T0182117

Progress in Mathematical Physics

Volume 73

More information about this series at http://www.springer.com/series/4813

Lalaonirina R. Rakotomanana

Covariance and Gauge Invariance in Continuum Physics

Application to Mechanics, Gravitation, and Electromagnetism

 Birkhäuser

Lalaonirina R. Rakotomanana
Institut de Recherche Mathématique
Université de Rennes
Rennes CX, France

ISSN 1544-9998   ISSN 2197-1846 (electronic)
Progress in Mathematical Physics
ISBN 978-3-030-06298-9   ISBN 978-3-319-91782-5 (eBook)
https://doi.org/10.1007/978-3-319-91782-5

Mathematics Subject Classification (2010): 53C820, 70H33, 70S05, 70S10, 70S15, 70S20, 74-02, 74A30, 74F15, 74J20, 78A25, 78A48, 78M30, 83C22, 83C35, 83D05

Printed on acid-free paper

This book is published under the imprint Birkhäuser, www.birkhauser-science.com by registered company Springer International Publishing AG part of Springer Nature.
The registered company address is: Gewerbestrasse 11, 6330 Cham, Switzerland

To Oly, Rindra, Herinarivo, and Haga

Preface

This monograph is designed for scientists and graduate students in physics, theoretical physics, mechanics, and mathematic physics. It may be considered as a second part of a previous book (*A geometric approach to thermomechanics of dissipating continua*, Birkhaüser) dealing with the thermomechanics of second gradient continuum bodies; in the same way the present book is also essentially based on a differential geometry of Riemann–Cartan. Variational approaches which were voluntarily omitted in the previous book constitute the main methods applied for deriving governing equations in the present monograph. We mainly consider the physics of continuum which can be described by a Lagrangian function. Three aspects of continuum physics are investigated: gradient continuum mechanics, relative gravitation, and some introduction to coupling with electromagnetism. Various aspects of invariance of the Lagrangian, as covariance of the formulation, as gauge invariance of the conservation equations, are the main topics esquissed in the monograph.

Different aspects of invariance are treated all along the manuscript. In short, covariance (i.e. passive diffeomorphism invariance) is the invariance of the shape of balance laws and constitutive laws equations under arbitrary change of coordinate system. Conversely, active diffeomorphism invariance is related to gauge invariance and Noether's theorems. First, we consider the covariance of Lagrangian density $\mathscr{L}$, which is assumed to explicitly depend on metric tensor and on affine connection. Following the method of Lovelock and Rund, we apply the covariance principle to the Lagrangian density $\mathscr{L}$. It is shown that the arguments of $\mathscr{L}$ are necessarily the torsion and/or the curvature associated with the connection, in addition to metric. In a second part, we consider the active diffeomorphism invariance by using local Poincaré gauge theory according to Utiyama method in relativistic gravitation. Most development is based on the Lie derivative of metric, torsion, and curvature, to obtain both the expression of the energy-momenta and the conservation laws for general metric-affine continuum. The invariance results we obtained are then applied to Lagrangian of second gradient continuum and to the spacetime Lagrangian in the framework of relative gravitation. Some problems of coupling between spacetime and matter are also addressed. The arguments of the Lagrangian are then the metric, the torsion, and the curvature of the spacetime, and for the matter, the strain, the contortion (or the torsion), and the covariant derivative of the contortion (or the torsion) based on the Levi-Civita connection associated with the metric of the matter. Some illustrations on gravitation, electromagnetism, and continuum mechanics in the framework of Einstein–Cartan spacetime are given.

I am grateful to the editing staff at Springer Verlag AG for their help and guidance during writing of the manuscript. I wish to thank all my colleagues and students with whom I have worked during these last 30 years. Among them, I would like to mention Nirmal Antonio Tamarasselvame who contributes to the formulation of the main covariance theorem in the Chap. 3. I would also like to mention Gregory Futhazar, a former PhD student, and my colleague Loïc Le Marrec for their contribution on the investigation of nonhomogeneous wave propagation with the help of Riemann–Cartan manifolds. Some of the illustrations in Chap. 5 were obtained during a common research work.

Last but not least, I particularly want to thank my wife Oly, my children Rindra, Herinarivo, and Haga, to whom this book is dedicated, for their understanding and their constant encouragement.

Cesson-Sévigné, France   Lalaonirina R. Rakotomanana
July 2017

Contents

Chapter 1
General Introduction

1.1 Classical Physics, Lagrangian, and Invariance

The motion and in a general manner the evolution of a physical system must be described with respect to an arbitrary coordinate system, and the laws of physics should be independent on the choice of the reference frame. These are the two most basic invariance requirements for any physics model. For non relativistic physics, namely the classical mechanics, these requirements may be classified into Galilean Invariance and covariance e.g. Rosen (1972). Galilean invariance is related to the notion of class of inertial frames of reference which move with constant relative velocity each other, and related to the homogeneity of time. The covariance means an invariance of the shape of the model equations under general coordinate transformations. Covariance is a mathematical condition and does not impose any restrictions on the contents of the laws of physics.

In the framework of special relativity, two invariance properties, namely invariance with respect to a change of reference frame and covariance, also constitute the basic requirements of the theory. In such a case, Galilean transformations are merely replaced by Lorentz transformations in a four-dimensional spacetime. In that way, mechanics is compatible with electromagnetism and, additionally, absolute constant light speed is assumed. In a broad sense, physical quantities of special relativity are modeled by scalars, four-vectors, and four-tensors. In most cases, they may be considered as arguments of a Lagrangian function $\mathscr{L}$ which defines the overall mathematical model. Given the concept of reference frame, the key property of such equations is that if they hold in one inertial frame, then they should hold in any inertial frame. In the same way, if all the components of a tensor vanish in one inertial frame, they should vanish in every inertial frame.

For general relativity theory, or gravitation theory, the covariance still means the invariance of the governing equations (say same shape for governing equations) with respect to any change of coordinate system, following a diffeomorphism (a group of transformations larger than translation and rotation group). The invariance with

© Springer International Publishing AG, part of Springer Nature 2018
L. R. Rakotomanana, *Covariance and Gauge Invariance in Continuum Physics*,
Progress in Mathematical Physics 73, https://doi.org/10.1007/978-3-319-91782-5_1

respect to any reference frame is also the second invariance requirement. The second invariance condition leads to gauge invariance.

Most of field equations governing theoretical physics are deduced from a variational principle after defining a suitable Lagrangian density $\mathscr{L}$ and its arguments. For relativistic gravity and cosmology see e.g. Clifton et al. (2012) for recent and quasi-exhaustive review. To start with, let $dq^i := F^i_\alpha \, dx^\alpha$ be a local coordinate change (or a local map), where F^i_α are called tetrads or triads for three dimensional geometry e.g. Marsden and Hughes (1983). In fact, a frame of reference may be identified with a set of four vectors $\{\mathbf{e}_\alpha(x^\mu) := F^i_\alpha(x^\mu) \, \hat{\mathbf{e}}_i\}$, where $\hat{\mathbf{e}}_i$ is an given set of orthonormal base, and $\{\mathbf{e}_\alpha(x^\mu), \alpha = 0, 1, 2, 3, 4\}$ is defined in any given point of the spacetime. In a Minkowskian spacetime or an Euclidean space, an induced metric is defined as $g_{\alpha\beta} := \hat{g}_{ij} F^i_\alpha F^j_\beta$ where $\hat{g}_{ij}$ are components of space or spacetime metric. Local map $dq^i := F^i_\alpha \, dx^\alpha$ cannot be always integrated. The use on multivalued tetrads as local transformations was done in Kleinert (2000) on the basis of the so called nonholonomic mapping principle to generate twisted and curved spacetime. The method was inspired from the plastic deformation of crystals where dislocations and disclinations are present as defects. In such a case the transformations are not single-valued (Kleinert 2008), almost everywhere $F^i_\alpha(x^\mu)$ and its inverse $F^\alpha_i(x^\mu)$ are elements of $\mathbb{GL}^+_n(\mathbb{R})$ at each point x^μ. It is therefore usual to define connection coefficients, as extension of the usual definition $\Gamma^\gamma_{\alpha\beta} := F^\gamma_i \, \partial_\beta F^i_\alpha$ associated to the tetrads. In relativistic gravitation, several theories were built upon various forms of the Lagrangian density function e.g. Carter (1973), Clifton et al. (2012), Taub (1954). The arguments of Lagrangian density function $\mathscr{L}$ in Einstein gravitation is essentially based on the curvature of the spacetime, which expresses the strong relation between gravitation and geometry. In the original paper of Einstein, the curvature is obtained solely with metric components and their first and second derivatives. Later this theory of curvature gravitation is improved in some sense to include the torsion for capturing quantum effects in the presence of high energies e.g. Hammond (2002), Olmo and Rubiera-Garcia (2013), or to model vorticity in fluid flows e.g. Garcia de Andrade (2005), and continuum with microstructure e.g. Kleinert (2008). Torsion of connection is a variable independent of the metric since it is purely obtained by means of an affine connection $\hat{\Gamma}^\gamma_{\alpha\beta}$ of the space (resp. spacetime), which is a priori not related to the metric components $\hat{g}_{\alpha\beta}$. Introduction of the metric and the affine connection as independent arguments of the Lagrangian constitutes the basics for the Palatini formulation of relativistic gravitation e.g. Koivisto (2011). Earlier as 1928, Cartan proposed spacetime with torsion and curvature to support the spacetime geometry e.g. Cartan (1986). The first results were obtained by Cartan in the formulation of Newtonian gravitation in this pure affinely framework e.g. Ruedde and Straumann (1997), which extended the well-known special relativity theory. Indeed original special relativity does not include gravitation effects but rather electromagnetic waves. The coupling of spacetime and matter (assumed to be continuous or not) then becomes another challenge in the formulation of electromagnetic fields e.g. Dias and Moraes (2005), Plebanski (1960), Prasanna (1975a), of relativistic gravitation e.g. Anderson (1981), Hehl et al. (1976), and particularly of the so-called extended bodies under gravitation e.g. Dixon (1975), Ehlers and Geroch (2004), Papapetrou (1951). Spacetime and matter have their own geometry backgrounds. To separate spacetime gravity (G) and matter (M), the following form was proposed $\mathscr{L} := \mathscr{L}_G(\hat{g}_{\alpha\beta}, \hat{\Gamma}^{\gamma}_{\alpha\beta}) + \mathscr{L}_M(\hat{g}_{\alpha\beta}, g_{\alpha\beta}, \Gamma^{\gamma}_{\alpha\beta}, \Phi)$, where $g_{\alpha\beta}$, $\Gamma^{\gamma}_{\alpha\beta}$ (with hat for spacetime or without hat for matter), and Φ are metric tensor, connection—possibly independent of the metric—and some matter field respectively e.g. Forger and Römer (2004), Hehl et al. (1995), Petrov and Lompay (2013), Sotiriou and Faraoni (2010), Utiyama (1956), Vitagliano et al. (2011). Variable Φ is a priori introduced to analyze multi-physics phenomenon as electromagnetic fields when necessary, but the presence of metric $\hat{g}_{\alpha\beta}$ in the matter Lagrangian is necessary for the theory to be viable e.g. Lehmkuhl (2011). Connections are not necessary the same for matter and for space (or for spacetime) e.g. Amendola et al. (2011), Defrise (1953) as implicitly assumed in classical elasticity e.g. Marsden and Hughes (1983). On the one hand, classical elastic material may evolve in a spacetime with gravity, meaning that the matter connection $\Gamma^{\gamma}_{\alpha\beta}$ is affinely equivalent to Euclidean connection (with zero torsion and curvature), whereas the spacetime connection $\hat{\Gamma}^{\gamma}_{\alpha\beta}$ may have non zero torsion and curvature (Defrise 1953; Manoff 1999). On the other hand, continuum with dislocations and disclinations density is endowed with connection $\Gamma^{\gamma}_{\alpha\beta}$ with nonzero torsion and/or curvature but evolves within a classical Newtonian spacetime with Euclidean connection e.g. Rakotomanana (2003). For short, analysis of continuum motion always requires the simultaneous consideration of two manifolds: the spacetime and the body (Defrise 1953). This suggests that in principle the metric and the connection which are arguments of the gravity Lagrangian density $\mathscr{L}_G$ are not the same as the metric and the connection, arguments of the matter Lagrangian density $\mathscr{L}_M$ e.g. Amendola et al. (2011), Manoff (1999). Generalized continuum in the presence of dislocations and disclinations field may be related to relativistic mechanics in e.g. Baldacci et al. (1979), where the concept of micro universe (microcosms) was introduced to model plastic behavior due to dislocations. Some authors also related defects of crystal to the Einstein universe as a Cosserat continuum with defects e.g. Kleinert (1987). Early papers and recent works described the defects with anholonomic configuration spaces by using the multiplicative decomposition e.g. Bilby et al. (1955), Clayton et al. (2004). This opens the question of generalized transformations of continuum with evolving local topology e.g. Verçyn (1990). As alternative formulation of continuum mechanics where strain and stress tensors are related by constitutive laws, often derived from a strain energy function, which is merely a "static" case of matter Lagrangian density $\mathscr{L}_M$. For generalized continuum with micro-structure, strain energy density is assumed function of strain, first gradient, and second gradient of strain e.g. Mindlin (1964, 1965). They are based on the torsion and curvature of intermediate non Euclidean manifolds. A brief overview of Lagrangian involved in relativistic gravitation and in strain gradient continuum shows that they commonly take the form: $\mathscr{L}_G = \mathscr{L}_G(g_{\alpha\beta}, \partial_\lambda g_{\alpha\beta}, \partial_\mu \partial_\lambda g_{\alpha\beta})$, where $g_{\alpha\beta}$ are the metric components and $\partial_\lambda g_{\alpha\beta}$, and $\partial_\lambda \partial_\mu g_{\alpha\beta}$ are their first and second derivatives with respect to space (or spacetime) coordinates. To satisfy formulation invariance, it is necessary to review the mathematical requirement and physical principle underlying the derivation of all these constitutive laws and corresponding fields equations for either gradient elasticity or relativistic gravitation.

1.2 General Covariance, Gauge Invariance

The general covariance of field equations means rigorously invariance with respect to the action of the group of all spacetime diffeomorphisms. Concept of general invariance then includes two aspects in physics theory: a passive diffeomorphism which is a change of coordinates, and an active diffeomorphism which is a gauge transformation acting on tensor fields of the theory e.g. Manoff (1999). Passive diffeomorphism (covariance) is a trivial requirement stating that it should be possible to use different coordinates to describe one physical situation. Covariance dictates that the laws of physics (conservation laws, and constitutive laws) keep the same form, regardless of the coordinate system. Invariance with respect to an active diffeomorphism implies that any solution of the field equations can be transformed and still satisfies the same, untransformed field equations. Accordingly, these two aspects of invariance apply to both relativistic and classical mechanics e.g. Anderson (1971), Frewer (2009): invariance of Lagrangian function $\mathscr{L}$ with respect to passive diffeomorphism e.g. Antonio and Rakotomanana (2011), and invariance with respect to active diffeomorphism one e.g. Ali et al. (2009), McKellar (1981). The gauge aspect arises when deriving the physical laws with respect to a class of reference frames: Lorentzian transformations for relativistic gravitation and Galilean transformations for classical mechanics. An action $\mathscr{S}$ defined from a Lagrangian $\mathscr{L}$ is proposed and its invariance under some continuous set of spacetime transformations leads to local and/or global conservation laws e.g. Forger and Römer (2004). In this way, Noether's theorem classically establishes correspondence between the symmetries and the conservation laws of physics theory based a variational principle e.g. Lovelock and Rund (1975), Pons (2011). Historical analysis of these two aspects of invariance may be found in e.g. Brading and Ryckman (2008), Frewer (2009), Norton (1993). The confusion between gauge invariance with respect reference frames—physical requirement—and the covariance with respect to coordinate change—mathematical requirement—was a source of long lasted dispute, in the scope of relativistic gravity (Earman 1974). As earlier as 1956 Utiyama formulated relativistic gravitation as a gauge theory (Utiyama 1956). Since then, numerous extended models have been proposed. In 1961, Kibble started with the invariance of a Lagrangian function with respect to Lorentz transformations (in flat Minkowskian spacetime) (Kibble 1961), and showed that the arguments of the free Lagrangian could be geometrically interpreted as affine connection coefficients. He stated that existence of gravitational field is obtained from the Lorentz invariance of the Lagrangian. It was also shown in Kibble (1961) that translational invariance (in addition to Lorentz invariance) imposes that the Lagrangian can not depend explicitly of the coordinates (x^μ). Application of

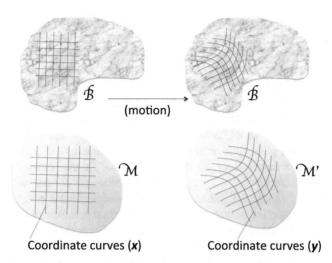

Fig. 1.1 Motion of a body $\mathscr{B}$ is defined with respect to a reference (either $\mathscr{R}$ or $\mathscr{R}'$). In each of these references, a coordinate system associated to coordinate curves (either x^{μ} or y^{μ}) may be used to define the position of the body with respect to the reference. Links (1) schematize either the motion of the body $\mathscr{B} \to \mathscr{B}'$ or the change of reference frame $\mathscr{R} \to \mathscr{R}'$. Links (2) represent the relativity of the motion of body $\mathscr{B}$ with respect to $\mathscr{R}$ or to $\mathscr{R}'$. Links (3) represent the change of coordinate system for either the body $\mathscr{B}$ or the reference frame $\mathscr{R}$. The goal is to investigate the invariance of the formulation of the physics of such a body $\mathscr{B}$ independently on the choice of either the reference or the coordinate system within each reference

translational invariance for elastic continuum matter with dislocations may be found in e.g. Lazar and Anastassiadis (2008), Malyshev (2000). The idea is to search for Lagrangian functions which are gauge invariant, in the sense that Lagrangian is invariant under spacetime diffeomorphisms, (Fig. 1.1) and additionally under local Lorentzian transformations e.g. Bruzzo (1987), Hehl et al. (1976), Kleinert (2008), Utiyama (1956). The Lorentz transformation is replaced by Galilean transformation for non relativistic mechanics e.g. Havas (1964). In any case, the complete Poincaré transformations group is considered e.g. Cho (1976c), Pons (2011), Zeeman (1964). In this later reference, Zeeman demonstrated that the causality principle induced the Lorentz group of invariance.

For classical continuum mechanics and more generally for classical continuum physics, three aspects of invariance hold e.g. Rosen (1972), Svendsen and Betram (1999): (a) Euclidean Frame Indifference means invariance with respect to Euclidean observers; (b) Covariance is the previous passive diffeomorphism invariance; (c) Rigid Motion Indifference is the invariance with respect to superimposed rigid body motions. Covariance is related to a mathematical function whereas the term "indifference" is related to Euclidean observers (see Svendsen and Betram 1999). For gradient material evolving within a Newtonian spacetime, the material frame indifference was further investigated, where they extended the concept of material isomorphism of Noll for strain gradient materials and the multiplicative decomposition e.g. Le and Stumpf (1996). For continuum moving in Euclidean space, Svendsen and Betram (1999) show that any two of previous principles automatically imply the third. Consequently, covariance condition is a necessary to ensure the material frame-indifference of constitutive laws. Covariance and frame-indifference have been compared in Kempers (1989) where the classical frame-indifference principle was shown to follow from covariance when reduced to non relativistic limit, and when the inertial terms may be neglected (Frewer 2009). Then for situation where inertia is absent or can be neglected, the passive diffeomorphism invariance of either relativistic or non relativistic continuum mechanics includes the frame-indifference principle e.g. Kempers (1989). For the sake of the completeness, on the one hand for relativistic gravitation, diffeomorphism invariance applies to four dimensional spacetime instead of only three-dimensional space (subclass of diffeomorphisms applied only to space is called internal transformations e.g. Krause 1976). On the other hand, the Euclidean-Frame-Indifference should be replaced by Lorentz transformations e.g. Bruzzo (1987), McKellar (1981), Utiyama (1956) to be compatible with electromagnetism. Applying the invariance on a 2-covariant tensor-valued function, depending on metric, its first and second derivatives (linear with respect to this later), Cartan deduced the Einstein's gravitation equations in terms of curvature (Cartan 1922). This is a key point for understanding 80 decades of debate on the general covariance propounded by Einstein for with the diffeomorphism is applied in the spacetime manifold but not on the space only e.g. Norton (1993), Rosen (1972). For completeness, the concept of spacetime is actually related to the concept of affine connection which endows the spacetime manifold (Kadianakis 1996). One of the main problems is to define how the twisted and curved spacetime is coupled with the gradient continuum matter in general gravitation. Some aspects of this coupling are still open and development of Lagrangian function if any together with their associated conservation laws requires further studies e.g. Kleinert (2000). The geometric background would be the Riemann–Cartan manifold e.g. Nakahara (1996). The accounting of torsion and curvature in the derivation of constitutive and conservations laws of some domains of the continuum physics are developed in the present book.

1.3   Objectives and Planning

Conversely to usual continuum physics, strain gradient continuum physics uses the metric and the gradient of metric as geometric variables. Most of the strain gradient continuum models use an Euclidean connection which derives from metric as Levi-Civita connection to calculate the derivatives of various tensor variables involved in the models. However, there exist some approaches based on Cartan geometry with a connection which does not derive from the metric and for which the associated torsion and curvature do not necessarily vanish. The main goal of the present book is to investigate the invariance aspects and their consequences on the derivation of the continuum mechanics and the relativistic gravitation laws beyond the Riemann spacetime and the classical continuum mechanics. We aim to apply the passive and active diffeomorphisms on Lagrangian function $\mathscr{L}$ to obtain a set of arguments satisfying both the covariance and gauge invariance. First, we apply the covariance on Lagrangian function of the type $\mathscr{L}(g_{\alpha\beta}, \nabla_\gamma g_{\alpha\beta}, \nabla_\gamma \nabla_\lambda g_{\alpha\beta})$, in an arbitrary coordinate system (x^α) utilizing the covariant derivative with respect to an affine connection ∇. Covariance deals with either $\mathscr{L}_G$ or $\mathscr{L}_M$. Second we consider the active diffeomorphism-invariance requirement to derive the field equations resulting from the local Poincaré gauge invariance, by means of Lie derivative along an arbitrary vector field ξ^α. The adopted method for conservation laws derivation will be based on the additional decomposition of the Lagrangian variation of a primal variables as metric, torsion, and curvature into a Eulerian variation and the Lie derivative variation of these variables. We accordingly deduce the corresponding conservations laws. Some preliminary examples such as gravitational and electromagnetic waves and waves within nonhomogeneous twisted continuum are treated in this paper particularly devoted on the coupling of spacetime and microstructured continuum matter.

After general introduction on Lagrangian function and general purposes on the types of invariance such as covariance and gauge invariance in this chapter, Chap. 2 is devoted with the definition of Lagrangian function. The geometric background of the continuum mechanics and relativistic gravitation is reminded, namely metric, torsion and curvature within a continuum bodies and within spacetimes. Relation between the invariance of Lagrangian and the Euler–Lagrange equations is introduced. Some examples of Lagrangian in the framework of continuum mechanics and gravitation are given.

Chapter 3 contains an essential part of this book since it deals with the covariance of Lagrangian depending on the metric and its two first derivatives. The main result of this chapter shows that the covariance of such a Lagrangian induces that the Lagrangian should depend on the metric, the torsion and the curvature of the connection. It clearly shows that the arguments of the Lagrangian defined by a restrictive theorem due to invariance with respect to the choice of metric, and on the choice of the connection on the manifold.

Gauge invariance is treated in Chap. 4. The Lagrangian is assumed to depend on the metric, the torsion and the curvature of the continuum body and of the ambient spacetime. Poincaré's gauge invariance is used to obtain conservation laws of both continuum mechanics and relativistic gravitation. The concept of metric strain, torsion strain, and curvature strain is introduced. They are mainly based on the contortion tensor for metric compatible connection. Lie derivative constitutes the main tool for deriving the conservation laws.

We give some examples and applications in the Chaps. 5 and 6. Introductory aspects of gravitational are reminded in Chap. 5 where we give applications in the domain of wave propagation within non homogeneous continuum, and some recall on the gravitational waves. The extension of the geodesic deviation in Riemannian spacetime to auto parallel deviation for gravitation field within Einstein–Cartan spacetime is obtained allowing us to consider the influence of torsion on a particle motion. Some examples of twisted and curved continuum body are treated in details.

A large part of the Chap. 6 is devoted to the coupling between electromagnetism and gravitation, namely in the framework of Einstein–Cartan theory. Various forms of the Maxwell's equations are developed on either connection-based formulation or form-like formulation. The coupling of gravitation and electromagnetism is effective by choosing a Lagrangian function involving gravitation and electromagnetic field. A symmetric energy-momentum tensor field is obtained by applying Lagrangian variation, then the Poincaré's invariance allows us to derive the conservation equations including Maxwell's equations, energy conservation and momentum equation. Our approach is heavily based on the Lagrangian formulation and then, the coupling of gravitation and electromagnetism is considered by investigating the Einstein–Maxwell's equations of fields. Some relations between torsion filed and the cosmological constant are developed in this chapter (Anti-de Sitter spacetime).

Chapter 2
Basic Concepts on Manifolds, Spacetimes, and Calculus of Variations

2.1 Introduction

In the present book, we are interested in continuum physics namely mechanics, gravitation, electromagnetism and their mutual interaction. Most of field equations governing theoretical physics are deduced from a variational principle after defining a suitable Lagrangian density $\mathscr{L}$ and its arguments. Three steps are considered for deriving the field equations governing their evolution and mutual interaction. The first focus on the continuum geometry and by the way the spacetime with the concept of reference frame where physics happen.

Extension of the Galilean Principle of Relativity stating that the laws of mechanics have the same form in all inertial frames, leads the two postulates of special relativity of Einstein conciliating mechanics and electromagnetism: (a) Physical laws have the same form in all inertial frames (covariance), and (b) the light speed is finite and is the same in all inertial frames (causality). The two postulates induce the concept of space-time, and to group of transformations of frames of reference.

For accounting for the presence of gravitation within the previous flat spacetime, Einstein arrives to the principle of equivalence stating that any non-inertial frame is equivalent to a some gravitational field, and this is true locally for non homogeneous field. The spacetime becomes curved with an event-dependent metric tensor. More generally, the derivation of more and more sophisticated spacetimes may be done by considering first the deformation of the flat Minkowski spacetime $\mathbb{R}^4$ by defining a time and position-dependent pseudo-metric tensor to give a curved Einstein–Riemann spacetime. This spacetime is curved with a non zero curvature tensor entirely defined by the metric and its derivatives. The second step would be considering in addition the deformation of the connection of the Einstein–Riemann spacetime to give an Einstein–Cartan spacetime which is twisted and curved. In such a case the metric and the connection are two independent variables.

© Springer International Publishing AG, part of Springer Nature 2018   9
L. R. Rakotomanana, *Covariance and Gauge Invariance in Continuum Physics*,
Progress in Mathematical Physics 73, https://doi.org/10.1007/978-3-319-91782-5_2

The goal of this chapter is to give some introductory elements of differential geometry and of variational calculus to on spacetime, and more generally on manifolds.

2.2 Space-Time Background

Space and time may be considered as the most fundamental concepts of physics and even of natural sciences.

2.2.1 Basics on Flat Minkowski Spacetime

For classical mechanics' theory, the Newton's laws of gravitation are based on the assumption on the existence of an absolute three dimensional space and an absolute monodimensional time. In other words, classical mechanics of Newton considers that space is distinct from body and that time flows uniformly without regard to whether anything else happens in the world. Accordingly, for each time t, Newton's laws of gravitation assume an absolute distance function between two points defined with an Euclidean metric of the space. The spacetime of classical mechanics is defined by a four-dimensional manifold defined as a product of a monodimensional Euclidean time with a three-dimensional Euclidean space $\mathscr{T} \times \mathscr{E}$. Absolute space $\mathscr{E}$ and absolute time $\mathscr{T}$ do not depend upon physical events happening inside. The spacetime coordinates can be written as (t, x^1, x^2, x^3). Both of them have their proper metric for measuring the flow of time $|t - t_0|$ and the distance $\|\mathbf{MM_0}\|^2 :=$ $(x^1 - x_0^1)^2 + (x^2 - x_0^2)^2 + (x^3 - x_0^3)^2$ between points respectively and connection which relates the values of local field, namely the velocity field say $\mathbf{v}(t, M)$, at two arbitrary neighboring points $M(t, x^i)$ and $\tilde{M}(t, x^i + dx^i)$ of this spacetime.

A first major modification of the concept of space and time was the relativization of time. Special relativity thus lies upon two cornerstones: (a) the extended relativity principle which assumes first the existence of inertial or Galilean frames $\mathscr{G}$, just as in classical mechanics, and second that all respect to these frames, all physics laws (especially electromagnetism) should be invariant; (b) the light speed axiom assuming that the same value of the light speed holds in all Galilean frames, irrespective of the emission properties of the source. Usually, $c := 299,'792.458$ [km s^{-1}] may be taken as the light speed. Implicitly it then assumes, as in classical mechanics, that there is an absolute spacetime, set of events with a spacetime coordinates (x^μ) interpreted as coordinates of a four-dimensional affine space. However, space and time merge into a spacetime concept. Spacetime is described as a four-dimensional continuum where any event can be described by coordinates $(x^0 = ct, x^1, x^2, x^3)$ of $\mathbb{R}^4$, where index 0 stands for time.

Definition 2.1 (Minkowski Space) A Minkowski space $\mathscr{M}$ is the vector space $\mathbb{R}^4$ endowed with an indefinite inner product given by:

$$< \mathbf{x}, \mathbf{y} >:= x^0 y^0 - x^1 y^1 - x^2 y^2 - x^3 y^3 \tag{2.1}$$

where x^μ are the coordinates of $\mathbf{x} \in \mathbb{R}^4$, and y^ν the coordinates of $\mathbf{y} \in \mathbb{R}^4$.

To avoid lengthy equations including the light speed c, it is sometimes worthwhile to define the spacetime coordinates $(x^0 := ct, x^1, x^2, x^3)$, and therefore to have previous metric $\hat{g}_{\alpha\beta} := \mathrm{diag}\,\{+1, -1, -1, -1\}$) for Minkowskian spacetime $\mathscr{M}$. Minkowski spacetime is the basic geometric background of flat (absence of gravitation) space and time. For ambient spacetime, a uniform metric holds for the entire Euclidean space $\hat{g}_{ij} := -\mathrm{diag}\,\{+1, +1, +1\}$ and $\hat{g}_{\alpha\beta} := \mathrm{diag}(\{+1, -1, -1, -1\})$ for Minkowski spacetime.[1]

Remark 2.1 To separate the space and the time, and then to point out the asymptotic behavior of the spacetime for small velocity of matter compared to that of the light, in a (flat) Minkowskian spacetime, one can define a global coordinate system $\{x^0 := t, x^1, x^2, x^3\}$ in which the metric tensor $\hat{g}_{\alpha\beta}$, and its inverse $\hat{g}^{\alpha\beta}$ have (diagonal) matrix components e.g. Havas (1964):

$$\hat{g}^{\alpha\beta} := \mathrm{diag}\left(1/c^2, -1, -1, -1\right), \quad \hat{g}_{\alpha\beta} := \mathrm{diag}\left(c^2, -1, -1, -1\right) \tag{2.2}$$

Havas introduced two separated space and time metrics by considering the following limits (Havas 1964):

$$\hat{h}^{\alpha\beta} := \lim_{c \to \infty} \hat{g}^{\alpha\beta}(c) \cdot c^2 \simeq \mathrm{diag}\,(0, -1, -1, -1), \qquad c >> 1 \tag{2.3}$$

$$\hat{t}_\alpha \hat{t}_\beta = \hat{t}_{\alpha\beta} := \lim_{c \to \infty} \hat{g}_{\alpha\beta}(c)/c^2 = \mathrm{diag}\,(1, 0, 0, 0) \tag{2.4}$$

Reference manifold (a four dimensional "spacetime continuum") is a priori assumed e.g. Amendola et al. (2011) for relativistic mechanics, and e.g. Kadianakis (1996) for classical mechanics. The particular choice of reference body with Euclidean (resp. Minkowskian) metric assumes the existence of an implicit flat configuration and constitutes an underlying constitutive assumption e.g. Clayton et al. (2004), Le and Stumpf (1996). Consider two events $\mathbf{x} := (x^\mu)$, and $\mathbf{y} := (y^\mu)$ of the spacetime $\mathscr{M}$. It is possible to define a partial ordering on the Minkowski manifold $\mathscr{M}$, denoted $\mathbf{x} < \mathbf{y}$ if the event $\mathbf{x}$ can influence the event $\mathbf{y}$. We remind that $\mathbf{x} < \mathbf{y}$ if $\mathbf{y} - \mathbf{x}$ is a time vector (see below) $< \mathbf{y} - \mathbf{x}, \mathbf{y} - \mathbf{x} > \in \mathbb{R}^{+*}$ is strictly positive oriented to the future e.g. Zeeman (1964).

---

[1] It is worth noting that covariant components of event hold as $(x_0 = ct, x_1 = -x^1, x_2 = -x^2, x_3 = -x^3)$ in this flat Minkowskian spacetime $\mathscr{M}$.

2.2.2   Twisted and Curved Spacetimes

In the presence of gravitation or external forces for material continuum mechanics, spacetime/material continuum can be divided into small pieces called "microcosms" e.g. Gonseth (1926), and these microcosms are glued together smoothly (without voids). In the special relativity theory, the spacetime has the structure of Minkowskian space $\mathcal{M}$. In the general relativity theory, the spacetime is a set of (infinite number of) flat microcosms in which the special relativity holds (principle of equivalence). Global properties are different from those of small microcosms to give the need of differential manifold as underlying background geometry e.g. Nakahara (1996). The results are that spacetime with gravitation is endowed with time and position (event) dependent metric tensor $g_{\alpha\beta}(x^\mu)$. In a more systematic method, the flat Minkowski spacetime may be twisted and curved by using a multi-valued coordinate transformations e.g. Kleinert (2000). This method called Nonholonomic Mapping Principle (NMP) is analogous to the method of Volterra dislocations usually applied to model the creation and evolution of dislocations and disclinations of crystalline solids e.g. Bilby et al. (1955), Ross (1989). Details for these spacetimes will be developed later in the present book, together with an introduction to the variational procedure for obtaining the governing equations.

2.3   Manifolds, Tensor Fields, and Connections

Relativity principle cannot be replaced by the covariance. Some confusion may appear between change of reference frames and (passive) change of coordinate systems e.g. Bain (2004). This latter does not play role in the spacetime geometry for neither classical mechanics, nor Riemannian/Cartan/ or Riemann–Cartan spacetime in relativity. Coordinate systems have no physical meaning whereas the frame of reference is a fundamental concept in physics theory (Principle of Relativity).

2.3.1   Coordinate System, and Group of Transformations

Continuous bodies and spacetimes are often described by three or four dimensional continua respectively. Mathematical models of continuum mechanics and relativistic gravitation then dwell upon the concept of neighborhoods and continuous mappings. This naturally leads to the concept of manifold e.g. Lovelock and Rund (1975), Wang (1967). Namely, in Galilean view, space is a tri-dimensional manifold and time is a one dimensional manifold, they exist separately. In the Minkowskian point of view, for relativistic physics, spacetime is a four-dimensional manifold. In both theories, manifolds are differentiable. After introducing manifolds which is the background of continuum geometry and kinematics, two variables constitute the basis of spacetime and continuous material model: the metric tensor and the connection.

2.3.1.1 Manifolds, Tangent Space, Cotangent Space

We remind some basic elements of differentiable manifold. Extensive developments of manifold may be seen elsewhere e.g. Nakahara (1996). Briefly, we fixe once and for all the n-dimensional vector space E (reference body). An n-dimensional manifold $\mathscr{B}$, embedded onto E, is a point set which is covered completely by a countable set of neighborhoods $U_1, U_2, \ldots$, such that each point $P \in \mathscr{B}$ belongs to at least one of these neighborhoods. U_k is also called coordinate patch. The manifold $\mathscr{B}$ is said differentiable if there is at least one covering of $\mathscr{B}$ by a set of coordinate patches $\{U_k, \}$ such that in each overlap of the patches, the coordinates are mutually related by differentiable functions. For two given manifolds, say $\mathscr{B}_1$ and $\mathscr{B}_2$, a diffeomorphism is an invertible mapping $\varphi : \mathscr{B}_1 \rightarrow \mathscr{B}_2$, such that φ and its inverse φ^{-1} are both C^∞, meaning that the mappings are continuous and infinitely differentiable. For the case where $\mathscr{B}_1 = \mathscr{B}_2 = \mathscr{B}$, any mapping φ is invertible and provided that $\varphi \in C^\infty$, then it is a diffeomorphism. A coordinate system is defined on each U_k such that one may assign in a unique manner n real numbers $x^1, \ldots, x^n$ to each point $P \in U_k$. As P ranges over U_k, the corresponding numbers $x^1, \ldots, x^n$ range over an open domain D_k of E. It thus exists a one-to-one mapping of each U_k onto D_k, this mapping will be assumed continuous. The numbers $x^1, \ldots, x^n$ are called the coordinates of P. In the case where $U_1 \cap U_2 \neq \emptyset$ there are two sets of coordinates associated to a same point $P \in U_1 \cap U_2$ (Fig. 2.1).

In sum, the manifold $\mathscr{B}$ is then provided with a family of pairs $\{(U_k, \varphi_k\}$ such that for any two coordinate patches U_n and U_m such that $U_n \cap U_m \neq \emptyset$, the map $\varphi_n (\varphi_m)^{-1} : \varphi_m (U_n \cap U_m) \rightarrow \varphi_n (U_n \cap U_m)$ is infinitely differentiable. φ_n are called coordinate functions (or coordinates), and for any point $P \in U_k$, represented by $(x^1(P), \cdots, x^n(P))$. In practise, the differentiability of manifold may be approached by means of change of coordinate system.

Fig. 2.1 Overlapping of coordinate patches. The point $P \in \mathscr{B}$ of the manifold is reported as being associated to two overlapping coordinate patches U_1 and U_2. In the overlap, there will be points that have coordinates from each coordinate patches. Sets $D_1 := \varphi_1(U_1)$ and $D_2 := \varphi_2(U_2)$ belong to E

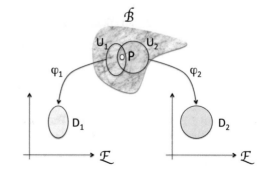

Fig. 2.2 Function f on
manifold $\mathscr{B}$. The function
$f : P \in \mathscr{B} \to \mathbb{R}$ is defined
by its coordinate presentation
$f\varphi^{-1} : (x^1, \cdots, x^n) \in$
$\mathbb{R}^n \to f(P) \in \mathbb{R}$

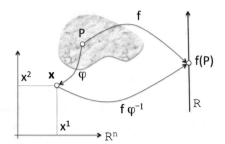

Calculus on manifold of dimension n starts with the function and tangent space. A function on manifold $\mathscr{B}$ denoted $f : \mathscr{B} \to \mathbb{R}$ is a (smooth) map from $\mathscr{B}$ to real $\mathbb{R}$. It is represented on Fig. 2.2 together with its coordinate presentation. Let consider now a curve on the manifold defined by the function $\mathscr{C} : [a, b] \in \mathbb{R} \to \mathscr{B}$ and a function $f : \mathscr{B} \to \mathbb{R}$. Say a real $t = 0 \in [a, b]$. The tangent vector of the curve at $\mathscr{C}(0)$ is the directional derivative of a function $f[\mathscr{C}(t)]$ along the curve $\mathscr{C}$ at the point $t = 0$:

$$\frac{d}{dt} f[\mathscr{C}(t)]|_{t=0} = \frac{\partial \left(f\varphi^{-1}(x^\mu)\right)}{\partial x^\alpha} \frac{dx^\alpha \{\mathscr{C}(t)\}}{dt}|_{t=0} \tag{2.5}$$

where the "partial derivative" of f in the composition of functions is practically expressed in terms of the partial derivative of its coordinate presentation. It is worth to define the tangent vector $\mathbf{u}$ at P of $\mathscr{B}$ as:

$$u^\alpha := \frac{dx^\alpha \{\mathscr{C}(t)\}}{dt} \tag{2.6}$$

and use a synthetic expression of the relation (2.5):

$$\frac{d}{dt} f[\mathscr{C}(t)]|_{t=0} = \mathbf{u}[f] = u^\alpha \frac{\partial f}{\partial x^\alpha} \tag{2.7}$$

Definition 2.2 (Tangent Space) All the equivalent classes of the curves $\mathscr{C}_k$, and by the way all the tangent vectors $\mathbf{u}_k$ at point P, from a vector space called the tangent space of $\mathscr{B}$ at point P, it is denoted $T_P\mathscr{B}$.

The base of the tangent space of dimension n is denoted by:

$$\mathbf{f}_\alpha := u^\alpha \frac{\partial}{\partial x^\alpha}, \qquad \alpha = 1, \cdots, n \tag{2.8}$$

Definition 2.3 (Cotangent Space) The cotangent space at P of the manifold $\mathscr{B}$ is the set of linear function from $T_P\mathscr{B} \to \mathbb{R}$. It is denoted $T_P^*\mathscr{B}$.

An arbitrary element of the cotangent space $\omega \in T_P^*\mathcal{B}$ is also called dual vector. the basic example of dual vector is the differential df of a real function:

$$df = \frac{\partial f}{\partial x^\alpha} dx^\alpha \tag{2.9}$$

Then the action of the linear map $df \in T_P^*\mathcal{B}$ on any vector $\mathbf{u} \in T_P\mathcal{B}$ is defined by inner product:

$$(df, \mathbf{u}) = \mathbf{u}[f] = u^\alpha \frac{\partial f}{\partial x^\alpha} \tag{2.10}$$

The base of the cotangent space is denoted here $\{dx^\alpha\}$ allowing us to decompose any 1-form ω on the manifold $\mathcal{B}$ as $\omega = \omega_\beta dx^\beta$, where ω_β are the covariant components of ω.

2.3.1.2   Change of Coordinate System

Let (y^i) *(Latin indexes)* and (x^α) *(Greek indexes)* corresponding coordinate systems respectively in U_1 and U_2. Latin indexes and Greek indexes will be used to distinguish two different coordinate systems. The transformation between coordinates (y^i) and (x^α) is diffeomorphism, more precisely a passive diffeomorphism,

$$y^i = y^i(x^\alpha), \qquad x^\alpha = x^\alpha(y^i) \tag{2.11}$$

$$J_i^\alpha = \frac{\partial x^\alpha}{\partial y^i}, \qquad J_{ij}^\alpha = \frac{\partial J_i^\alpha}{\partial y^j} = \frac{\partial^2 x^\alpha}{\partial y^i \partial y^j}, \qquad J_{ijk}^\alpha = \frac{\partial J_{ij}^\alpha}{\partial y^k} = \frac{\partial^3 x^\alpha}{\partial y^i \partial y^j \partial y^k}$$

We also have that for any coordinate system (y^i), and for any permutation $\sigma \in \mathfrak{S}_n$

$$\frac{\partial^n}{\partial y^1 \dots \partial y^n} = \frac{\partial^n}{\partial y^{\sigma(1)} \dots \partial y^{\sigma(n)}}$$

Einstein Convention   When a lowercase index and a upper case one such as j, k, $l, \dots$ appear twice in a term then summation over that index is applied. The range of summation is $1, \dots, n$, the letter n is exceptionally excluded from the summation. We have

$$\frac{\partial x^\alpha}{\partial y^i} \frac{\partial y^i}{\partial x^\beta} = J_i^\alpha A_\beta^i = \delta_\beta^\alpha, \qquad \frac{\partial y^i}{\partial x^\alpha} \frac{\partial x^\alpha}{\partial y^j} = A_\alpha^i J_j^\alpha = \delta_j^i$$

with respectively summation (Einstein) over i and α from 1 to n.

Let U an open subset of $\mathbb{R}^3$, $U := \{(r, \theta, \varphi) \in \mathbb{R}^3 : 0 < r, 0 < \theta < \pi, 0 < \varphi < 2\pi\}$ As illustration, let consider the transformation of spherical to Cartesian coordinates (see Fig. 2.9) for the definition of coordinates:

$$\begin{cases} x^1 = r \sin\theta \cos\varphi \\ x^2 = r \sin\theta \sin\varphi \\ x^3 = r \cos\theta \end{cases}, \quad J_\alpha^i = \begin{bmatrix} \sin\theta\cos\varphi & r\cos\theta\cos\varphi & -r\sin\theta\sin\varphi \\ \sin\theta\sin\varphi & r\cos\theta\sin\varphi & r\sin\theta\cos\varphi \\ \cos\theta & -r\sin\theta & 0 \end{bmatrix}$$

(2.12)

where J_α^i is the Jacobian matrix associated to the change of coordinates. This transformation is not onto but it is a diffeomorphism onto its image. The determinant is $\mathrm{Det}(J_\alpha^i) = r^2 \sin\theta > 0$, positive for $r > 0$ and $0 < \theta \leq \pi$.

2.3.1.3 Examples of Group of Transformations

Both the body and the spacetime are described by manifolds: the continuum of dimension three, and the spacetime of dimension four e.g. Defrise (1953), Kadianakis (1996). Here some fundamental examples for coordinate system changes where change of coordinates may include change of reference frames e.g. Bain (2004). For the sake of simplicity, let us denote $x^0 := ct$ the time variable and $x^i = (x^1, x^2, x^3)$ the space coordinates. The spacetime coordinates hold as $x^\alpha = (x^0, x^1, x^2, x^3)$.

The *Leibniz group* that is the group of transformations: $y^j := R_i^j(x^0)x^i + u^j(x^0)$ for $i, j = 1, 2, 3$, and $y^0 := x^0 + u^0$ where $R_i^j(x^0)$ is an orthogonal transformation depending on time variable, and $u^\alpha = (u^0, u^1, u^2, u^3)$. These are transformations between rigid Euclidean reference frames with arbitrary rotating and arbitrary accelerating with respect each other e.g. Bernal and Sanchez (2003) (Fig. 2.3).

The second is the *Maxwell group* represented by the transformations: $y^j := R_i^j x^i + u^j(x^0)$ for $i, j = 1, 2, 3$, and $y^0 := x^0 + u^0$ where R_i^j is a uniform orthogonal transformation, independent of time variable. Maxwell group includes fixed rotating frames but arbitrary accelerating frames.

The third example consists of *Newton–Cartan transformations group*: $y^j := R_i^j x^i + v^j x^0 + u^j$, for $i, j = 1, 2, 3$, and $y^0 := x^0 + u^0$ where $v^j = (v^1, v^2, v^3)$ is a constant velocity. This is the *Galilean* group (or Galilean Lie group; Rosen 1972) with 10 parameters. The basic root of the design of Galilean transformation comes from the "intuitive" observation that the velocity of a material point as seen by observers at rest in two reference frames differs and depend on the relative constant velocity of the two frames, linearly.

Role of these groups in regards of constitutive laws may be found in e.g. Rakotomanana (2003) when considering the principle of material indifference. An extension of the Newton–Cartan group is the local Galilean group $R_i^j(x^k)$

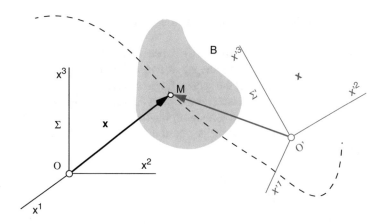

Fig. 2.3 Change of spacetime coordinate system $x^\mu \rightarrow x'^\alpha(x^\mu)$. The interpretation is a "passive diffeomorphism" on this figure, where the material point M moves with respect to the reference system Σ—along the dotted trajectory—with coordinates (x'^α) with respect to moving reference Σ' (in red) (time origin change, translation, and rotation)

and $u^j(x^k)$, where the linear transformation and the translation depend on space coordinates but not on the time variable x^0.

Another important group is that of *internal transformations*. We can attach numerous coordinate systems to one reference frame. For that purpose, it is worth to consider the internal transformations e.g. Krause (1976):

$$y^0 = y^0(x^0, x^1, x^2, x^3), \qquad y^j = y^j(x^1, x^2, x^3), \quad j = 1, 2, 3$$

where x^0 and y^0 correspond to time variable in each coordinate system. Leibniz and Maxwell groups are particular cases. The first equation expresses a synchronization of the time measure (clock) within frame of reference, whereas the second equation one expresses a change of coordinate system within the three-dimensional space of the spacetime. Internal transformations are nonlinear extension of Galilean transformations, but they do not allow us to derive the adequate transformations of spacetime. Moreover, if we assume that inertial forces are present in a given frame, they cannot be removed by means of internal transformations. This basically constitutes the subtle difference between coordinate systems and frames of reference.

2.3.1.4 Lorentz Invariance

Concepts of Galilean transformation and absolute space decoupled from absolute time are valid for most mechanics of material body. However, these concepts are found to be inadequate when the velocity of the body approaches the light speed, and namely for electrodynamic phenomenon where body velocity is near the light propagation speed. This aspect should be accounted for in electrodynamics when dealing with Maxwell's equations. Galilean transformation is no more adequate for building the physics theory.

In that way, the special relativity theory is based on two postulates: (1) physics laws are the same (have the same shape) in two reference frames in relativistic constant motion (no rotation); (2) the speed of light c is finite and independent of the motion of its source in any reference frame e.g. Ryder (2009). Now, to introduce the difference between covariance and diffeomorphism invariance, let consider the wave equation of any scalar field ϕ in Cartesian coordinates system ($x^0 := ct, x^1, x^2, x^3$), with the time t:

$$\frac{\partial^2 \phi}{\partial (x^1)^2} + \frac{\partial^2 \phi}{\partial (x^2)^2} + \frac{\partial^2 \phi}{\partial (x^3)^2} - \frac{\partial^2 \phi}{\partial (x^0)^2} = 0 \qquad (2.13)$$

c being the light velocity. Based on the two postulates, it is thus classically shown that the change of coordinate system (called Lorentz transformations) according to the relation $y^\alpha = J^\alpha_\mu x^\mu$ where (see appendix for the determination of this transformation):

$$J^i_j = \delta^i_j + \frac{(\gamma - 1) v^i v_j}{|\mathbf{v}|^2}, \qquad J^0_i = J^i_0 = -\gamma v^i, \qquad J^0_0 = \gamma \qquad (2.14)$$

where v^i are three real parameters satisfying $|\mathbf{v}|^2 := (v^1)^2 + (v^2)^2 + (v^3)^2 < 1$ (any particle has a velocity lower than the light speed which was set to $c = 1$), and $\gamma := (1 - |\mathbf{v}|^2)^{-1/2}$ is the invariance group extending the Galilean invariance of classical Newtonian mechanics to Lorentz invariance of special relativity. The coordinate transformation associated to (2.14) is called Poincaré-Lorentz boost (see (2.15) for the proof), recall that Lorentz boost is simply a Lorentz transformation which does not involve rotation,

$$\begin{cases} y^0 = \gamma \left(x^0 - \mathbf{v} \cdot \mathbf{x} \right) \\ \mathbf{y} = \mathbf{x} + (\gamma - 1) \frac{1}{\|\mathbf{v}\|^2} \mathbf{v} \otimes \mathbf{v} (\mathbf{x}) - \gamma \mathbf{v} x^0 \end{cases} \qquad (2.15)$$

where $(y^0, \mathbf{y}) := (y^0, y^1, y^2, y^3)$ are the new spacetime coordinates. It should be stressed that the Lorentz transformation mixes the time-coordinate x^0 and the spatial coordinates $x^i, i = 1, 2, 3$ between the two reference frames. They cannot be dissociated to extract an absolute time. The Lorentz transformation (2.14) is the minimal invariance requirement to physics theory, including mechanics and electromagnetism e.g. Kibble (1961) and associated to special relativity. For the scalar function $\Phi(y^\alpha) = \phi(x^\mu) = \phi(A^\mu_\alpha x^\alpha)$, rewriting the wave equation (2.13) in terms of the new coordinates gives (Fig. 2.4):

$$\frac{\partial^2 \Phi}{\partial (y^1)^2} + \frac{\partial^2 \Phi}{\partial (y^2)^2} + \frac{\partial^2 \Phi}{\partial (y^3)^2} - \frac{\partial^2 \Phi}{\partial (y^0)^2} = 0 \qquad (2.16)$$

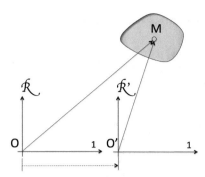

Fig. 2.4 Lorentz invariance: The change of reference frame is defined by a translation of $\mathscr{R} \to \mathscr{R}'$ such that the relative velocity v of the second reference $\mathscr{R}'$ is constant. The Lorentz transformation is defined as a transformation verifying that the four-dimensional length of an infinitesimal vector remains constant: $ds^2 = (dx^0)^2 - [(dx^1)^2 + (dx^2)^2 + (dx^3)^2] = (dy^0)^2 - [(dy^1)^2 + (dy^2)^2 + (dy^3)^2]$ where the coordinate system in $\mathscr{R}$ is denoted (x^μ) whereas the coordinate system in $\mathscr{R}'$ is (y^μ)

The two wave equations (2.13) and (2.16) have exactly the same shape showing that the wave equation is invariant under Lorentz transformation e.g. Westman and Sonego (2009). Both of them may be written in a concise formulation as follows:

$$\hat{g}^{\alpha\beta} \nabla_\alpha \nabla_\beta \Phi = 0, \qquad \hat{g}^{\alpha\beta} := \mathrm{diag}\{+1, -1, -1, -1\}, \qquad \nabla_\alpha := \partial_\alpha \qquad (2.17)$$

and in which we have defined the differential operator ∇_α. This operator will be extended later to the notion of affine connection in order to ensure the covariance property. However, between Cartesian coordinates and spherical coordinates, we do not obtain the same shape for wave equations as (2.13) and (2.16). This assesses that explicit shape of the wave equation is not invariant under general change of coordinate system of the spacetime. This should not be confused with covariance, as we will see hereafter. Conversely the wave equation (2.17) has exactly the same shape for any coordinate system if the metric and the connection is chosen worthily. It is said covariant with respect to a diffeomorphism. A further step would be the extension of the global invariance concept, which may be associated to the change of reference body, to a local gauge invariance that is more appropriate to gradient continuum and to relativistic gravitation theory as performed in e.g. Utiyama (1956).

2.3.2 Elements on Spacetime and Invariance for Relativity

We remind in this subsection some mathematical basis for the relativity, namely some tensor analysis, Minkowski spacetime, and Poincaré's transformations.

2.3.2.1 Forms, Tensors and (Pseudo)-Riemannian Manifolds

We notice $T_P \mathscr{B}$ the tangent space of a manifold $\mathscr{B}$ at point P and $T_P \mathscr{B}^*$ its dual space. Let $\{\mathbf{e}_1, \ldots, \mathbf{e}_n\}$ be a base of $T_P \mathscr{B}$ (contravariant vectors with the lower index) and its dual base $\{\mathbf{f}^1, \ldots, \mathbf{f}^n\}$ in $T_P \mathscr{B}^*$ (covariant vectors with upper index) such that $\langle \mathbf{f}^i, \mathbf{e}_j \rangle \equiv \delta^i_j$.

Definition 2.4 (Tensor) Let $\{\mathbf{u}_1, \ldots, \mathbf{u}_q\} \in T_P \mathscr{B}$ and $\{\mathbf{v}^1, \ldots, \mathbf{v}^p\} \in T_P \mathscr{B}^*$ be some arbitrary vectors. A p-contravariant and q-covariant tensor field $\mathcal{T}$ on $\mathscr{B}$ is a multilinear form defined at each point $P \in \mathscr{B}$ by $\mathcal{T} : (\mathbf{v}^1, \ldots, \mathbf{v}^p, \mathbf{u}_1, \ldots, \mathbf{u}_q) \in (T_P \mathscr{B}^*)^p \times (T_P \mathscr{B})^q \longrightarrow \mathcal{T}(\mathbf{v}^1, \ldots, \mathbf{v}^p, \mathbf{u}_1, \ldots, \mathbf{u}_q) \in \mathbb{R}$. The sum $(p+q)$ is called the rank of the tensor field. The couple (p, q) is its type.

For any two tensors $\mathcal{T}$ and $\mathcal{S}$, respectively q and q' covariant on $\mathscr{B}$, we define the tensor product $\mathcal{T} \otimes \mathcal{S}$ as a linear application such that:

$$\mathcal{T} \otimes \mathcal{S} \left(\mathbf{u}_1, \ldots, \mathbf{u}_q, \mathbf{w}_1, \ldots, \mathbf{w}_{q'} \right) \equiv \mathcal{T} \left(\mathbf{u}_1, \ldots, \mathbf{u}_q \right) \mathcal{S} \left(\mathbf{w}_1, \ldots, \mathbf{w}_{q'} \right) \qquad (2.18)$$

This definition may be extended to contravariant and mixed tensors e.g. Rakotomanana (2003).

Definition 2.5 A differential form of order q, or a q-form, is a totally antisymmetric tensor of type $(0, q)$.

A type $(1, 0)$ tensor, denoted $\mathbf{u}$, is called vector field and a type $(0, 1)$ tensor, denoted ω a co-vector field or a 1-form field on the manifold $\mathscr{B}$. Let give some examples of tensor products of vectors and 1-forms:

1. $\mathbf{e}_{i_1} \otimes \mathbf{e}_{i_2}, \mathbf{e}_{i_1} \otimes \mathbf{e}_{i_2} \otimes \mathbf{e}_{i_3}, \ldots, \mathbf{e}_{i_1} \otimes \mathbf{e}_{i_2} \otimes \mathbf{e}_{i_3} \otimes \cdots \otimes \mathbf{e}_{i_r}$
2. $\mathbf{f}^{j_1} \otimes \mathbf{f}^{j_2}, \mathbf{f}^{j_1} \otimes \mathbf{f}^{j_2} \otimes \mathbf{f}^{j_3}, \ldots, \mathbf{f}^{j_1} \otimes \mathbf{f}^{j_2} \otimes \mathbf{f}^{j_3} \otimes \cdots \otimes \mathbf{f}^{j_r}$
3. $\mathbf{e}_{i_1} \otimes \mathbf{f}_{j_1}, \ldots, \mathbf{e}_{i_1} \otimes \mathbf{f}^{j_1} \otimes \mathbf{f}^{j_2} \otimes \cdots \mathbf{e}_{i_r}$.

where the indexes $i_1, i_2, \cdots, i_r$ and $j_1, i_2, \cdots, j_r$ may take any value of $1, 2, \cdots, n$. Number of these indexes is not limited.

Definition 2.6 Let $\left\{ \mathbf{f}^{i_1}, \cdots, \mathbf{f}^{i_r} \right\}$ a set of r one-forms on the manifold $\mathscr{B}$ of dimension n, with $r \leq n$. Indexes $(i_1, \cdots, i_r)$ take the values $(1, \cdots, n)$. Consider all permutations (even or odd) $\sigma : (i_1, \cdots, i_r) \rightarrow (\sigma(i_1), \cdots, \sigma(i_r))$ with its sign:

$$\text{sign}(\sigma) := \epsilon^{\sigma(i_1), \cdots, \sigma(i_r)}_{i_1, \cdots, i_r} = \begin{cases} +1 \text{ for even permutations} \\ -1 \text{ for odd permutations} \end{cases}$$

The wedge product is defined by the totally antisymmetric tensor product:

$$\mathbf{f}^{i_1} \wedge \mathbf{f}^{i_2} \wedge \cdots \wedge \mathbf{f}^{i_r} := \sum_\sigma \text{sign}(\sigma) \, \mathbf{f}^{\sigma(i_1)} \otimes \mathbf{f}^{\sigma(i_2)} \otimes \cdots \otimes \mathbf{f}^{\sigma(i_r)} \qquad (2.19)$$

where summation runs over all permutations σ.

Fig. 2.5 Transformation of a square to a parallelogram (in a tridimensional Euclidean space). Vectors **a** and **b** are sides of the square whereas 1-forms α and β are normal vectors of the square

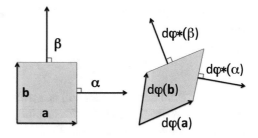

Typical example of differential form of order 1 on an open subset of $\mathbb{R}^n$ is given as follows: Consider a scalar $\mathscr{C}^1$ function $\varphi : (\xi^1, \cdots, \xi^n) \in \mathbb{R}^n \to \varphi(\xi^1, \cdots, \xi^n) \in \mathbb{R}$ then its total differential (cf. exterior derivative in appendix) is a differential 1-form:

$$d\varphi = \partial_i \varphi \, d\xi^i$$

where the set $\{d\xi^1, \cdots, , d\xi^n\}$ constitute a covariant independent 1-forms. Let give some examples of wedge product of 1-forms:

1. $d\xi^{i_1} \wedge d\xi^{i_2} = d\xi^{i_1} \otimes d\xi^{i_2} - d\xi^{i_2} \otimes d\xi^{i_1}$
2. $d\xi^{i_1} \wedge d\xi^{i_2} \wedge d\xi^{i_3} = d\xi^{i_1} \otimes d\xi^{i_2} \otimes d\xi^{i_3} + d\xi^{i_2} \otimes d\xi^{i_3} \otimes d\xi^{i_1} + d\xi^{i_3} \otimes d\xi^{i_1} \otimes d\xi^{i_2} - d\xi^{i_1} \otimes d\xi^{i_3} \otimes d\xi^{i_2} - d\xi^{i_2} \otimes d\xi^{i_1} \otimes d\xi^{i_3} - d\xi^{i_3} \otimes d\xi^{i_2} \otimes d\xi^{i_1}$
3. $\ldots, d\xi^{i_1} \wedge d\xi^{i_2} \wedge d\xi^{i_3} \wedge \cdots \wedge d\xi^{i_r}$, with $r \leq n$

where the indexes $i_1, i_2, \cdots, i_r$ may take any value of $1, 2, \cdots, n$.

Remark 2.2 In an Euclidean space, it is usual to represent a 1-form by a "vector". However, it may induce confusion namely when we consider large deformation of a continuum $\mathscr{B}$ (manifold) undergoing an arbitrary transformation even if it is a diffeomorphism $\varphi : \mathscr{B} \to \varphi(\mathscr{B})$. Indeed, let consider the linear tangent transformation $d\varphi$ (whose components are triads F_μ^i) with the Cartesian components $\partial_\mu \varphi^i$ (also usually called deformation gradient in the framework of continuum mechanics). Figure 2.5 reports an Illustration of tangent and cotangent space. During the deformation of a square to a parallelogram, the two vectors **a** and **a** become $d\varphi(\mathbf{a})$ and $d\varphi(\mathbf{b})$ respectively, while the 1-forms α and β are transformed into $d\varphi^*(\alpha)$ and $d\varphi^*(\beta)$ where the components of φ^* are $\mathrm{Det}(F_\nu^j)\left(F_\mu^i\right)^{-1}$. For a tridimensional Euclidean manifold endowed with metric tensor $g_{\mu\nu}$ and a velum-form ω_0 (determinant), 1-forms α and β may be represented by vectors (homeomorphism between vector and normal vectors of surface). Deformation of vectors and those of 1-forms are different.

Definition 2.7 (Components of Tensor) If $\mathfrak{T}$ is a tensor field of type (p, q) then the scalars $\mathfrak{T}^{j_1 \ldots j_q}_{l_1 \ldots l_p}$ are the components of $\mathfrak{T}$ projected onto the base formed by $\{\mathbf{e}_{j_1}, \ldots, \mathbf{e}_{j_q}\}$ and $\{\mathbf{f}^{l_1}, \ldots, \mathbf{f}^{l_p}\}$, defined by $\mathfrak{T}^{j_1 \ldots l_p}_{j_1 \ldots j_q} := \mathfrak{T}\left(\mathbf{f}^{l_1}, \ldots, \mathbf{f}^{l_p}, \mathbf{e}_{j_1}, \ldots, \mathbf{e}_{j_q}\right)$.

Any (p, q) tensor $\mathcal{T}$ may thus be decomposed along the tensor base according to the tensor product (2.18):

$$\mathcal{T} := \mathcal{T}^{j_1 \ldots j_q}_{l_1 \ldots l_p} \, \mathbf{f}^{l_1} \otimes \cdots \otimes \mathbf{f}^{l_p} \otimes \mathbf{e}_{j_1} \otimes \ldots \otimes \mathbf{e}_{j_q}$$

In the present book, a tensor will be assimilated to its components, as soon as the vector bases are defined. If $\mathcal{T}^{h_1 \ldots h_r}_{k_1 \ldots k_s}$ constitute the components of a type (r, s) tensor then, under the transformation (2.11)

$$\mathcal{T}^{j_1 \ldots j_r}_{l_1 \ldots l_s} = A^{j_1}_{\alpha_1} \ldots A^{j_r}_{\alpha_r} J^{\beta_1}_{l_1} \ldots J^{\beta_s}_{l_s} \, \mathcal{T}^{\alpha_1 \ldots \alpha_r}_{\beta_1 \ldots \beta_s} \tag{2.20}$$

and the corresponding inverse formulation

$$\mathcal{T}^{\alpha_1 \ldots \alpha_r}_{\beta_1 \ldots \beta_s} = J^{\alpha_1}_{j_1} \ldots J^{\alpha_r}_{j_r} A^{l_1}_{\beta_1} \ldots A^{l_s}_{\beta_s} \, \mathcal{T}^{j_1 \ldots j_r}_{l_1 \ldots l_s} \tag{2.21}$$

Properties 2.3.1 *According to the previous results:*

1. *A scalar ψ is a type $(0, 0)$ tensor which has the same form in any coordinate system: ψ in (y^i), $\overline{\psi}$ in (x^α) and $\psi = \overline{\psi}$.*
2. *If all tensor components vanish in a coordinate system then they vanish in any other coordinate system.*

Definition 2.8 (Metric) Let $\mathcal{B}$ be a differentiable manifold. A metric $\mathbf{g}$ is a type $(0, 2)$ tensor on $\mathcal{B}$ which satisfies the following axioms at each point $P \in \mathcal{B}$:

1. $\mathbf{g}(\mathbf{u}, \mathbf{v}) = \mathbf{g}(\mathbf{v}, \mathbf{u})$,
2. $\mathbf{g}(\mathbf{u}, \mathbf{u}) \geq 0$ where the equality holds only when $\mathbf{u} = 0$.

Projected onto a base $\{\mathbf{e}_1, \ldots, \mathbf{e}_n\}$ of $T_P \mathcal{B}$, components of $\mathbf{g}$ are g_{ij}, components of its inverse $\mathbf{g}^{-1}$ are g^{ij} with $g^{ik} g_{kj} = \delta^i_j$. Since $g_{ij} = g_{ji}$ the eigenvalues of the matrix are real and strictly positive (or negative depending on the convention) and their eigenvectors orthogonal. We say that $\mathbf{g}$ is a pseudo-Riemannian metric if it satisfies the first axiom and the second one is replaced by: $\mathbf{g}(\mathbf{u}, \mathbf{v}) = 0$ for any $\mathbf{u} \in T_P \mathcal{B}$, then $\mathbf{v} = 0$. For pseudo-Riemannian metric, some of the eigenvalues are negative. Say j the number of negative (or positive depending on the convention) eigenvalues.

Definition 2.9 (Riemannian Manifold) A differential manifold $\mathcal{B}$ endowed with a (pseudo-) metric $(\mathcal{B}, \mathbf{g})$ is a (pseudo-) Riemannian manifold.

For $j = 1$, the pseudo-Riemannian manifold $(\mathcal{B}, \mathbf{g})$ is called Lorentzian. On such a manifold, the tangent four-vectors $\mathbf{u} \in T_P \mathcal{B}$ can be divided into three classes (with the convention $j = 1$ is the number of positive eigenvalues):

1. $g_{\alpha\beta} u^\alpha u^\beta < 0$, vector $\mathbf{u}$ is spacelike
2. $g_{\alpha\beta} u^\alpha u^\beta = 0$, vector $\mathbf{u}$ is lightlike
3. $g_{\alpha\beta} u^\alpha u^\beta > 0$, vector $\mathbf{u}$ is timelike.

Definition 2.10 (Galilean Structure) Let $\mathcal{M}$ be a four-dimensional differentiable manifold, endowed with a $(0, 2)$ tensor field $\mathbf{g}$ and a nonsingular 1-form τ. The triplet $(\mathcal{M}, \mathbf{g}, \tau)$ constitutes a Galilean structure provided the two following conditions are satisfied:

1. Tensor $\mathbf{g}$, called space metric, is symmetric and degenerate of rank 3 (three non zero eigenvalues).
2. There exists a vector field $\mathbf{u} \neq 0$ on $\mathcal{M}$ such that $\mathbf{g}(\mathbf{u}, \mathbf{v}) \equiv 0$ for any vector field $\mathbf{v} \in \mathcal{T}_\mathbf{x}\mathcal{M}$, and $\tau(\mathbf{u}) = 1$.

This definition introduces the lightlike vector field $\mathbf{u} \in \mathcal{T}_\mathbf{x}\mathcal{M}$, which is an element of the Galilean structure. The 2-covariant symmetric tensor $\tau \otimes \tau$, degenerate of rank 1 (one non zero eigenvalue) is called time metric. The spacetime metric is therefore defined as $\hat{\mathbf{g}} := \tau \otimes \tau - \mathbf{g}$ (hereafter, the hat on the spacetime metric will be dropped if there is no ambiguity). In mechanics, it is worth to introduce a local coordinate system $(x^0 := ct, x^1, x^2, x^3)$ in the neighborhood of any event (spacetime point) in such a way that $\tau := cdt$ and for any index i, $\tau(\partial_i) = 0$, meaning that the space coordinates x^i are independent variables with respect to time t e.g. Dixon (1975). c denotes the light speed.

2.3.2.2 Hilbert's Causality Principle

The causality principle constitutes the most fundamental principle that should be satisfied by any physics model. In the framework of general relativity, this means that each event (point) x^μ of the spacetime $\mathcal{M}$ should admit an intrinsic valid notion of the past, present, and future. These states should not depend on the mathematical description (invariant notion). When considering dynamical evolution of metric on a manifold or change of spacetime coordinates, Hilbert's conditions must be satisfied to respect the causality e.g. Brading and Ryckman (2008). Arbitrariness of coordinates change (or dynamics of metric evolution) should be limited if we want to stand on the point of view that two points of the same timeline are in a cause-effect relationship. The following four inequalities must be satisfied besides the condition $\det\left[g_{\alpha\beta}\right] < 0$ e.g. Brading and Ryckman (2008), Havas (1964):

$$g_{00} > 0, \quad g_{11} < 0, \quad \det\begin{bmatrix} g_{11} & g_{12} \\ g_{21} & g_{22} \end{bmatrix} > 0, \quad \det\begin{bmatrix} g_{11} & g_{12} & g_{13} \\ g_{21} & g_{22} & g_{23} \\ g_{31} & g_{32} & g_{33} \end{bmatrix} < 0 \qquad (2.22)$$

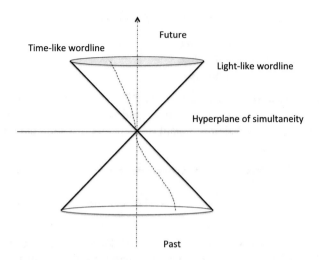

Fig. 2.6 Minkowski spacetime. Diagram showing the light cone for a worldline in Minkowski spacetime. The observer is at the junction between two light cones, the past and the future. The space is represented hypersurface of simultaneity

Minkowski spacetime is as a four-dimensional spacetime used to describe the consequences of special relativity theory on the finiteness of light speed and for the time-like worldline $g_{\alpha\beta}\, u^{\alpha} u^{\beta} := c^2 dt - g_{ij} dx^i dx^j > 0$ and for the light-like wordline $g_{\alpha\beta}\, u^{\alpha} u^{\beta} := c^2 dt - g_{ij} dx^i dx^j = 0$ (Fig. 2.6). These conditions ensure that change of spacetime coordinates will respect to the flow of time $x^0 := ct$ (they allow us to define a so-called "proper time coordinate"). Recently, the application of Hilbert's causality conditions provides some restriction on the formation of black hole cosmology after analysis of the famous Schwarzschild solutions of Einstein's equations of relativistic gravitation e.g. Logunov and Mestivirishvili (2012). Application of causality condition in strain gradient continuum mechanics may be found in e.g. Metrikine (2006), which is based on a conjecture extending Einstein's causality on the second-order differential equations. Broadly speaking, Hilbert's causality conditions involve two problems in relativistic gravitation and electromagnetism: the causal ordering and the univocal determination of metric components. They are strongly tight with the invariance of Lagrangian function and its arguments e.g. Brading and Ryckman (2008), and even beyond the covariance, related with the search of invariance group of spacetimes e.g. Zeeman (1964), Williams (1973).

Definition 2.11 (Isometry) Let $(\mathscr{B}, \mathbf{g})$ be a (pseudo-) Riemannian manifold. A diffeomorphism $\mathbf{f} : \mathscr{B} \to \mathscr{B}$ is an isometry if it preserves the metric:

$$A^i_{\alpha}\, A^j_{\beta}\, g_{ij}\left(y^k\right) = g_{\alpha\beta}(x^{\mu}) \tag{2.23}$$

where y^k and x^{μ} are the coordinates of $\mathbf{f}(M) \in \mathbf{f}(\mathscr{B})$, and $M \in \mathscr{B}$ respectively.

The orthogonality group $\mathcal{O}(n), n = 3, 4$ is the group of $n \times n$ real matrices whose transpose is equal to their inverse. In an Euclidean n-dimensional manifold, the linear transformation of (tangent) vectors as $\hat{\mathbf{e}}_j = R_{jh}\mathbf{e}_h$, provided that the coefficients R_{jh} satisfy the orthogonality condition $R_{jh}\ R_{kh} \equiv \delta_{jk}$, implies that $\det\mathbf{R} = \pm 1$. These passive diffeomorphisms are orthogonal transformations, and for the subgroup $\det\mathbf{R} = 1$ rotations (proper orthogonal). Isometries form a group which includes the identity map. Since an isometry preserves the length of a (tangent) vector, it characterizes a rigid motion in continuum mechanics. In the framework of relativistic gravitation, the background is the Minkowskian spacetime, a four-dimensional real vector space $\mathbb{R}^{(1,3)}$ endowed with an inner product, which is a nondegenerate bilinear form with signature $(1, 3)$, where x^0 is the time. But the associated metric is not positive-definite. Adopting the notation that $\mathcal{O}(n) := \mathcal{O}(0, n)$, the orthogonal group of the Minkowskian spacetime is denoted $\mathcal{O}(1, 3)$. This group $\mathcal{O}(1, 3)$ is called Lorentzian group of transformations.

Definition 2.12 (Lorentz Group) The Lorentz group $\mathcal{O}(1, 3)$ is the group of linear transformations that preserve the Minkowski spacetime inner product (Eq. (2.1)) on $\mathbb{R}^4$.

Let now remind briefly the necessary and sufficient condition for a 4×4 matrix— representing a linear transformation—$\mathbf{A} = (A^\alpha_\beta)$ to leave the inner product (Eq. (2.1)) of any two four-vectors $\mathbf{u}$ and $\mathbf{v}$ invariant. Suppose the transformed vectors:

$$u'^\mu = A^\mu_\alpha u^\alpha, \qquad u'^\nu = A^\nu_\beta u^\beta, \qquad \mathbf{u}' \cdot \mathbf{v}' = g_{\mu\nu} A^\mu_\alpha u^\alpha\, A^\nu_\beta u^\beta \equiv g_{\alpha\beta} u^\alpha u^\beta$$

In order for the condition $\mathbf{u}' \cdot \mathbf{v}' \equiv \mathbf{u} \cdot \mathbf{v}$ to hold for two arbitrary vectors $\mathbf{u}$ and $\mathbf{v}$, we deduce the necessary and sufficient condition (then a definition) of Lorentz transformations, which resembles to the condition (2.23):

$$g_{\mu\nu} A^\mu_\alpha\, A^\nu_\beta \equiv g_{\alpha\beta} \tag{2.24}$$

These transformations constitute a group (closure, associativity, identity and inverse conditions). This group defined as such is any larger as the group defined hereafter (boosts and rotations). The subgroup which preserves the time orientation is denoted $\mathcal{O}^+(1, 3)$. The subgroup which preserves both the time and the space orientation is the group of proper Lorentz transformations $\mathcal{SO}^+(1, 3)$.

2.3.2.3 Euclidean Spacetime and Isometries

An Euclidean spacetime is a quadruplet, called affine fiber bundle, $(\mathcal{M}, \mathbf{g}, T, \tau)$, where $\mathcal{M}$ is a four dimensional differentiable manifold endowed with a metric $\mathbf{g}$, whose elements $(x^\alpha, \alpha = 0, 1, 2, 3) \in \mathcal{M}$ are called events. T is a one dimensional ordered Euclidean space and having underlying vector space on the real set $\mathbb{R}$. Projection map $\tau : \mathcal{M} \to T$ characterizes the absolute time $\tau(x^\alpha) = ct$ of the event

Fig. 2.7 Euclidean
spacetime. The $3D$ affine
Euclidean (sub)-spaces
$\mathcal{M}_t := \{x^\alpha \in$
$\mathcal{M}$, such that $\tau(x^\alpha) = ct\}$,
$t \in T$ are called
instantaneous spaces. The
structure group of the bundle
is the group of isometries of a
three dimensional Euclidean
space. Time T remains the
same for all points of the
spacetime (absolute time)

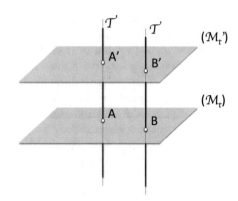

$x^\alpha \in \mathcal{M}$ e.g. Kadianakis (1996). At this step, there is no need of a supplementary
structure such as affine structure. According to classical Newton mechanics, there is
a absolute time τ and a absolute three-dimensional space. Hence the manifold $\mathcal{M}$,
set of the events, may be merely considered as the Cartesian product of the sets of
instants T and the set of all space points $\mathcal{S} \to \mathcal{M}_t$, say $\mathcal{M} := T \times \mathcal{M}_t$ e.g. Ehlers
(1973). Any instantaneous subspace $\mathcal{M}_t$ is then implicitly assumed Euclidean 3-
spaces (Fig. 2.7).

2.3.2.4   Minkowski Spacetime and Lorentz Transformations

Let $(\mathcal{M}, \mathbf{g})$ be a (pseudo)—Riemannian manifold, and consider the kinematics-
based Lorentz transformation, where the velocity v is along the axis x^1 for
simplifying. The metric is defined by $g_{\alpha\beta} := \mathrm{diag}(1, -1, -1, -1)$, to give the
Minkowski line element $ds^2 = g_{\alpha\beta}\, dx^\alpha dx^\beta = (dx^0)^2 - (dx^1)^2 - (dx^2)^2 - (dx^3)^2$.
The group of proper Lorentz transformations $\mathcal{O}(1, 3)$ are generated by six types
of transformations that are three simple spatial rotations, and three time-space
operators called boosts. Consider the usual form of Lorentz transformations (one
of the generator boost) $(x^0 := ct)$:

$$\begin{cases} y^0 = \dfrac{1}{\sqrt{1 - v^2}}\left(x^0 - vx^1\right) \\ y^1 = \dfrac{1}{\sqrt{1 - v^2}}\left(-vx^0 + x^1\right) \\ y^2 = x^2 \\ y^3 = x^3 \end{cases} \implies \begin{cases} x^0 = \dfrac{1}{\sqrt{1 - v^2}}\left(y^0 + vy^1\right) \\ x^1 = \dfrac{1}{\sqrt{1 - v^2}}\left(vy^0 + y^1\right) \\ x^2 = y^2 \\ x^3 = y^3 \end{cases}$$

By plugging the second system of equations into the Minkowski line element, we
obtain $ds^2 = (dy^0)^2 - (dy^1)^2 - (dy^2)^2 - (dy^3)^2 = g_{\alpha\beta}\, dy^\alpha dy^\beta$, stating that the
Minkowski line element is invariant with respect to Lorentz transformations (2.14),
assessing that they are isometries. Defining $\gamma := (1 - v^2)^{-1/2}$, and $\theta :=$

$\ln\left[(1+v^2)/\sqrt{1-v^2}\right]$, the three boosts generator B_i, $i = 1, 2, 3$ are given by:

$$\begin{bmatrix} \cosh\theta & \sinh\theta & 0 & 0 \\ \sinh\theta & \cosh\theta & 0 & 0 \\ 0 & 0 & 1 & 0 \\ 0 & 0 & 0 & 1 \end{bmatrix}, \begin{bmatrix} \cosh\theta & \sinh\theta & 0 & 0 \\ 0 & 1 & 0 & 0 \\ \sinh\theta & 0 & \cosh\theta & 0 \\ 0 & 0 & 0 & 1 \end{bmatrix}, \begin{bmatrix} \cosh\theta & 0 & 0 & \sinh\theta \\ 0 & 1 & 0 & 0 \\ 0 & 0 & 1 & 0 \\ \sinh\theta & 0 & 0 & \cosh\theta \end{bmatrix}$$

Boosts specify the change from one observer's reference frame to another one. More classically, three space rotations generator R_i, $i = 1, 2, 3$ are given by (θ is the rotation angle):

$$\begin{bmatrix} 1 & 0 & 0 & 0 \\ 0 & 1 & 0 & 0 \\ 0 & 0 & \cos\theta & -\sin\theta \\ 0 & 0 & \sin\theta & \cos\theta \end{bmatrix}, \begin{bmatrix} 1 & 1 & 0 & 0 \\ 0 & \cos\theta & 0 & \sin\theta \\ 0 & 0 & 1 & 0 \\ 0 & -\sin\theta & 0 & \cos\theta \end{bmatrix}, \begin{bmatrix} 1 & 0 & 0 & 1 \\ 0 & \cos\theta & -\sin\theta & 0 \\ 0 & \sin\theta & \cos\theta & 0 \\ 0 & 0 & 0 & 1 \end{bmatrix}$$

By introducing the dimension of the light velocity, the term v is replaced by v/c, and we recover the Galilean group of transformations when $v/c \to 0$ e.g. Ehlers (1973). Field equations of Newtonian mechanics are invariant under any member of the Galilean group, whereas those of the special relativity are invariant under the Lorentz group $v/c \neq 0$. Physical quantities in special relativity theory are always referred to a fixed Minkowski spacetime. The Lorentz transformations built with change of reference, (boosts and rotations) together with time reversal and space inversion equals to the entire Lorentz group.

Remark 2.3 Components of the metric tensor $\{g_{ij}\}$ constitute a symmetric matrix, the eigenvalues are real. For Riemannian manifold, all the eigenvalues are strictly positive. For a pseudo-Riemannian manifold, some of them may be strictly negative. The index of the metric is the pair (i, j) where i is the number of positive eigenvalues and j the number of negative eigenvalues. For $j = 1$, the metric is called Lorentzian.

Causality of spacetime events and invariance with respect to Lorentz group are strongly tight. Consider a Minkowskian spacetime $\mathscr{M}$ endowed with the metric $g_{\alpha\beta}$. In 1964, Zeeman's theorem states that the most general point (event) transformation of $\mathscr{M}$ to itself that preserves the Hilbert's causality relation within the spacetime is an 11-parameters group, called $\mathscr{G}$ generated by the orthochronous Lorentz transformations, the spacetime translation, and dilations (Zeeman 1964). In other words, the Lorentz group naturally appears as the symmetry group of the spacetime $\mathscr{M}$ if a single principle of causality is admitted a priori.

2.3.2.5 Global Poincaré Transformations

Indeed, common continuous symmetries of special relativistic theory is Lorentz invariance, meaning that the dynamics is the same in any Lorentz frame. The group of Lorentz transformations (2.14) (which is not compact) can be decomposed into two parts: (a) Boosts, where we go from one Lorentz frame to another, i.e., we change the velocity v; (b) Rotations, where we change the orientation of the coordinate frame. Lorentz transformations are not the only elements of the symmetry group of the Minkowskian spacetime. In the framework of special relativity with the metric $\hat{g}_{\alpha\beta} = \text{diag}\{+1, -1, -1, -1\}$, the Poincaré-Minkowski symmetry group of the spacetime is obtained by adding translation in space and time u^{ν}, $\nu = 0, 1, 2, 3$:

$$y^{\mu} = A^{\mu}_{\nu} x^{\nu} + u^{\nu}, \qquad \hat{g}_{\alpha\beta} A^{\alpha}_{\mu} A^{\beta}_{\nu} = \hat{g}_{\mu\nu} \qquad (2.25)$$

constitutes the symmetry group of the physics laws.[2] In the language of group theory, we have the definition.

Definition 2.13 The Poincaré's group is $\mathcal{SO}^{+}(1, 3) \ltimes \mathbb{R}^{1,3}$, the semi-direct product of the group of translations of the Minkowski spacetime $\mathbb{R}^{1,3}$ and the proper Lorentz transformations $\mathcal{SO}^{+}(1, 3)$.

Zeeman (1964) showed that the causality group of a Minkowskian spacetime $\mathcal{M}$ was the group generated by the orthochronous Lorentz group, the translation of $\mathcal{M}$ and dilations of $\mathcal{M}$. Later, Williams (1973) extended this result to show that the group of automorphisms $\mathcal{G}$ of $\mathcal{M}$ that preserved the norms of timelike vectors, was in fact the complete Poincaré group. In that way, it is needless to use the principle of relativity since the invariance of the light velocity c leads to the invariance with respect to Poincaré group e.g. Williams (1973). Poincaré group is a larger group of symmetries than Lorentz group, obtained when we add translations to the set of symmetries (space and time translation). This is the main idea behind the Einstein's theory of special relativity. The set of conserved quantities associated with Poincaré group is larger. Translational and boost invariance implies conservation of four momentum, and rotational invariance implies conservation of angular momentum. The choice of a metric tensor components $g_{\alpha\beta}$ may be explained as follows. At any point $P \in \mathcal{B}$, the tangent space $T_P\mathcal{B}$ is spanned by $\{\mathbf{e}_1, \cdots, \mathbf{e}_n\}$, and the dual space $T_P\mathcal{B}^*$ by $\{\mathbf{f}^1, \cdots, \mathbf{f}^n\}$. There is an alternative choice by taking a particular base defined as $\hat{\mathbf{e}}_a := \mathbf{F}^{-1}(\mathbf{e}_a)$, where $\det\mathbf{F} > 0$, and in such a way that $\mathbf{g}(\hat{\mathbf{e}}_a, \hat{\mathbf{e}}_b) := \hat{g}_{ab}$. The quantity $\mathbf{F}$ is called gradient of transformation in classical continuum mechanics

---

[2]Given the two groups $\mathbb{R}^{1,3}$ and $\mathcal{SO}^{+}(1, 3)$ such $\mathbb{R}^{1,3} \cap \mathcal{SO}^{+}(1, 3) = \{\mathbb{I}\}$, the semi-direct product of $\mathbb{R}^{1,3}$ and $\mathcal{SO}^{+}(1, 3)$ is a group denoted and with its element:

$$\mathcal{SO}^{+}(1, 3) \ltimes \mathbb{R}^{1,3} := \left\{ g = l\,k \in \mathcal{SO}^{+}(1, 3) \ltimes \mathbb{R}^{1,3}, \text{where } k \in \mathcal{SO}^{+}(1, 3), l \in \mathbb{R}^{1,3} \right\}.$$

e.g. Marsden and Hughes (1983), and called tetrads in relativistic gravitation theory e.g. Nakahara (1996). $\mathbf{F}$ is not necessarily a gradient of a mapping (triad/tetrad approach). In $3D$ continuum mechanics, it is usual to define the Green-Lagrange strain $\varepsilon_{\alpha\beta} := (1/2)(g_{\alpha\beta} - \hat{g}_{\alpha\beta})$. However, introducing this tensor as a primal strain variable necessarily induces a coupling between continuum matter and the ambient spacetime. Its extension in relativistic gravitation needs further development as we will see later. Arguments of any scalar field $\mathscr{L}$ may be chosen as the components of the metric $g_{\alpha\beta}$ onto a "deformed base" $\{\mathbf{e}_\gamma\}$. The definition of strain ε does not require the introduction a priori of displacement field and its gradient. It expresses the difference of shape between two configurations of a continuum. While the physical variables (modeled by tensors) may be calculated in arbitrary coordinate system, the displacement field components are naturally referred to the initial Cartesian frame. It is therefore physically significant to distinguish the Cartesian reference frame and arbitrary coordinate system.

2.3.3 Volume-Form

Governing equations of continuum physics mainly lie on the conservation laws of some physical quantities such as mass, energy and information to name but a few. The main uses of differential forms concern the integration of these physical extensive quantities on either within manifolds $\mathscr{B}$ or at their boundary $\partial\mathscr{B}$. For that purpose, a coordinate-free theory of integration is therefore necessary on manifolds.

A manifold $\mathscr{B}$ of dimension n is orientable if there exists a n-form, denoted ω_n which vanishes nowhere e.g. Lovelock and Rund (1975).

Definition 2.14 (Volume-Form) A volume-form on a manifold $\mathscr{B}$ of dimension n, is a nowhere vanishing n-form ω_n on $\mathscr{B}$.

For example, consider a surface by (ξ^1, ξ^2) immersed in a tridimensional space $\mathbb{R}^3$ defined by the mapping: $\varphi : (\xi^1, \xi^2) \in \mathbb{R}^3 \to \varphi(\xi^1, \xi^2) \in \mathbb{R}^3$. The tangent plane at point $\varphi(\xi^1, \xi^2)$ is the plane through the point φ spanned by the two (independent) vectors:

$$\mathbf{u}_1 := \frac{\partial_1\varphi}{\|\partial_1\varphi\|} \quad \text{and} \quad \mathbf{u}_3 := \frac{\partial_1\varphi \times \partial_2\varphi}{\|\partial_1\varphi \times \partial_2\varphi\|}, \quad \text{and} \quad \mathbf{u}_2 := \mathbf{u}_3 \times \mathbf{u}_1$$

The surface being immersed within a tridimensional space, the vector basis $\{\mathbf{u}_1, \mathbf{u}_2, \mathbf{u}_3\}$ is called adapted moving orthonormal frame. Consider the differential form $d\varphi := \partial_1\varphi d\xi^1 + \partial_2\varphi d\xi^2$ which obviously belongs to the tangent plane, and therefore may be decomposed onto the adapted moving reference frame as:

$$d\varphi := \partial_1\varphi\, d\xi^1 + \partial_2\varphi\, d\xi^2 = \theta^1\mathbf{u}_1 + \theta^2\mathbf{u}_2$$

where we define two 1-forms:

$$\theta^1 := \partial_1\varphi \cdot \mathbf{u}_1 \, d\xi^1 + \partial_2\varphi \cdot \mathbf{u}_1 \, d\xi^2$$
$$\theta^2 := \partial_1\varphi \cdot \mathbf{u}_2 \, d\xi^1 + \partial_2\varphi \cdot \mathbf{u}_2 \, d\xi^2$$

We deduce that:

$$d\varphi \wedge d\varphi = \left(\partial_1\varphi \cdot \mathbf{u}_1 \, d\xi^1\right) \wedge \left(\partial_2\varphi \cdot \mathbf{u}_2 \, d\xi^2\right) + \left(\partial_2\varphi \cdot \mathbf{u}_1 \, d\xi^2\right) \wedge \left(\partial_1\varphi \cdot \mathbf{u}_2 \, d\xi^1\right) \quad (2.26)$$

$$= [(\partial_1\varphi \cdot \mathbf{u}_1)(\partial_2\varphi \cdot \mathbf{u}_2) - (\partial_2\varphi \cdot \mathbf{u}_1)(\partial_1\varphi \cdot \mathbf{u}_2)] \, d\xi^1 \wedge d\xi^2$$

$$= \mathrm{Det} \begin{pmatrix} \partial_1\varphi \cdot \mathbf{u}_1 & \partial_1\varphi \cdot \mathbf{u}_2 \\ \partial_2\varphi \cdot \mathbf{u}_1 & \partial_2\varphi \cdot \mathbf{u}_2 \end{pmatrix} d\xi^1 \wedge d\xi^2 \qquad (2.27)$$

which is called area 2-form of the surface.

More generally, a volume-form on a n-dimensional manifold $\mathscr{M}$ is a nowhere vanishing n-form on $\mathscr{M}$. A particular n-form is $\mathbf{f}^1 \wedge \mathbf{f}^2 \cdots \wedge \mathbf{f}^n$. If $\{\mathbf{e}_1, \mathbf{e}_2, \ldots, \mathbf{e}_n\}$ is the vector base dual to $\{\mathbf{f}^1, \mathbf{f}^2, \ldots, \mathbf{f}^n\}$ then we have: $\left(\mathbf{f}^1 \wedge \mathbf{f}^2 \cdots \wedge \mathbf{f}^n\right)(\mathbf{e}_1, \mathbf{e}_2, \ldots, \mathbf{e}_n) = 1$. For spacetime or continuum with a metric structure, various volume-form may be designed as $\omega := f(\mathbf{x}) \, \omega_n$ provided any strictly positive function $f : \mathbf{x} \in \mathscr{M} \to f(\mathbf{x}) \in \mathbb{R}_+^*$. We can construct a volume-form by using an linearly independent n-forms $\mathbf{f}^1 \wedge \mathbf{f}^2 \cdots \wedge \mathbf{f}^n \neq 0$, and define $\omega_n := f \, \mathbf{f}^1 \wedge \mathbf{f}^2 \cdots \wedge \mathbf{f}^n$, where $f(\mathbf{x}) \in \mathbb{R}$ is a positive non-vanishing C^∞ scalar function. For (pseudo-) Riemannian manifold $(\mathscr{B}, \mathbf{g})$, there exists a natural volume-form which is invariant under coordinate transformation, say in a coordinate system $(x^1, \ldots, x^n)$, $\omega_n := \sqrt{\mathrm{Detg}} \, dx^1 \wedge dx^2 .. \wedge dx^n$ where $n = 3$, or $n = 4$ e.g. Nakahara (1996).

Remark 2.4 The volume form $\omega_n := \sqrt{\mathrm{Detg}} \, dx^1 \wedge dx^2 .. \wedge dx^n$ is invariant under the passive diffeomorphism (2.11). In terms of the y-coordinates, the volume-form holds as (the metric is written more explicitly) the extension of the area form (2.26):

$$\omega_n = \sqrt{\left|\mathrm{Det}\left(J_i^\alpha J_j^\beta g_{\alpha\beta}\right)\right|} \, dy^1 \wedge dy^2 .. \wedge dy^n \qquad (2.28)$$

Owing that $dy^i := A_\alpha^i dx^\alpha$, we obtain the volume-form in terms of y-coordinates $\omega_n := \pm\sqrt{\mathrm{Detg}} \, dy^1 \wedge dy^2 .. \wedge dy^n$ where $\sqrt{|\mathrm{Detg}|}$ is the determinant with the coordinates y^i. The sign $\pm$ may be dropped if we consider a passive diffeomorphism with strictly positive Jacobian everywhere.

Remark 2.5 Two volume-forms ω_n and ω_n' on the manifold $\mathscr{B}$ are said to be equivalent if $\omega_n = f \, \omega_n'$ for some scalar function on $\mathscr{B}$, say $f \in \mathscr{C}^\infty$ with $f > 0$ at every point $M \in \mathscr{B}$. For example $dx \wedge dy \wedge dz$ is not equivalent to $dy \wedge dx \wedge dz$, and $dx \wedge dy \wedge dz$ (Cartesian coordinate system) is not equivalent to $dr \wedge d\varphi \wedge d\theta$ where (r, θ, φ) is the standard spherical coordinate system. Conversely $dx \wedge dy \wedge dz$ is equivalent to $dr \wedge d\theta \wedge d\varphi$ for $r \neq 0$, and $\theta \in]0, \pi[$ since $dx \wedge dy \wedge dz = r^2 \sin\theta dr \wedge d\theta \wedge d\varphi$.

Definition 2.15 On an oriented metric manifold $\mathscr{B}$, the Lagrangian (n-form) is defined by Lagrangian density $\mathscr{L}$ multiplied by the volume form ω_n, and thus defines the action:

$$\mathscr{S} := \int_{\mathscr{B}} \mathscr{L} \omega_n = \int_{\mathscr{B}} \mathscr{L} \sqrt{\mathrm{Det}\mathbf{g}}\, dx^1 \wedge dx^2.. \wedge dx^n \tag{2.29}$$

This allows us to integrate the Lagrangian density $\mathscr{L}$ over the oriented manifold $\mathscr{B}$. By the way, the scalar field $\mathscr{L}$ behaves like a tensor density e.g. Lovelock and Rund (1975) and not as a scalar (tensor), but if we include the term $\sqrt{\mathrm{Det}\mathbf{g}}$ in the Lagrangian density, we can define a scalar (tensor) field $\mathscr{L}\sqrt{\mathrm{Det}\mathbf{g}} \to \mathscr{L}$ where the factor $\sqrt{\mathrm{Det}\mathbf{g}}$ is worthily included once for all in order to have the proper weight. The presence of this factor is essential to recover the Einstein's field equations in vacuo. The Lagrangian density can not explicitly depend on the coordinates (x^μ) by virtue of the Lorentz translational invariance as shown by Kibble in e.g. Kibble (1961).

2.3.4   Affine Connection

A connection is a structure which specifies how tensors are transported along a curve on a manifold. Space connection provides the geometric structure to derive laws of motion which constitutes the originality of Newton's approach. Newton's laws are fundamentally based on the connecting of instantaneous spaces to calculate the acceleration e.g. Appleby (1977), Kadianakis (1996). By analogy, for deriving equations of continuum mechanics it is necessary to introduce the matter connection (Fig. 2.8). Space and matter connections are different for either classical mechanics e.g. Rakotomanana (2003), or in relativistic gravity theory e.g. Tamanini (2012), except for classical elasticity e.g. Marsden and Hughes (1983).

2.3.4.1   Affine Connection, Affinely Connected Manifold

To start with, a local tangent base is associated to a coordinate system: the contravariant base $\{\mathbf{e}_\alpha\}$ is associated to the system $\{y^\alpha\}$.

Definition 2.16 (Affine Connection) An affine connection ∇ on a manifold $\mathscr{B}$ is a map defined by

$$\nabla : (\mathbf{u}, \mathbf{v}) \in T_P\mathscr{B} \times T_P\mathscr{B} \longrightarrow \nabla_\mathbf{u}\mathbf{v} \in T_P\mathscr{B} \tag{2.30}$$

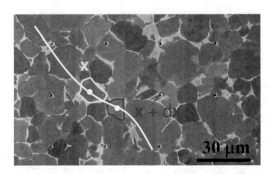

Fig. 2.8 Two connected material points x^μ and $x^\mu + dx^\mu$ along a material line. At this length scale $\ell = 30\,\mu\text{m}$, the metric alone is not sufficient for modeling the shape change due to relativistic motions of grains. Grain interfaces are source of dislocations during plastic flow of polycrystalline solids. Dislocations may multiply within grains by a multiple cross-slip process. The same dissection of spacetime holds in relativistic gravitation to obtain local inertial microcosms—attached to local flat inertial frames of reference. From these microcosms, in which the special relativity theory holds, the global structure of gravitational field is reconstructed after introducing curvature field to obtain continuous bound of the whole. For Einstein–Cartan spacetime, torsion field should be also introduced

which satisfies the following conditions (λ and μ are scalars, ϕ is scalar field)

1. $\nabla_{\lambda\mathbf{u}_1+\mu\mathbf{u}_2}\mathbf{v} = \lambda\nabla_{\mathbf{u}_1}\mathbf{v} + \mu\nabla_{\mathbf{u}_2}\mathbf{v}$
2. $\nabla_{\mathbf{u}}(\lambda\mathbf{v}_1 + \mu\mathbf{v}_2) = \lambda\nabla_{\mathbf{u}}\mathbf{v}_1 + \mu\nabla_{\mathbf{u}}\mathbf{v}_2$
3. $\nabla_{\phi\mathbf{u}}\mathbf{v} = \phi\nabla_{\mathbf{u}}\mathbf{v}$
4. $\nabla_{\mathbf{u}}(\phi\mathbf{v}) = \phi\nabla_{\mathbf{u}}\mathbf{v} + \mathbf{u}(\phi)\mathbf{v}$

The connection coefficients Γ^c_{ab} are implicitly defined by $\nabla_{\mathbf{e}_a}\mathbf{e}_b := \Gamma^c_{ab}\mathbf{e}_c$. $\nabla_{\mathbf{u}}$ represents a covariant derivative along the direction $\mathbf{u} \in T_P\mathscr{B}$. It generalizes the derivative of tensor fields on manifold. The covariant derivative of a type (p, q) tensor is a type $(p, q + 1)$ tensor. For example, the covariant derivative of a scalar field ϕ and a vector field $\mathbf{w}$ along the vector $\mathbf{e}_k$ (a vector of the tangent space $T_P\mathscr{B}$) may be expressed in terms of their components on the base $\{\mathbf{e}_a\}$ associated to coordinate system (y^a) : $\nabla_{\mathbf{e}_k}\phi = \partial_k\phi$, and $\nabla_{\mathbf{e}_k}\mathbf{w} = (\partial_k w^a + \Gamma^a_{kc}w^c)\mathbf{e}_a$. Of course, by definition, the covariant derivative of a scalar field is a 1-form $\nabla\phi := \partial_k\phi\,\mathbf{e}^k = \nabla_k\phi$. The covariant derivative of a vector field is a $(1, 1)$ type tensor:

$$\nabla\mathbf{w} = \nabla_{\mathbf{e}_k}(w^a\mathbf{e}_a) \otimes \mathbf{e}^k = \left[(\nabla_{\mathbf{e}_k}w^a)\mathbf{e}_a + w^c\nabla_{\mathbf{e}_k}\mathbf{e}_c\right] \otimes \mathbf{e}^k$$
$$= (\partial_k w^a + \Gamma^c_{kc}w^c)\mathbf{e}_a \otimes \mathbf{e}^k := \nabla_k w^a\,\mathbf{e}_a \otimes \mathbf{e}^k$$

For tensor components, we adopt whenever needed the notations $\partial_k(..) = (..)_{,k}$ and $\nabla_{\mathbf{e}_k}(..) = (..)_{|k}$. This allows us to define the components of the covariant derivative of a vector field $\mathbf{w}$, and that of a 1-form field ω on a manifold $\mathscr{B}$, as: $\nabla_k w^a := \partial_k w^a + \Gamma^a_{kc}\,w^c$, and $\nabla_k\omega_a := \partial_k w_a - \Gamma^c_{ka}\,\omega_c$ by using the reciprocity relation $\omega_a(\mathbf{w}^b) := \delta^b_a$ for the two dual fields.

Definition 2.17 (Affinely Connected Manifold) A differential manifold $\mathscr{B}$ endowed with an affine connection ∇ is an affinely connected manifold.

Definition 2.18 (Metric Compatible Connection) On a manifold $\mathscr{B}$ endowed with connection ∇, and a metric **g**, connection ∇ is metric compatible connection if and only if $\nabla \mathbf{g} \equiv 0$.

For either classical or relativistic mechanics, there is a correspondence between the set of reference frames of spacetime and the set of compatible affine connections e.g. Kadianakis (1996). The existence of infinite number of reference frames is related to the possible existence of infinite number of affine connections on the spacetime manifold. For the Galilean structure (see definition below), the metric compatibility of the connection takes the form of e.g. Bain (2004), Dixon (1975), Ruedde and Straumann (1997):

$$\nabla_\gamma h_{\alpha\beta} \equiv 0, \qquad \nabla_\beta \tau_\alpha \equiv 0 \qquad (2.31)$$

where **h** and τ are metric and 1-form on the manifold $\mathscr{B}$, that will be defined later. However, the connection coefficients $\Gamma^\gamma_{\alpha\beta}$ of ∇ are not uniquely determined by these metric compatibility (Eq. (2.31)). This allows us to define:

Definition 2.19 (Galilean Connection) Consider a Galilean structure $(\mathscr{M}, \mathbf{h}, \tau)$ where **h** and τ are metric and 1-form on the manifold $\mathscr{M}$. An affine connection ∇ is a Galilean connection if and only if its is compatible with the metric say: $\nabla_\gamma h_{\alpha\beta} \equiv 0$, and $\nabla_\beta \tau_\alpha \equiv 0$.

From the compatibility of the 1-form τ, we deduce that the exterior derivative vanishes $d\tau \equiv 0$, meaning an integrability of the distribution associated to τ. First, the maximal integral manifolds are the spatial sections of constant time. Second, there is no unique symmetric connection on a Galilean structure (manifold). Any connection with coefficients $\Gamma^\gamma_{\alpha\beta} + 2\left(\tau_\alpha \omega_{\beta\lambda} - \tau_\beta \omega_{\alpha\lambda}\right) h^{\lambda\gamma}$, where $\Gamma^\gamma_{\alpha\beta}$ is Galilean and $\omega_{\alpha\beta}$ are components of arbitrary closed 2-form $(d\omega \equiv 0)$, is also Galilean e.g. Ruedde and Straumann (1997).

Definition 2.20 (Metric-Affine Manifold) A manifold $\mathscr{B}$ endowed with a metric **g** and an affine (symmetric) connection ∇ is called metric-affine (connected) manifold.

Metric and connection need not depend on each other (Hehl and Kerlick 1976). However, the metric alone determines a particular connection $\overline{\nabla}$. Connection plays also central role in continuum mechanics since it represents the evolution of material manifold topology during non holonomic deformation in a $3D$ body e.g. Rakotomanana (2003). It supports dynamical concepts in the four-dimensional spacetime manifold, such as the concept of frame of reference e.g. Defrise (1953). To conform with either classic or relativistic gravity theory, it is postulated from now and hereafter that the metric components $g_{\alpha\beta}$ and the coefficients of the affine connection $\Gamma^\gamma_{\alpha\beta}$ should be single-valued and smooth enough to be of class $\mathscr{C}^2$.

2.3.4.2 Example: Spherical Coordinate System

As first illustration of connection, consider the Euclidean space defined by the origin O the orthogonal axes $Ox_1x_2x_3$. It is obvious that the connection coefficients of the Euclidean space are zero for the Cartesian coordinate system. Let us calculate the connection coefficients of the Euclidean three-dimensional space with the spherical coordinate system (r, θ, φ) (Fig. 2.9). The local base for the spherical coordinate system is defined from the vector position $\mathbf{OM} = r\mathbf{e}_r$:

$$
\begin{cases}
\mathbf{f}_r := \partial_r \mathbf{OM} = \sin\theta\cos\varphi\,\mathbf{e}_1 + \sin\theta\sin\varphi\,\mathbf{e}_2 + \cos\theta\,\mathbf{e}_3 \\
\mathbf{f}_\theta := \partial_\theta \mathbf{OM} = r\cos\theta\cos\varphi\,\mathbf{e}_1 + r\cos\theta\sin\varphi\,\mathbf{e}_2 - r\sin\theta\,\mathbf{e}_3 \\
\mathbf{f}_\varphi := \partial_\varphi \mathbf{OM} = -r\sin\theta\sin\varphi\,\mathbf{e}_1 + r\sin\theta\cos\varphi\,\mathbf{e}_2
\end{cases}
$$

where we observe that vectors of the local base are orthogonal each other but they have different norm. From the definition of the reciprocal (dual) basis $\mathbf{f}^i\left(\mathbf{f}_j\right) \equiv \delta^i{}_j$, we easily obtain:

$$
\begin{cases}
\mathbf{f}^r = \sin\theta\cos\varphi\,\mathbf{e}_1 + \sin\theta\sin\varphi\,\mathbf{e}_2 + \cos\theta\,\mathbf{e}_3 \\
\mathbf{f}^\theta = \dfrac{1}{r}\left(\cos\theta\cos\varphi\,\mathbf{e}_1 + \cos\theta\sin\varphi\,\mathbf{e}_2 - \sin\theta\,\mathbf{e}_3\right) \\
\mathbf{f}^\varphi = \dfrac{1}{r\sin\theta}\left(\cos\varphi\,\mathbf{e}_1 - \sin\varphi\,\mathbf{e}_2\right)
\end{cases}
$$

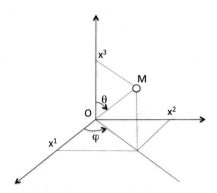

Fig. 2.9 System of spherical coordinates (r, θ, φ): Coordinates are basically defined from the vector position of the point M as $\mathbf{OM} = r\left(\sin\theta\cos\varphi\,\mathbf{e}_1 + \sin\theta\sin\varphi\,\mathbf{e}_2 + \cos\theta\,\mathbf{e}_3\right)$. The mesh of coordinate curves are formed by the intersection of the surfaces: $\sum_{i=1}^{3}(x^i)^2 = r^2$ (spheres), $(x^1)^2 + (x^2)^2 = \tan^2\theta\, x^3$ (cones), and $x^2 = x^1\tan\varphi$ (planes)

We deduce the Christoffel's symbols by writing $\mathbf{f}^k \left(\partial_i \mathbf{f}_j \right) := \Gamma_{ij}^k$:

$$\begin{cases} \overline{\Gamma}_{\theta\theta}^r = -r \\ \overline{\Gamma}_{\varphi\varphi}^r = -r \sin^2\theta \\ \overline{\Gamma}_{\varphi\varphi}^\theta = -\sin\theta\cos\theta \end{cases} \qquad \begin{cases} \overline{\Gamma}_{r\theta}^\theta = \overline{\Gamma}_{\theta r}^\theta = 1/r \\ \overline{\Gamma}_{r\varphi}^\varphi = \overline{\Gamma}_{\varphi r}^\varphi = 1/r \\ \overline{\Gamma}_{\theta\varphi}^\varphi = \overline{\Gamma}_{\varphi\theta}^\varphi = \cos\theta/\sin\theta \end{cases} \qquad (2.32)$$

The other coefficients of the connection are equal to zero. The only non vanishing metric components projected onto the local base, endowed from the scalar product of the space as:

$$g_{rr} := \|\mathbf{f}_r\|^2 = 1, \qquad g_{\theta\theta} := \|\mathbf{f}_\theta\|^2 = r^2, \qquad g_{\varphi\varphi} := \|\mathbf{f}_\varphi\|^2 = r^2 \sin^2\theta$$

which allow us to calculate the line element $ds^2 := dr^2 + r^2 d\theta^2 + r^2 \sin^2\theta d\varphi^2$.

2.3.4.3 Example: Elliptic-Hyperbolic Coordinate System

Connection is a concept beyond the elementary introduction based on the consideration of a curvilinear coordinate system. For the sake of clarity, let us consider a particular example of elliptic-hyperbolic coordinate system in a plane. It is defined by the mapping:

$$\begin{cases} x^1 = \cosh y^1 \cos y^2 \\ x^2 = \sinh y^1 \sin y^2 \end{cases}$$

where (x^1, x^2) are the classical Cartesian coordinates in the plane whereas (y^1, y^2) such that $y^1 > 0$ and $0 < y^2 < 2\pi$ are called elliptic-hyperbolic coordinates in the plane. It is easily checked that the couple (y^1, y^2) constitutes an admissible system of coordinates (except at the origin O). The local base associated to (x^1, x^2) and associated (y^1, y^2) are denoted respectively $(\mathbf{e}_1, \mathbf{e}_2)$ and $(\mathbf{f}_1, \mathbf{f}_2)$ such that:

$$\begin{cases} \mathbf{f}_1 = \sinh y^1 \cos y^2 \, \mathbf{e}_1 + \cosh y^1 \sin y^2 \, \mathbf{e}_2 \\ \mathbf{f}_2 = -\cosh y^1 \sin y^2 \, \mathbf{e}_1 + \sinh y^1 \cos y^2 \, \mathbf{e}_2 \end{cases}$$

The connection coefficients are defined as $\Gamma_{ij}^k := \mathbf{f}^k \left(\partial_i \mathbf{f}_j \right)$ in which $\mathbf{f}^k$ is the dual vector defined as $\mathbf{f}^k \left(\mathbf{f}_j \right) \equiv \delta_j^k$. For illustrating the calculus, consider the example Γ_{11}^1:

$$\Gamma_{11}^1 = \begin{pmatrix} \dfrac{\cosh y^1 \sin y^2}{\cosh^2 y^1 - \cos^2 y^2} \\ \dfrac{\sinh y^1 \cos y^2}{\cosh^2 y^1 - \cos^2 y^2} \end{pmatrix} \cdot \partial_1 \begin{pmatrix} \sinh y^1 \cos y^2 \\ \cosh y^1 \sin y^2 \end{pmatrix} = \frac{\cosh y^1 \sinh y^1}{\cosh^2 y^1 - \cos^2 y^2}$$

where the dual vector is uniquely calculated as $\mathbf{f}^1 (\mathbf{f}_1) = 1$, and $\mathbf{f}^1 (\mathbf{f}_2) = 0$. The non vanishing other components are obtained accordingly:

$$\Gamma^1_{11} = \Gamma^2_{12} = \Gamma^2_{21} = -\Gamma^1_{22} = \frac{\cosh y^1 \sinh y^1}{\cosh^2 y^1 - \cos^2 y^2}$$

$$\Gamma^2_{11} = -\Gamma^1_{12} = -\Gamma^1_{21} = -\Gamma^2_{22} = -\frac{\cos y^2 \sin y^2}{\cosh^2 y^1 - \cos^2 y^2}$$

Remark 2.6 Connections associated to curvilinear coordinates does not constitute the general case but they illustrate the fact that connection is not a tensor. These examples show that the connection coefficients associated to Cartesian system are identically equal to zero although coefficients associated to elliptic-hyperbolic do not vanish.

2.3.4.4 Practical Formula for Covariant Derivative

There are numerous index convention for the component formulation of the covariant derivative. We remind the covariant derivative of general tensor $\mathbf{T} = (T^{\alpha_1 \cdots \alpha_p}_{\beta_1 \cdots \beta_q})$:

$$\nabla_\gamma T^{\alpha_1 \cdots \alpha_p}_{\beta_1 \cdots \beta_q} = \partial_\gamma T^{\alpha_1 \cdots \alpha_p}_{\beta_1 \cdots \beta_q} + \sum_{s=1}^{s=p} \Gamma^{\alpha_s}_{\gamma\mu} T^{\alpha_1 \cdots \alpha_{s-1} \, \mu \, \alpha_{s+1} \cdots \alpha_p}_{\beta_1 \cdots \beta_q} - \sum_{s=1}^{s=q} \Gamma^{\mu}_{\gamma\beta_s} T^{\alpha_1 \cdots \alpha_p}_{\beta_1 \cdots \beta_{s-1} \, \mu \, \beta_{s+1} \cdots \beta_q}$$

$$(2.33)$$

where the index γ is the first covariant index of connection coefficients for the second $\Gamma^{\alpha_s}_{\gamma\mu}$ and third $\Gamma^{\mu}_{\gamma\beta_s}$ terms. Some authors use the transpose covariant derivative and thus put the γ index at the second place e.g. Lovelock and Rund (1975). As usual examples, the covariant derivative of tensor $T^{\alpha_1\alpha_2}$, $T_{\beta_1\beta_2}$, $T^{\alpha_1}_{\beta_1\beta_2}$, and $T^{\alpha_1}_{\beta_1\beta_2\beta_3}$ are respectively:

$$\nabla_\gamma T^{\alpha_1\alpha_2} = \partial_\gamma T^{\alpha_1\alpha_2} + \Gamma^{\alpha_1}_{\gamma\mu} T^{\mu\alpha_2} + \Gamma^{\alpha_2}_{\gamma\mu} T^{\alpha_1\mu}$$

$$\nabla_\gamma T_{\beta_1\beta_2} = \partial_\gamma T_{\beta_1\beta_2} - \Gamma^{\mu}_{\gamma\beta_1} T_{\mu\beta_2} - \Gamma^{\mu}_{\gamma\beta_2} T_{\beta_1\mu}$$

$$\nabla_\gamma T^{\alpha_1}_{\beta_1\beta_2} = \partial_\gamma T^{\alpha_1}_{\beta_1\beta_2} + \Gamma^{\alpha_1}_{\gamma\mu} T^{\mu}_{\beta_1\beta_2} - \Gamma^{\mu}_{\gamma\beta_1} T^{\alpha_1}_{\mu\beta_2} - \Gamma^{\mu}_{\gamma\beta_2} T^{\alpha_1}_{\beta_1\mu}$$

$$\nabla_\gamma T^{\alpha_1}_{\beta_1\beta_2\beta_3} = \partial_\gamma T^{\alpha_1}_{\beta_1\beta_2\beta_3} + \Gamma^{\alpha_1}_{\gamma\mu} T^{\mu}_{\beta_1\beta_2\beta_3} - \Gamma^{\mu}_{\gamma\beta_1} T^{\alpha_1}_{\mu\beta_2\beta_3} - \Gamma^{\mu}_{\gamma\beta_2} T^{\alpha_1}_{\beta_1\mu\beta_3} - \Gamma^{\mu}_{\gamma\beta_3} T^{\alpha_1}_{\beta_1\beta_2\mu}$$

2.3.4.5 Torsion and Curvature

Say a manifold $\mathscr{B}$ endowed with an affine connection ∇. Covariant derivative of a vector field with components (w^a) (considered as $(1, 0)$ tensor) is a $(1, 1)$ tensor field with components $\nabla_k w^a$. The second covariant derivative leads to a $(1, 2)$ tensor with components $\nabla_l \nabla_k w^a$ which may be different of $\nabla_k \nabla_l w^a$ even if $w^a(\mathbf{x})$ is twice differentiable with respect to components x^b. Torsion and curvature fields are introduced for that purpose.

Definition 2.21 Let $(\mathscr{B}, \nabla)$ be an affinely connected manifold. The Lie-Jacobi brackets and Cartan coefficients of structure are defined by: $[\mathbf{u}, \mathbf{v}](\psi) := \mathbf{u}\mathbf{v}(\psi) - \mathbf{v}\mathbf{u}(\psi)$, and then $[\mathbf{e}_\alpha, \mathbf{e}_\beta] := \aleph^\gamma_{0\alpha\beta} \mathbf{e}_\gamma$, where ψ is a scalar field on $\mathscr{B}$, and $\mathbf{u}$ and $\mathbf{v}$ vector fields.

Considering two vector bases on the tangent space $T_M\mathscr{B}$, there are two possibilities: (a) Case where $(\mathbf{e}_1, \cdots, \mathbf{e}_n)$ is a base associated to a global coordinate system $(x^1, \cdots, x^n)$ then $\aleph^\gamma_{0\alpha\beta} \equiv 0$; (b) Case where $(\mathbf{e}_1, \cdots, \mathbf{e}_n)$ is a base which cannot be associated to a global coordinate system then $\aleph^\gamma_{0\alpha\beta} \neq 0$. The basis $\{\mathbf{e}_\alpha\}$ is said anholonomic when the coefficients of structure do not vanish, the frame $\{\partial_\alpha\}$ are holonomic because its members commute each other.

Definition 2.22 (Torsion) The torsion tensor $\aleph$ is a type $(1, 2)$ tensor

$$\begin{cases} \aleph(\mathbf{f}^k, \mathbf{e}_i, \mathbf{e}_j) = \mathbf{f}^k \left(\nabla_{\mathbf{e}_i} \mathbf{e}_j - \nabla_{\mathbf{e}_j} \mathbf{e}_i - [\mathbf{e}_i, \mathbf{e}_j] \right) \\ \aleph^k_{ij} = \Gamma^k_{ij} - \Gamma^k_{ji} - \aleph^k_{0ij} \end{cases} \tag{2.34}$$

where $\mathbf{e}_i$ and $\mathbf{f}^i$ are vector and 1-form on $\mathscr{B}$ (Fig. 2.10).

To highlight the role of torsion on a manifold, consider a scalar field $\phi(x^\mu)$ twice differentiable on $\mathscr{B}$ ($\mathscr{C}^2$). Consider the difference between second covariant derivatives, assuming that we use a coordinate vector base,

$$\nabla_\alpha \nabla_\beta \phi - \nabla_\beta \nabla_\alpha \phi = (\partial_\alpha \partial_\beta \phi - \partial_\beta \partial_\alpha \phi) - (\Gamma^\gamma_{\alpha\beta} - \Gamma^\gamma_{\beta\alpha})\partial_\gamma \phi$$

$$= -\aleph^\gamma_{\alpha\beta} \partial_\gamma \phi = -\aleph^\gamma_{\alpha\beta} \nabla_\gamma \phi \tag{2.35}$$

showing that on a manifold with torsion, the commutativity of second covariant derivatives is not ensured even for a smooth $\mathscr{C}^2$ scalar field. This demonstrates that

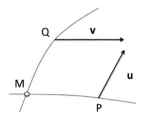

Fig. 2.10 Cartan parallelogram. Let M be a point on the manifold $\mathscr{M}$ and (MP) and (MQ) two set of curves intersecting at M. Both **MP** and **MQ** are assumed infinitesimal and then constitute two vectors of the tangent space $T_M\mathscr{M}$. Vectors **u** and **v** are the images of **MQ** and **MP** by parallel transport along (MP) an (MQ) respectively. The torsion measures the disclosure of the parallelogram made up of infinitesimal vectors **MP** and **MQ** and their parallel transport **v** and **u**

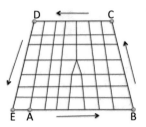

Fig. 2.11 Cartan parallelogram in a discrete crystal lattice. The non closure of the path $ABCDE$ measures the number of defect inside the parallelogram. Only one defect (triangle) is represented here. The opening **EA** define the so-called Bürgers vector

torsion is a characteristic of the manifold rather than of the field ϕ. Even for a smooth field ϕ on the continuum, the second (covariant) derivative may be non commutative. If $\aleph^\gamma_{\alpha\beta}$ vanish in some coordinate system, they will vanish in any other coordinate system. The skew symmetry does not depend on the choice of coordinate system. Torsion tensor may be associated to defects in a continuum and crystal lattice. It is well known that defects are often due to the non integrability of displacement and of gradient of displacement as sketched on Fig. 2.11. Torsion measures the density of defects on this figure, which reports the discrete lattice version of the continuum Cartan parallelogram disclosure. The Cartan disclosure may be associated to defects in matter, namely for crystalline solids with dislocations e.g. Bilby et al. (1955), but also for the spacetime when considering geometrization of quantum effects e.g. Ross (1989).

Definition 2.23 (Curvature) The curvature tensor $\mathfrak{R}$ is a type $(1, 3)$ tensor

$$\begin{cases} \mathfrak{R}(\mathbf{f}^k, \mathbf{e}_i, \mathbf{e}_j, \mathbf{e}_l) = \mathbf{f}^k(\nabla_{\mathbf{e}_i}\nabla_{\mathbf{e}_j}\mathbf{e}_l - \nabla_{\mathbf{e}_j}\nabla_{\mathbf{e}_i}\mathbf{e}_l - \nabla_{[\mathbf{e}_i,\mathbf{e}_j]}\mathbf{e}_l) \\ \qquad \mathfrak{R}^k_{ijl} = (\Gamma^k_{jl,i} + \Gamma^m_{jl}\Gamma^k_{im}) - (\Gamma^k_{il,j} + \Gamma^m_{il}\Gamma^k_{jm}) - \aleph^m_{0ij}\Gamma^k_{ml} \end{cases} \quad (2.36)$$

where $\mathbf{e}_i$ and $\mathbf{f}^i$ are vector and 1-form on $\mathscr{B}$.

Curvature tensor is also a characteristic of the manifold $\mathscr{B}$, and is skew-symmetric in its first pair of lower indices (i, j). In the following, we assume that the base is always associated to a coordinate system, say $\aleph^m_{0ij} \equiv 0$. In a n-dimensional manifold there are $n^2(n^2-1)/12$ independent components. Because of its symmetry properties, only one tensor can be constructed from the curvature tensor, namely the Ricci curvature tensor $\mathfrak{R}_{\beta\lambda} := \mathfrak{R}^\alpha_{\alpha\beta\lambda}$ with six independent components on a three dimensional manifold e.g. Nakahara (1996).

Remark 2.7 The existence of torsion and curvature tensors does not require the existence of a metric on the manifold $\mathscr{B}$. The theorem of Frobenius, e.g. Nakahara (1996), allows us to simplify the expression of the torsion and the curvature projected onto the vector base associated to a coordinate system.

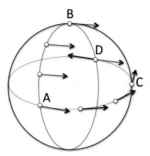

Fig. 2.12 Curvature of a sphere. A vector **w** is defined at point A. This vector is (parallel) transported along the path $\widehat{ABD}$ of the vertical circle by maintaining **w** parallel to itself and represented by a dotted arrow. The same vector **w** is transported along the path $\widehat{ACD}$ along the horizontal circle by maintaining **w** tangent to the circle. At point D the two transported vectors initially the same are opposite. This happens due to the non vanishing curvature of the sphere

Remark 2.8 To a connection $\Gamma^{\gamma}_{\alpha\beta}$ on a differentiable manifold $\mathscr{M}$ we have associated torsion $\aleph^{\gamma}_{\alpha\beta}$ and curvature $\Re^{\gamma}_{\alpha\beta\lambda}$ tensors and then their respective all higher order covariant derivatives. Conversely, the knowledge of torsion and curvature on a neighborhood of each point of the manifold also allows us to determine the connection. At least locally, the manifold geometry with connection is entirely defined by means of torsion and curvature.

Remark 2.9 Let $(\mathbf{u}, \mathbf{v}, \mathbf{w})$ be three arbitrary vectors belonging to the tangent space $T_{\mathbf{x}}\mathscr{B}$. Torsion and curvature may be also defined as operators on vectors by writing:

$$\begin{cases} \aleph(\mathbf{u}, \mathbf{v}) := \nabla_{\mathbf{u}}\mathbf{v} - \nabla_{\mathbf{v}}\mathbf{u} - [\mathbf{u}, \mathbf{v}] \\ \Re(\mathbf{u}, \mathbf{v}, \mathbf{w}) := \nabla_{\mathbf{u}}\nabla_{\mathbf{v}}\mathbf{w} - \nabla_{\mathbf{v}}\nabla_{\mathbf{u}}\mathbf{w} - \nabla[\mathbf{u}, \mathbf{v}]\mathbf{w} \end{cases} \tag{2.37}$$

By using the definition of the connection ∇, it is easy to show that for three scalar real functions (f, g, h) on the manifold $\mathscr{B}$, we have $\Re(f\mathbf{u}, g\mathbf{v}, h\mathbf{w}) = fgh\Re(\mathbf{u}, \mathbf{v}, \mathbf{w})$ proving that $\Re$ is a multilinear object. So is the case for $\aleph$ (Fig. 2.12).

Properties 2.3.2 *Let consider a vector field* **u** *defined on the connected manifold* $\mathscr{B}$, $\{\mathbf{e}_1, \cdots, \mathbf{e}_n\}$ *and* $\{\mathbf{f}^1, \cdots, \mathbf{f}^n\}$ *are vector base of* $\mathfrak{T}_M\mathscr{B}$ *and dual vector base on* $\mathfrak{T}^*_M\mathscr{B}$. *Then the non commutativity of the second covariant derivatives of the vector field holds:*

$$\nabla_{\alpha}\nabla_{\beta}u^{\lambda} - \nabla_{\beta}\nabla_{\alpha}u^{\lambda} = -\aleph^{\gamma}_{\alpha\beta}\nabla_{\gamma}u^{\lambda} + \Re^{\lambda}_{\alpha\beta\gamma}u^{\gamma} \tag{2.38}$$

Proof It suffices to write first its definition and then its distributive property:

$$(\nabla_{\alpha}\nabla_{\beta}u^{\lambda} - \nabla_{\beta}\nabla_{\alpha}u^{\lambda})\mathbf{e}_{\lambda} := \nabla_{\mathbf{e}_{\alpha}}\nabla_{\mathbf{e}_{\beta}}(u^{\lambda}\mathbf{e}_{\lambda}) - \nabla_{\mathbf{e}_{\beta}}\nabla_{\mathbf{e}_{\alpha}}(u^{\lambda}\mathbf{e}_{\lambda})$$

$$= \left[\nabla_{\mathbf{e}_{\alpha}}\left(\nabla_{\mathbf{e}_{\beta}}u^{\lambda}\right)\mathbf{e}_{\lambda} + \nabla_{\mathbf{e}_{\alpha}}\left(\nabla_{\mathbf{e}_{\beta}}\mathbf{e}_{\lambda}\right)u^{\lambda}\right]$$

$$- \left[\nabla_{\mathbf{e}_\beta} \left(\nabla_{\mathbf{e}_\alpha} u^\lambda \right) \mathbf{e}_\lambda + \nabla_{\mathbf{e}_\beta} \left(\nabla_{\mathbf{e}_\alpha} \mathbf{e}_\lambda \right) u^\lambda \right]$$

$$= \left(-\aleph_{\alpha\beta}^\gamma \nabla_\gamma u^\lambda + \Re_{\alpha\beta\gamma}^\lambda u^\gamma \right) \mathbf{e}_\lambda$$

thanks to the definition of the torsion and the curvature tensor, and Eq. (2.35). This relation is valid in any coordinate base □. For a metric compatible connection, it is obvious to derive the relation:

$$\nabla_\alpha \nabla_\beta u_\lambda - \nabla_\beta \nabla_\alpha u_\lambda = -\aleph_{\alpha\beta}^\gamma \nabla_\gamma u_\lambda + g_{\lambda\mu} g^{\gamma\nu} \Re_{\alpha\beta\gamma}^\mu u_\nu$$

Remark 2.10 For the sake of the simplicity, we have considered that $\{\mathbf{e}_1, \cdots, \mathbf{e}_n\}$ and $\{\mathbf{f}^1, \cdots, \mathbf{f}^n\}$ are coordinate basis and coordinate dual basis. It should be also mentioned that even components $u^\lambda(x^\alpha) \in \mathscr{C}^2$ are twice differentiable, the non commutativity of the second covariant derivatives comes from the twisting and the curving of the manifold $\mathscr{B}$.

Remark 2.11 In view of the property (2.38) the curvature and the torsion tensor may induce independently the non-commutativity of covariant derivatives of a vector field **u** on a curved and twisted manifold $\mathscr{B}$.

Torsion and curvature tensors are intrinsic properties of a connection, and then of the metric-affine continuum manifold $(\mathscr{M}, \mathbf{g}, \nabla)$. In relativistic gravitation, affine connection may be interpreted from two different point of views. Torsion and curvature may represent force field (usually the gravitation field for the curvature) and the geodesics of connections define the classical inertial path of free particles in standard relativistic gravitation. The second utilization of affine connection is the curved (and/or twisted) spacetime where material body is evolving e.g. Havas (1964). For a torsionless metric-affine manifold $\mathscr{M}$, the curvature tensor $\Re_{\alpha\beta\lambda}^\gamma \equiv 0$ identically vanishes if and only if the manifold is flat. It means that Cartesian coordinate system with $g_{\alpha\beta} = \delta_{\alpha\beta}$ (Kronecker symbols) and connection coefficients $\Gamma_{\alpha\beta}^\gamma = 0$ can be introduced throughout $\mathscr{M}$. It means that the parallel transport of vectors on $\mathscr{M}$ is independent of path, and the covariant derivatives commute.

2.3.4.6 Newtonian Spacetime

In the absence of gravitation, the spacetime of classical (Newtonian) physics is described by a symmetric tensor $g^{\alpha\beta}$, a 1-form τ_α, and a symmetric affine connection $\Gamma_{\alpha\beta}^\gamma$ satisfying the relationships e.g. Bain (2004), Dixon (1975), Goenner (1974):

$$\begin{cases} g^{\alpha\beta} \tau_\beta = 0 \\ \nabla_\gamma g^{\alpha\beta} = 0 \quad \text{and} \quad \nabla_\alpha \tau_\beta = 0 \\ \Re_{\alpha\beta\lambda}^\gamma = 0 \end{cases} \tag{2.39}$$

where the first line means the orthogonality condition of the space and time, the second the metric compatibility of the connection, and the third line expresses the absence of gravitation forces. The tensor $g^{\alpha\beta}$ is required to be (negative) semi-definite, and of matrix rank 3. We intentionally introduce the contravariant components of the metric since it is clearly related to the relativistic metric. These conditions induce that there is a family of coordinate systems in which:

$$
\begin{cases}
g^{\alpha\beta} = \operatorname{diag}\{0, -1, -1, -1\} & \text{as} \quad \left(= \lim_{c\to\infty} \operatorname{diag}\left\{1/c^2, -1, -1, -1\right\}\right) \\
\tau_\alpha = \operatorname{diag}\{1, 0, 0, 0\} \\
\Gamma^\gamma_{\alpha\beta} \equiv 0
\end{cases}
$$

Coordinate systems are related to one another by group of Galilean transformations. They are identified as inertial frames of references. Equation (2.39) and particularly the metric compatibility is not sufficient to define a unique inertial frame but rather a family of frames. Compatibility (2.39) also ensures that there exists a scalar function t which satisfies $\tau_\alpha := \partial_\alpha t$, meaning that the Newtonian spacetime possesses an absolute time.

Remark 2.12 Spacetime flatness (third line of Eq. (2.39)) $\mathfrak{R}^\gamma_{\alpha\beta\lambda}(x^\mu) \equiv 0$ prohibits relativistic rotation between inertial frames. This condition also imposes at most linearity in the time-dependence of translation between inertial frames, the relativistic acceleration between them is thus excluded e.g. Bain (2004). This allows us to define the Galilean group of transformations associated to classical gravitation e.g. Bain (2004). Previous studies e.g. Andringa et al. (2011) have shown that the Bargmann group, which is an extension of the Galilean group in the presence of gravitation, is the gauging group of the Newton–Cartan gravitation. In this way, the metric compatibility and orthogonality conditions in the relations (2.39) of $g_{\alpha\beta}$ an τ_α allows us to obtain the most general compatible connection

$$
\begin{aligned}
\Gamma^\gamma_{\alpha\beta} &= \tau^\gamma \left(\partial_\alpha \tau_\beta - \partial_\beta \tau_\alpha\right) + (1/2)g^{\gamma\lambda}\left(\partial_\alpha g_{\lambda\beta} + \partial_\beta g_{\alpha\lambda} - \partial_\lambda g_{\alpha\beta}\right) \\
&\quad + g^{\gamma\lambda}\left(\omega_{\lambda\alpha}\tau_\beta - \omega_{\lambda\beta}\tau_\alpha\right)
\end{aligned}
\tag{2.40}
$$

where ω is a 2-form.

In the scope of continuum mechanics, the six independent components of the curvature constitute six compatibility conditions if equal to zero, ensuring that given a strain state of the deformed configuration, there are three single-valued components of displacement from an initial torsionless and curvature-free configuration (Marsden and Hughes 1983). To any affine connection, we can associate a torsion and a curvature tensors. Conversely, the knowledge of the torsion and curvature at any point of the affinely manifold $\mathscr{B}$ suffices to determine the connection, at least locally. This further means that the local geometric structure of the manifold is contained, at least locally, in the knowledge of $\aleph$ and $\mathfrak{R}$. For a manifold where $\aleph(\mathbf{x}) = 0$ and $\mathfrak{R}(\mathbf{x}) = 0$, the connection is called flat and the manifold is locally equivalent to the embedding affine space e.g. Nakahara (1996). In continuum mechanics, the non twisted, and non curved manifold $\mathscr{B}$ constitutes the geometric background of strongly continuum model e.g. Marsden and Hughes (1983), Rakotomanana (1997).

Remark 2.13 (Transposition Invariance) On an affinely connected manifold $\mathscr{M}$, ∇ with coefficients $\Gamma^{\gamma}_{\alpha\beta}$, we can define new connection $\check{\nabla}$ such that $\check{\Gamma}^{\gamma}_{\alpha\beta} := \Gamma^{\gamma}_{\alpha\beta} - \aleph^{\gamma}_{\alpha\beta} = \Gamma^{\gamma}_{\beta\alpha}$. This transformation is called transposition. This leads to new covariant derivative. As illustration, for a type $(1, 1)$ tensor A^{μ}_{ν}, these two derivatives hold:

$$\nabla_{\rho} A^{\mu}_{\nu} := \partial_{\rho} A^{\mu}_{\nu} + \Gamma^{\mu}_{\rho\lambda} A^{\lambda}_{\nu} - \Gamma^{\lambda}_{\rho\nu} A^{\mu}_{\lambda} \neq \check{\nabla}_{\rho} A^{\mu}_{\nu} := \partial_{\rho} A^{\mu}_{\nu} + \Gamma^{\mu}_{\lambda\rho} A^{\lambda}_{\nu} - \Gamma^{\lambda}_{\nu\rho} A^{\mu}_{\lambda}$$

Consequently, two different curvatures also exist. Curvature is not transposition invariant since $\check{\mathfrak{R}}^{\gamma}_{\alpha\beta\lambda} \neq \mathfrak{R}^{\gamma}_{\alpha\beta\lambda}$. Transposition invariance plays role in the non zero torsion gravitational theories, particularly when investigating the deviation acceleration of two close curves in an Einstein–Cartan spacetime.

2.3.4.7 Levi-Civita Connection

The Levi-Civita connection is an example of Euclidean connection (derived from the metric) and introduced by the following theorem:

Theorem 2.1 (Fundamental Theorem of (Pseudo-) Riemannian Geometry) *On any (pseudo-) Riemannian manifold $(\mathscr{B}, \mathbf{g})$, there exists a unique connection compatible with the (pseudo-) metric and free-torsion ($\aleph = 0$), called Levi-Civita connection and denoted $\overline{\nabla}$.*

Proof See in e.g. Nakahara (1996). Coefficients of $\overline{\nabla}$ reduce to symbols of Christoffel $\partial_a \mathbf{e}_b = \overline{\Gamma}^{c}_{ba} \mathbf{e}_c$, calculated in terms of metric $\mathbf{g}$ e.g. Nakahara (1996)

$$\overline{\Gamma}^{c}_{ab} = (1/2) \, g^{cd} \left(\partial_b g_{ad} + \partial_a g_{db} - \partial_d g_{ab} \right) \tag{2.41}$$

By the way, the Einstein theory of relativistic gravitation is built upon the Levi-Civita connection $\overline{\Gamma}^{c}_{ab}$ associated to a metric, with zero torsion but nonzero curvature. For example, if $\mathbf{h}$ is a type $(0, 2)$ tensor then the covariant derivative with respect to $\mathbf{e}_k$ (a vector of the base) is a type $(0, 2 + 1)$ tensor for which the coordinates are noticed (care should be taken for place of indices if the connection is not symmetric)

$$\overline{\nabla}_{\mathbf{e}_k} h_{ij} = h_{ij|k} = h_{ij,k} - \Gamma^{a}_{ik} h_{aj} - \Gamma^{a}_{jk} h_{ia} \tag{2.42}$$

and then with respect to $\mathbf{e}_l$, the second covariant derivative is $h_{ij|k|l}$ with

$$
\begin{aligned}
h_{ij|k|l} = {} & h_{ij,kl} - \Gamma^a_{ik} h_{aj,l} - \Gamma^a_{jk} h_{ia,l} - \Gamma^a_{ik,l} h_{aj} - \Gamma^a_{jk,l} h_{ia} - \Gamma^b_{il} h_{bj,k} \\
& + \Gamma^b_{il}(\Gamma^c_{bk} h_{cj} + \Gamma^c_{jk} h_{bc}) - \Gamma^b_{jl} h_{ib,k} + \Gamma^b_{jl}(\Gamma^c_{ik} h_{cb} + \Gamma^c_{bk} h_{ic}) \\
& - \Gamma^b_{kl} h_{ij,b} + \Gamma^b_{kl}(\Gamma^c_{ib} h_{cj} + \Gamma^c_{jb} h_{ic})
\end{aligned}
\tag{2.43}
$$

where we have denoted $\overline{\Gamma} = \Gamma$ for the sake of the simplicity. On a metric-affine manifold, the metric compatibility of the connection holds as $\nabla \mathbf{g} \equiv 0$. Metricity guarantees a local Minkowskian structure in the relativistic gravitation, and a local Euclidean structure in gradient continuum mechanics e.g. Hehl et al. (1976). This may be violated when electromagnetic is involved, and this was investigated by Weyl e.g. Weyl (1929).

For completeness, the components of the curvature tensor associated to the Levi-Civita connection are derived from (2.41) as:

$$
\begin{aligned}
\overline{\mathfrak{R}}^\lambda_{\alpha\beta\mu} = {} & (1/2)g^{\lambda\sigma}\left(\partial_\mu\partial_\alpha g_{\sigma\beta} - \partial_\mu\partial_\beta g_{\sigma\alpha} + \partial_\sigma\partial_\beta g_{\mu\alpha} - \partial_\sigma\partial_\alpha g_{\mu\beta}\right) \\
& + (1/4)g^{\lambda\sigma}\left(\partial_\alpha g_{\sigma\gamma} + \partial_\gamma g_{\alpha\sigma} - \partial_\sigma g_{\alpha\gamma}\right)g^{\gamma\kappa}\left(\partial_\beta g_{\kappa\mu} + \partial_\mu g_{\beta\kappa} - \partial_\kappa g_{\beta\mu}\right) \\
& - (1/4)g^{\lambda\sigma}\left(\partial_\beta g_{\sigma\gamma} + \partial_\gamma g_{\beta\sigma} - \partial_\sigma g_{\beta\gamma}\right)g^{\gamma\kappa}\left(\partial_\alpha g_{\kappa\mu} + \partial_\mu g_{\alpha\kappa} - \partial_\kappa g_{\alpha\mu}\right)
\end{aligned}
$$

Remark 2.14 Apart from Levi-Civita connection $\overline{\nabla}$, there exists Cartan connection ∇ metric compatible $\nabla_a g_{bc} \equiv 0$ with non symmetric coefficients $\Gamma^c_{ab} - \Gamma^c_{ba} \neq 0$. It should be pointed out that $\overline{\nabla}_{\mathbf{e}_k} h_{ij} \neq \nabla_{\mathbf{e}_k} h_{ij}$. The condition that metric is covariantly constant with respect to ∇ means that length measuring "rods" and angle between two arbitrary rods remain constant along parallel transport ∇ e.g. Verçyn (1990).

Metric compatibility allows us to have local Euclidean structure at each point P of the metric-affine manifold $(\mathscr{B}, \mathbf{g}, \nabla)$. Moreover, metric compatibility permits to decompose the connection coefficients e.g. Nakahara (1996): $\Gamma^\gamma_{\alpha\beta} = \overline{\Gamma}^\gamma_{\alpha\beta} + D^\gamma_{\alpha\beta} + \Omega^\gamma_{\alpha\beta}$, where the skew symmetric term $\Omega^\gamma_{\alpha\beta}$ is considered as non holonomic rotation, and symmetric term $D^\gamma_{\alpha\beta}$ stands for non holonomic strain e.g. Rakotomanana (2005):

$$
\Omega^\gamma_{\alpha\beta} = (1/2)(\Gamma^\gamma_{\alpha\beta} - \Gamma^\gamma_{\beta\alpha}), \qquad D^\gamma_{\alpha\beta} = g^{\gamma\lambda}g_{\alpha\mu}\,\Omega^\mu_{\lambda\beta} + g^{\gamma\lambda}g_{\mu\beta}\,\Omega^\mu_{\lambda\alpha}
\tag{2.44}
$$

The sum $\mathfrak{T}^\gamma_{\alpha\beta} := \Omega^\gamma_{\alpha\beta} + D^\gamma_{\alpha\beta}$ is called contortion tensor whereas $\Omega^\gamma_{\alpha\beta}$ is often called the object of the anholonomy. In the general case, curvature does not vanish, say $\overline{\mathfrak{R}}^\lambda_{\alpha\beta\mu} \neq 0$.

Remark 2.15 In relativistic gravitation, metric compatibility can be also visualized as the condition the spacetime to be a set of microcosms e.g. Gonseth (1926) glued by the affine connection within a Minkowskian spacetime. The presence of the anholonomy supplies the spacetime with additional rotational degrees of freedom e.g. Hehl and von der Heyde (1973). Another practical advantage of metric compatibility is the possibility of raising and lowering of indices of tensors very easily. For generalized continuum mechanics, and particularly for crystals with defects, the non compatibility is necessary to extend the model to include Somigliana dislocations e.g. Clayton et al. (2005).

Properties 2.3.3 *Let $\mathscr{B}$ be a Riemannian manifold and $P \in \mathscr{B}$ any point. In orthonormal base associated to a normal coordinate system centered on P, the symbols of Christoffel vanish:* $\overline{\Gamma}^{k}_{ij}(P) = 0$ *for $i, j, k = 1, \ldots, n$.*

2.3.4.8 Normal Coordinate System and Inertial Frame

A normal coordinate system on a Riemannian manifold $\mathscr{B}$ centered at P may be also defined by local relations: $g_{\alpha\beta}(P) := \delta_{\alpha\beta}$, and $\overline{\Gamma}^{\gamma}_{\alpha\beta}(\xi)\,\xi^{\alpha}\xi^{\beta} \equiv 0$, where $(\xi^1, \cdots, \xi^n)$ are local coordinates of points $P' := P + \xi$ about the center P. This should be related to the local form of the equivalence principle in the theory of relativity (Krause 1976). Riemannian normal coordinates are standard tool to demonstrate various differential geometry theorems, their use to derive equations allows us to simplify resolution of most problems. The use of normal coordinates may enlighten the introduction of the curvature tensor as arguments of the Lagrangian density. Let $\overline{\nabla}$ be a metric compatible and torsion free connection, if we assume that $\mathscr{L} = \mathscr{L}(g_{\alpha\beta}, \partial_{\alpha}\overline{\Gamma}^{\kappa}_{\beta\lambda} + \overline{\Gamma}^{\xi}_{\beta\lambda}\overline{\Gamma}^{\kappa}_{\alpha\xi})$ then it is worth to consider $\mathscr{L} = \mathscr{L}(g_{\alpha\beta}, \overline{\mathfrak{R}}^{\gamma}_{\alpha\beta\lambda})$. Indeed, this could be related to Taylor expansion of metric about a point x^{μ}:

$$g\left(x^{\mu}; \xi\right) = \underbrace{g_{\alpha\beta}\left(x^{\mu}\right)}_{Simple\ material} - \underbrace{\frac{1}{3}\overline{\mathfrak{R}}_{\alpha\beta\gamma\lambda}\left(x^{\mu}\right)\xi^{\beta}\xi^{\lambda}}_{2^{rd}\ strain\ gradient} - \underbrace{\frac{1}{6}\overline{\nabla}_{\sigma}\overline{\mathfrak{R}}_{\alpha\beta\gamma\lambda}\left(x^{\mu}\right)\xi^{\beta}\xi^{\lambda}\xi^{\sigma}}_{3^{nd}\ strain\ gradient}$$

$$(2.45)$$

where we have calculated the perturbation of the length $ds^2 = g_{\alpha\beta}dx^{\alpha}dx^{\beta}$ when a small term ξ^{μ} is superimposed on dx^{α} (Cartan's theorem). Therefore it may exist a form-invariant Lagrangian density depending on the metric and on the curvature associated to $\overline{\nabla}$. Torsion does not appear since we a priori assume torsionless connection. No confusion will be done between the partial differentiation of a tensor field with respect to y^k (in the sense of Levi-Civita) and the differentiation with respect to the affine connection $\nabla_{\mathbf{e}_k}$. It has been shown that in a flat spacetime the vanishing of the connection coefficients (here reduced to the Chistoffel's symbols) $\Gamma^{\gamma}_{\alpha\beta}(x^{\mu}) \equiv 0$ is the necessary and sufficient condition for the system of coordinates $\{x^{\mu}\}$ define an inertial frame e.g. for either classical mechanics (Kadianakis 1996), or relativistic mechanics (Krause 1976; Manoff 2001a) (at least locally). In a locally inertial frame where $\Gamma^{\gamma}_{\alpha\beta}(x^{\mu}) \equiv 0$, the components of the curvature can be calculated by means of the Christoffel's symbols to give, without going into details: $\mathfrak{R}^{\gamma}_{\alpha\beta\lambda} = (1/2)g^{\gamma\sigma}\left(\partial_{\lambda}\partial_{\alpha}g_{\sigma\beta} - \partial_{\lambda}\partial_{\beta}g_{\sigma\alpha} + \partial_{\sigma}\partial_{\beta}g_{\lambda\alpha} - \partial_{\sigma}\partial_{\alpha}g_{\lambda\beta}\right)$, which is valid only in the local flat coordinate system centered at x^{μ}. This relation is not a tensorial relation. From this relation we calculate Riemann curvature $\mathfrak{R}_{\alpha\beta\lambda\mu} := \mathfrak{R}^{\gamma}_{\alpha\beta\lambda}\,g_{\gamma\mu}$. Conversely, even in a curved spacetime or space, it is possible to make the Christoffel symbols $\overline{\Gamma}^{\gamma}_{\alpha\beta}$ vanish at any one point, but not in its neighborhood. This is related in relativistic gravitation to the finding of a locally inertial frame (local equivalence principle).

Remark 2.16 The equivalence principle between inertia and gravitation is correct only at a point in a strict sense. Extended to a finite region surrounding the point, the equivalence principle no more holds (Shen and Moritz 1996). For non zero torsion field on a metric-affine manifold $\mathscr{M}$, connection coefficients $\Gamma^{\gamma}_{\alpha\beta}$ cannot be transformed to zero any longer.

2.3.5   Tetrads and Affine Connection: Continuum Transformations

Starting from a flat Euclidean space or from a flat Minkowski spacetime $\mathscr{M}$, we may obtain a twisted and/or curved space or spacetime $\mathscr{B}$ by using multi-valued transformations and tetrads.

Consider a manifold $(\mathscr{B}, \mathbf{g})$ embedded in a n-dimensional Euclidean space $\mathscr{E}$. Let $\mathbf{X} \in \mathscr{B}$ a point of the manifold, and let consider a mapping φ which associates $\mathbf{X}$ to a point of the Euclidean space $\mathbf{x} \in \mathscr{E}$. We denote the mapping $\mathbf{x}(\mathbf{X})$ for simplifying and recall that the image $\varphi\,(\mathscr{B})$ is a manifold. This section reminds that defects theory may be also modeled within the framework of Einstein–Cartan theory of gravitation e.g. Ruggiero and Tartaglia (2003). Both of them have similar equations and geometric background. Here we shortly present the tetrad (triad) approach by locally defining set of vectors $\mathbf{e}_{\alpha}\,(\mathbf{X})$, $\alpha = 1, n$, (n=3 or 4) rather than directly mapping $\mathbf{x}\,(\mathbf{X})$. Extension to four-dimensional spacetimes is obvious.

2.3.5.1   Transformation of a Continuum

Basically, a continuum material body is a three-dimensional differentiable manifold $\mathscr{B}$ such that there exists a global orientation-preserving diffeomorphisms $\psi : \mathscr{B} \to \psi\,(\mathscr{B}) \in \mathscr{E}$. The image of $\mathscr{B}$ is a connected subset of the Euclidean space. Configurations of the continuum are defined by the pair $(\psi, \psi(\mathscr{B}))$. For classical mechanics, time parameter t can be defined with an absolute manner independently on the space $\mathscr{E}$. Given two configurations of the continuum matter corresponding to two time parameters t_r (reference) and t_c (current), a transformation of the body with respect to the reference configuration $(\psi_r, \psi_r(\mathscr{B}))$ is a mapping (Fig. 2.13):

$$\varphi := \psi_c\,\psi_r^{-1} \tag{2.46}$$

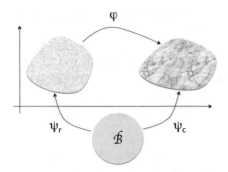

Fig. 2.13 Transformation of a continuum $\mathscr{B}$ with respect to a reference configuration $(\psi_r, \psi_r(\mathscr{B}) \in \mathscr{E})$. Both the current configuration $(\varphi, \psi_c(\mathscr{B}) \in \mathscr{E})$ and the reference configurations belong to the Euclidean space. Most studies consider the reference configuration as the initial configuration of the body and also assume that $\mathscr{B}$ and $\psi_r(\mathscr{B})$ are merged

where $(\varphi, \psi_c(\mathscr{B}))$ is the current configuration with respect to $(\psi_r, \psi_r(\mathscr{B}))$. Different cases may occur by considering, for the sake of the simplicity, that the set $\mathscr{B}$ is a flat manifold (no torsion and no curvature). The first case is when the current configuration is defect-free in the sense that the mapping ψ_r is a diffeomorphism ensuring that the reference configuration has zero torsion $\aleph_r \equiv 0$ and zero curvature $\mathfrak{R}_r \equiv 0$. In such a case, If φ is a diffeomorphism, then the current configuration is also defect-free. Conversely if such is not the case, torsion and/or curvature may appear at the current configuration meaning creation of defect during the transformation. The second case occurs when the reference configuration is not defect-free in the sense that ψ_r is not a diffeomorphism, then the torsion and/or the curvature are not equal to zero in this reference configuration, $\aleph_r \neq 0$ and zero curvature $\mathfrak{R}_r \neq 0$. In the same way the current configuration has the following internal defects $\aleph_c \neq 0$ and zero curvature $\mathfrak{R}_c \neq 0$. In such a case, two cases may occur, the first where there are additional creation of defects then $\aleph_c \neq \aleph_i$ and $\mathfrak{R}_c \neq \mathfrak{R}_i$ (φ is not a diffeomorphism), and the second one where $\aleph_c \equiv \aleph_i$ and $\mathfrak{R}_c \equiv \mathfrak{R}_i$. The present section will give some details about all these possibilities.

2.3.5.2 Holonomic Mapping

Let consider a smooth and single-valued mapping (it is a homeomorphism and we call it holonomic mapping e.g. Rakotomanana 2003). It is usual to define the deformation gradient which is also called basis triads (rigorously it is not a gradient e.g. Marsden and Hughes 1983) in components form, together with its dual basis triads: $F_\alpha^i(\mathbf{X}) := \partial_\alpha x^i(\mathbf{X})$, and $F_j^\beta(\mathbf{x}) := \partial_j X^\beta(\mathbf{x})$. The triads satisfy orthogonality and completeness relationships: $F_\alpha^i(\mathbf{X}) F_i^\beta[\mathbf{x}(\mathbf{X})] = \delta_\alpha^\beta$, and $F_\alpha^i(\mathbf{X}) F_j^\alpha[\mathbf{x}(\mathbf{X})] = \delta_j^i$. We may write the vector transformation and the metric components, where $\hat{\mathbf{e}}_i$ is a vector rigidly attached to the Euclidean space

$\mathscr{E}: \mathbf{e}_\alpha = F_\alpha^i \, \hat{\mathbf{e}}_i$, and $g_{\alpha\beta} = \mathbf{g}\left(\mathbf{e}_\alpha, \mathbf{e}_\beta\right)$. On the one hand, if the transformation $\mathbf{x}(\mathbf{X})$ is smooth and single valued, it is integrable, i.e. its derivatives commute, due to the Schwarz's integrability conditions: $\partial_\beta F_\alpha^i - \partial_\alpha F_\beta^i = \partial_\beta \partial_\alpha x^i - \partial_\alpha \partial_\beta x^i = 0$. On the other hand, we can differentiate the vector base, and implicitly define the affine connection coefficients (called Weitzenböck connection): $\partial_\alpha \mathbf{e}_\beta := \Gamma_{\alpha\beta}^\gamma \, \mathbf{e}_\gamma = \partial_\alpha F_\beta^i \, \hat{\mathbf{e}}_i = F_i^\gamma \, \partial_\alpha F_\beta^i \, \mathbf{e}_\gamma$ with $\Gamma_{\alpha\beta}^\gamma = F_i^\gamma \, \partial_\alpha F_\beta^i$. Torsion tensor is zero during an holonomic transformation $\aleph_{\alpha\beta}^\gamma := (\partial_\alpha F_\beta^i - \partial_\beta F_\alpha^i) F_i^\gamma = 0$ and the curvature vanishes whenever $\Gamma_{\alpha\beta}^\gamma$ are of class $\mathscr{C}^1$. Weitzenböck connection implicitly assumes that there is a Minkowskian spacetime as ambient spacetime. Otherwise, one has to introduce a slightly general definition as $\Gamma_{\alpha\beta}^\gamma := F_i^\gamma \, \hat{\Gamma}_{j\beta}^i \, F_\alpha^j + F_i^\gamma \, \partial_\alpha F_\beta^i$, where F_α^i and $\hat{\Gamma}_{j\beta}^i$ can be considered a Poincaré gauge potentials e.g. Cho (1976b), Malyshev (2000), Obukhov et al. (1989).

Remark 2.17 Tetrads satisfy the condition $+\infty \geq \mathrm{Det}[F_\alpha^i(x^\mu)] > 0$ at each point $\mathbf{x} \in \mathscr{M}$. They constitute a group of isomorphisms, a Lie group of dimension n^2 called general linear group and denoted $\mathbb{G}L_n(\mathbb{R})$. Nevertheless, tetrads may have an infinite value for determinant. Its inverse may thus not be defined.

Remark 2.18 Usual continuum transformation preserves its topology, meaning that close points remain close. Accordingly there are two equivalent descriptions: (a) material or Lagrangian description (with respect to a reference configuration, usually initial configuration); and (b) spatial or Eulerian description (with respect to a current state, position of the points of the continuum). Transformations are diffeomorphism, and bodies are differential manifolds e.g. Marsden and Hughes (1983).

2.3.5.3 Nonholonomic Mapping and Torsion e.g. Kleinert (2000)

Let now consider transformations which do not preserve the topology of bodies. The idea behind generalized continua models, and particularly for strain gradient continua is to define an extended geometric structure for bodies by including additional degrees of freedom. For these continua "macroscopic" mapping is still diffeomorphism whereas evolution of geometric structure modifies the bodies internal topology e.g. Rakotomanana (1997). We recall in the following how to describe such transformations in the scope of generalized continuum e.g. Pettey (1971) by using of differential geometric and affinely connected manifolds. Commonly, a generalized continuum is a connected, locally connected, and locally compact metric manifold. Manifold $\mathscr{B}$ is locally compact, if every point of $\mathscr{B}$ has a compact neighborhood. $\mathscr{B}$ is connected if it cannot be written as an union $\mathscr{B} = \mathscr{B}_1 \cup \mathscr{B}_2$ where $\mathscr{B}_1$ and $\mathscr{B}_2$ are both open and $\mathscr{B}_1 \cup \mathscr{B}_2 = \emptyset$. It is locally connected, if every point of $\mathscr{B}$ has a connected neighborhood e.g. Nakahara (1996).

Say a mapping $\mathbf{X} \to \mathbf{x}$ that is not smooth and/or not single valued. In such a case, the triads $F_\alpha^i(X^\mu)$ are not integrable. Of course, it is possible to map the

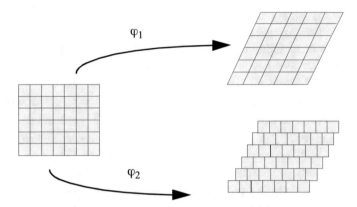

Fig. 2.14 Shear deformation: homeomorphic mapping or holonomic transformation $\varphi_1 \in \mathcal{C}^1$ and $\varphi_1^{-1} \in \mathcal{C}^1$, and nonholonomic transformation $\varphi_2 \notin \mathcal{C}^0$, here including Volterra translational dislocations (inducing "fragmentation" to obtain brick elements) here $+\infty \geq \mathrm{Det} F_\alpha^i > 0$

tangent vector $d\mathbf{X}$ to the vector $d\mathbf{x}$ via an linear tangent transformation defined by the triads: $\mathbf{e}_\alpha = F_\alpha^i \, \hat{\mathbf{e}}_i$, and $dx^i = F_\alpha^i \, dX^\alpha$ in which the non integrability of $F_\alpha^i(\mathbf{X})$ means that: $\partial_\beta F_\alpha^i - \partial_\alpha F_\beta^i = \partial_\beta \partial_\alpha x^i - \partial_\alpha \partial_\beta x^i \neq 0$. This mapping is called non holonomic. It is worth to modify the previous development to give ($\{\hat{\mathbf{e}}_1, \cdots, \hat{\mathbf{e}}_n\}$ is assumed attached to the Euclidean space): $\partial_\alpha \mathbf{e}_\beta - \partial_\beta \mathbf{e}_\alpha = F_i^\gamma (\partial_\alpha F_\beta^i - \partial_\beta F_\alpha^i)\mathbf{e}_\gamma := \aleph_{\alpha\beta}^\gamma \, \mathbf{e}_\gamma \neq 0$ showing that the torsion $\aleph_{\alpha\beta}^\gamma$ is not zero for such a mapping. It does not lead to a single-valued mapping $\mathbf{x}(\mathbf{X})$. This transformation captures translational dislocations of Volterra e.g. Maugin (1993). It is well-known, see Fig. 2.14 that a macroscopic shear deformation may be modeled by either a homeomorphism $\varphi_1 \in \mathcal{C}^1$ from an initial non defected configuration, or a non homeomorphic transformation (here not even single-valued) $\varphi_2 \notin \mathcal{C}^0$ inducing the same macroscopic shear, but the nonholonomic map φ_2 includes "perturbed" transformations at another length scale. Rather than directly considering the discontinuity of the displacement field $[\mathbf{u}]$, we prefer to extend the continuum model to weakly continuous model where both the strain (change of the metric) $\delta g_{\alpha\beta}$ and connection change $\delta \Gamma_{\alpha\beta}^\gamma$ are considered: in sum, use of the Riemann–Cartan manifold e.g. Rakotomanana (1997).

Remark 2.19 Although the knowledge of the 9 triads (resp. 16 tetrads) $F_\alpha^i(\mathbf{X})$ allows us to define the 6 (resp. 10) metric components $g_{\alpha\beta}$, the inverse is not true. Triads (resp. tetrads) are determined only up to Galilean (resp. Lorentzian) transformation which possibly vary with point location $\mathbf{X}$ e.g. Utiyama (1956).

Typical example of torsion in the continuum theory is given by edge dislocation where multi-valued transformation of continuum is defined together with its basis tetrads (b is the norm of the Burgers vector) e.g. Kleinert (2000, 2008):

$$\begin{cases} x^1 = X^1 \\ x^2 = X^2 - (b/2\pi) \ \arctan(X^2/X^1) \end{cases} \qquad F_\alpha^i = \frac{b}{2\pi} \frac{1}{(X^1)^2 + (X^2)^2} \begin{pmatrix} 1 & 0 \\ X^2 & -X^1 \end{pmatrix}$$

It is easy to calculate the torsion induced by this multi-valued transformation from a non defected initial continuum: $\partial_1 F_2^2 - \partial_2 F_1^2 = -b \ \delta^{(2)}(\mathbf{x}) \neq 0$, with other components equal to zero. Since the tetrads are single-valued, then the curvature tensor is identically equal to zero. Although easily visualized, the usual Bürgers vector may present drawback from the topological point of view (Yang et al. 1998). Indeed, the integral $b^\gamma := \oint \aleph_{\alpha\beta}^\gamma \ dx^\alpha \wedge dx^\beta$ around a closed path in a twisted manifold $\mathscr{B}$ is not diffeomorphism invariant in a such way that it violates the coordinate invariance. Under non uniform coordinate transform $\tilde{x}^\mu(x^\alpha)$ it is straightforward to show that the components of the Bürgers vector in this new coordinate system hold $\tilde{b}^\lambda := \oint \tilde{\aleph}_{\mu\nu}^\lambda \ d\tilde{x}^\mu \wedge d\tilde{x}^\nu = \oint J_\gamma^\lambda \ \aleph_{\alpha\beta}^\gamma \ dx^\alpha \wedge dx^\beta$ is different of $J_\gamma^\lambda \ b^\gamma$. Indeed $J_\gamma^\lambda(x^\kappa) := \partial\tilde{x}^\lambda/\partial x^\gamma$ is not necessarily uniform over the manifold (Yang et al. 1998). The density of dislocation is conversely a good candidate for characterizing the manifold twist. There are three rectilinear infinite Volterra dislocations, when considering a dislocation line along a given axis, say the screw, edge and climb dislocations (Kleman and Friedel 2008). The non integrable part of the tetrads of all of these kinds of dislocations could be expressed as $\mathbf{F}^*(\mathbf{x}) := \mathbf{b} \otimes \mathbf{n}(\mathbf{x}) \ \delta_\Omega(\mathbf{x})$, where δ_Ω is a Dirac distribution localized on the discontinuity surface Ω having the unit normal $\mathbf{n}(\mathbf{x})$ at point $\mathbf{x}$. Expressions of the displacement fields corresponding with the three linear dislocations may be developed in a rigorous way, accounting for the distributional components within the dynamical framework e.g. Pellegrini (2012).

Remark 2.20 In the scope of relativistic gravitation, the previous nonholonomic deformation together with non zero torsion is better approached by considering the spacetime absolute transport as follows. Let two distinct events (positions) of the spacetime $\mathbf{v} := v^\alpha \ \mathbf{e}_\alpha$ and $\mathbf{w} := v^\alpha \ \mathbf{e}_\alpha$. The two vectors are parallel if their components are equal, owing that the base vectors $\mathbf{e}_\alpha(x^\mu)$ and $\mathbf{e}_\alpha(x^\mu + dx^\mu)$ are respectively functions of the events where they are defined in the spacetime. The affine connection $\Gamma_{\alpha\beta}^\gamma := F^\gamma \partial_\alpha F_\beta^i$ is then easily obtained in the Weitzenböck spacetime (zero curvature) e.g. Hayashi (1979). The Weitzenböck spacetime reduces to Minkowski spacetime when the torsion is equal to zero, meaning that the tetrad $F_\alpha^i(\mathbf{X})$ is integrable.

2.3.5.4 Nonholonomic Transformation and Curvature

Consider a transformation $\mathbf{X} \rightarrow \mathbf{x}(\mathbf{X})$ which may be not integrable, more precisely a multi-valued function. It is known that such a mapping is not sufficient to describe all topological defects in the framework of crystalline solids e.g. Kleinert (2000), Rakotomanana (1997). To go further we must also add the multi-valuedness of the tetrads F_α^i themselves, meaning a possible rotation degree of freedom of the tetrads basis too.

More precisely, first derivatives of vectors $\{e_1, \cdots, e_n\}$, where $e_\alpha := F_\alpha^i \hat{e}_i$, are not integrable, then second-order derivatives do not commute (cf. Schwartz's theorem): $\partial_\alpha \partial_\beta\, e_\lambda - \partial_\beta \partial_\alpha\, e_\lambda = \partial_\alpha (\Gamma_{\lambda\beta}^\kappa e_\kappa) - \partial_\beta (\Gamma_{\lambda\alpha}^\kappa e_\kappa) := \mathfrak{R}_{\alpha\beta\lambda}^\gamma\, e_\gamma$. Therefore, Cartan curvature does not vanish: $\mathfrak{R}_{\alpha\beta\lambda}^\gamma = (\partial_\alpha \Gamma_{\beta\lambda}^\gamma + \Gamma_{\beta\lambda}^\mu \Gamma_{\alpha\mu}^\gamma) - (\partial_\beta \Gamma_{\alpha\lambda}^\gamma + \Gamma_{\alpha\lambda}^\mu \Gamma_{\beta\mu}^\gamma) \neq 0$. As for the Weitzenböck connection, essentially defined in a flat spacetime, a more rigorous method would be the introduction of additional gauge fields $\hat{\Gamma}_{j\beta}^i(x^\mu)$ to give the (not torsionless) connection $\Gamma_{\alpha\beta}^\gamma := F_i^\gamma \hat{\Gamma}_{j\beta}^i F_\alpha^j + F_i^\gamma \partial_\alpha F_\beta^i$ e.g. Obukhov et al. (1989). This means that we create curvature for gradient continuum (or for the spacetime in the relativistic gravitation framework e.g. Cho 1976b) by introducing quantities $\hat{\Gamma}_{j\beta}^i(x^\mu)$ to generate non zero curvature. Such a transformation may be related to the process of rotational dislocations e.g. Maugin (1993), or some plastic deformation of matter e.g. Kobelev (2010). Typical example for the presence of non vanishing curvature is given by the multivalued transformation: $x^i := \delta_\mu^i [X^\mu + \Omega\, \epsilon_\nu^\mu\, X^\mu \arctan(X^2/X^1)]$, where $\Omega \in \mathbb{R}$ is a rotation angle, and ϵ_ν^μ denotes the antisymmetric Levi-Civita tensor. We check that the torsion tensor is equal to zero and that only one Riemann curvature component does not vanish: $\mathfrak{R}_{1212} = -2\pi\, \Omega\, \delta^{(2)}(\mathbf{x}) \neq 0$. This transformation is associated to edge disclination e.g. Kleinert (2008), where angle Ω is assumed small e.g. Ruggiero and Tartaglia (2003).

In sum, connection in a twisted and curved spacetime cannot be obtained only in terms of triads/tetrads. It is necessary to introduce a spin connection as follows: $\hat{\Gamma}_{\alpha\beta}^\gamma := F_i^\gamma \hat{\nabla}_\alpha F_\beta^i$ where the symbol $\hat{\nabla}$ denotes the spacetime connection (not necessarily torsionless) which is different from the matter connection ∇ e.g. Cho (1976a,c) (Kibble suggested to introduce additional gauge potentials to create spacetime curvature; Kibble 1961). The explicit formulation takes the form of $\Gamma_{\alpha\beta}^\gamma := F_i^\gamma \partial_\alpha F_\beta^i - F_i^\gamma \hat{\Gamma}_{\alpha\beta}^\lambda F_\lambda^i$ where the first term reduces to the Weitzenböck connection with non zero torsion but zero curvature. The second term $\tilde{\gamma}_{\alpha\beta}^\gamma := F_i^\gamma \hat{\Gamma}_{\alpha\beta}^\lambda F_\lambda^i$ has a role of spin connection, with possibly nonzero torsion (twisted spacetime) and/or non zero curvature (curved spacetime) e.g. Cho (1976b), in generalized teleparallel gravitation theory (Sotiriou et al. 2011). Such a connection is metric compatible.

Remark 2.21 By definition, the multi-valued tetrads form parallel field in the sense that $\nabla_\beta F_\alpha^i := \partial_\beta F_\alpha^i - \Gamma_{\beta\alpha}^\lambda F_\lambda^i = \partial_\beta F_\alpha^i - F_j^i (\partial_\beta F_\alpha^j) F_\lambda^i \equiv 0$. Therefore the induced metric is also a parallel tensor showing that the connection is metric compatible $\nabla_\lambda g_{\alpha\beta} = 0$.

2.3.5.5 Torsion, Curvature, and Smoothness of Tensor Fields

The role of torsion and curvature in continuum mechanics are not limited to describe translation dislocations and rotation disclinations. Experimental testings have shown

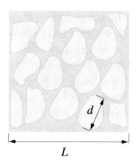

L

Fig. 2.15 The Hall-Peach effects are characterized by the increase of the strength of polycrystalline metals when the grain size decreases. The adimensional number $\lambda := d/L$ is defined for accounting for scale length. For samples with $\lambda \simeq 10^{-6} \to 0$ it is usual to consider plasticity using macroscopic continuum models. Gradient models are worth for samples with intermediate $\lambda \simeq 10^{-3} \to 10^{-1}$, whereas micro-mechanics theories as discrete crystalline plasticity is better for $\lambda \simeq 10^{-1} \to 1$

the existence of a material length scale for microcracking and plasticity of materials. Roughly speaking, continuum plasticity is used at macroscopic level ($\geq 300\,\mu$m) whereas microscopic slip models constitute the basic tool for crystalline plasticity ($\simeq 10^{-4}\,\mu$m). In between, gradient continuum models have been proposed to model mesoscopic plasticity e.g. Rakotomanana (2005). More precisely, four regions of plastic behavior usually exist. The first occurs at lengthscale below approximately 200 Å (corresponding to roughly 10^4 atoms). The second kind of plasticity behavior appears at $2\,\mu$m (corresponding to nearly 10^8 atoms). The third spatial domain ranges between 2 and $300\,\mu$m, where the influence of lengthscale still pertains. Thus the influence of the length scale cannot be neglected, suggesting to use the Riemann–Cartan manifold as geometric background for generalized continuum model. On Fig. 2.15, we sketch the influence of grain size (microcosms e.g. Gonseth 1926) on the continuum modeling by defining appropriate length scale λ. We now remind result obtained in a previous work on weakly continua (Rakotomanana 1997), where discontinuity of field is assumed to be diffuse and to continuously vary within matter. We have considered a class of continua for which both scalar field and vector field may be discontinuous.

Theorem 2.2 (Theorem on Discontinuous Fields) *Consider a metric-affine manifold $\mathscr{B}$. Let θ be a scalar field on $\mathscr{B}$. If the variation of θ from one point to another point depends on the path then the torsion field in $\mathscr{B}$ does not vanish $\aleph \neq 0$ and it characterizes the discontinuity of field θ on $\mathscr{B}$. Let $\mathbf{w}$ be a vector field on $\mathscr{B}$. If the variation of $\mathbf{w}$ from one point to another point depends on the path then $\aleph$ or $\mathfrak{R}$ do not vanish on $\mathscr{B}$ and they characterize the discontinuity of field $\mathbf{w}$ on $\mathscr{B}$.*

Proof See in e.g. Rakotomanana (1997).   □

The generalized continuum model presented here is based on the Cartan's circuit concept where the continuum $\mathscr{B}$ does not have any lattice structure (Noll 1967).

A continuum $\mathscr{B}$ is said affinely equivalent to the ambient Euclidean space $\mathscr{E}$ if and only if the torsion and curvature tensor fields are identically null at every point of $\mathscr{B}$. We observe at least two classes when considering transformations of continuum with an initial defect-free configuration: (a) the actual configuration remains defect-free, for which the mapping from initial and actual configurations is a diffeomorphism; and (b) the current configuration presents defects, for which the mapping is not a diffeomorphism. For this second case, there is no global coordinate system in the current configuration which can be associated to the initial configuration by the relation $dx^i = F^i_\alpha \, dX^\alpha$, since $F^i_\alpha(X^\mu)$ is not integrable on the whole continuum $\mathscr{B}$.

Torsion is not only associated to translational dislocations but also to local discontinuity of scalar field on a continuum $\mathscr{B}$. Curvature is not only associated to rotational dislocations but also to local discontinuity of vector field on continuum e.g. Katanaev and Volovich (1992), Rakotomanana (1997), Ruggiero and Tartaglia (2003). The geometric structure behind these continuum models are the following: (a) metric-affine geometry for arbitrary metric $\mathbf{g}$ and connection with torsion $\aleph \neq 0$ and curvature $\mathfrak{R} \neq 0$; (b) Riemann–Cartan geometry if additionally the connection is compatible with the metric $\nabla\mathbf{g} \equiv 0$ and $(\aleph \neq 0, \mathfrak{R} \neq 0)$; (c) Riemann geometry for compatible connection and $(\aleph \equiv 0, \mathfrak{R} \neq 0)$; (d) Weitzenböck geometry for compatible metric and $(\aleph \neq 0, \mathfrak{R} \equiv 0)$; and (e) Euclidean geometry for compatible metric and $(\aleph \equiv 0, \mathfrak{R} \equiv 0)$.

Strain gradient continuum may undergo inelastic deformation. The existence of sharp gradients of field in adiabatic matter shear bands during plastic deformations after high velocity impact, constitutes a typical strain gradient continuum. Such is also the case for matter constituted of grains, at an appropriate length scale. In the scope of crystalline solid mechanics, Le and Stumpf have shown in e.g. Le and Stumpf (1996) that elastic and plastic deformation fields, as well as the local associated rotations can be uniquely calculated from both the metric components $g_{\alpha\beta}$, and the torsion components $\aleph^\gamma_{\alpha\beta}$, if the curvature associated to the connection identically vanishes $\mathfrak{R}^\lambda_{\alpha\beta\mu} \equiv 0$. In the scope of strain gradient continuum and classical mechanics, we consider the product of manifold $\mathscr{M} := \mathfrak{T} \times \mathscr{B}$, where $\mathscr{B}$ reduces to a three-dimensional Riemann–Cartan manifold, endowed with a connection with torsion and curvature. We first recall the definition of the *Riemann curvature* $\mathfrak{R}_{\alpha\beta\lambda\mu} := \mathfrak{R}^\gamma_{\alpha\beta\lambda} \, g_{\gamma\mu}$ with n^4 components on a n-dimension manifold. Various symmetries of Riemannian curvature reduce the number of independent components to $n^2(n^2 - 1)/12$. Second, the skew-symmetry of the torsion tensor with respect to the two lower indices also diminishes the number of independent components of $\aleph^\gamma_{\alpha\beta}$. On a $3D$ continuum, the curvature field on a 3-manifold is entirely characterized by the Ricci curvature $\mathcal{R}_{\alpha\beta} := \mathfrak{R}^\lambda_{\lambda\alpha\beta}$ with 9 components.

Remark 2.22 From a topological point of view, a generalized continuum may be defined by a locally connected and locally compact manifold. This class of generalized continuum is metrizable if it has a one-to-one continuous image in a metric after Proizvolov theorem e.g. Pettey (1971) (resp. semi-metric) space

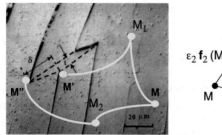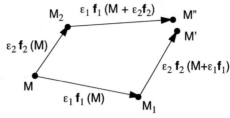

Fig. 2.16 (Left) Illustration of a shear band formation resulting from a displacement discontinuity at points M' and M''. (Right) Corresponding Cartan's circuit with area $\epsilon_1 \times \epsilon_2$. At the macroscopic level, nucleation and migration of a great number of discontinuities (defects) appear on the macroscopic continuum boundary, and give rise to macroscopic plastic strain (Lüders-Hartmann bands)

(Euclidean ambient space) in classical continuum mechanics (resp. Lorentzian spacetime for relativistic gravitation).

For establishing the covariance of constitutive laws of strain gradient continuum, we limit to coordinate systems (x^α) and (y^i) that are all diffeomorphically equivalent to each other. This involves a larger group than orthogonal transformations and also larger than the class of linear transformations, but it remains a small fraction of all possible change of coordinate systems e.g. Nakahara (1996), Rakotomanana (2003). Further studies in this direction still hold as great challenge, by considering the concept of path-dependent integration method e.g. Kleinert (2008). The path dependence is illustrated by the Cartan circuit together with discontinuities of fields (Fig. 2.16). Dislocations may be considered as discontinuity of displacement field (Burgers vector) and the density of dislocations may be calculated by torsion tensor field on continuum manifolds. Discontinuity of geometrical variables are not only well sound from mathematics point of view, they also have real physical meaning. Some experimental studies have shown the possibility of measuring an abrupt change of density by means of High Resolution Transmission Electron Microscopy e.g. Barra et al. (2009).

Say an affinely connected manifold $\mathscr{B}$. Let consider two vectors $\mathbf{f}_1$ and $\mathbf{f}_2$ at a point M, they define two paths of length ϵ_1 and ϵ_2. Non zero torsion and curvature fields induce the following relationships (Rakotomanana 1997):

$$\begin{cases} \lim\limits_{(\epsilon_1,\epsilon_2)\to 0} (\theta' - \theta'')/\epsilon_1\epsilon_2 = \aleph\,(\mathbf{f}_1, \mathbf{f}_2)\,[\theta] \\ \lim\limits_{(\epsilon_1,\epsilon_2)\to 0} [\mathbf{f}^\gamma\left(\mathbf{w}'\right) - \mathbf{f}^\gamma\,(\mathbf{w}'')]/\epsilon_1\epsilon_2 = \Re\,(\mathbf{f}_1, \mathbf{f}_2, \mathbf{w}, \mathbf{f}^\gamma) - \mathbf{f}^\gamma\left(\nabla_{\aleph(\mathbf{f}_1,\mathbf{f}_2)}\mathbf{w}\right) \end{cases}$$

$$(2.47)$$

where $\theta\,(\mathbf{x})$, and $\mathbf{w}\,(\mathbf{x})$ are respectively a scalar field and a vector field on $\mathscr{B}$. We notice $\theta' := \theta\left(M'\right)$, and $\theta'' := \theta\left(M''\right)$; and $\mathbf{w}' := \mathbf{w}\left(M'\right)$, and $\mathbf{w}'' := \mathbf{w}\left(M''\right)$. The system of equations (2.47) are the continuum extension of the discrete dislocation loop which is characterized by a Burgers vector $\mathbf{b} := [\mathbf{u}]$, defined by the discontinuity of the displacement field within continuum (otherwise strongly continuous elsewhere) $\mathbf{b} := - \oint_{\mathcal{C}} d\mathbf{u}$. The integral is calculated along a closed curve $\mathcal{C}$ around the isolated dislocation. For multivalued fields, the concept of generalized continuum allows us to define locally connected (and locally compact) manifolds rather than directly accounting for discontinuity of scalar and vector fields, $\theta(\mathbf{x})$ and $\mathbf{w}(\mathbf{x})$. If the curvature vanishes, vector field is however necessarily not single-valued as shown by the second equation (2.47). These relations show the interdependence of translation and rotation dislocations and therefore their practical co-existence. The twisted and curved continuum manifold offers a representation of a material (resp. spacetime) manifold that is everywhere dislocated (resp. singular or with continuously distributed spins). Moreover, these relations may be read from right to left direction, and state that whenever $\aleph$ or $\mathfrak{R}$ are not equal to zero then any scalar and vector fields on the affinely connected continuum $\mathcal{B}$ are necessarily multi-valued fields. For Riemannian continua, only vector fields may be multivalued, whereas for Weitzenböck continua e.g. Hayashi (1979), only scalar field and some particular vector field may be multi-valued, since $\mathfrak{R} \equiv 0$. Within the class of the Weitzenböck continua, Le and Stumpf analyzed the existence and uniqueness of an anholonomic crystal reference in the domain of nonlinear dislocations and similarly for elastic plastic deformation of continua (Le and Stumpf 1996). They showed that elastic and plastic deformations fields, with corresponding rotation fields, are uniquely determined from the metric $g_{\alpha\beta}$ and the torsion $\aleph^{\gamma}_{\alpha\beta}$, if the curvature tensor of crystal reference identically vanishes (it constitutes the integrability condition). For a larger class of Riemann–Cartan continua, there is influence of both torsion and curvature e.g. Bruzzo (1987), Capoziello and de Laurentis (2011), Clayton et al. (2004), Shapiro (2002), Utiyama (1956).

Remark 2.23 The analogy between continuum with defects in three dimensions and extension of the Minkowski spacetime (flat, smooth) to spacetime with curvature and then to spacetime with curvature and torsion seems to be straightforward e.g. Ruggiero and Tartaglia (2003). Indeed it is tempting to state that the extended spacetimes can be considered as Minkowski spacetime with distributions of defects such as discontinuity of scalar and vector field. In three dimensional dislocations theory, the link between the discontinuity of tensor fields (scalar and vector here) and the presence of torsion and curvature fields over a manifold (2.47) is illustrated by the Volterra processes which can be described by a well known "cut and paste" scheme. Mathematically, these two tensors also represent the non integrability of differential du^{α} meaning that the displacement field is multivalued. Volterra processes lead to three relativistic translations of two regarding lips and three relativistic rotations of the same lips e.g. Rakotomanana (2003). When passing from continuum mechanics in three dimensions to relativistic gravitation in four dimensions, there is a difference because the curvature tensor is entirely defined by means of Ricci curvature (six independent components) whereas in four dimension, this is not the case.

2.4 Invariance for Lagrangian and Euler–Lagrange Equations

Most physics phenomenae within continuum may be described by a Lagrangian function $\mathscr{L}$ defined on the continuum manifold. Equations governing the fields inside the continuum may be derived by means of this Lagrangian. One common point of the most theories of physics is the spacetime where motion and more generally evolution of fields happens. Therefore, any Lagrangian function should have at least the geometric arguments: metric components and their partial derivative, connection coefficients and their partial derivatives.

The two invariance conditions namely the covariance and the gauge invariance of Lagrangian function are first introduced here by means of well known simple system, a material point in motion within a reference frame and subject to a potential field. Then it will be extended to derive the Euler–Lagrange equations in continuum physics. Roughly speaking, covariance helps to rigorously define the list of arguments of the Lagrangian function, whereas gauge invariance helps to define the conservation laws of fields describing physics inside continuum.

Overview of these two invariance condition is sketched in this subsection, and will be treated more in details in the next chapters.

2.4.1 Covariant Formulation of Classical Mechanics of a Particle

Let M be a material point of mass m in motion with a velocity $\mathbf{v}$ with respect to a Galilean reference $\mathscr{R}$. The material point is submitted to an external force deriving from a potential $\mathscr{U}$. The Newton's law is written accordingly:

$$m\mathbf{a} = -\nabla\mathscr{U} := \mathbf{F} \tag{2.48}$$

where $\mathbf{a}$ and $\nabla\mathscr{U}$ denote the acceleration with respect to $\mathscr{R}$ and the gradient of the potential respectively. For example, a uniform gravitation $\mathbf{g}$ the potential takes the form of $\mathscr{U} := -m\mathbf{g} \cdot \mathbf{OM}$ whereas for the gravitation field generated by a massive body of mass M, the potential takes the form of $\mathscr{U} = -GMm/r$ where r denotes the distance between the test particle of mass m and the mass center of the body. Once the reference is chosen, we have to define a coordinate system and its associate local base to project onto the motion equation (2.48). For the sake of the simplicity, the three standard coordinate system are reported on Fig. 2.17: The projection of the Newton's equation (2.48) onto the local base does not induce a covariant formulation of the particle motion. To overcome this difficulty, a covariant formulation is obtained by first defining the Lagrangian of the material point as $\mathscr{L} := (m/2)\|\mathbf{v}\|^2 - \mathscr{U} = \mathscr{L}(t, q^\alpha, \dot{q}^\alpha)$ where q^α denotes any admissible coordinate system and $\dot{q}^\alpha$ its derivative with respect to t. Second

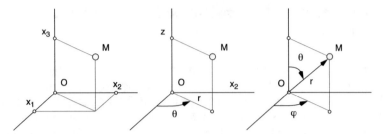

Fig. 2.17 Coordinate systems and their associate orthonormal local bases: (a) Cartesian (x^1, x^2, x^3) with $(\mathbf{e}_1, \mathbf{e}_2, \mathbf{e}_3)$, (b) Cylindrical (r, θ, z) with $(\mathbf{e}_r, \mathbf{e}_\theta, \mathbf{e}_z)$, and (c) spherical (r, θ, φ) with $(\mathbf{e}_r, \mathbf{e}_\theta, \mathbf{e}_\varphi)$

we deduce the Euler–Lagrange equation by varying the Lagrangian $\delta\mathscr{L}$ along a small variation δq^α. A straightforward and classical calculus allows us to derive the so-called Euler–Lagrange equation:

$$\frac{\partial \mathscr{L}}{\partial q^\alpha} - \frac{d}{dt}\left(\frac{\partial \mathscr{L}}{\partial \dot{q}^\alpha}\right) = 0, \qquad \alpha = 1, 3 \qquad (2.49)$$

Independently on the choice of coordinate system $(q^\alpha = (x^1, x^2, x^3)$, $q^\alpha = (r, \theta, z)$, or $q^\alpha = (r, \theta, \varphi))$, the system of equations (2.49) keeps the same shape, it is said covariant motion equation of the material point. For a material point of mass m in a plane with the velocity $\mathbf{v}$ and subject to a potential $\mathscr{U}$, let us derive the motion equation by using two different formulations. We consider a Cartesian and a polar coordinate systems.

1. Newton's formulation of the point motion holds $m\mathbf{a} = -\nabla\mathscr{U}$

$$\begin{cases} m\ddot{x} = -\dfrac{\partial \mathscr{U}}{\partial x}(x, y) \\[2mm] m\ddot{y} = -\dfrac{\partial \mathscr{U}}{\partial y}(x, y) \end{cases} \qquad \begin{cases} m\left(\ddot{r} - r\dot{\theta}^2\right) = -\dfrac{\partial \mathscr{U}}{\partial r}(r, \theta) \\[2mm] m\left(r\ddot{\theta} + 2\dot{r}\dot{\theta}\right) = -\dfrac{1}{r}\dfrac{\partial \mathscr{U}}{\partial r}(r, \theta) \end{cases}$$

Obviously the formulation of Newton for the mechanics of a point and more generally of any arbitrary system is not covariant. The shape changes after a coordinate transformation.

2. Lagrange's formulation of the point motion writes with the Lagrangian $\mathscr{L} := (m/2)|\mathbf{v}|^2 - \mathscr{U}$ and the Poincarés invariance $\delta \int_{t_i}^{t_f} \mathscr{L}\,dt = 0$ with worth boundary conditions:

$$\begin{cases} \dfrac{d}{dt}\left(\dfrac{\partial \mathscr{L}}{\partial \dot{x}}\right) - \dfrac{\partial \mathscr{L}}{\partial x} = 0 \\[3mm] \dfrac{d}{dt}\left(\dfrac{\partial \mathscr{L}}{\partial \dot{y}}\right) - \dfrac{\partial \mathscr{L}}{\partial y} = 0 \end{cases} \qquad \begin{cases} \dfrac{d}{dt}\left(\dfrac{\partial \mathscr{L}}{\partial \dot{r}}\right) - \dfrac{\partial \mathscr{L}}{\partial r} = 0 \\[3mm] \dfrac{d}{dt}\left(\dfrac{\partial \mathscr{L}}{\partial \dot{\theta}}\right) - \dfrac{\partial \mathscr{L}}{\partial \theta} = 0 \end{cases}$$

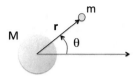

Fig. 2.18 In the framework of Newton mechanics, the mass m is subject to a force exerted by the mass M by considering that motion evolves within a rigid and absolute spacetime

This constitutes the simplest example illustrating the covariance of the Lagrange formulation. Of course, we arrive to the Newton's equations when the Lagrangian $\mathscr{L}$ is explicitly formulated. In the next section, we extend this concept.

By the way, we remind some basic notions on the orbits of a test particle submitted to a central gravitational force due to a mass M. The Lagrangian of a test particle of mass m (test particle means that its mass m does engender a gravitation filed since $m << M$): $\mathscr{L} = (m/2)\left(\dot{r}^2 + r^2\dot{\theta}^2\right) - \mathscr{U}(r)$. Considering the motion in the plane containing M and m is sufficient. Since $\partial_\theta\mathscr{L} \equiv 0$, it is then straightforward to show that the total mechanical energy in any (three-dimensional) central potential is conserved and takes the form of:

$$\mathscr{E} = \frac{m}{2}\dot{r}^2 + \frac{\ell^2}{2mr^2} + \mathscr{U}(r), \qquad \ell := mr^2\dot{\theta}$$

where ℓ is the magnitude of the angular momentum vector of the particle, which is conserved (first invariant). For Newtonian gravity, the particular form of total energy holds (Fig. 2.18):

$$\mathscr{E} = \frac{m}{2}\dot{r}^2 + \mathscr{U}_{eff}, \qquad \mathscr{U}_{eff} := \frac{\ell^2}{2mr^2} - G\frac{Mm}{r} \tag{2.50}$$

The energy $\mathscr{E}$ is the second invariant of the motion. $\mathscr{U}_{eff}$ is called effective potential. Expressions of the invariants give the orbits differential equations as follows:

$$\begin{cases} \dot{r} = \sqrt{\dfrac{2}{m}\left(\mathscr{E} - \mathscr{U}_{eff}(r)\right)} \\ \dot{\theta} = \dfrac{\ell}{mr^2} \end{cases} \tag{2.51}$$

The combination of the two invariants ℓ and $\mathscr{E}$ allows us to express the parametric equation of the orbits as connecting r and θ:

$$\frac{d\theta}{dr} = \frac{\ell/(mr^2)}{\sqrt{2/m[\mathscr{E} - \mathscr{U}_{eff}(r)]}} \tag{2.52}$$

Two types orbits are possible for this effective potential: (a) bound orbits with energy $\mathscr{E} < 0$, and (b) unbound orbits with energy $\mathscr{E} \geq 0$. We will investigate some orbits corresponding to these Newtonian orbits (2.52) in the presence of curved and twisted gravitation.

2.4.2   Basic Elements for Calculus of Variations

We give some basic ingredients for deriving the Euler–Lagrange equations by starting with the invariance under change of coordinate system. The previous subsection is extended by considering an arbitrary Lagrangian function on a manifold $\mathscr{M}$. Let first consider a curve $\mathscr{C}$ of $\mathscr{M}$, assumed to be a class $\mathcal{C}^2$, which is represented parametrically by:

$$\mathscr{C} : t \in \mathbb{R}^* \to x^\alpha(t) \in \mathscr{M} \tag{2.53}$$

where t is a parameter, not necessarily the time variable, and x^α is a point of the manifold (Fig. 2.19). The derivative of the coordinate with respect to t is denoted $\dot{x}^\alpha$. Say a Lagrangian function $\mathscr{L}(t, x^\alpha, \dot{x}^\alpha)$ of $2n + 1$ independent variables of class $\mathcal{C}^2$. We assume that the curve $\mathscr{C}$ is delimited by two points $x^\alpha(t_1)$ and $x^\alpha(t_2)$ respectively defined by A and B on the figure. The action associated to this Lagrangian function on the curve $\mathscr{C}$ is defined as the integral:

$$\mathscr{S} := \int_{t_1}^{t_2} \mathscr{L}\left(t, x^\alpha(t), \dot{x}^\alpha(t)\right) dt \tag{2.54}$$

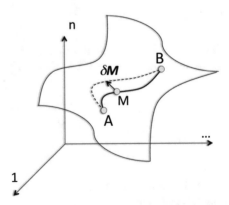

Fig. 2.19 Configuration space of the Lagrangian for a discrete system with coordinates (x^α, $\alpha = 1, \cdots, n$). The two extremities A and B of the curve $\mathscr{C}$ are the initial and the final points, the curve lying on the manifold represented by the surface. At the current point M the virtual displacement (virtual velocity) $\delta\mathbf{M}$ has the components δx^α. The real motion is represented by the plain line curve $\mathscr{C}$ and the virtual trajectory is represented by the dotted curve

Consider now the coordinate transformations $y^i = y^i(x^\alpha)$ which does not affect the parameter t. The elementary definition of the covariance is that the Lagrangian function is conserved under an arbitrary coordinate transformations (it is also called diffeomorphism invariance since the coordinate transformations constitute a diffeomorphism). The goal of this subsection is to recover the standard Euler–Lagrange equations associated to this Lagrange function by means of covariance.

The invariance of the action under the coordinate transformation stipulates that the Lagrangian function $\mathscr{L}$ is a scalar. By denoting that the Lagrangian function in terms of the coordinates y^i is $\overline{\mathscr{L}}$, the covariance of the action is satisfied if:

$$\overline{\mathscr{L}}\left(t, y^i, \dot{y}^i\right) = \mathscr{L}\left(t, x^\alpha(y^i), \dot{x}^\alpha(y^i, \dot{y}^i)\right) \tag{2.55}$$

and this invariance should be verified for all values of $(y^i, \dot{y}^i)$. To obtain the invariance condition, we have thus to differentiate this relation of invariance with respect to y^i and $\dot{y}^i$ respectively. For the sake of the clarity, we have to do some preliminary calculus. The inverse of the coordinate transformations allows us to derive the following relationships:

$$x^\alpha = x^\alpha(y^i), \qquad \dot{x}^\alpha = J_i^\alpha(y^i)\,\dot{y}^i = \dot{x}^\alpha\left(y^i, \dot{y}^i\right) \tag{2.56}$$

From these equations, we obtain:

$$\frac{\partial \dot{x}^\alpha}{\partial \dot{y}^i} = J_i^\alpha, \qquad \frac{\partial \dot{x}^\alpha}{\partial y^i} = J_{ij}^\alpha\,\dot{y}^j \tag{2.57}$$

Differentiation of the Lagrangian function $\overline{\mathscr{L}}$ with respect to y^i and $\dot{y}^i$ gives:

$$\frac{\partial \overline{\mathscr{L}}}{\partial y^i} = J_i^\alpha \frac{\partial \mathscr{L}}{\partial x^\alpha} + J_{ij}^\alpha \frac{\partial \mathscr{L}}{\partial \dot{x}^\alpha}\,\dot{y}^j, \qquad \frac{\partial \overline{\mathscr{L}}}{\partial \dot{y}^i} = J_i^\alpha \frac{\partial \mathscr{L}}{\partial \dot{x}^\alpha} \tag{2.58}$$

assessing that the derivatives $\partial \mathscr{L}/\partial \dot{y}^i$ are the components of a covariant vector whereas the derivatives $\partial \mathscr{L}/\partial y^i$ are not. Owing that the partial derivative of a scalar function is by definition the covariant derivative, care should be taken by observing that here the dependence of the Lagrangian on the derivative $\dot{x}^\alpha$ introduces difficulties. Variables are in fact not independent. The construction of a covariant vector associated to the derivative $\partial \mathscr{L}/\partial x^\alpha$ is done by eliminating the right-hand-side term of the first equation of the system (2.58). For this purpose, total differentiation of the second equation of (2.58) with respect to the parameter t e.g. Lovelock and Rund (1975) gives:

$$\frac{d}{dt}\left(\frac{\partial \overline{\mathscr{L}}}{\partial \dot{y}^i}\right) = J_i^\alpha \frac{d}{dt}\left(\frac{\partial \mathscr{L}}{\partial \dot{x}^\alpha}\right) + J_{ij}^\alpha \frac{\partial \mathscr{L}}{\partial \dot{x}^\alpha}\dot{x}^j \tag{2.59}$$

Therefore, we deduce the covariance requirement of the Lagrangian function

$$\frac{d}{dt}\left(\frac{\partial \overline{\mathscr{L}}}{\partial \dot{y}^i}\right) - \frac{\partial \overline{\mathscr{L}}}{\partial y^i} = J_i^\alpha\left[\frac{d}{dt}\left(\frac{\partial \mathscr{L}}{\partial \dot{x}^\alpha}\right) - \frac{\partial \mathscr{L}}{\partial x^\alpha}\right] \tag{2.60}$$

The covariant vector:

$$\mathscr{E}_\alpha := \frac{d}{dt}\left(\frac{\partial \mathscr{L}}{\partial \dot{x}^\alpha}\right) - \frac{\partial \mathscr{L}}{\partial x^\alpha} \tag{2.61}$$

is interpreted as the Lagrangian derivative (or generalized gradient) of the Lagrangian function $\mathscr{L}$. Let now consider the action $\mathscr{S}$ and we assume that it is stationary with respect to the evolution of the parameter t. Then its variation holds:

$$\delta\mathscr{S} = \delta\int_{t_1}^{t_2}\mathscr{L}\left(t, x^\alpha(t), \dot{x}^\alpha(t)\right)dt = 0 \tag{2.62}$$

with the additional assumptions that the two extremities are fixed, say $\delta x^\alpha(t_1) = 0$ and $\delta x^\alpha(t_2) = 0$. Straightforward calculus using integration by parts leads to the classical equation:

$$\delta\mathscr{S} = \delta\int_{t_1}^{t_2}\left[\frac{d}{dt}\left(\frac{\partial \mathscr{L}}{\partial \dot{x}^\alpha}\right) - \frac{\partial \mathscr{L}}{\partial x^\alpha}\right]\delta x^\alpha \, dt = 0 \tag{2.63}$$

Owing that coordinates x^α are independent, the conditions of stationarity of the action are the so-called Euler–Lagrange equations:

$$\mathscr{E}_\alpha = 0, \quad \alpha = 1, n \tag{2.64}$$

and these equations are covariant. In the following we extend this concept to obtain the Euler–Lagrange equations associated to a Lagrangian function

2.4.3 Extended Euler–Lagrange Equations

We remind the concept of Lagrangian functions and its invariance, which is a common tool for both relativistic gravity e.g. Kibble (1961), strain gradient continuum e.g. Lazar and Anastassiadis (2008), and more generally most of physics theory. We first give a preliminary definition of a Lagrangian function e.g. Lovelock and Rund (1975) and the notation δ here corresponds to what we will call Lagrangian variation in the second part of the work.

Definition 2.24   Consider an open subset $\mathcal{M} \subset \mathbb{R}^n$, and a scalar (real) function on $\mathcal{M}$ depending on coordinates, $(x^\mu, \mu = 0, n - 1)$, and on the set of m scalar functions Φ^i, $i = 1, m$ with their derivatives:

$$\mathcal{L}\left(x^\mu, \Phi^i(x^\mu), \Phi^i_{\mu_1}(x^\mu), \cdots, \Phi^i_{\mu_1\cdots\mu_k}(x^\mu)\right) \tag{2.65}$$

where:

$$\Phi^i_{\mu_1}(x^\mu) := \partial_{\mu_1}\Phi^i(x^\mu), \cdots, \Phi^i_{\mu_1\cdots\mu_k}(x^\mu) := \partial_{\mu_1}\cdots\partial_{\mu_k}\Phi^i(x^\mu) \tag{2.66}$$

$\mathcal{L}$ is called Lagrangian function of order k, and:

$$\Sigma_i^{\mu_1\cdots\mu_k} := \partial\mathcal{L}/\partial\Phi^i_{\mu_1\cdots\mu_k} \tag{2.67}$$

are the Lagrangian momenta (or *currents*).

As arguments of the Lagrangian function, metric components $g_{\alpha\beta}$ and coefficients connection $\Gamma^\gamma_{\alpha\beta}$ play the role of Φ^i in the scope of relativistic gravitation theory. The diffeomorphism invariance is required for both the gravity Lagrangian density $\mathcal{L}_G$ and the matter Lagrangian density $\mathcal{L}_M$. Both of them may be submitted to the following transformations e.g. Capoziello and de Laurentis (2011), Obukhov and Puetzfeld (2014):

$$\begin{cases} x^\mu \to \tilde{x}^\mu = \tilde{x}^\mu(x^\alpha) \\ \Phi^i \to \tilde{\Phi}^i := \Phi^i + \delta\Phi^i \end{cases} \tag{2.68}$$

Variation $\delta x^\mu := \tilde{x}^\mu - x^\mu$ of the domain $\mathcal{M}$ is called external variation (pictured as arbitrary small transformations of the spacetime coordinates), whereas variation of fields $\delta\Phi^i$ are called internal or substantial (arbitrary small variation of matter fields such as the metric, the torsion, and the curvature tensor fields). In view of deriving equations from covariance and gauge invariance, it is important to consider a worth set of variables. For instance considering the set of metric and connection may induce some interpretation errors, as we will see later, whereas considering metric and torsion, and curvature is worth (conforming to covariance requirement). Considering the displacement and the connection is worth. Some mistakes might appear when wrong set of variables is chosen.

   We consider a generic Lagrangian $\mathcal{L}$ hereafter. The change of Lagrangian due to these two transformations is given by:

$$\mathcal{L}(\tilde{x}^\mu, \tilde{\Phi}^i, \tilde{\Phi}^i_{\mu_1}, \cdots, \tilde{\Phi}^i_{\mu_1\cdots\mu_k}) = \mathcal{L}(x^\mu, \Phi^i, \Phi^i_{\mu_1}, \cdots, \Phi^i_{\mu_1\cdots\mu_k}) \tag{2.69}$$

$$+ \frac{\partial\mathcal{L}}{\partial x^\mu}(x^\mu, \Phi^i, \Phi^i_{\mu_1}, \cdots, \Phi^i_{\mu_1\cdots\mu_k})\delta x^\mu \tag{2.70}$$

$$+\dots$$

$$+\frac{\partial \mathscr{L}}{\partial \Phi^i_{\mu_1\dots\mu_k}}(x^\mu, \Phi^i, \Phi^i_{\mu_1}, \dots, \Phi^i_{\mu_1\dots\mu_k})\delta\Phi^i_{\mu_1\dots\mu_k}$$

$$(2.71)$$

The Lagrangian must satisfy three conditions: (a) the same shape of the function should be retained independently on the coordinate system (covariance) e.g. Antonio and Rakotomanana (2011); (b) the Lagrangian should satisfy the gauge invariance under the group of Lorentz transformations (which is an extension of the Galilean invariance) e.g. Kibble (1961), and (c) the gauge invariance of the Lagrangian with respect to active diffeomorphisms applied on fields. We deduce respectively:

1. The *covariance* requirement of the Lagrangian should be satisfied, since physical contents do not depend on the choice of coordinate system,

$$\mathscr{L}(\tilde{x}^\mu, \tilde{\Phi}^i, \tilde{\Phi}^i_{\mu_1}, \dots, \tilde{\Phi}^i_{\mu_1,\dots\mu_k}) = \mathscr{L}(x^\mu, \Phi^i, \Phi^i_{\mu_1}, \dots, \Phi^i_{\mu_1\dots\mu_k}) \qquad (2.72)$$

The shape of the Lagrange remains the same after change of coordinate system.
2. The gauge invariance with respect to the existence of a (locally) *homogeneous* ambient Minkowskian spacetime induces:

$$\frac{\partial \mathscr{L}}{\partial x^\mu}\left(x^\mu, \Phi^i, \Phi^i_{\mu_1}, \dots, \Phi^i_{\mu_1\dots\mu_k}\right)\delta x^\mu = 0, \qquad \forall \delta x^\mu \qquad (2.73)$$

stating that Lagrangian cannot depend explicitly on the coordinate x^μ e.g. Capoziello and de Laurentis (2011). This is called (local) translation invariance in Kibble (1961), Lovelock and Rund (1975). Partial derivatives mean that all physics fields and their derivatives are held constant.
3. The *gauge invariance* with respect to active diffeomorphisms (variations of the fields) then gives e.g. Pons (2011) :

$$\frac{\partial \mathscr{L}}{\partial \Phi^i}\left(x^\mu, \Phi^i, \Phi^i_{\mu_1}, \dots, \Phi^i_{\mu_1\dots\mu_k}\right)\delta\Phi^i$$

$$+\frac{\partial \mathscr{L}}{\partial \Phi^i_{\mu_1}}\left(x^\mu, \Phi^i, \Phi^i_{\mu_1}, \dots, \Phi^i_{\mu_1\dots\mu_k}\right)\delta\Phi^i_{\mu_1}$$

$$+\dots$$

$$+\frac{\partial \mathscr{L}}{\partial \Phi^i_{\mu_1,\dots\mu_k}}\left(x^\mu, \Phi^i, \Phi^i_{\mu_1}, \dots, \Phi^i_{\mu_1\dots\mu_k}\right)\delta\Phi^i_{\mu_1,\dots\mu_k} = 0, \qquad \forall \delta\Phi^i$$

for any independent variations of the fields (but not their derivatives). After integrating by parts and worthily transferring boundary terms at $\partial\mathscr{M}$, we obtain classically the *Euler–Lagrange equation* (arguments are omitted for the sake of simplicity) e.g. Pons (2011), Sharma and Ganti (2005):

$$\frac{\partial \mathscr{L}}{\partial \Phi^i} - \frac{\partial}{\partial x^{\mu_1}} \left(\frac{\partial \mathscr{L}}{\partial \Phi^i_{\mu_1}} \right) + \frac{\partial^2}{\partial x^{\mu_1} \partial x^{\mu_2}} \left(\frac{\partial \mathscr{L}}{\partial \Phi^i_{\mu_1 \mu_2}} \right) \pm \ldots \equiv 0 \qquad (2.74)$$

From now and hereafter, the Lagrangian is assumed to depend not on the coordinates x^μ explicitly (conforming to the Kibble result Kibble 1961; Lovelock and Rund 1975) and we only consider geometric variables as fields. Typically, the variations of the domain, or the variations of the fields include the following terms ($\xi^\alpha, \xi^\alpha_\beta x^\beta, \partial_\alpha \Phi^i_{\mu_1 \cdots \mu_n} \xi^\alpha$), where ξ^α, and ξ^α_β are the variations of vector and tensor fields (gauge) e.g. Pons (2011). As we will see later, the gauge variations ξ^α_β imply the symmetry of hypermomenta $\Sigma^{\mu_1 \cdots \mu_n}_i$, whereas the partial derivatives $\partial_\alpha \Phi^i_{\mu_1 \cdots \mu_n} \xi^\alpha$ are related to the local conservation laws. In this work, central focus is on mechanics of gradient continuum and on relativistic gravitation where the physical variables for matter are matter metric and matter connection in addition to spacetime metric and spacetime connection.

Before embarking into details, as we will see in the second part of this work, it is worth to distinguish various kinds of variations when dealing with a continuum matter in motion in a continuum spacetime e.g. Carter and Quintana (1977). A Lagrangian variation of a tensor field $\Delta\mathbf{T}$ is a comoving variation of this quantity with the material point (small) displacement ξ. The Euler variation $\delta\mathbf{T}$ is the variation of a tensor field is the variation of this quantity at a fixed point in the spacetime. We will see that the Lie derivative (cf. Appendix A.3) of this tensor field along the vector field ξ defines the difference between the Lagrangian variation and the corresponding Eulerian variation $\Delta\mathbf{T} = \delta\mathbf{T} + \mathcal{L}_\xi \mathbf{T}$.

The variational procedure, that is reminded in this chapter, does not explicitly mention the torsion and curvature variation. In a general manner, the previous tensor $\mathbf{T}$ may be an arbitrary tensor namely metric, torsion and curvature. Interestingly, in the next chapters, torsion and curvature tensors constitute part of the geometrical background of the continuum (or spacetime) and then constitute in any case some "hidden" arguments of Lagrangian. $\mathscr{L}$. In the next chapters, we will show for instance that an action involving only metric and curvature can generate conservations equations containing torsion variable as in e.g. Futhazar et al. (2014), Kleinert (2000), Noll (1967), Rakotomanana (1997). This is essentially due to the presence of torsion as intrinsic part of the connection.

2.5   Simple Examples in Continuum and Relativistic Mechanics

2.5.1   Particles in a Minkowski Spacetime

Let use temporarily the coordinates $(x^0 := t, x^1, x^2, x^3)$ to highlight the importance of the light speed in relativistic theory. Minkowskian spacetime is a Lorentzian manifold endowed with a metric $\hat{g}_{\alpha\beta} := \text{diag}\{c^2, -1, -1, -1\}$ at every point P allowing to define arc length $ds^2 := g_{\alpha\beta}dx^\alpha dx^\beta = c^2dt^2 - dx^2 - dy^2 - dz^2$. Let $x^\mu := (x^0, x^1, x^2, x^3)$ be the coordinates of a mass point evolving within a Minkowskian spacetime. The path of this mass point is called worldline and defined as $x^\mu := (x^0(t), x^1(t), x^2(t), x^3(t))$ where $x^0 := t$ is a time parameter. The four-vector tangent to this worldline is a tangent vector of the manifold:

$$\frac{dx^\mu}{dt} = \left(v^0 := \frac{dx^0}{dt} = 1, v^i := \frac{dx^i}{dt}\right), \qquad i = 1, 2, 3 \qquad (2.75)$$

Remark that v^i are components of three-dimensional vector, but (v^μ) do not behave like a four dimensional vector. In the four dimensional spacetime, we can not use this definition because the time t is not invariant under Lorentz transformations (2.14), as we will check below. Indeed, the first principle of special relativity theory imposes that the velocity of any particle within the space is lower than the light speed then $c^2 > (v^1)^2 + (v^2)^2 + (v^3)^2$. To overcome the difficulty of time invariance, we use the concept of proper time τ as independent variable.

Definition 2.25 (Proper Time) We call the proper time parameter τ such that $d\tau^2 := ds^2/c^2$. This gives, where γ is called the Lorentz factor,

$$d\tau := \sqrt{1 - (v^2/c^2)}dt, \qquad \gamma := \left(1 - (v^2/c^2)\right)^{-1/2} \qquad (2.76)$$

In the particle's own rest frame, such that $dx^i = 0$, the proper time τ coincides on t. Physically, the proper time τ is merely the measure of the recorded time on the particle own's clock.

Definition 2.26 (Four-Vector, Number-Flux Four-Vector) The 4-velocity vector u^μ in the special relativity theory is defined as:

$$u^\mu := \left(u^0 := \frac{dx^0}{d\tau}, u^i := \frac{dx^i}{d\tau}\right) = \gamma\left(1, v^i\right), \qquad i = 1, 2, 3 \qquad (2.77)$$

where we check that $g_{\alpha\beta}u^\alpha u^\beta = \gamma^2(c^2 - v^2) = c^2$. The number-flux four-vector is defined as, n is the number density of particles in the rest frame,

$$n^\mu := n\, u^\mu \qquad (2.78)$$

In Galilean physics, the number density is a scalar with is the same independently on the reference frames whereas the three-dimensional flux nv^i is frame-dependent since velocities are frame-dependent. The number-flux four-vector nu^μ is a frame-independent four-vector. The previous relations allow us to deduce that $\mathbf{g}(\mathbf{u}, \mathbf{u}) = n^2c^2$. If we adopt the coordinate system $x^\mu = (x^0 := ct, x^i)$, the result would be $\mathbf{g}(\mathbf{u}, \mathbf{u}) = n^2$. For completeness, the momentum of a mass point is equal to $\mathbf{p} := \gamma \, m_0 \mathbf{v}$, where the "static mass" m_0 (rest mass) is multiplied by the factor γ. Therefore, the necessary work for accelerating the point from rest to a 3-velocity $\mathbf{v}$ is: $\mathfrak{T} = m_0c^2[1 - (v/c)^2]^{-1/2} - m_0c^2 = (\gamma - 1) \, m_0c^2$, because the material point at rest must have energy content equivalent to m_0c^2. The Lagrangian function of a material point in a relativistic free fall state is defined as, following the original idea of Einstein about electron motion,

$$\mathscr{L} := \mathfrak{T} = (\gamma - 1) \, m_0c^2 \tag{2.79}$$

For sufficiently low speed, the previous expended work (which equals to the kinetic energy) reduces to: $\mathfrak{T} = (\gamma - 1) \, m_0c^2 \simeq (1/2)m_0v^2 + (3/8)m_0v^4/c^2 + \cdots$.

2.5.2   Some Continua Examples

2.5.2.1   Energy-Momentum Tensor

In the following, we give some basic elements to build the energy-momentum tensor by considering a matter action $\mathscr{S}_M := \int_\mathcal{M} \mathscr{L}_M \left(g_{\alpha\beta}, \partial_\gamma \, g_{\alpha\beta}, \cdots ; u_\alpha, \cdots \right) \omega_n$, where the arguments of the matter Lagrangian $\mathscr{L}_M$ may include metric and its partial derivatives, four-velocity (for dust) and other physical variables if necessary. It should be however noticed that the presence of partial derivatives of the metric as arguments of the matter Lagrangian induces some misunderstandings and requires to go further to the concept of covariance. Starting with Eq. (4.98), the variation of the Einstein–Hilbert action gives:

$$\delta \mathscr{S}_M = \int_\mathcal{M} \delta \left(\mathscr{L}_M \sqrt{\mathrm{Detg}} \right) dx^0 \wedge \cdots dx^3 = \int_\mathcal{M} \left(\frac{\partial \mathscr{L}_M}{\partial g^{\alpha\beta}} - \frac{\mathscr{L}_M}{2} g_{\alpha\beta} \right) \delta g^{\alpha\beta} \, \omega_n$$

where the presence of the metric $\mathbf{g}$ in the integrand is essential. The boundary conditions terms are (voluntarily) omitted for the sake of the clarity. At this stage, we recall the definition of the stress-energy momentum e.g. Carter and Quintana (1977), Pons (2011):

$$T_{\alpha\beta} := \frac{\partial \mathscr{L}_M}{\partial g^{\alpha\beta}} - \frac{\mathscr{L}_M}{2} g_{\alpha\beta} \tag{2.80}$$

The second term of the stress-energy $T_{\alpha\beta}$ appears because of the choice of a Lagrangian scalar density rather than the scalar Lagrangian. Indeed, if we express the derivative by means of the scalar Lagrangian, the following relation holds:

$$\sqrt{\text{Detg}}\, T_{\alpha\beta} := 2\, \frac{\partial \left(\sqrt{\text{Detg}}\, \mathscr{L}_M\right)}{\partial g^{\alpha\beta}} \tag{2.81}$$

Relations (2.80) and (2.81) define the same stress-energy tensor. This second relation could be naively deduced from the variation equation $\sqrt{\text{Detg}}\, T_{\alpha\beta}\, \delta g^{\alpha\beta} := \delta \left(\sqrt{\text{Detg}}\, \mathscr{L}_M\right)$ meaning that the stress-energy tensor is obtained from the variation of the scalar Lagrangian.

Einstein equations for relativistic matter fields (derived in its extended form in the second part of this paper) take the form of: $\Re_{\alpha\beta} - (1/2)\,\mathcal{R}g_{\alpha\beta} = \chi\, T_{\alpha\beta}$, where $T_{\alpha\beta}$ is the energy-momentum tensor characterizing the matter (we include it into the Lagrangian hereafter). Multiplying this equation with $g^{\mu\nu}$ allows us to derive: $\Re_{\alpha\beta} = \chi \left[T_{\alpha\beta} - (1/2)\, T\, g_{\alpha\beta} \right]$ with, $T := g^{\alpha\beta}\, T_{\alpha\beta}$; This allows us to obtain the field equations for special relativity at the lowest order in ε, and by taking the first dominant order in $2\varepsilon_{\alpha\beta} := g_{\alpha\beta} - \hat{g}_{\alpha\beta}$ we have gravitational equation:

$$\chi \left[T_{\alpha\beta} - (T/2)g_{\alpha\beta} \right] = \hat{g}^{\lambda\sigma} \left(\partial_\mu \partial_\alpha \varepsilon_{\sigma\beta} - \partial_\mu \partial_\beta \varepsilon_{\sigma\alpha} + \partial_\sigma \partial_\beta \varepsilon_{\mu\alpha} - \partial_\sigma \partial_\alpha \varepsilon_{\mu\beta} \right) \tag{2.82}$$

where the constitutive formulations $T_{\alpha\beta}(\hat{g}_{\mu\nu}, \varepsilon_{\mu\nu}, \partial_\lambda \partial_\gamma \varepsilon_{\mu\nu})$ highlight the problem of including metric and its derivatives as arguments of the Lagrangian. It is via the dependence of matter Lagrangian on the metric $g_{\mu\nu}$ that we can derive the matter mass-energy-momentum tensor field $T_{\alpha\beta}$ e.g. Dixon (1975), Lehmkuhl (2011), Maugin (1978), Williams (1989). Result based on the Lagrangian dependence only on curvature and on metric was shown in e.g. Lovelock (1971) for Riemannian manifold and its extension constitutes a main expectation in this paper. Another research direction was developed with metric-affine theory where metric and affine connection, and their derivatives were given a priori as independent arguments of Lagrangian (Palatini method). The basic idea is to consider multivalued mappings e.g. Kleinert (2008) to merge to the Palatini–Einstein method, where the Lagrangian takes the form of $\mathscr{L}(g_{\alpha\beta}, \Gamma^\gamma_{\lambda\mu}, \partial_\nu \Gamma^\gamma_{\lambda\mu})$, depending on connection coefficients and their partial derivatives. Conversely to original relativistic gravitation theory, curvature is here associated to connection and its partial derivatives e.g. Exirifard and Sheikh-Jabbari (2008), Forger and Römer (2004), Lovelock and Rund (1975). In the present paper, we are interested in invariance of both matter Lagrangian and spacetime Lagrangian (covariance and gauge invariance).

Some examples of Lagrangian function of basic material models are given in this subsection. More detailed development will be conducted in following chapters.

2.5.2.2 Dust in Relativistic Mechanics

It is also known that not every material system can be given a Lagrangian formulation, and sometimes there are materials for which Lagrangian does not explicitly depend on the metric tensor. Such is the case for a model of incoherent matter called "dust" whose particles do not interact. In special relativity, from a slightly different point of view (following a dust-matter along a worldline) the action of a free dust-matter is defined as $\mathscr{S} := -m_0c^2 \int_{\tau_1}^{\tau_2} d\tau$, where τ is the proper time of the particle. This induces:

$$\mathscr{S} = -m_0c \int_{s_1}^{s_2} ds = -m_0c \int_{s_1}^{s_2} \sqrt{\hat{g}_{\mu\nu}dx^\mu dx^\nu} = -\int_{\tau_1}^{\tau_2} m_0c\sqrt{\hat{g}_{\mu\nu}u^\mu u^\nu}\, d\tau$$
(2.83)

In general relativity, the Minkowskian metric is merely replaced by a dynamic metric $g_{\mu\nu}$ to give:

$$\mathscr{S} = -\int_{\tau_1}^{\tau_2} m_0c\sqrt{g_{\mu\nu}u^\mu u^\nu}\, d\tau$$
(2.84)

which gives exactly the dust-matter action Eq. (2.85). Such is the case when the matter Lagrangian is assumed to depend explicitly on the metric. The associated energy-momentum tensor is defined as $T_{\alpha\beta} := \rho u_\alpha u_\beta$ where ρ is the proper density of matter and u_α its four-velocity field e.g. Dirac (1974). To derive the variational formulation, let first consider the relativistic (Hamiltonian) action for a single dust-matter associated to Lagrangian $\mathscr{L}$:

$$\mathscr{S} := \int_{\tau_1}^{\tau_2} (\gamma - 1)\, m_0c^2\, d\tau = \int_{t_1}^{t_2} (1 - 1/\gamma)\, m_0c^2\, dt \rightarrow \mathscr{S}$$

$$:= -\int_{t_1}^{t_2} m_0c^2\sqrt{1 - (v/c)^2}\, dt$$

by dropping the rest energy m_0c^2 which is constant with respect to t. Then we can write:

$$\mathscr{S} := -\int_{t_1}^{t_2} m_0c\sqrt{c^2 - v^2}\, dt = -\int_{\tau_1}^{\tau_2} m_0c\sqrt{g_{\mu\nu}u^\mu u^\nu}\, d\tau$$
(2.85)

by using the relation between Eqs. (2.75) and (2.77), and owing that (Fig. 2.20):

$$\sqrt{c^2 - v^2} = (1/\gamma)\sqrt{g_{\mu\nu}u^\mu u^\nu} \qquad \text{and} \qquad d\tau = dt/\gamma$$

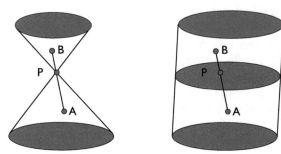

Fig. 2.20 Local motion of a dust matter P. A is the past position, P the present position, and B its future in the spacetime continuum $\mathcal{M}$. (Left) in relativity, and (right) in Newtonian physics. For relativity, it is asserted that there exists an intrinsic time-order only for events occurring on the worldline APB for an observer, the dust particle. The present $\{P\}$ is a three-dimensional space in Newtonian physics. This means that the Newtonian physics asserts that there is an intrinsic time-order for arbitrary pair of events, due to the assumption of an absolute time with any event

By extension, the action for relativistic dust is thus given by the integral of all dust-matter by replacing number $m := \rho \omega_n$ e.g. Dirac (1974)[3]:

$$\mathscr{S} := -\int_{\mathscr{M}} \rho\, c \sqrt{u_\alpha u^\alpha}\; \omega_n, \qquad \omega_n := \sqrt{\mathrm{Detg}}\; dx^0 \wedge dx^1 \wedge dx^2 \wedge dx^3 \qquad (2.86)$$

where we observe in fact an implicit dependence on the metric tensor $\mathbf{g}$ for either the norm $u_\alpha u^\alpha := g_{\alpha\beta} u^\alpha u^\beta$ or in the term "Det $\mathbf{g}$" we will introduce soon after. By a standard variation in the framework of general relativity it is straightforward to deduce that $T_{\alpha\beta} = \rho\, u_\alpha u_\beta$. Indeed, we can derive the variation of this action as (Dirac 1974):

$$\delta\mathscr{S} = -\frac{1}{2}\int_{\mathscr{M}} \rho\, u_\alpha u_\beta\, \delta g^{\alpha\beta}\; \omega_n$$

For particular case, it is suggested to define the constant $\chi := 8\pi G/c^4$. Combining the gravitation field and the dust, we arrive at the Dirac action for relativistic dust:

$$\mathscr{S} := \int \left(\frac{c^4}{16\pi G}\, \overline{\mathcal{R}} - \rho\, c \sqrt{u_\alpha u^\alpha} \right) \omega_n \qquad (2.87)$$

A dust is a typical example of a test-particle whose dimension is very small compared to that of all other moving bodies within the spacetime. The gravitational field induced by dust is therefore neglected, and the curvature $\overline{\mathcal{R}}$ is considered as solely due to an "external field".

---

[3]In relativistic mechanics, dust is usually defined as a set of particles forming a perfect fluid with zero pressure and with no interaction between them. Particles move independently each other.

2.5.2.3 Perfect Fluids in Relativistic Mechanics

In relativistic mechanics, a perfect fluid is a fluid that has no viscosity and no heat conduction e.g. Minazzoli and Karko (2012). In the same way, the stress-energy tensor for perfect fluid is given by: $T_{\alpha\beta} := (\rho + p)u_\alpha u_\beta + p\ g_{\alpha\beta}$. Same comments as for dust could be applied. Indeed, perfect fluid is defined as continuum in which all forces against slipping are zero. The only force between neighboring fluid element is pressure (normal component of stress tensor). This model extends the ideal gas of classical thermodynamics. It extends the concept of dust matter by adding pressure interaction. A recent example of barotropic perfect fluid is given in e.g. Minazzoli and Karko (2012) where the action of the fluid within a gravitational field is given by:

$$\mathscr{S} := \int_{\mathscr{B}} \left[\frac{c^4}{16\pi G} \overline{\mathcal{R}} - \rho \left(c^2 + \int \frac{P(\rho)}{\rho^2} d\rho \right) \right] \omega_n \tag{2.88}$$

where ρ is the energy density of the fluid (the rest energy is the number particles of fluid per unit volume in the mean frame of the reference of these particles), $P(\rho)$ the pressure, and c the light speed. The corresponding stress-energy tensor can be derived by variation to give:

$$T^{\alpha\beta} = -\left\{ \rho \left[c^2 + \Pi(\rho) \right] + P(\rho) \right\} u^\alpha u^\beta + P(\rho) g^{\alpha\beta}$$

which leads to the field equation:

$$\mathfrak{R}^{\alpha\beta} - (1/2)\mathcal{R}\ g^{\alpha\beta} = -\left\{ \rho \left[c^2 + \Pi(\rho) \right] + P(\rho) \right\} u^\alpha u^\beta + P(\rho) g^{\alpha\beta} \tag{2.89}$$

In relations defining the Lagrangian function of the dust (2.87) and the perfect fluids (2.88) the scalar curvature is calculated with its classical components (4.98) in the framework of Einstein relativistic gravitation.

2.5.2.4 Strain Gradient Continuum

For $3D$ elastic continuum (in this paragraph Greek indices run for $(1, 2, 3)$), it is usual to take a sufficiently smooth mapping so that $x^i := x^i(X^\alpha)$ and its inverse $X^\alpha := X^\alpha(x^i)$ have at least smooth second derivatives. Associated to these coordinates, we define the deformation gradient (tangent linear application) $F^i_\alpha := \partial x^i / \partial X^\alpha$ and metric components $g_{\alpha\beta} := F^i_\alpha F^j_\beta \hat{g}_{ij}$. This relation, showing that the material metric is induced by the spacetime metric, means that the matter is coupled with the spacetime e.g. Lehmkuhl (2011), Verçyn (1990). Motions of classical continuum preserve topology of bodies. Mathematical models deduced from these assumptions do not depend on scale lengths. In early 1960s, the interest for gradient continuum grew due to the need of accounting for small scale dependence of material response. Toupin developed elastic continuum models with couple-stress (Toupin 1962) and observed that some components of the gradient of strain were not accounted for. Later he pointed out the correspondence of strain gradient elastic models and discrete lattice models of solids with nevertheless some flaws when considering centro-symmetric materials. In 1964, Mindlin proposed an enhanced generalized elasticity theory to describe elastic material with microstructure effects by adding the strain gradient as additional variables (Mindlin 1964). To overcome the centro-symmetric problem, Mindlin (1965) have proposed a strain gradient theory in which the strain energy function is assumed to depend on both strain, strain gradient, and second gradient of strain $\mathscr{L}_M(\varepsilon_{ij}, \partial_k \varepsilon_{ij}, \partial_l \partial_k \varepsilon_{ij})$ where ε is the strain tensor. Owing that $g_{\alpha\beta} := \hat{g}_{\alpha\beta} + 2\varepsilon_{\alpha\beta} = \hat{g}_{\alpha\beta} + \partial_\alpha u_\beta + \partial_\beta u_\alpha + \partial_\alpha u^\gamma \partial_\beta u_\gamma$, Lagrangian may be formulated as function of derivatives of displacement field u_α, say $\mathscr{L}(\partial_\alpha u_\beta, \partial_\lambda \partial_\alpha u_\beta, \partial_\mu \partial_\lambda \partial_\alpha u_\beta)$ e.g. Agiasofitou and Lazar (2009). For example, the most known model of gradient elastic continuum is defined by Mindlin Lagrangian function as $\mathscr{L} := \mathscr{L}_G - \mathscr{L}_M$ with:

$$\mathscr{L}_M = (\lambda/2)\, \varepsilon_{\alpha\alpha}\varepsilon_{\beta\beta} + \mu\, \varepsilon_{\alpha\beta}\varepsilon_{\alpha\beta} + a_1\, \partial_\alpha \varepsilon_{\alpha\gamma}\, \partial_\gamma \varepsilon_{\beta\beta} + a_2\, \partial_\alpha \varepsilon_{\beta\beta}\, \partial_\alpha \varepsilon_{\gamma\gamma}$$

$$+ a_3\, \partial_\alpha \varepsilon_{\alpha\gamma}\, \partial_\beta \varepsilon_{\beta\gamma} + a_4\, \partial_\alpha \varepsilon_{\beta\gamma}\, \partial_\alpha \varepsilon_{\beta\gamma} + a_5\, \partial_\alpha \varepsilon_{\beta\gamma}\, \partial_\gamma \varepsilon_{\alpha\beta} \qquad (2.90)$$

$$\mathscr{L}_G = (\rho/2)\, \partial_0 u_\alpha\, \partial_0 u_\alpha + (\rho/2)\ell_G^2\, \partial_0 \partial_\beta u_\alpha\, \partial_0 \partial_\beta u_\alpha$$

where λ and μ are the usual Lamé parameters and the various a_i are additional elastic constitutive constants e.g. Mindlin (1964). Here u_α stands for the three-dimensional displacement components rather than for four-vector velocity in relativistic mechanics. The additional (higher order) term of inertia in $\mathscr{L}_G$ captures the time derivative of the displacement gradient e.g. Polizzotto (2012). If only the skew symmetric part of the displacement gradient is retained then this is similar to the rotational inertia for Timoshenko beam e.g. Bideau et al. (2011) which is a particular Cosserat continuum model. $x^0 := t$ stands for time, and ρ and ℓ_G are respectively the matter density and a characteristic lengthscale. On the one hand, derivatives of the matter Lagrangian $\mathscr{L}_M$ with respect to the strain $\varepsilon_{\alpha\beta}$ and its partial derivatives $\partial_\gamma \varepsilon_{\alpha\beta}$ allow us to obtain the matter constitutive laws (it should be stressed that in such operation the strain and its derivatives are independent on the displacement and its derivatives in $\mathscr{L}$). On the other hand, the gauge invariance of the entire Lagrangian $\mathscr{L}$ gives the conservations laws provided that the components of strain $\varepsilon_{\alpha\beta}$ and its derivatives are expressed in terms of the displacement u_α. A simple illustration of $\mathscr{L}$ (2.90) is given by the Lagrangian density of an elastic beam with line density ρ, Young's modulus $\mathbb{E}$ (Poisson ratio $\nu \equiv 0$), moving in an Newtonian spacetime,

$$\mathscr{L} = (1/2)\rho\, (\partial_t u)^2 + (1/2)\rho\, \ell_G^2\, (\partial_x \partial_t u)^2 - (1/2)\mathbb{E}\left[(\partial_x u)^2 + \ell_M^2\, (\partial_x \partial_x u)^2\right]$$

where $u(x, t)$ is the axial displacement along the beam of each point x, and with another lengthscale parameter for the matter $\ell_M := 2(a_1 + a_2 + a_3 + a_4 + a_5)/\mathbb{E}$

e.g. Mindlin (1964), Polyzos and Fotiadis (2012). For completeness, conservation laws associated to this Lagrangian hold e.g. Challamel et al. (2009):

$$\partial_x \partial_x \left(u - \ell_M^2 \, \partial_x \partial_x u \right) = (1/c^2) \partial_t \partial_t \left(u - \ell_G^2 \, \partial_x \partial_x u \right), \qquad c^2 := \mathbb{E}/\rho$$

The nonlocality of this model includes both the strain gradient effect and the Eringen's integral effect. This is a typical form of kinetic and strain energy density proposed in e.g. Askes and Aifantis (2011) where review of recent models, deduced from Mindlin Lagrangian, has been done in statics and dynamics. See e.g. Polizzotto (2012) for discussion about problems arising from boundary conditions combined with higher order inertia terms. Gradient theories of elastic continuum mechanics are nowadays used to address various problems in which the effects of the inhomogeneities of matter cannot be disregarded. In the 1960s e.g. Mindlin (1965), Mindlin (1964), as well as in quite recent literature and/or reviews e.g. Askes and Aifantis (2011), Cordero et al. (2016), Javili et al. (2013), these gradient theories assume strain and gradient of strain and even higher gradients as primal variables for establishing constitutive laws. In a general manner, Lagrangian (2.90) defines second/third grade continuum models, and its covariance constitutes one of the issue we would like to address in this book.

For the sake of the clarity, the classical approach for strain gradient continuum, namely the strain gradient elastic model, may be slightly modified as follows. First, the arguments of the strain energy density function are considered $\mathscr{U} = \mathscr{U}(F_\alpha^i, F^\gamma \partial_\alpha F_\beta^i, \cdots)$ where the argument F_α^i may be replaced easily by $g_{\alpha\beta} := F_\alpha^i F_\beta^j \, \hat{g}_{ij}$ for invariance requirements. The second argument $\Gamma_{\alpha\beta}^\gamma := F^\gamma \partial_\alpha F_\beta^i$ is preferred to $\partial_\alpha F_\beta^i$. It should be stressed that it does not any utilization restriction since the tetrads F_β^i cannot be singular. Higher derivatives of the connection may be also suggested as arguments of the energy $\mathscr{U}$, or more generally the Lagrangian $\mathscr{L}$, to define higher gradient elastic models. The next chapter is devoted to the covariance of Lagrangian under an arbitrary diffeomorphism (change of coordinate system).

Chapter 3
Covariance of Lagrangian Density Function

3.1 Introduction

During the last two centuries, the concept of absolute spacetime has been extended in two main directions, and accordingly the definition of a continuum has following more or less these evolutions. Galilean physics, and namely the Newton mechanics, is mainly based on the existence of an absolute rigid space and time. For both special and general relativistic physics, Einstein and numerous other authors which were involved in, revised the concept of space and time into spacetime by relativizing the time (Minkowski spacetime) and by transforming of absolute and rigid space into a variable and dynamical four-metric to model the interaction of Einstein spacetime and matter. Further extension of the relativistic continuum physics was obtained when Cartan added the torsion as dynamical variable to obtain the Einstein–Cartan spacetime. More generally, the basic geometry underlying any physics theory may thus be proposed to include metric and affine connection, not necessarily associated to metric (metric affine-manifold). Lagrangian we are interested in are function defined on such a manifold.

This chapter is devoted to the so-called "axiom of general invariance of Hilbert" e.g. Brading and Ryckman (2008). In 1905, Hilbert proposed two axioms for setting the basis for the covariant derivation of both relativistic gravitation and electromagnetism. The two axioms were proposed as follows:

1. (Mie axiom on Lagrangian) The Lagrangian density $\mathscr{L}$ depends on the spacetime metric $g_{\mu\nu}$ (10 components), their first and second derivatives, and the electromagnetic potential A_μ (4 components) and their first derivatives;
2. (Covariance) The Lagrangian function $\mathscr{L}$ is an invariant with respect to arbitrary transformations of the point-event coordinates x^μ.

The present chapter deals with the second Hilbert's axiom. For historical context, we remind the Cartan theorem. Let $\mathscr{B}$ a four dimensional manifold endowed with a metric tensor $g_{\alpha\beta}\,(x^\nu)$. Assume that the manifold is orientable with a volume-form

© Springer International Publishing AG, part of Springer Nature 2018
L. R. Rakotomanana, *Covariance and Gauge Invariance in Continuum Physics*,
Progress in Mathematical Physics 73, https://doi.org/10.1007/978-3-319-91782-5_3

$\omega_n(x^\nu)$. It is worth to remind the theorem of Cauchy (1850)–Weyl (1939) stating that for any scalar-valued function $\mathscr{L}$ with vector arguments $\mathbf{e}_\alpha := F_\alpha^i \hat{\mathbf{e}}_i$, $\alpha = 1, n$, elements of the tangent space $T_\mathbf{x}\mathscr{M}$, the Lagrangian function $\mathscr{L}$ is invariant under the rotation group $\mathcal{O}^+$ if and only if it exists a representation of the function $\tilde{\mathscr{L}}$ such that: $\mathscr{L} = \tilde{\mathscr{L}}\left(g_{\alpha\beta}, \det(g_{\alpha\beta})\right)$. Proof may be found in e.g. Rakotomanana (2003) (annex E). This allows us to assume a priori the metric components as arguments of the Lagrangian density of the action: $\mathscr{S} := \int_{\mathscr{B}} \mathscr{L}\, \omega_n$, where $\mathscr{L}$ depends on the fields and their derivatives (of finite order), and ω_n a volume-form on $\mathscr{B}$. Let consider spacetime transformations. Passive diffeomorphism views transformations acting on the coordinates leaving the fields unchanged (covariance), whereas active diffeomorphisms assume transformations to act on fields and leave the coordinates unchanged (gauge invariance) Anderson (1971). We first consider passive diffeomorphisms by studying in the following section the invariance of Lagrangian function $\mathscr{L}$ under the group of diffeomorphisms Diff($\mathscr{B}$). Three theorems are particularly involved for the covariance analysis.

3.2 Some Basic Theorems

We report here three theorems that are useful to the derivation of the Lagrangian covariance: Cartan's theorem in the framework of relativistic gravitation, Lovelock's theorem extending the Cartan's theorem to nonlinear Lagrangian in the framework of Riemann geometry, and the Quotient theorem necessary for establishing the main result of this chapter.

3.2.1 Theorem of Cartan

After the introduction of Minkowski spacetime, the second major modification of the spacetime is the recognizing that the gravity is not an a priori existing force field but the acceptation that is an aspect of geometry which is captured by the metric and by the affine connection. Derivation of the gravitation equations consists in searching for 2-covariant tensor $T_{\alpha\beta}$ with the conditions: first, $T_{\alpha\beta}$ is linear with respect to the second derivatives of the metric components; and second, the tensor should be divergence free for satisfying the conservation laws. In 1922, Cartan derived the Einstein's equations of gravity on affinely connected manifold. He showed in Cartan (1922) the uniqueness of the Einstein tensor "$G_{\alpha\beta} := \mathfrak{R}_{\alpha\beta} - (1/2)\mathcal{R}\, g_{\alpha\beta}$" assessing the necessary (and sufficient) condition to use Einstein's equations of gravity within usual assumptions.

Theorem 3.1 (Cartan (1922)) *Let tensor $T_{\alpha\beta}\left(g_{\alpha\beta}, \partial_\gamma g_{\alpha\beta}, \partial_\lambda \partial_\gamma g_{\alpha\beta}\right)$ to be linear with respect to the second derivatives of $g_{\alpha\beta}$ and diffeomorphism invariant in the sense that it has a form invariant with respect to all diffeomorphisms of the* *manifold—change of coordinates* $y^\mu = y^\mu (x^\nu)$. *Then the unique solution takes the form of:*

$$T_{\alpha\beta} = a \, \mathfrak{R}_{\alpha\beta} + b \, \mathcal{R} \, g_{\alpha\beta} + c \, g_{\alpha\beta} \qquad (3.1)$$

where $\mathfrak{R}_{\alpha\beta}$ and $\mathcal{R}$ are the Ricci and the scalar curvatures respectively. a, and b are real arbitrary constants and c are scalar field.

Proof See Cartan (1922). □

The theorem of Cartan shows that at any point P of a pseudo-Riemannian (and Lorentzian) manifold $\mathcal{M}$, it is possible to find a coordinate system (x^μ) such that the metric tensor writes $g_{\alpha\beta}$ and the first derivatives of the components $\partial_\gamma g_{\alpha\beta}$, with respect to the chosen coordinates, are zero. However, certain combinations of second-order derivatives $\partial_\lambda \partial_\gamma g_{\alpha\beta}$ which from the Riemann curvature $\mathfrak{R}^\gamma_{\alpha\beta\lambda}$ cannot be eliminated. This is a fundamental result in gravitation theory and constitutes the mathematical basis for gravitation equations $\chi \, T_{\alpha\beta} := \mathfrak{R}_{\alpha\beta} - (1/2) \, \mathcal{R} \, g_{\alpha\beta} + \Lambda \, g_{\alpha\beta}$, where $\chi := 8\pi G/c^4$ is the cosmological constant introduced to be compatible with the universe expansion according to the finding of Hubble in 1929. $G \simeq 6.67 \times 10^{-28} [\mathrm{m}^3 \, \mathrm{kg}^{-1} \, \mathrm{s}^{-2}]$ and $c \simeq 3 \times 10^8 [\mathrm{ms}^{-1}]$ are the gravity constant and the light velocity respectively. Cartan theorem states that at every point of the four dimensional Riemannian manifold (spacetime), among the 100 second derivatives of the metric $\partial_\lambda \partial_\gamma g_{\alpha\beta}$, the invariance with respect to passive diffeomorphisms allows us to eliminate 80 of them to retain only 20, corresponding to the components of curvature $\mathfrak{R}^\gamma_{\alpha\beta\lambda}$. Cartan theorem was extended in the seventies by using a scalar field rather than a 2-covariant tensor (Lovelock and Rund 1975). Since then numerous geometrized theories of gravitation including nonlinear dependence on curvature have been proposed, called fourth-order gravitation theory. For short, they were mainly introduced to prevent the bing bang singularity in relativistic gravitation, to account for the inflationary cosmological model, and most interestingly to attempt to unify gravitation and quantization of matter fields e.g. Schimdt (2007). Cartan theorem can be applied both to the relativistic gravity theory and to the strain gradient continuum theory. Following the idea of Cartan, Anderson provided the structure of divergence-free contravariant $(2, 0)$ type tensor (Anderson 1978). In this paper, the separation of the influence of the metric components and their second partial derivatives was shown. Different steps of proof are analogous to that of Lovelock and Rund and details may be found in Antonio and Rakotomanana (2011).

3.2.2 Theorem of Lovelock (1969)

Consider a set of variables $\Phi^i (x^\mu)$. Let assume that the Lagrangian is function of the $\Phi^i (x^\mu)$ and their first and second derivatives with respect to x^μ. Metric components $g_{\alpha\beta}(x^\mu)$ play the role of $\Phi^i (x^\mu)$ in the gravitation theory. The associated Euler–Lagrange equations are expected to be of fourth-order as we previously show.

However, in the general theory of relativity, the Einstein equations of gravitation are of second-order in terms of metric components. Lovelock investigated the case of Lagrange density which is a function of the metric and its first and second derivatives with respect to x^μ. The following theorem was shown in Lovelock (1969).

Theorem 3.2 *Let $\mathcal{M}$ a four-dimensional torsionless manifold endowed with a metric $g_{\alpha\beta}$, and a metric compatible affine connection $\Gamma^\gamma_{\alpha\beta}$ with a curvature $\mathfrak{R}^\gamma_{\alpha\beta\lambda}$. The only second order Euler–Lagrange equations which can arise from a scalar density $\mathcal{L}(g_{\alpha\beta}, \partial_\gamma g_{\alpha\beta}, \partial_\lambda \partial_\gamma g_{\alpha\beta})$ are:*

$$\mathfrak{R}_{\alpha\beta} - (1/2)\mathcal{R}g_{\alpha\beta} + \Lambda g_{\alpha\beta} = 0 \tag{3.2}$$

are i.e. the Einstein's gravitational equations with the cosmological constant Λ.

Theorem of Lovelock constitutes one of the major results on the uniqueness of the Einstein equations on manifold with a metric tensor with signature $(+, -, -, -)$. In the same way, motivated by the fact that gravitational field equations should be second-order equations in terms of metric components, these equations being obtained from a Lagrangian density $\mathcal{L}(g_{\alpha\beta}, \partial_\gamma g_{\alpha\beta}, \partial_\lambda \partial_\gamma g_{\alpha\beta})$, it was also shown in Anderson (1981) that in a four-dimensional manifold, the Lagrangian density must take the form of: $\mathcal{L} = a\,\mathcal{R} + b$ where a and b are constants. The resulting Euler–Lagrange equations are deduced and can be re-arranged to be exactly (3.2). This result is obtained under very general conditions without the assumption that the Lagrangian is linear in curvature $\mathcal{R}$. In Anderson (1981), Anderson extended this result to include the interaction of gravitation field and external sources due to matter immersed in the spacetime $\mathcal{M}$.

3.2.3 Theorem of Quotient

For self consistency, we remind some technical theorems in tensor analysis (Lovelock and Rund 1975). The most important is the quotient law saying that a set of real numbers ($\mathbb{R}$) form the components of a tensor of a certain rank, if and only if its scalar product with another arbitrary tensor is again a tensor (practically, we attempt to obtain a scalar by a worth choice). This can be used as a criterial test whether a set of numbers form a tensor or not.

Lemma 3.1 *Locally at point P, let the both $(n \times n)$ quantities Σ^{ij} and $\overline{\Sigma}^{ij}$ then the both $(n \times n \times n)$ quantities $\Sigma^{ij,k}$ and $\overline{\Sigma}^{ij,k}$ (the comma "," doesn't represent partial derivative). If, for any symmetric type $(0,2)$ tensor $\mathbf{h}$, for i, j, $k = 1, \ldots, n$,*

$$\Sigma^{ij}h_{ij} + \Sigma^{ij,k}h_{ij|k} = \overline{\Sigma}^{ij}h_{ij} + \overline{\Sigma}^{ij,k}h_{ij|k} \tag{3.3}$$

then

$$(\Sigma^{ij} + \Sigma^{ji}) = (\overline{\Sigma}^{ij} + \overline{\Sigma}^{ji}) \tag{3.4}$$

and

$$(\Sigma^{ij,k} + \Sigma^{ji,k}) = (\overline{\Sigma}^{ij,k} + \overline{\Sigma}^{ji,k}) \qquad (3.5)$$

Proof The equality (3.3) being valid for any symmetric tensor **h**, it is thus valid for a non null constant tensor (locally). The covariant derivative vanishes and from (3.3) we have the following equality $(\Sigma^{ij})h_{ij} = (\overline{\Sigma}^{ij})h_{ij}$. The term h_{ij} cannot simplify because of the summation in i and j. For i and j fixed, we choose $h_{ij} = h_{ji} = 1$ and the other components null. For i and j range over $\{1, \dots, n\}$, we obtain

$$\begin{cases} \Sigma^{11} = \overline{\Sigma}^{11} \\ \Sigma^{12} + \Sigma^{21} = \overline{\Sigma}^{12} + \overline{\Sigma}^{21} \\ \Sigma^{13} + \Sigma^{31} = \overline{\Sigma}^{13} + \overline{\Sigma}^{31}, \qquad \dots \end{cases} \qquad (3.6)$$

thus we deduce (3.4). In the same way, locally at point P, a null tensor **h** with a non null constant covariant derivative can be chosen too. In such a case from (3.3), we get $(\Sigma^{ij,k})h_{ij|k} = (\overline{\Sigma}^{ij,k})h_{ij|k}$. The term $h_{ij|k}$ cannot simplified because of the summation in i, j and k. For i, j and k fixed, we choose $h_{ij|k} = h_{ji|k} = 1$ and the other components null. For i, j and k range over $\{1, \dots, n\}$, we obtain

$$\begin{cases} \Sigma^{11,1} = \overline{\Sigma}^{11,1} \\ \Sigma^{11,2} = \overline{\Sigma}^{11,2} \end{cases}, \qquad \begin{cases} \Sigma^{12,1} + \Sigma^{21,1} = \overline{\Sigma}^{12,1} + \overline{\Sigma}^{21,1} \\ \Sigma^{12,2} + \Sigma^{21,2} = \overline{\Sigma}^{12,2} + \overline{\Sigma}^{21,2} \\ \Sigma^{13,1} + \Sigma^{31,1} = \overline{\Sigma}^{13,1} + \overline{\Sigma}^{31,1} \\ \Sigma^{13,2} + \Sigma^{31,2} = \overline{\Sigma}^{31,2} + \overline{\Sigma}^{31,2}, \qquad \dots \end{cases} \qquad (3.7)$$

thus we obtain (3.5).   □

We have also the version with the partial derivative:

Lemma 3.2 *Locally at point P, let the both $(n \times n)$ quantities Σ^{ij} and $\overline{\Sigma}^{ij}$ then the both $(n \times n \times n)$ quantities $\Sigma^{ij,k}$ and $\overline{\Sigma}^{ij,k}$ (the comma "," doesn't represent partial derivative). If, for any symmetric type $(0, 2)$ tensor **h**, for i, j, $k = 1, \dots, n$:*

$$\Sigma^{ij}h_{ij} + \Sigma^{ij,k}h_{ij,k} = \overline{\Sigma}^{ij}h_{ij} + \overline{\Sigma}^{ij,k}h_{ij,k} \qquad (3.8)$$

then

$$(\Sigma^{ij} + \Sigma^{ji}) = (\overline{\Sigma}^{ij} + \overline{\Sigma}^{ji}) \qquad (3.9)$$

and

$$(\Sigma^{ij,k} + \Sigma^{ji,k}) = (\overline{\Sigma}^{ij,k} + \overline{\Sigma}^{ji,k}) \qquad (3.10)$$

Now the following quotient theorem holds e.g. Lovelock and Rund (1975).

Theorem 3.3 (Quotient Law) *Locally at point P, if the $(n \times n)$ quantities Σ^{ij} and the $(n \times n \times n)$ quantities $\Sigma^{ij,k}$ (the comma "," doesn't represent the partial derivative) are such that the quantities $\Sigma^{ij} h_{ij} + \Sigma^{ij,k} h_{ij|k}$ represent a scalar field for any symmetric type (0,2) tensor **h**, then the quantities $(\Sigma^{ij} + \Sigma^{ji})$ and $(\Sigma^{ij,k} + \Sigma^{ji,k})$ respectively represent the components of a tensor type (2,0) and the components of a type (3, 0) tensor.*

Proof Let define the scalar $\psi = \Sigma^{ij} h_{ij} + \Sigma^{ij,k} h_{ij|k}$ and $\overline{\psi} = \Sigma^{\alpha\beta} h_{\alpha\beta} + \Sigma^{\alpha\beta,\gamma} h_{\alpha\beta|\gamma}$ in the system (y^i) and (x^α) respectively for any symmetric tensor type (0,2) **h**. The equality $\psi = \overline{\psi}$ becomes

$$\Sigma^{ij} h_{ij} + \Sigma^{ij,k} h_{ij|k} = \Sigma^{\alpha\beta} A_\alpha^i A_\beta^j h_{ij} + \Sigma^{\alpha\beta,\gamma} A_\alpha^i A_\beta^j A_\gamma^k h_{ij|k} \qquad (3.11)$$

According to the Lemma 3.1

$$\Sigma^{ij} + \Sigma^{ji} = \Sigma^{\alpha\beta} A_\alpha^i A_\beta^j + \Sigma^{\alpha\beta} A_\alpha^j A_\beta^i \qquad (3.12)$$

and

$$\Sigma^{ij,k} + \Sigma^{ji,k} = \Sigma^{\alpha\beta,\gamma} A_\alpha^i A_\beta^j A_\gamma^k + \Sigma^{\alpha\beta,\gamma} A_\alpha^j A_\beta^i A_\gamma^k \qquad (3.13)$$

then a permutation between i and j gives

$$\begin{cases} \Sigma^{ij} + \Sigma^{ji} = \left(\Sigma^{\alpha\beta} + \Sigma^{\beta\alpha} \right) A_\alpha^i A_\beta^j \\ \Sigma^{ij,k} + \Sigma^{ji,k} = \left(\Sigma^{\alpha\beta,\gamma} + \Sigma^{\beta\alpha,\gamma} \right) A_\alpha^i A_\beta^j A_\gamma^k \end{cases} \qquad (3.14)$$

Therefore $(\Sigma^{ij} + \Sigma^{ji})$ and $(\Sigma^{ij,k} + \Sigma^{ji,k})$ are respectively components of type (2, 0) tensor and (3, 0) according to the definitions (2.20) and (2.21). □

3.3 Invariance with Respect to the Metric

Consider a scalar field $\mathscr{L}$ depending on the metric components and their partial derivatives $\mathscr{L} = \mathscr{L}(g_{ij}, \partial_k g_{ij}, \partial_l \partial_k g_{ij})$ and $\mathscr{L} = \mathscr{L}(g_{\alpha\beta}, \partial_\gamma g_{\alpha\beta}, \partial_\lambda \partial_\gamma g_{\alpha\beta})$ respectively in coordinate system (y^i) and (x^α). The metric components $g_{\alpha\beta}(x^\mu)$ (respectively $g_{ij}(y^k)$ in terms of new coordinates) are the unknown functions. We consider general covariance by imposing the invariance of Lagrangian shape under coordinate transformations $y^i(x^\alpha)$. The corresponding partial derivatives are denoted

$$\Lambda^{ij} = \frac{\partial \mathscr{L}}{\partial g_{ij}}, \qquad \Lambda^{ij,k} = \frac{\partial \mathscr{L}}{\partial g_{ij,k}}, \qquad \Lambda^{ij,kl} = \frac{\partial \mathscr{L}}{\partial g_{ij,kl}}. \qquad (3.15)$$

They should not be confused with the hypermomenta or currents, which are physical reactions due to the variations of kinematics, and deduced from the variation of the Lagrangian function. Care should be taken since relativistic gravitation and gradient continuum models need hypermomenta and currents to be viable theory e.g. Sotiriou (2008). Due to the symmetry of **g**, we have the major properties of symmetry

$$\Lambda^{ij} = \Lambda^{ji}, \qquad \Lambda^{ij,k} = \Lambda^{ji,k}, \qquad \Lambda^{ij,kl} = \Lambda^{ji,kl} = \Lambda^{ij,lk}. \tag{3.16}$$

Minor symmetry property $\Lambda^{ij,kl} = \Lambda^{kl,ij}$ is also satisfied but they are not used here. The covariance of $\mathscr{L}$ (which is one of the necessary conditions to ensure the indifference of the constitutive laws with respect to Superimposed Rigid Body Motions e.g. Betram and Svendsen 2001) takes the form of

$$\mathscr{L}(g_{\alpha\beta}, \partial_\gamma g_{\alpha\beta}, \partial_\lambda \partial_\gamma g_{\alpha\beta}) = \mathscr{L}(g_{ij}, \partial_k g_{ij}, \partial_l \partial_k g_{ij}). \tag{3.17}$$

Lagrangian with first and second derivatives of the metric components as additional arguments plays an keyrole for the extension of Einstein tensor in relativistic gravitation e.g. Cartan (1922), Exirirfard and Sheikh-Jabbari (2008), Lovelock and Rund (1975). In 1922, Cartan showed the uniqueness of the Einstein equations of relativistic gravity (Cartan 1922). Result on diffeomorphism invariance (covariance) we are now proving extends this Cartan's result, to a nonlinear dependence of Lagrangian on metric second derivatives $\partial_\lambda \partial_\mu g_{\alpha\beta}$. The corresponding Euler–Lagrange equations deduced from the covariant Lagrangian can be used for that purpose.

Remark 3.1 It is necessary to clarify the link between invariance of the equations (their form) and the physical relativity principle e.g. Betram and Svendsen (2001), Westman and Sonego (2009). Coordinates (x^μ) in relativistic gravitation are merely mathematical parameters to label spacetime point (event) and do not have any operational property. Such is not the case for Newtonian mechanics and special relativity.

For remind, because no observer is distinguished, laws in physics have to be observer-invariant (frame indifference). The keypoint is to define the group of invariance. Each physics theory has its invariance group. Galilean invariance is required for classical mechanics whereas Minkowskian invariance is the basis for special Relativistic mechanics, the later is imposed by the need of compatibility of mechanics and electromagnetic fields and propagation.

Anyhow, this should be completed by the fact that matter Lagrangian functions $\mathscr{L}$ should have at least the same shape for arbitrary coordinate systems. The link between the two concepts was not very clear e.g. Bain (2004). Interestingly, Ryskin gave an accurate formulation of objectivity (more recent term defining the material frame indifference) as *"Any physical law must be expressible in a form-independent of coordinate system"* (Ryskin 1985). This is called covariance in the Einstein language. Further, it was extended to four dimensional formalism, thus including the time variable, and suggests that every physical law verifying the diffeomorphism-invariance seems also material frame indifferent. However, this is not true.

3.3.1 Transformation Rules for the Metric and Its Derivatives

According to the definitions (2.20) and (2.21), the metric components and its derivatives satisfy the following rules of transformations

$$
\begin{cases}
g_{ij} = J_i^\alpha J_j^\beta g_{\alpha\beta} \\
g_{ij,k} = (J_{ik}^\alpha J_j^\beta + J_i^\alpha J_{jk}^\beta) g_{\alpha\beta} + J_i^\alpha J_j^\beta J_k^\gamma g_{\alpha\beta,\gamma} \\
g_{ij,kl} = (J_{ikl}^\alpha J_j^\beta + J_{ik}^\alpha J_{jl}^\beta + J_{il}^\alpha J_{jk}^\beta + J_i^\alpha J_{jkl}^\beta) g_{\alpha\beta} \\
\qquad + (J_{ik}^\alpha J_j^\beta J_h^\gamma + J_i^\alpha J_{jk}^\beta J_l^\gamma + J_{il}^\alpha J_j^\beta J_k^\gamma + J_i^\alpha J_{jl}^\beta J_k^\gamma + J_i^\alpha J_j^\beta J_{kl}^\gamma) g_{\alpha\beta,\gamma} \\
\qquad + (J_i^\alpha J_j^\beta J_k^\gamma J_l^\lambda) g_{\alpha\beta,\gamma\lambda}
\end{cases}
\tag{3.18}
$$

According to the symmetry $J_{pq}^\mu = J_{qp}^\mu$ (the transformation is assumed of class C^2), one can write $J_{pq}^\mu = (1/2)(J_{pq}^\mu + J_{qp}^\mu)$, that induces $(\partial J_{ij}^\alpha / \partial J_{pq}^\mu) = (1/2)\delta_\mu^\alpha(\delta_i^q \delta_j^p + \delta_j^q \delta_i^p)$, $(\partial J_{ijk}^\alpha / \partial J_{pq}^\mu) = 0$ and $(\partial J_i^\alpha / \partial J_{pq}^\mu) = 0$. If one does not consider the symmetric part then there is a loss of some terms in the derivation. Now, introducing the expressions (3.18) into the equality (3.17), and differentiating with respect to J_{pq}^μ give

$$
0 = \Lambda^{ij,k} \left[\delta_\mu^\alpha(\delta_k^q \delta_i^p + \delta_i^q \delta_k^p) J_j^\beta + \delta_\mu^\beta(\delta_k^q \delta_j^p + \delta_j^q \delta_k^p) J_i^\alpha \right] g_{\alpha\beta}
$$

$$
+ \Lambda^{ij,kl} \left[J_i^\alpha J_j^\beta \delta_\mu^\gamma(\delta_k^q \delta_l^p + \delta_l^q \delta_k^p) + J_i^\alpha J_k^\gamma \delta_\mu^\beta(\delta_j^q \delta_l^p + \delta_l^q \delta_j^p) \right] g_{\alpha\beta,\gamma}
$$

$$
+ \Lambda^{ij,kl} \left[J_j^\beta J_k^\gamma \delta_\mu^\alpha(\delta_i^q \delta_l^p + \delta_l^q \delta_i^p) + J_i^\alpha J_l^\gamma \delta_\mu^\beta(\delta_j^q \delta_k^p + \delta_k^q \delta_j^p) \right] g_{\alpha\beta,\gamma}
$$

$$
+ \Lambda^{ij,kl} \left[J_j^\beta J_l^\gamma \delta_\mu^\alpha(\delta_i^q \delta_k^p + \delta_k^q \delta_i^p) \right] g_{\alpha\beta,\gamma}
$$

$$
+ \Lambda^{ij,kl} \left[J_{jk}^\beta \delta_\mu^\alpha(\delta_i^q \delta_l^p + \delta_l^q \delta_i^p) + J_{il}^\alpha \delta_\mu^\beta(\delta_j^q \delta_k^p + \delta_k^q \delta_j^p) \right] g_{\alpha\beta}
$$

$$
+ \Lambda^{ij,kl} \left[J_{jl}^\beta \delta_\mu^\alpha(\delta_i^q \delta_k^p + \delta_k^q \delta_i^p) + J_{ik}^\alpha \delta_\mu^\beta(\delta_j^q \delta_l^p + \delta_l^q \delta_j^p) \right] g_{\alpha\beta}
$$

Previous equation is valid for arbitrary coordinate transformation, in particular for the identity transformation: $x^\alpha = y^i$, $J_i^\alpha = \delta_i^\alpha$, $J_{ij}^\alpha = 0$. In such a case, we simplify

$$
0 = \Lambda^{ij,k} \left[\delta_\mu^\alpha(\delta_k^q \delta_i^p + \delta_i^q \delta_k^p)\delta_j^\beta + \delta_\mu^\beta(\delta_k^q \delta_j^p + \delta_j^q \delta_k^p)\delta_i^\alpha \right] g_{\alpha\beta}
$$

$$
+ \Lambda^{ij,kl} \left[\delta_i^\alpha \delta_j^\beta \delta_\mu^\gamma(\delta_k^q \delta_l^p + \delta_h^t \delta_k^p) + \delta_i^\alpha \delta_k^\gamma \delta_\mu^\beta(\delta_j^q \delta_l^p + \delta_l^q \delta_j^p) \right] g_{\alpha\beta,\gamma}
$$

$$
+ \Lambda^{ij,kl} \left[\delta_j^\beta \delta_k^\gamma \delta_\mu^\alpha(\delta_i^q \delta_l^p + \delta_l^q \delta_i^p) + \delta_i^\alpha \delta_l^\gamma \delta_\mu^\beta(\delta_j^q \delta_k^p + \delta_k^q \delta_j^p) \right] g_{\alpha\beta,\gamma}
$$

$$
+ \Lambda^{ij,kl} \left[\delta_j^\beta \delta_l^\gamma \delta_\mu^\alpha(\delta_i^q \delta_k^p + \delta_k^q \delta_i^p) \right] g_{\alpha\beta,\gamma}
$$

Further simplifications and symmetry of Λ induce[1]

$$2\Lambda^{q\beta,\gamma p}g_{\mu\beta,\gamma} + 2\Lambda^{p\beta,\gamma q}g_{\mu\beta,\gamma} + \Lambda^{\alpha\beta,pq}g_{\alpha\beta,\mu} + \Lambda^{p\beta,q}g_{\mu\beta} + \Lambda^{q\beta,p}g_{\mu\beta} = 0.$$

$$(3.19)$$

In the particular case of a normal coordinate system, these reduce to $\Lambda^{p\mu,q} + \Lambda^{q\mu,p} = 0$. According to the symmetry of $\Lambda^{ij,k}$, for arbitrary indexes i, j, k, we have

$$\Lambda^{ji,k} + \Lambda^{ki,j} = 0$$

$$\Lambda^{kj,i} + \Lambda^{ki,j} = 0 \quad (i \longleftrightarrow k)$$

$$\Lambda^{ji,k} + \Lambda^{kj,i} = 0 \quad (i \longleftrightarrow j)$$

then $\Lambda^{ji,k} = -\Lambda^{ki,j} = \Lambda^{kj,i} = -\Lambda^{ji,k}$, and finally

$$\Lambda^{ji,k} = \Lambda^{ij,k} = 0.$$

$$(3.20)$$

Nevertheless Eq. (3.20) are only valid in normal coordinate system.

3.3.2   Introduction of Dual Variables

Introducing the expressions (3.18) into (3.17) and differentiating respectively with respect to $g_{\alpha\beta,\gamma\lambda}$, $g_{\alpha\beta,\gamma}$, $g_{\alpha\beta}$, allow to write

$$
\begin{cases}
\Lambda^{\alpha\beta,\gamma\lambda} = \Lambda^{ij,kl} J_i^\alpha J_j^\beta J_k^\gamma J_l^\lambda \\
\Lambda^{\alpha\beta,\gamma} = \Lambda^{ij,kl} \dfrac{\partial g_{ij,kl}}{\partial g_{\alpha\beta,\gamma}} + \Lambda^{ij,k} \dfrac{\partial g_{ij,k}}{\partial g_{\alpha\beta,\gamma}} \\
\Lambda^{\alpha\beta} = \Lambda^{ij,kl} \dfrac{\partial g_{ij,kl}}{\partial g_{\alpha\beta}} + \Lambda^{ij,k} \dfrac{\partial g_{ij,k}}{\partial g_{\alpha\beta}} + \Lambda^{ij} \dfrac{\partial g_{ij}}{\partial g_{\alpha\beta}}
\end{cases}
\tag{3.21}
$$

The first equation shows that $\Lambda^{ij,kl}$ are components of a type $(4, 0)$ tensor. Conversely the two other equations show that $\Lambda^{ij,k}$ and Λ^{ij} are not components of tensor. Thus we should introduce two tensorial quantities instead of $\Lambda^{ij,k}$ and Λ^{ij} respectively. Let $\mathbf{h}$ an arbitrary symmetric type $(0, 2)$ tensor (i.e. $\mathbf{h}$ follows the same rule of transformation (3.18) as $\mathbf{g}$), and Π^{ij} and $\Pi^{ij,k}$ (the comma "," does not represent partial derivative) two unknown quantities that verify the following equation

$$\Lambda^{ij,kl}h_{ij,kl} + \Lambda^{ij,k}h_{ij,k} + \Lambda^{ij}h_{ij} = \Lambda^{ij,kl}h_{ij|k|l} + \Pi^{ij,k}h_{ij|k} + \Pi^{ij}h_{ij} \tag{3.22}$$

---

[1]Latin indices and Greek indices mix since the transformation is the identity.

The covariant derivatives (2.42) and (2.43) are introduced into (3.22), that reduce to an equality without covariant derivative terms. It depends on $\{\Gamma_{ij}^k, \Gamma_{ij,l}^k\}$, $\{h_{ij}, h_{ij,k}, h_{ij,kl}\}$, $\{\Lambda^{ij}, \Lambda^{ij,k}, \Lambda^{ij,kl}\}$ and $\{\Pi^{ij}, \Pi^{ij,k}\}$. Then, the Lemma 3.2 is applied to Eq. (3.22) in order to identify the coefficients of $h_{ij,k}$ and h_{ij} (coefficients of $h_{ij,kl}$ are the same in both hand sides of equation). Some worth permutations between the indices are necessary. For the sake of the clarity, details of calculus are reported in appendix. Remind that Λ^{ij}, $\Lambda^{ij,k}$ and $\Lambda^{ij,kl}$ have symmetry properties. Such is not necessary the case for Π^{ij} and $\Pi^{ij,k}$. Thus we obtain the following equations

$$
\begin{cases}
\Pi_{(S)}^{ij,k} = \Lambda^{ij,k} + 2\Gamma_{al}^i \Lambda^{aj,kl} + 2\Gamma_{al}^j \Lambda^{ia,kl} + \Gamma_{bl}^k \Lambda^{ij,bl} \\
\Pi_{(S)}^{ij} = \Lambda^{ij} + \Gamma_{ak,l}^i \Lambda^{aj,kl} + \Gamma_{ak,l}^j \Lambda^{ia,kl} \\
\quad - \Gamma_{al}^b \Gamma_{bk}^i \Lambda^{aj,kl} - \Gamma_{cl}^b \Gamma_{bk}^j \Lambda^{ic,kl} \\
\quad - \Gamma_{bl}^i \Gamma_{ck}^j \Lambda^{bc,kl} - \Gamma_{bl}^j \Gamma_{ck}^i \Lambda^{bc,kl} \\
\quad - \Gamma_{kl}^b \Gamma_{cb}^i \Lambda^{cj,kl} - \Gamma_{kl}^b \Gamma_{cb}^j \Lambda^{ci,kl} \\
\quad + (1/2)\Gamma_{ak}^i (\Pi^{aj,k} + \Pi^{ja,k}) + (1/2)\Gamma_{ak}^j (\Pi^{ia,k} + \Pi^{ai,k})
\end{cases}
\tag{3.23}
$$

where $\Pi_{(S)}^{ij,k} = (1/2)(\Pi^{ij,k} + \Pi^{ji,k})$ and $\Pi_{(S)}^{ij} = (1/2)(\Pi^{ij} + \Pi^{ji})$.

Lemma 3.3 $F := \Lambda^{ij,kl} h_{ij,kl} + \Lambda^{ij,k} h_{ij,k} + \Lambda^{ij} h_{ij}$ is a scalar field.

Proof Let us notice $F = \Lambda^{ij,kl} h_{ij,kl} + \Lambda^{ij,k} h_{ij,k} + \Lambda^{ij} h_{ij}$ and $\overline{F} = \Lambda^{\alpha\beta,\gamma\lambda} h_{\alpha\beta,\gamma\lambda} + \Lambda^{\alpha\beta,\gamma} h_{\alpha\beta,\gamma} + \Lambda^{\alpha\beta} h_{\alpha\beta}$ respectively in coordinate system (y^i) and (x^α). According to (3.21), we get

$$
\Lambda^{\alpha\beta,\gamma\lambda} h_{\alpha\beta,\gamma\lambda} = \left[\Lambda^{ij,kl} J_i^\alpha J_j^\beta J_k^\gamma J_l^\lambda\right] h_{\alpha\beta,\gamma\lambda}
$$

$$
\Lambda^{\alpha\beta,\gamma} h_{\alpha\beta,\gamma} = \left[\Lambda^{ij,kl} \frac{\partial g_{ij,kl}}{\partial g_{\alpha\beta,\gamma}} + \Lambda^{ij,k} \frac{\partial g_{ij,k}}{\partial g_{\alpha\beta,\gamma}}\right] h_{\alpha\beta,\gamma}
$$

$$
\Lambda^{\alpha\beta} h_{\alpha\beta} = \left[\Lambda^{ij,kl} \frac{\partial g_{ij,kl}}{\partial g_{\alpha\beta}} + \Lambda^{ij,k} \frac{\partial g_{ij,k}}{\partial g_{\alpha\beta}} + \Lambda^{ij} \frac{\partial g_{ij}}{\partial g_{\alpha\beta}}\right] h_{\alpha\beta}
$$

Factorization of the coefficients of $\Lambda^{ij,kl}$, $\Lambda^{ij,k}$ and Λ^{ij} gives $\overline{F} = (a) + (b) + (c)$ with

$$
(a) = \Lambda^{ij,kl} \left[\frac{\partial g_{ij,kl}}{\partial g_{\alpha\beta,\gamma\lambda}} h_{\alpha\beta,\gamma\lambda} + \frac{\partial g_{ij,kl}}{\partial g_{\alpha\beta,\gamma}} h_{\alpha\beta,\gamma} + \frac{\partial g_{ij,kl}}{\partial g_{\alpha\beta}} h_{\alpha\beta}\right]
$$

$$
(b) = \Lambda^{ij,k} \left[\frac{\partial g_{ij,k}}{\partial g_{\alpha\beta,\gamma}} h_{\alpha\beta,\gamma} + \frac{\partial g_{ij,k}}{\partial g_{\alpha\beta}} h_{\alpha\beta}\right]
$$

$$
(c) = \Lambda^{ij} \left[\frac{\partial g_{ij}}{\partial g_{\alpha\beta}} h_{\alpha\beta}\right]
$$

According to the relations (3.18), the quantities in square brackets are simplified, (a) $= \Lambda^{ij,kl}\left[h_{ij,kl}\right]$, (b) $= \Lambda^{ij,k}\left[h_{ij,k}\right]$, (c) $= \Lambda^{ij}\left[h_{ij}\right]$ and thus $\overline{F} = F$. $\qquad\square$

Lemma 3.4 $(F - \Lambda^{ij,kl}h_{ij|k|l})$ *is a scalar.*

Proof Let us notice $G := \Lambda^{ij,kl}h_{ij|k|l}$ and $\overline{G} := \Lambda^{\alpha\beta,\gamma\lambda}h_{\alpha\beta|\gamma|\lambda}$. By (3.21) we have

$$\Lambda^{\alpha\beta,\gamma\lambda}h_{\alpha\beta|\gamma|\lambda} = \Lambda^{ij,kl}J_i^\alpha J_j^\beta J_k^\gamma J_l^\lambda h_{\alpha\beta|\gamma|\lambda} \tag{3.24}$$

h being a type $(0,2)$ tensor, the second covariant derivative $h_{\alpha\beta|\gamma|\lambda}$ forms the components of a type $(0,4)$ tensor. Consequently $J_i^\alpha J_j^\beta J_k^\gamma J_l^\lambda h_{\alpha\beta|\gamma|\lambda} = h_{ij|k|l}$, thus $\overline{G} = G$. By previous lemma, we have $\overline{F} - \overline{G} = F - G$. $\qquad\square$

Remark 3.2 A direct and simple proof may be obtained by observing that Λ^{ijkl} is in fact a type $(4,0)$ tensor and $h_{ij|k|l}$ is a type $(0,4)$ tensor, then their contraction is a scalar.

Consequently, $\Pi^{ij,k}h_{ij|k} + \Pi^{ij}h_{ij}$ is a scalar for an arbitrary type $(0,2)$ tensor **h**. Using the quotient Theorem 3.3, $\Pi_{(S)}^{ij,k}$ and $\Pi_{(S)}^{ij}$ are the components of type $(3,0)$ and $(2,0)$ tensor respectively, these tensors are also symmetric.

3.3.3 Theorem

Expressions of $\Pi_{(S)}^{ij,k}$ in Eq. (3.23) hold in an arbitrary coordinate system. In normal coordinate system, the Christoffel symbols vanish, e.g. Nakahara (1996) and we have, from (3.23), $\Pi_{(S)}^{ij,k} = \Lambda^{ij,k}$. However, from (3.20) $\Lambda^{ij,k} = 0$ in normal coordinate system, thus $\Pi_{(S)}^{ij,k} = 0$ in normal coordinate system and this is also true for any other coordinate system, because $\Pi_{(S)}^{ij,k}$ are components of tensor. The first equation in (3.23) is simplified in any coordinate system

$$0 = \Lambda^{ij,k} + 2\Gamma_{al}^i\Lambda^{aj,kl} + 2\Gamma_{al}^j\Lambda^{ia,kl} + \Gamma_{bl}^k\Lambda^{ij,bl}. \tag{3.25}$$

We can establish the following theorem:

Theorem 3.4 *Let a scalar field* $\mathscr{L} = \mathscr{L}\left(g_{ij}, g_{ij,k}, g_{ij,kl}\right)$ *defined on a Riemannian manifold. If* $\dfrac{\partial\mathscr{L}}{\partial g_{ij,kl}} = 0$ *then* $\dfrac{\partial\mathscr{L}}{\partial g_{ij,k}} = 0$.

Proof $\forall\, i, j, k, l$, the condition $\Lambda^{ij,kl} = 0$ is introduced into (3.25). $\qquad\square$

Consequently, from the second equality of (3.23), we deduce $\dfrac{\partial\mathscr{L}}{\partial g_{ij}} = \Pi_{(S)}^{ij}$. An equivalent formulation of Theorem 3.4 may be found in Lovelock and Rund (1975).

Theorem 3.5 *On a Riemannian manifold $\mathscr{B}$, there does not exist a scalar density such $\mathscr{L} = \mathscr{L}(g_{ij}, \partial_k g_{ij})$ that only depends on the metric components g_{ij} and their first partial derivatives $g_{ij,k}$.*

In any coordinate system, we have the following decomposition:

$$
\begin{cases}
\Pi^{ij} = (1/2)(\Pi^{ij} + \Pi^{ji}) + (1/2)(\Pi^{ij} - \Pi^{ji}) \\
\Pi^{ij,k} = (1/2)(\Pi^{ij,k} + \Pi^{ji,k}) + (1/2)(\Pi^{ij,k} - \Pi^{ji,k})
\end{cases}
\tag{3.26}
$$

In Lovelock and Rund (1975), the quantities Π^{ij} and $\Pi^{ij,k}$ are assumed to be symmetric with respect to the indices i and j: $\Pi^{ij} = \Pi^{ji}$, $\Pi^{ij,k} = \Pi^{ji,k}$ and it is proven that the quantities $\Pi^{ij,k}$ are always zero. In the present study it has been proven that only the symmetric part $\Pi^{ij,k}_{(S)}$ is null: the quantities $\Pi^{ij,k}$ are skew-symmetric $\Pi^{ij,k} = -\Pi^{ji,k}$. The present study is slightly more general than result presented in Lovelock and Rund (1975). To study the fields in physics, the arguments have to be tensors, as for $\mathscr{L}(\mathbf{g})$ depending on metric tensor. To extend the arguments of $\mathscr{L}$, defined on Riemannian manifold endowed with an affine connection ∇, the new form is then $\mathscr{L}(\mathbf{g}, \nabla\mathbf{g}, \nabla^2\mathbf{g})$ where all the arguments are tensors. The corresponding form in the coordinate system (x^α) is $\mathscr{L}(g_{\alpha\beta}, \nabla_\gamma g_{\alpha\beta}, \nabla_\lambda \nabla_\gamma g_{\alpha\beta})$. If the connection is Euclidean then we have $\nabla_\gamma g_{\alpha\beta} \equiv \partial_\gamma g_{\alpha\beta}$. Another motivation is that, according to the Lemma of Ricci e.g. Nakahara (1996), the covariant derivative of the metric tensor $\mathbf{g}$, in the sense of Levi-Civita connection, is identically equal to zero.

3.4 Invariance with Respect to the Connection

Metric of matter is induced by ambient Euclidean space for classical mechanics and Minkowskian spacetime for relativistic mechanics. Conversely there exist many possibilities of choice for the affine connection. We demand that the metric be covariantly constant: $\nabla\mathbf{g} \equiv 0$ to be compatible with the metric.[2] This means that $\nabla\mathbf{g}$ cannot be an explicit argument of the Lagrangian $\mathscr{L}$. This is the reason why we should consider ∇ as an argument rather than $\nabla\mathbf{g}$. We labeled the biconnection $\nabla^2 = \nabla \circ \nabla$. In order to extend the list of arguments of $\mathscr{L}$, we will consider the following forms: $\mathscr{L}(\mathbf{g}, \nabla)$, $\mathscr{L}(\mathbf{g}, \nabla^2)$ and $\mathscr{L}(\mathbf{g}, \nabla, \nabla^2)$ e.g. Antonio and Rakotomanana (2011). However connection is not tensor, thus we aim to obtain tensorial arguments built upon the connection and/or the bi-connection. The

---

[2]Although we do not deal with the derivation of Euler–Lagrange equations associated to these Lagrangian density functions $\mathscr{L}$, it is worth to mention that an appropriate divergence theorem generalized Gauss formula, necessary to derive the equations of motion, is mandatory to the definition of a compatible volume-form on the affine manifold (Saa 1995). For short, the condition on the metric compatibility allows us to obtain such a volume-form easily.

introduction of ∇, as advocated by Palatini, as Lagrangian function arguments is also related to the affine variational method in relativistic gravitation. It is used to obviate non covariance properties which arises when matter Lagrangian depends upon metric derivatives.[3] Beyond the compatibility of ∇ with the metric, we also should consider the compatibility of the volume-form with the connection e.g. Saa (1995). There is some arbitrariness in the choice of the volume-form on a affine manifold but we can take $\sqrt{\text{Det}\mathbf{g}}$ within Lagrangian density $\mathscr{L}$, and then can skip this aspect.

Let now consider the link between local frame and affine connection. Let $\mathscr{B}$ be continuum—a metric-affine manifold—endowed with a metric $\mathbf{g}$, and an affine connection ∇. Every neighbor of a point $\mathbf{x}$ can be considered as a microcosm which has its own local Galilean group $\{\mathbf{Q}(x^i), \mathbf{v}_0(x^i)\}$ defined by the transformation (a local Newton–Cartan transformations): $\mathbf{OM'}(t) = \mathbf{Q}(x^i)\,[\mathbf{OM}(t)] + \mathbf{v}_0(x^i)t$ where $x^0 := t$, and where Latin indices are used for space coordinates $i = (1, 2, 3)$. Let two microcosms $\mathbf{x} := (x^0, x^i)$ and $\mathbf{x} + d\mathbf{x} := (x^0 + dx^0, x^i + dx^i)$ with the line element $ds^2 = g_{\alpha\beta}dx^\alpha dx^\beta$, Greek indices hold $\alpha, \beta = 0, 1, 3$, say $x^\mu(s)$. The change of the local Galilean group from one microcosm to another is then obtained with parallel transport (affine connection):

$$dv_0 = \nabla_{d\mathbf{x}}\mathbf{v}_0 = \left(\partial_\beta v_0^\alpha + \Gamma^\alpha_{\beta\gamma} v_0^\gamma\right) dx^\beta \mathbf{e}_\alpha$$

$$d\mathbf{Q} = \nabla_{d\mathbf{x}}\mathbf{Q} := \left(\partial_\gamma Q^\alpha_\beta + \Gamma^\alpha_{\gamma\lambda} Q^\lambda_\beta - \Gamma^\lambda_{\gamma\beta} Q^\alpha_\lambda\right) dx^\gamma \mathbf{e}_\alpha \otimes \mathbf{e}^\beta$$

with coefficients $\Gamma^\gamma_{\alpha\beta} = \mathbf{e}^\gamma\left(\nabla_{\mathbf{e}_\alpha}\mathbf{e}_\beta\right)$. It is then obvious that affine connection defines change of local Galilean reference frames e.g. Kadianakis (1996). The method may be extended to local Minkowskian spacetime and even to generalized change of frames of references. In the following, we consider the invariance with respect to connection (Fig. 3.1).

3.4.1 Preliminary

A scalar field of any sort which is not changed by the diffeomorphisms group is called invariant (the shape of the function is not changed). For the sake of

---

[3]In the scope of relativistic gravitation, the underlying idea of Palatini e.g. Capoziello and de Laurentis (2011) is to consider the connection $\Gamma^\gamma_{\alpha\beta}$ necessary to link two microcosms, independent from the spacetime metric $g_{\alpha\beta}$ as independent argument for the Lagrangian density function. There is no reason to limit the connection to Levi-Civita connection deduced from the metric. Some studies on relativistic gravitation introduce the concept of bi-metric theory by considering two slightly different metrics as $g_{\alpha\beta}$ and $\tilde{g}_{\alpha\beta} := f[\mathcal{R}(\Gamma^\gamma_{\alpha\beta})]\,g_{\alpha\beta}$. Nevertheless we limit to a rather general purpose by considering the metric, the connection and bi-connection as independent arguments of $\mathscr{L}$.

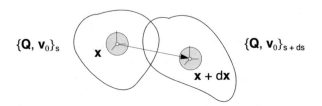

Fig. 3.1 Local Galilean frames of references, characterized by an orthogonal transformation **Q** and a velocity translation of frame $\mathbf{v}_0$, at two microcosms **x**, and $\mathbf{x} + d\mathbf{x}$ depending on the line element $ds^2 := g_{\alpha\beta} dx^\alpha dx^\beta$. For relativistic gravitation, the principle of equivalence applied for two microcosms **x**, and $\mathbf{x} + d\mathbf{x}$ allows us to reconstruct the four-dimensional global structure of the spacetime with gravitation by means of connection ∇

the simplicity, we define the invariance of any scalar field (more precisely the diffeomorphism-invariance): $\mathscr{L}(X, Y) = \mathscr{L}(X', Y')$ with both formal arguments X and Y defined in any two coordinate systems (with and without).

Lemma 3.5 *Let us consider arbitrary constants* K_1, K_2, C_1, C_2, C_3 *and the variables* x, x', y, y', p, p', q, q' *which follow the transformations*

$$\begin{cases} x' = K_1 x + K_2 \\ y' = K_1 y \\ p' = C_1 p + C_2 (x + y) + C_3 \\ q' = C_1 q \end{cases} \tag{3.27}$$

Now let us consider a scalar function $\mathscr{L}$ *which satisfies the equations (diffeomorphism-invariance)*

1 $\mathscr{L}(x, y) = \mathscr{L}(x', y')$
2 $\mathscr{L}(p, q) = \mathscr{L}(p', q')$
3 $\mathscr{L}(x, y, p, q) = \mathscr{L}(x', y', p', q')$.

If $(\partial K_1/\partial K_2) = 0$, $(\partial K_1/\partial C_3) = 0$, $(\partial K_2/\partial C_3) = 0$, $(\partial C_1/\partial C_3) = 0$ *then, from equation (1) we have* $\mathscr{L}(y) = \mathscr{L}(y')$, *from equation (2) we have* $\mathscr{L}(q) = \mathscr{L}(q')$, *from equation (3) we have* $\mathscr{L}(y, q) = \mathscr{L}(y', q')$.

Proof Equation 1: According to (3.27) we have $\mathscr{L}(x, y) = \mathscr{L}(K_1 x + K_2, K_1 y)$. We differentiate this equation with respect to K_2, to find

$$0 = \frac{\partial \mathscr{L}}{\partial x'} \frac{\partial x'}{\partial K_2} + \frac{\partial \mathscr{L}}{\partial y'} \frac{\partial y'}{\partial K_2}$$

$$0 = \frac{\partial \mathscr{L}}{\partial x'} \left[\frac{\partial K_1}{\partial K_2} x + 1 \right] + \frac{\partial \mathscr{L}}{\partial y'} \frac{\partial K_1}{\partial K_2} y$$

which involves that $\partial \mathscr{L}/\partial x' = 0$ if $\partial K_1/\partial K_2 = 0$. Then, according to $(\partial \mathscr{L}/\partial x) = (\partial \mathscr{L}/\partial x')(\partial x'/\partial x)$, we prove that $\partial \mathscr{L}/\partial x = 0$. *Equation 2* According to (3.27) we have $\mathscr{L}(p, q) = \mathscr{L}(C_1\, p + C_2\, (x + y) + C_3, C_1\, q)$. We differentiate this equation with respect to C_3, to find

$$0 = \frac{\partial \mathscr{L}}{\partial p'} \frac{\partial p'}{\partial C_3} + \frac{\partial \mathscr{L}}{\partial q'} \frac{\partial q'}{\partial C_3}$$

$$0 = \frac{\partial \mathscr{L}}{\partial p'} \left[\frac{\partial C_1}{\partial C_3} p + \frac{\partial C_2}{\partial C_3} (x + y) + 1 \right] + \frac{\partial \mathscr{L}}{\partial q'} \frac{\partial C_1}{\partial C_3} q$$

which implies that $\partial \mathscr{L}/\partial p' = 0$ if $\partial C_1/\partial C_3 = 0$, the term in square brackets not vanishing. Then, according to $(\partial \mathscr{L}/\partial p) = (\partial \mathscr{L}/\partial p')(\partial p'/\partial p)$, we prove that $\partial \mathscr{L}/\partial p = 0$. Finally we obtain $\mathscr{L}(q) = \mathscr{L}(q')$. *Equation 3* According to (3.27) we have $\mathscr{L}(x, y, p, q) = \mathscr{L}(K_1\, x + K_2, K_1\, y, C_1\, p + C_2\, (x + y) + C_3, C_1\, q)$. We differentiate this equation with respect to C_3, to find

$$0 = \frac{\partial \mathscr{L}}{\partial x'} \frac{\partial x'}{\partial C_3} + \frac{\partial \mathscr{L}}{\partial y'} \frac{\partial y'}{\partial C_3} + \frac{\partial \mathscr{L}}{\partial p'} \frac{\partial p'}{\partial C_3} + \frac{\partial \mathscr{L}}{\partial q'} \frac{\partial q'}{\partial C_3}$$

$$0 = \frac{\partial \mathscr{L}}{\partial x'} \left[\frac{\partial K_1}{\partial C_3} x + \frac{\partial K_2}{\partial C_3} \right] + \frac{\partial \mathscr{L}}{\partial y'} \frac{\partial K_1}{\partial C_3} y + \frac{\partial \mathscr{L}}{\partial p'} \left[\frac{\partial C_1}{\partial C_3} p + \frac{\partial C_2}{\partial C_3} (x + y) + 1 \right]$$
$$+ \frac{\partial \mathscr{L}}{\partial q'} \frac{\partial C_1}{\partial C_3} q$$

which implies that $\partial \mathscr{L}/\partial p' = 0$ if $\partial K_1/\partial C_3 = 0$, $\partial K_2/\partial C_3 = 0$, $\partial C_1/\partial C_3 = 0$. Then we have too $\partial \mathscr{L}/\partial p = 0$. According to $(\partial \mathscr{L}/\partial x) = (\partial \mathscr{L}/\partial p')(\partial p'/\partial x)$, we prove that $\partial \mathscr{L}/\partial x = 0$. Finally, according to $(\partial \mathscr{L}/\partial x) = (\partial \mathscr{L}/\partial x')(\partial x'/\partial x) = (\partial \mathscr{L}/\partial x')\, K_1$, we show that $\partial \mathscr{L}/\partial x' = 0$. Then we obtain $\mathscr{L}(y, q) = \mathscr{L}(y', q')$.

$\square$

Remark 3.3 The previous proof is based on the principle of fields invariance introduced by Lovelock and Rund (1975). It is equivalent to the form-invariance, a term borrowed from Svendsen and Betram (1999).

In an arbitrary coordinate system (y^i), the metric components, the connection and the bi-connection coefficients are respectively g_{ij}, Γ^k_{ij} and $\Gamma^k_{ij,l} + \Gamma^m_{ij}\Gamma^k_{lm}$. The forms $\mathscr{L}(g, \nabla)$, $\mathscr{L}(g, \nabla^2)$ and $\mathscr{L}(g, \nabla, \nabla^2)$ are then explicitly written as $\mathscr{L}(g_{ij}, \Gamma^k_{ij})$, $\mathscr{L}(g_{ij}, \Gamma^k_{ij,l} + \Gamma^m_{ij}\Gamma^k_{lm})$ and $\mathscr{L}(g_{ij}, \Gamma^k_{ij}, \Gamma^k_{ij,l} + \Gamma^m_{ij}\Gamma^k_{lm})$, respectively. Let (x^α) an other coordinate system, let us assume the diffeomorphism-invariance of the scalar field $\mathscr{L}$ (three cases):

$$\begin{cases} \mathscr{L}(\Gamma^k_{ij}) = \mathscr{L}(\Gamma^\gamma_{\alpha\beta}) \\ \mathscr{L}(\Gamma^k_{ij,l} + \Gamma^m_{ij}\Gamma^k_{lm}) = \mathscr{L}(\Gamma^\gamma_{\alpha\beta,\lambda} + \Gamma^\mu_{\alpha\beta}\Gamma^\gamma_{\lambda\mu}) \\ \mathscr{L}(\Gamma^k_{ij}, \Gamma^k_{ij,l} + \Gamma^m_{ij}\Gamma^k_{lm}) = \mathscr{L}(\Gamma^\gamma_{\alpha\beta}, \Gamma^\gamma_{\alpha\beta,\lambda} + \Gamma^\mu_{\alpha\beta}\Gamma^\gamma_{\lambda\mu}) \end{cases} \qquad (3.28)$$

where the metric components will be omitted for the sake of simplicity. For further applications, let us introduce the following components:

$$
\begin{cases}
\mathbb{T}^k_{ij} = (1/2)\left(\Gamma^k_{ij} - \Gamma^k_{ji}\right) \\
\mathbb{S}^k_{ij} = (1/2)\left(\Gamma^k_{ij} + \Gamma^k_{ji}\right) \\
\mathbb{B}^k_{lij} = (1/2)\left(\Gamma^k_{ij,l} + \Gamma^m_{ij}\Gamma^k_{lm} - \Gamma^k_{lj,i} - \Gamma^m_{lj}\Gamma^k_{im}\right) \\
\mathbb{A}^k_{lij} = (1/2)\left(\Gamma^k_{ij,l} + \Gamma^m_{ij}\Gamma^k_{lm} + \Gamma^k_{lj,i} + \Gamma^m_{lj}\Gamma^k_{im}\right)
\end{cases}
\tag{3.29}
$$

According to (3.29), permutation between i and j allows the decomposition $\Gamma^k_{ij} = \mathbb{S}^k_{ij} + \mathbb{T}^k_{ij}$. A permutation between i and l allows the decomposition $\Gamma^k_{ij,l} + \Gamma^m_{ij}\Gamma^k_{lm} = \mathbb{A}^k_{lij} + \mathbb{B}^k_{lij}$. A first permutation between i and j then another successive permutation between i and l allow the simultaneous decompositions $\Gamma^k_{ij} = \mathbb{S}^k_{ij} + \mathbb{T}^k_{ij}$ and $\Gamma^k_{ij,l} + \Gamma^m_{ij}\Gamma^k_{lm} = \mathbb{A}^k_{lij} + \mathbb{B}^k_{lij}$. Details of calculus are given in appendix. Thanks to these decompositions, covariance of $\mathscr{L}$ (3.28) becomes

$$
\begin{cases}
\mathscr{L}(\mathbb{S}^k_{ij}, \mathbb{T}^k_{ij}) = \mathscr{L}(\mathbb{S}^\gamma_{\alpha\beta}, \mathbb{T}^\gamma_{\alpha\beta}) \\
\mathscr{L}(\mathbb{A}^k_{lij}, \mathbb{B}^k_{lij}) = \mathscr{L}(\mathbb{A}^\gamma_{\lambda\alpha\beta}, \mathbb{B}^\gamma_{\lambda\alpha\beta}) \\
\mathscr{L}(\mathbb{S}^k_{ij}, \mathbb{T}^k_{ij}, \mathbb{A}^k_{lij}, \mathbb{B}^k_{lij}) = \mathscr{L}(\mathbb{S}^\gamma_{\alpha\beta}, \mathbb{T}^\gamma_{\alpha\beta}, \mathbb{A}^\gamma_{\lambda\alpha\beta}, \mathbb{B}^\gamma_{\lambda\alpha\beta})
\end{cases}
\tag{3.30}
$$

3.4.2 Application: Covariance of $\mathscr{L}$

Let us consider the following identification of variables $x, x', y, y', p, p', q, q'$:

$$
x = \mathbb{S}^k_{ij}, \quad y = \mathbb{T}^k_{ij}, \quad p = \mathbb{A}^k_{lij}, \quad q = \mathbb{B}^k_{lij},
\tag{3.31}
$$

$$
x' = \mathbb{S}^\gamma_{\alpha\beta}, \quad y' = \mathbb{T}^\gamma_{\alpha\beta}, \quad p' = \mathbb{A}^\gamma_{\lambda\alpha\beta}, \quad q' = \mathbb{B}^\gamma_{\lambda\alpha\beta}.
\tag{3.32}
$$

The transformation laws between the above variables take the form of (3.27) with (see appendix)

$$
\begin{cases}
K_1 = J^i_\alpha J^j_\beta A^\gamma_k \\
K_2 = J^j_{\alpha\beta} A^\gamma_j
\end{cases}, \qquad
\begin{cases}
C_1 = J^i_\alpha J^j_\beta J^l_\lambda A^\gamma_k \\
C_2 = J^i_{\alpha\lambda} J^j_\beta A^\gamma_k + J^i_\lambda J^j_{\alpha\beta} A^\gamma_k + J^i_\alpha J^j_{\beta\lambda} A^\gamma_k \\
C_3 = J^i_\mu J^l_\lambda J^j_{\alpha\beta} A^\gamma_j A^\mu_{il} + J^i_{\mu\lambda} J^j_{\alpha\beta} A^\mu_i A^\gamma_j
\end{cases}
\tag{3.33}
$$

We have $\partial K_1/\partial K_2 = 0$, $\partial K_1/\partial C_3 = 0$, $\partial K_2/\partial C_3 = 0$, $\partial C_1/\partial C_3 = 0$. According to Lemma 3.5, the covariance of $\mathscr{L}$ (3.30) means

$$\begin{cases} \mathscr{L}(\mathbb{T}_{ij}^k) = \mathscr{L}(\mathbb{T}_{\alpha\beta}^\gamma) \\ \mathscr{L}(\mathbb{B}_{lij}^k) = \mathscr{L}(\mathbb{B}_{\lambda\alpha\beta}^\gamma) \\ \mathscr{L}(\mathbb{T}_{ij}^k, \mathbb{B}_{lij}^k) = \mathscr{L}(\mathbb{T}_{\alpha\beta}^\gamma, \mathbb{B}_{\lambda\alpha\beta}^\gamma) \end{cases} \implies \begin{cases} \mathscr{L}(\aleph_{ij}^k) = \mathscr{L}(\aleph_{\alpha\beta}^\gamma) \\ \mathscr{L}(\mathfrak{R}_{lij}^k) = \mathscr{L}(\mathfrak{R}_{\lambda\alpha\beta}^\gamma) \\ \mathscr{L}(\aleph_{ij}^k, \mathfrak{R}_{lij}^k) = \mathscr{L}(\aleph_{\alpha\beta}^\gamma, \mathfrak{R}_{\lambda\alpha\beta}^\gamma) \end{cases} \quad (3.34)$$

since we can identify the torsion by $\aleph_{ij}^k = 2\mathbb{T}_{ij}^k$ and the curvature by $\mathfrak{R}_{lij}^k = 2\mathbb{B}_{lij}^k$.

3.4.3 Summary for Lagrangian Covariance

For the sake of the simplicity we have temporarily omitted the argument g_{ij}. Adding this argument does not change the proof. The overall result then includes the metric, the connection, and the bi-connection as arguments of the Lagrangian function $\mathscr{L}$. We also have the covariance:

$$\begin{cases} \mathscr{L}(g_{ij}, \aleph_{ij}^k) = \mathscr{L}(g_{\alpha\beta}, \aleph_{\alpha\beta}^\gamma) \\ \mathscr{L}(g_{ij}, \mathfrak{R}_{lij}^k) = \mathscr{L}(g_{\alpha\beta}, \mathfrak{R}_{\lambda\alpha\beta}^\gamma) \\ \mathscr{L}(g_{ij}, \aleph_{ij}^k, \mathfrak{R}_{lij}^k) = \mathscr{L}(g_{\alpha\beta}, \aleph_{\alpha\beta}^\gamma, \mathfrak{R}_{\lambda\alpha\beta}^\gamma) \end{cases} \quad (3.35)$$

All the arguments of $\mathscr{L}$ are components of tensors, they are invariant under the action of the diffeomorphism (in the sense that they transform covariantly according to usual tensor transformations depending of their type). Therefore, the Lagrangian function is covariant.

Remark 3.4 Obviously, under change of coordinate system it is not expected to generate any physical laws. In the framework of general relativity, the application of the Minimal Coupling Procedure would be a conversion of all partial derivatives $\partial_\gamma g_{\alpha\beta}$, and $\partial_\lambda \partial_\gamma g_{\alpha\beta}$ in the Minkowskian flat spacetime/continuum $\mathcal{M}$ into covariant derivatives $\nabla_\gamma g_{\alpha\beta}$, and $\nabla_\lambda \nabla_\gamma g_{\alpha\beta}$ and also choosing an appropriate volume-form ω_n in the underlying curved and possibly spacetime/continuum with torsion. For metric compatible connection, the standard MCP procedure would lead to $\nabla_\gamma g_{\alpha\beta} \equiv 0$, and $\nabla_\lambda \nabla_\gamma g_{\alpha\beta} \equiv 0$. The present approach is slightly different and can be considered as an extension of the MCP to Riemann–Cartan spacetime/continuum, since we assume the connection and the bi-connection as independent arguments.

The results are summarized in the following theorem:

Theorem 3.6 *Let a metric-affine manifold $(\mathscr{B}, \mathbf{g}, \nabla)$ where the affine connection is compatible with the metric $(\nabla \mathbf{g} = 0)$. To the connection are associated the torsion tensor $\aleph$ and the curvature tensor $\mathfrak{R}$. For any scalar function $\mathscr{L}$ defined on*

$\mathcal{B}$, depending on the metric, the connection and the bi-connection, the covariance induces:

$$\mathcal{L}(\mathbf{g}, \nabla) = \mathcal{L}(\mathbf{g}, \aleph), \quad \mathcal{L}(\mathbf{g}, \nabla^2) = \mathcal{L}(\mathbf{g}, \mathfrak{R}), \quad \mathcal{L}(\mathbf{g}, \nabla, \nabla^2) = \mathcal{L}(\mathbf{g}, \aleph, \mathfrak{R})$$
(3.36)

Equation (3.36) can be read in the two ways:

1. $\mathcal{L} = \mathcal{L}(\mathbf{g}, \nabla)$ *is covariant if and only if* $\mathcal{L} = \mathcal{L}(\mathbf{g}, \aleph)$
2. $\mathcal{L} = \mathcal{L}(\mathbf{g}, \nabla^2)$ *is covariant if and only if* $\mathcal{L} = \mathcal{L}(\mathbf{g}, \mathfrak{R})$
3. $\mathcal{L} = \mathcal{L}(\mathbf{g}, \nabla, \nabla^2)$ *is covariant if and only if* $\mathcal{L} = \mathcal{L}(\mathbf{g}, \aleph, \mathfrak{R})$

Covariance does not impose any restrictions on the spacetime theories (Newtonian for classical mechanics, Minkowskian for special relativity, Riemannian or Riemann–Cartan for relativistic gravitation). Covariance results conform to gravitational Utiyama theorem showing that the invariance under both the spacetime diffeomorphisms and the local Lorentz transformations impose the Lagrangian density to depend upon the tetrads, the connection and their derivatives only through the metric, torsion, and curvature e.g. Bruzzo (1987), Utiyama (1956). In the next section, we investigate the consequences of Lorentz invariance with respect to Lorentz transformations (2.14) of Lagrangian by means of gauge invariance. It should be reminded that the Lagrangian cannot explicitly depend on spacetime position (Bruzzo 1987; Kibble 1961).

3.4.4 Covariance of Nonlinear Elastic Continuum

A question would be the consequences of the covariance theorem in classical nonlinear elasticity, owing that classical nonlinear elasticity is based among other on the diffeomorphism assumption of the transformations of the body. For this purpose, let consider in this subsection a classical model for analyzing the elastic continuum deformation of a body $\mathcal{B}$.

3.4.4.1 Covariance of Strain Energy Density

Let then consider the three dimensional continuum $\mathcal{B}$ evolving within a Euclidean space $\mathscr{E}$ with metric $\hat{g}_{ij}$. Each material point of $\mathcal{B}$ is labelled by coordinates (t, X^α) in the initial configuration and by (t, x^i) in the actual deformed configuration. The basic assumption on the transformation of the body is the diffeomorphism property of the mapping $\varphi_t : X^\alpha \to x^i = \varphi_t^i(X^\alpha)$. The triads reduce to the deformation gradient $F_\alpha^i := \partial_\alpha \varphi_t^i$ allowing us to determine the components of the metric $\mathbf{g}$ of the space $\mathscr{E}$ as follows $g_{\alpha\beta} = C_{\alpha\beta} := F_\alpha^i \hat{g}_{ij} F_\beta^j$. We remind that $C_{\alpha\beta}$ denotes a quite common notation for the classical right Cauchy–Green strain tensor in the framework of nonlinear elasticity e.g. Marsden and Hughes (1983). It is worth to remind that the Cauchy-Green tensor (also called material metric tensor) may be interpreted as the components of the spatial metric tensor onto the deformed material base $\mathbf{f}_\alpha$, $\alpha = 1, 2, 3$ as follows:

$$C_{\alpha\beta} := \mathbf{f}_\alpha \cdot \mathbf{f}_\beta, \qquad \mathbf{f}_\alpha := F_\alpha^i \mathbf{e}_i$$

where $\mathbf{e}_i$, $i = 1, 2, 3$ is a vector base in the initial configuration, embedded in the continuum body and deforms with it. Then the induced metric in the continuum matter is decomposed along the dual base $\mathbf{f}^\alpha$, $\alpha = 1, 2, 3$ on the tangent space:

$$\mathbf{g} = C_{\alpha\beta} \, \mathbf{f}^\alpha \otimes \mathbf{f}^\beta \tag{3.37}$$

For strain gradient continuum models, partial derivatives of the metric components are often used as additional primal variables in the framework of gradient elasticity to give $\mathscr{L}(g_{\alpha\beta}, \partial_\gamma g_{\alpha\beta}, \partial_\lambda \partial_\gamma g_{\alpha\beta}, \cdots)$ e.g. Askes and Aifantis (2011), Metrikine (2006), Mindlin (1964). For the sake of the formulation invariance the previous Lagrangian function of strain gradient continuum models should be written in terms of covariant derivatives as follows $\mathscr{L}(C_{\alpha\beta}, \nabla_\gamma C_{\alpha\beta}, \nabla_\lambda \nabla_\gamma C_{\alpha\beta}, \cdots)$ where ∇ is a connection (to be defined).

Corollary 3.1 *Say an elastic continuum $\mathscr{B}$ evolving in a Euclidean space $\mathscr{E}$ with a Lagrangian function $\mathscr{L}$, owing that the transformation of $\mathscr{B}$ is assumed to be diffeomorphism. Then the Lagrangian function (the elastic potential) could not depend on the covariant derivative of the right Cauchy-Green tensor, inducing that:*

$$\mathscr{L} := (\rho/2) g_{ij} \partial_t \varphi_t^i \partial_t \varphi_t^j - \mathscr{U}(C_{\alpha\beta}, \nabla_\gamma C_{\alpha\beta}) \quad \rightarrow \quad \mathscr{L}$$

$$:= (\rho/2) g_{ij} \partial_t \varphi_t^i \partial_t \varphi_t^j - \mathscr{U}(C_{\alpha\beta}) \tag{3.38}$$

Proof Indeed, it is straightforward to show that the covariant derivative of the metric induced by the Euclidean space $\mathscr{E}$ in the continuum identically vanishes $\nabla_\gamma g_{\alpha\beta} \equiv 0$, or in a elasticity notation $\nabla_\gamma C_{\alpha\beta} \equiv 0$.

$$\nabla_\gamma C_{\alpha\beta} = \partial_\gamma C_{\alpha\beta} - \Gamma_{\gamma\alpha}^\mu C_{\mu\beta} - \Gamma_{\gamma\beta}^\mu C_{\alpha\mu}$$

with respectively the connection coefficients:

$$\Gamma_{\gamma\alpha}^\mu = (1/2) C^{\mu\sigma} \left(\partial_\gamma C_{\sigma\alpha} + \partial_\alpha C_{\gamma\sigma} - \partial_\sigma C_{\gamma\alpha} \right)$$

$$\Gamma_{\gamma\beta}^\mu = (1/2) C^{\mu\sigma} \left(\partial_\gamma C_{\sigma\beta} + \partial_\beta C_{\gamma\sigma} - \partial_\sigma C_{\gamma\beta} \right)$$

We (obviously) deduce the vanishing of the covariant derivative of the right Cauchy-Green strain tensor:

$$\nabla_\gamma C_{\alpha\beta} \equiv 0 \tag{3.39}$$

confirming that the usual connection that is used in classical elasticity is a metric compatible connection.   □

3.4.4.2 Examples of Nonlinear Elastic Material Models

Let denote the matrix $\mathbf{C} := [C_{\alpha\beta}]$ for the sake of the simplicity. The most usual models of nonlinear elastic material are:

1. the compressible Kirchhoff-St Venant model:

$$\mathcal{U} := \frac{\lambda}{2}\mathrm{Tr}^2\mathbf{C} + \mu\mathrm{Tr}\left(\mathbf{C}^2\right)$$

   where the real numbers λ and μ are called Lamé's elastic parameters;
2. the incompressible Mooney-Rivlin model:

$$\begin{cases} \mathcal{U} := C_1\,(\mathbb{I}_1 - 3) + C_2\,(\mathbb{I}_2 - 3) \\ \mathbb{I}_3 \equiv 1 \end{cases}$$

   where the real numbers C_1 and C_2 are elastic parameters, and $\mathbb{I}_1 := \mathrm{Tr}\mathbf{C}$ and $\mathbb{I}_2 := (1/2)\left[\mathrm{Tr}^2\mathbf{C} - \mathrm{Tr}\left(\mathbf{C}^2\right)\right]$ and $\mathbb{I}_3 := \mathrm{Det}\mathbf{C}$ are invariants of the matrix $\mathbf{C}$. The incompressibility of the material is defined by the second row.

Remark 3.5 This is in fact nothing more that a particular case of the theorem established by e.g. Lovelock and Rund (1975). The consequence of this corollary might be that the covariant gradient of the metric could not be used as primal variable for the strain energy density, and this is to be related with the previous theorem (3.6) to conclude that additional primal variables should be constructed by means of torsion and curvature. This is a basic requirement for the nonlinear elastic model to be covariant (Fig. 3.2).

As a remind, continuum $\mathcal{B}$ has compatible displacement field whenever the torsion and curvature vanish everywhere e.g. Maugin (1993), Rakotomanana (2003). Indeed, the displacement vector has three components whereas there are six relations between metric and (gradient of) displacement, meaning that the system is overconstrained. Satisfying compatibility of displacement and metric in three dimensional nonlinear elasticity results in the vanishing of torsion and curvature e.g. Rakotomanana (1997). The metric compatibility (3.39) constitutes another relation that is satisfied by the metric. Considering non metric compatible connection $\tilde{\nabla}$ might be possible but the geometric background of the continuum $\mathcal{B}$ should then be extended on a Weyl manifold rather than on a Riemann manifold by adding non-metricity $Q_{\alpha\beta}^\gamma := \tilde{\nabla}^\gamma C_{\alpha\beta} \neq 0$. Accordingly, the Lagrangian would take the form of:

$$\mathcal{L} := (\rho/2)g_{ij}\partial_t\varphi_t^i\partial_t\varphi_t^j - \mathcal{U}(C_{\alpha\beta}, Q_{\alpha\beta}^\gamma) \tag{3.40}$$

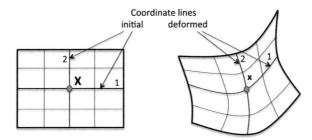

Fig. 3.2 In the initial configuration of continuum body $\mathcal{B}$ evolving within a Euclidean space $\mathscr{E}$ with metric $\hat{g}_{ij}$, a local vector base $(\mathbf{E}_i, i = 1, 2, 3)$ tangent to the coordinate lines (X^μ) is defined at each material point $\mathbf{X}$ which deforms to its position $\mathbf{x}$ in the deformed configuration $\varphi_t(\mathcal{B})$. The deformed local base tangent to the coordinate lines (x^i) in the deformed configuration is given by $\mathbf{f}_\alpha := \partial_\alpha \varphi_t^i (\mathbf{E}_i)$, and the components of the embedded metric tensor onto the deformed base are $g_{\alpha\beta} = C_{\alpha\beta} := \hat{\mathbf{g}}(\mathbf{f}_\alpha, \mathbf{f}_\beta) = \mathbf{f}_\alpha \cdot \mathbf{f}_\beta$ e.g. Rakotomanana (2003)

where the non-metricity tensor is an independent additional primal variable of the strain energy density e.g. Yavari and Goriely (2012). This may be related with some aspects of residual stress and Eshelbian inclusion within nonlinear elastic solids. In the following, we mainly focus on metric compatible connection.

Chapter 4
Gauge Invariance for Gravitation and Gradient Continuum

4.1 Introduction to Gauge Invariance

Geometrization of continuum physics that is the formulation of constitutive laws and conservation laws equations with respect to a reference spacetime involves some steps. First of all, physical measurable quantities should be identified with geometrical variables (metric, torsion, and curvature on the material manifold) and other additional variables if any. Second point, the spacetime is generally a dynamical background with its metric, torsion, and curvature, such is the case for general relativity. Then it is required to specify how all these geometrical variables are generated and modified by physical objects, namely material particle, material elements as line, surface, volume, defects, and how these physical objects evolve during the interaction of the continuum matter and the spacetime. The basic tool for deriving constitutive laws and conservation laws from a Lagrangian density lies on the concept of variation. Some aspects of variation calculus are introduced in the present chapter, namely the Lagrangian variation and the Eulerian variation (Poincaré invariance).

4.1.1 Transition from Covariance to Gauge Invariance

The first Lagrangian of Eq. (3.36) defines Weitzenböck continua which is well suited to model the tele parallel gravitation theory e.g. Hayashi (1979) and the theory of nonlinear elastic and plastic dislocations e.g. Lazar (2002), Le and Stumpf (1996). The second allows us to define the Lagrangian approach for Einstein gravitation e.g. Kleinert (2008). The third class extends the relativistic gravitation to Einstein–Cartan theory of gravitation, to metric-affine gravity e.g. Bruzzo (1987), Sotiriou and Liberati (2007), and to the concept of weakly continuous medium e.g. Rakotomanana (1997), or equivalently e.g. Ruggiero and Tartaglia (2003). A

© Springer International Publishing AG, part of Springer Nature 2018   95
L. R. Rakotomanana, *Covariance and Gauge Invariance in Continuum Physics*,
Progress in Mathematical Physics 73, https://doi.org/10.1007/978-3-319-91782-5_4

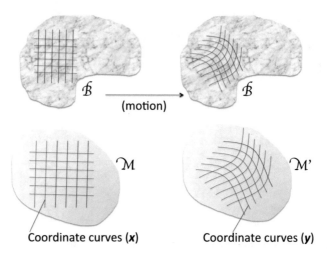

Fig. 4.1 Sketch of the various invariance concepts. The continuum matter $\mathscr{B}$ as well as the spacetime $\mathscr{M}$ or $\mathscr{M}'$ are modelled as Einstein–Cartan manifold having their own metric, torsion and curvature. The theorem on covariance is applied both on any Lagrangian function on the spacetime $\mathscr{M}$ or $\mathscr{M}'$, and on the continuum body $\mathscr{B}$ following a change of coordinate system $x^\mu \to y^\mu$. The Poincaré invariance concerns the motion of the $\mathscr{B}$ with respect to $\mathscr{M}$ (or $\mathscr{M}'$). The Poincaré's gauge invariance allows us to derive the conservation laws

spacetime (resp. continuum matter) where non metricity $\nabla_\gamma g_{\alpha\beta}$ vanishes can be pictured as a set of infinite number of Minkowskian (resp. Euclidean) "microcosms" glued together by means of affine connection $\Gamma^\gamma_{\alpha\beta}$ (Gonseth 1926; Pettey 1971). Depending on the affine connection (with/without torsion and/or curvature), sets of microcosms become the spacetime with gravitation for general relativity, and for strain gradient continuum. We consider in this section the invariance of Lagrangian with respect to the choice of affine connection (Kadianakis 1996). The basic question is to relate arbitrariness of connection with the invariance of Lagrangian with respect to a frame of reference (Fig. 4.1).

Different observers can only compare motions and transformations of a continuum if the reference frames of these observers are known along with laws of transformations: Galilean transformations for classical mechanics, and Lorentz transformations for special relativistic mechanics (Bernal and Sanchez 2003). It is interesting to relate these results with the fundamental findings in Kibble (1961) which studied the existence of free Lagrangian $\mathscr{L}$ in gravitation by starting with its invariance with respect to Lorentz transformations group in a flat spacetime. Later developments on affinely connected manifolds and their relation with mechanics show that there is correspondence between the set of reference frames of a spacetime manifold $\mathscr{M}$ and the set of compatible connections on $\mathscr{M}$ (Kadianakis 1996). Therefore, it strongly suggests that when we impose arbitrariness of affine connection, we implicitly impose, at least partially, invariance of Lagrangian density, and by the way the invariance of constitutive relations, with respect to change of reference frame: this is directly related to frame-indifference principle e.g. Betram and Svendsen

(2001), Ryskin (1985), Söderholm (1970) (in relativistic mechanics), Svendsen and Betram (1999), and as historical reference (Truesdell and Noll 1991). This may give new insights on the (old) debate on the covariance principle and the invariance of physical theories with respect to reference frames.

The variational formulation of relativistic gravitation allows us to deduce first the equations of gravitational fields, and the conservation laws in the same framework e.g. Carter (1973). To point out the different steps of the, let remind the work of Carter by starting to consider an action integral:

$$\mathscr{S} := \int_{\mathscr{B}} \left(\mathscr{U}(g_{\mu\nu}) - (1/2\chi)\mathscr{R} \right) \omega_n$$

modelling the motion of an elastic continuum with energy $\mathscr{U}$ depending on the metric within a curved spacetime $\mathscr{M}$. He has supposed that the displacement ξ vanishes at the boundary $\partial\mathscr{B}$. In such a case, the Lagrangian variation $\Delta\mathscr{S}$ and the Eulerian variation $\delta\mathscr{S}$ are the same, where the Lagrangian variation and the Eulerian variation are related by $\Delta = \delta + \mathcal{L}_\xi$ (Carter and Quintana 1977). Now the variation of the action takes the following form:

$$\Delta\mathscr{S} = \int_{\mathscr{B}} \left(\frac{\partial \mathscr{U}}{\partial g_{\mu\nu}} - \frac{1}{2\chi}\mathfrak{R}^{\mu\nu} \right) \Delta g_{\mu\nu}\, \omega_n - \frac{1}{2} \int_{\mathscr{B}} \left(\mathscr{U} - \frac{1}{2\chi}\mathscr{R} \right) g^{\mu\nu}\, \Delta g_{\mu\nu}\omega_n$$
$$+ \int_{\partial\mathscr{B}} \dots \tag{4.1}$$

where the last term consists in boundary contributions. By defining the stress-energy tensor, and by expressing the Lagrangian variation of the metric:

$$T^{\mu\nu} := 2\frac{\partial \mathscr{U}}{\partial g_{\mu\nu}} - \mathscr{U}\, g^{\mu\nu}, \qquad \Delta g_{\mu\nu} = \delta g_{\mu\nu} + \mathcal{L}_\xi g_{\mu\nu}$$

we deduce the both the gravitational field equation and the conservation laws due the arbitrariness of both the Eulerian variation of the gravitation field $\delta g_{\mu\nu}$ and the Lie derivative of the local vector field ξ (see appendix for the Lie derivative):

$$\begin{cases} \chi\, T^{\mu\nu} - (\mathfrak{R}^{\mu\nu} - (\mathscr{R}/2)\, g^{\mu\nu}) = 0 \\ \nabla_\nu \left[T^{\mu\nu} - (1/\chi)\, (\mathfrak{R}^{\mu\nu} - \mathscr{R}/2\, g^{\mu\nu}) \right] = 0 \end{cases}$$

where we have dropped the terms at the boundary and the divergence term. Now, provided we have the first Bianchi identity, the complete system of equations reduces to:

$$\begin{cases} \chi\, T^{\mu\nu} = (\mathfrak{R}^{\mu\nu} - (\mathscr{R}/2)\, g^{\mu\nu}) \\ \nabla_\nu T^{\mu\nu} = 0 \\ T^{\mu\nu} = 2\dfrac{\partial \mathscr{U}}{\partial g_{\mu\nu}} - \mathscr{U}\, g^{\mu\nu} \end{cases} \tag{4.2}$$

The field equation means the continuum matter bends the spacetime, whereas the second equation expresses the conservation laws of the matter, extending the linear and angular momentum equation. The third equation is the constitutive laws of the matter. One important aspect in the reference (Carter 1973) was the logical deduction of both three equations from one variational principle. In this section, we introduce the affine connection of the manifold as geometrical argument in additional to metric. Both the continuum body $\mathscr{B}$ and the spacetime $\mathscr{M}$ have their own geometrical variables. The idea is that invariance with respect to affine connection may be related to invariance of Lagrangian density with respect to reference frame. From another point of view any local frame of references may be identified with a triad (respectively tetrad) at each point of the manifold $\mathcal{E}$. This is the keypoint to extend the global invariance to local gauge invariance by considering vector field as local translation when applying Poincaré's gauge theory.

4.1.2 Mechanical Coupling of Matter and Spacetime

In this section, we consider the mechanical coupling meaning the interaction of the metric, torsion, and curvature of the continuum matter with those of the spacetime background. Other physical variables are omitted.

4.1.2.1 Spacetime Classification and Lagrangian

A metric-affine manifold $(\mathscr{M}, \mathbf{g}, \nabla)$ with the most general metric compatible connection is called Riemann–Cartan spacetime, often denoted U_4. If the torsion $\aleph^{\gamma}_{\alpha\beta}$ vanishes, a U_4 becomes a Riemannian spacetime, denoted V_4 in the domain of the General Relativity theory. If, alternatively, the curvature $\mathfrak{R}^{\lambda}_{\alpha\beta\mu}$ vanishes, then a U_4 manifold becomes Weitzenböck's teleparallel spacetime, denoted T_4 (Tele Parallel Gravity theory). Finally, the condition of zero curvature $\mathfrak{R}^{\lambda}_{\alpha\beta\mu} = 0$ transforms a V_4 into a Minkowskian spacetime M_4, and zero torsion $\aleph^{\gamma}_{\alpha\beta} = 0$ transforms a T_4 into an M_4. In the scope of gravitation, using of Riemann–Cartan manifold supports the idea to look for dynamical manifestation of spin-angular momentum of matter e.g. Hehl and von der Heyde (1973), where the torsion tensor is expected to describe additional rotational degrees of freedom. In the framework of continuum mechanics, as earlier as 1955, Riemann–Cartan manifold constitutes the geometric background for modeling continuous distributions of translational and rotational dislocations e.g. Bilby et al. (1955). Covariance is required for both the strain energy density of gradient continuum and the Lagrangian density involved in relativistic gravitation. In this section, we assume that both the material continuum and the spacetime are modeled by Riemann–Cartan manifolds with their own metric and connection respectively. The covariance requirement may be extended and applied to the theory of metric-affine gravity where Lagrangian density could be rewritten in a general manner to include strain gradient continuum[1]:

$$\mathcal{L} = \mathcal{L}(\underbrace{\hat{g}_{\alpha\beta}, \hat{\aleph}^{\gamma}_{\alpha\beta}, \hat{\mathfrak{R}}^{\gamma}_{\alpha\beta\lambda}}_{\text{spacetime}}, \underbrace{g_{\alpha\beta}, \aleph^{\gamma}_{\alpha\beta}, \mathfrak{R}^{\gamma}_{\alpha\beta\lambda}}_{\text{matter}}) \qquad (4.3)$$

clarify $\hat{\omega}_n$ or ω_n from which we can define various subclasses of a generalized continuum evolving within a more or less complex spacetime e.g. Clifton et al. (2012). The introduction of different connections for spacetime and for matter is now accepted in the literature (implicitly or explicitly) e.g. Appleby (1977), Bernal and Sanchez (2003), Defrise (1953), Kadianakis (1996), Petrov and Lompay (2013), Sotiriou and Liberati (2007), Tamanini (2012). Material metric and independent connection are also introduced to model macroscopic and mesoscopic mechanical interactions within matter. The great challenge of relativistic continuum mechanics concerns the accounting for the interaction of matter and spacetime. More precisely, modelling the interaction between the metric, torsion and curvature of the continuum matter $\mathcal{B}$ and those of the spacetime $\mathcal{M}$ remains an wide open topics research. Particularly, some authors have argued that the experimental testing of the presence of spacetime torsion might be induced by macroscopic rotating objects e.g. Acedo (2015), Mao et al. (2007). However, the measuring of spacetime torsion is only possible in presence of spinning particles since its coupling with intrinsic particle spin was argued very early e.g. Hehl and von der Heyde (1973), and even that its coupling with macroscopic rotating bodies is in contradiction the basic theory of Poincaré invariance. e.g. Hehl et al. (2013).

Remark 4.1 Consider a particular case of the Lagrangian (4.3) where the dependence is slightly modified as $\mathcal{L} = \mathcal{L}(\hat{g}_{\mu\nu}, F^{\mu}_{\alpha})$ where F^{μ}_{α} are tetrads, integrable or not. Classical results of continuum mechanics on the objectivity (covariance) induces that the Lagrangian should be expressed as $\mathcal{L} = \mathcal{L}(\hat{g}_{\mu\nu}, g_{\alpha\beta} := \hat{g}_{\mu\nu} F^{\mu}_{\alpha} F^{\nu}_{\beta})$ e.g. Marsden and Hughes (1983), Truesdell and Noll (1991), and it is a particular case of more tensorial functions e.g. Rakotomanana (2003).

4.1.2.2 Principle of Minimal Coupling

In order to formulate the variational principle of matter motion within a curved spacetime, it is necessary to account the interaction of the matter and the spacetime gravitation (Sciama 1964). The simplest interaction model would be the so-called minimal coupling procedure. The classical *Minimal Coupling Procedure* (MCP) for the function $\mathcal{L}$ is to replace the flat Lorentz spacetime metric $\hat{g}_{\alpha\beta}$ and connection $\hat{\nabla}$ by the matter metric $g_{\alpha\beta}$ and connection ∇ respectively, especially by using covariant derivatives instead of partial derivatives (for instance within a Cartesian axes of the flat space). Starting with a generic shape of a Lagrangian, depending on

---

[1]Index 0 stands for time variable, then the Lagrangian $\mathcal{L}$ also stands for dynamical situation.

the geometry of the continuum and some physical variables Φ, say:

$$\mathscr{L} = \mathscr{L}(g_{\alpha\beta}, \Gamma^{\gamma}_{\alpha\beta}, \partial_{\lambda}\Gamma^{\gamma}_{\alpha\beta}, \Phi^{\mu}, \partial_{\eta}\Phi^{\mu}, \cdots)$$

The previous result may be re-written as follows, by taking into account the covariance of the Lagrangian,

$$\mathscr{L} = \mathscr{L}(g_{\alpha\beta}, \aleph^{\gamma}_{\alpha\beta}, \mathfrak{R}^{\gamma}_{\alpha\beta\lambda}, \Phi^{\mu}, \nabla_{\eta}\Phi^{\mu}, \cdots)$$

in which replacement of the partial derivative by the covariant derivative may be considered as a minimal coupling procedure. However, thanks to the covariance theorem we previously established, the arguments of the Lagrangian are directly stated to be the metric, the torsion, and the curvature of matter. It should not be forgotten the relation between spacetime connection and matter connection that needs great cautious as we will see later. As extension, the MCP in the gravitation theory would consist in decomposing the overall Lagrangian $\mathscr{L}$ into a free gravitational Lagrangian $\mathscr{L}_G$ depending on the metric, the torsion and the curvature of the spacetime connection ($\hat{\nabla}$), and in adding to a suitable matter Lagrangian $\mathscr{L}_M$ depending both on the metric and connection of the spacetime and on the metric, the torsion and the curvature of the matter connection (∇).

Remark 4.2 Although there is a long last debate e.g. Hammond (2002), Hehl et al. (1995) on the worthiness of introducing or not the torsion as geometric variable of relativistic gravitation theory, there is at least no mathematical reason to exclude the torsion as primal variable of the Lagrangian e.g. Garcia de Andrade (2005). Einstein gravitation is based on the spacetime curvature. The main problem is the detection of the space torsion (if any) since usual matter model without microstructure such as matter torsion cannot detect torsion. Mao et al. have proposed a gyroscopic experiments using macroscopic rotating bodies (Gravity Probe B mission) to detect spacetime torsion (Mao et al. 2007), but it was shown that only bodies with microstructure such as intrinsic spin could be exploited for that purpose e.g. Hehl et al. (2013), Yasskin and Stoeger (1980). Nevertheless, results may differ if using models of minimal coupling or model of nonminimal coupling (Puetzfeld and Obukhov 2013b). The suggested general formulation of the Lagrangian (4.3) includes all possibilities in relativistic gravitation when restricted to continuum mechanics under gravitation.

The coupling of matter and spacetime needs great cautious e.g. Anderson (1981), Sotiriou (2008), particularly when matter action is introduced and linearly coupled with the classical Einstein–Hilbert action (theory based on scalar curvature of spacetime). The general form of Lagrangian function (4.3) may be shaped to fit some basic mathematical results in the domain of continuum immersed within a curved spacetime. In this way some fundamental results were obtained e.g. Anderson (1981) under some sound assumptions such as: (a) the spacetime $\mathscr{M}$ is four-dimensional, (b) the Lagrangian $\mathscr{L}$ is a scalar density, and (c) the fields (Euler–Lagrangian) equations of external variables (in our case the matter variables including the continuum metric $g_{\alpha\beta}$, torsion $\aleph^{\gamma}_{\alpha\beta}$ and curvature $\mathfrak{R}^{\gamma}_{\alpha\beta\lambda}$ components of material manifold) are of the first order in the spacetime metric and second order for external variables. In such a case, the principle of minimal gravitational coupling requires that the Lagrangian function $\mathscr{L}$ should be written as two additive parts: the gravitational Lagrangian $\mathscr{L}_G$ of the curved spacetime, and the Lagrangian corresponding to the material continuum $\mathscr{L}_M$ e.g. Anderson (1981). In this paper, Anderson worked within a pseudo-Riemannian spacetime with gravity, where torsion tensor identically vanishes $\aleph^{\gamma}_{\alpha\beta}$ (framework of Einstein relativistic gravitation). He established the following theorem:

Theorem 4.1 *Let $\mathscr{L}$ be a Lagrangian density $\mathscr{L}(\hat{g}_{\alpha\beta}, \partial_{\lambda}\hat{g}_{\alpha\beta}, \partial_{\lambda}\partial_{\mu}\hat{g}_{\alpha\beta}, \Phi_i, \partial_{\alpha}\Phi_i)$ which satisfies the condition that the Euler–Lagrange equations associated to the source variables Φ_i depend solely on arguments $(\hat{g}_{\alpha\beta}, \partial_{\lambda}\hat{g}_{\alpha\beta}, \Phi_i, \partial_{\alpha}\Phi_i, \partial_{\mu}\partial_{\lambda}\Phi_i)$. Then, the Lagrangian density $\mathscr{L}$ decomposes uniquely to the form:*

$$\mathscr{L} = \mathscr{L}_G(\hat{g}_{\alpha\beta}, \partial_{\lambda}\hat{g}_{\alpha\beta}, \partial_{\lambda}\partial_{\mu}\hat{g}_{\alpha\beta}) + \mathscr{L}_M(\hat{g}_{\alpha\beta}, \partial_{\lambda}\hat{g}_{\alpha\beta}, \partial_{\lambda}\partial_{\mu}\hat{g}_{\alpha\beta}; \Phi_i, \partial_{\alpha}\Phi_i) \quad (4.4)$$

where the scalar densities $\mathscr{L}_G$ and $\mathscr{L}_M$ satisfies: $\mathscr{L}_M(\hat{g}_{\alpha\beta}, \partial_{\lambda}\hat{g}_{\alpha\beta}, \partial_{\lambda}\partial_{\mu}\hat{g}_{\alpha\beta}; 0, 0) = 0$, and for which the Euler–Lagrange equations of only the part $\mathscr{L}_M$ and associated to the spacetime variables depend on the arguments $\{\hat{g}_{\alpha\beta}, \partial_{\lambda}\hat{g}_{\alpha\beta}, \partial_{\lambda}\partial_{\mu}\hat{g}_{\alpha\beta}; \Phi_i, \partial_{\alpha}\Phi_i\}$, and the analogous Euler–Lagrange associated to the source variables depend on the arguments:

$$\{\hat{g}_{\alpha\beta}, \partial_{\lambda}\hat{g}_{\alpha\beta}; \Phi_i, \partial_{\alpha}\Phi_i, \partial_{\mu}\partial_{\alpha}\Phi_i\}.$$

We deduce the corollary:

Corollary 4.1 *Let consider a Lagrangian $\mathscr{L}$ which is a scalar density of the type:*

$$\mathscr{L}(\hat{g}_{\alpha\beta}, \partial_{\lambda}\hat{g}_{\alpha\beta}, \partial_{\lambda}\partial_{\mu}\hat{g}_{\alpha\beta}, \Phi_i, \partial_{\alpha}\Phi_i),$$

and if the arguments of the associated Euler–Lagrange of the spacetime variables reduce to:

$$\{\hat{g}_{\alpha\beta}, \partial_{\lambda}\hat{g}_{\alpha\beta}, \partial_{\lambda}\partial_{\mu}\hat{g}_{\alpha\beta}; \Phi_i, \partial_{\alpha}\Phi_i\}.$$

Then the Lagrangian density necessarily takes the form of:

$$\mathscr{L} = a \, \mathcal{R}\sqrt{\mathrm{Detg}} + \Lambda\sqrt{\mathrm{Detg}} + \mathscr{L}_M(\hat{g}_{\alpha\beta}, \partial_{\lambda}\hat{g}_{\alpha\beta}, \partial_{\lambda}\partial_{\mu}\hat{g}_{\alpha\beta}; \Phi_i, \partial_{\alpha}\Phi_i) \quad (4.5)$$

where $a \in \mathbb{R}$, and $\Lambda \in \mathbb{R}$ are scalars, and where the scalar density $\mathscr{L}_M$ satisfies:

$$\mathscr{L}_M(\hat{g}_{\alpha\beta}, \partial_{\lambda}\hat{g}_{\alpha\beta}, \partial_{\lambda}\partial_{\mu}\hat{g}_{\alpha\beta}; 0, 0) = 0 \quad (4.6)$$

and for which the Euler–Lagrange equations of only the part $\mathscr{L}_M$ and associated to the spacetime variables depend on the arguments $\{\hat{g}_{\alpha\beta}, \partial_{\lambda}\hat{g}_{\alpha\beta}, \partial_{\lambda}\partial_{\mu}\hat{g}_{\alpha\beta}; \Phi_i, \partial_{\alpha}\Phi_i\}$, *and the analogous Euler–Lagrange associated to the source variables depend on the arguments:*

$$\left\{ \hat{g}_{\alpha\beta}, \partial_\lambda \hat{g}_{\alpha\beta}; \Phi_i, \partial_\alpha \Phi_i, \partial_\mu \partial_\alpha \Phi_i \right\}.$$

Proof See the paper of Anderson (1981) for detailed proof and related lemma and corollary.   □

Accordingly, in addition to the result obtained by Lovelock in the sixties, the general form of the Lagrangian must take the form of (adapted from a theorem in Anderson (1981), and provided the covariance theorem) e.g. Antonio and Rakotomanana (2011):

$$\mathscr{L} = \mathscr{L}(\hat{g}_{\alpha\beta}, \hat{\aleph}^\gamma_{\alpha\beta}, \hat{\mathfrak{R}}^\gamma_{\alpha\beta\lambda}; g_{\alpha\beta}, \aleph^\gamma_{\alpha\beta}, \mathfrak{R}^\gamma_{\alpha\beta\lambda}) \tag{4.7}$$

where both the spacetime and the matter may have their own metric and connection e.g. Koivisto (2011). In view of this function (4.7), the presence of the spacetime metric $\hat{g}_{\alpha\beta}$ in the Lagrangian $\mathscr{L}$ is essential to allow the mutual interaction of the spacetime and the matter (Lehmkuhl 2011). The presence of spacetime connection is sought by analogy. Indeed, spacetime metric and connection and its first derivatives are usually the arguments of spacetime Lagrangian $\mathscr{L}_G(\hat{g}_{\alpha\beta}, \hat{\aleph}^\gamma_{\alpha\beta}, \hat{\mathfrak{R}}^\gamma_{\alpha\beta\lambda})$. They may be different from arguments of matter Lagrangian density $\mathscr{L}_M$. In such a case, the coupling between spacetime and matter should be investigated more deeply. When dependence includes torsion and curvature, there is more complex coupling as we will see later. A rather common method is to introduce the Green-Lagrange strain tensor $\varepsilon_{\alpha\beta} := (1/2)(g_{\alpha\beta} - \hat{g}_{\alpha\beta})$ as argument of the matter Lagrangian e.g. Marsden and Hughes (1983). It is *de facto* a minimal coupling. Matter bends the spacetime and "forces" resulting from the spacetime curvature is source of deformation of matter. More generally, the description of the motion of a material body within spacetime necessarily involves two continua (pseudo-Riemannian and Riemannian manifolds for gravitation, and Euclidean and Riemannian manifolds for strain gradient continuum) e.g. Bernal and Sanchez (2003), Defrise (1953). Classical fields of physics can be unified by modelling spacetime as a four dimensional finite but unbounded elastic continuous medium, which can deform in presence of matter-energy fields. In this case, the material body (gradient or simple material) is regarded as the $3D$ boundary of a world-tube in such a way that the outside the world-tube (material) the region is empty.

4.2   Gravitation, Fields, and Matter

Recasting physics theory into the language of an action principle by means of Lagrangian density function is a cornerstone for obtaining invariant formulation (Sciama 1964). Lagrangian formalism over differentiable manifold $\mathscr{B}$ with an affine connection and a metric includes three basic structures e.g. Manoff (1999): (a) the Lagrangian density $\mathscr{L}$ and the choice of its arguments; (b) the Euler–Lagrange equations, obtained by variation procedure; and (c) the energy-momentum tensors, or constitutive relations for models. All of them lead to the field equations by means of gauge invariance.

4.2.1 Preliminaries

We give some known results about spacetimes and their hierarchical classification according to their metric and associated connection. Before embarking in the analysis of gravitation in various spacetimes, we get back for a while to the concept of covariance and the principle of relativity by reminding the free fall of a particle.

4.2.1.1 Free Fall of a Particle

The free fall of a particle is governed by the equation stating that the acceleration is equal to zero. Three possibilities ranging from Newton, Minkowski, and Einstein approach will be considered:

1. Let remind the Newton's first law for a particle of mass m (although the mass does not matter in the absence of "forces"). With respect to a given frame, the (three-dimensional) acceleration of this point is equal to zero:

$$\mathbf{a} = 0 \quad \text{with} \quad \mathbf{a} := \frac{d\mathbf{u}}{dx^0} \tag{4.8}$$

   where we use the coordinate system $x^\mu := \left(x^0 = ct, x^1, x^2, x^3\right)$ (practically, it means that we assume $c = 1$). Thus this free particle is at rest with respect to this reference frame or it moves with constant velocity along a straight line.

2. The second case deals with the free fall of a particle with respect to a reference with the velocity $\mathbf{u}$. Newton's law (4.8) is not invariant with respect to Lorentz transformation in the special relativistic mechanics. We should consider the four-acceleration (a^μ). For that purpose, we consider the four-velocity vector as defined in the relation (2.77) (with the convention $c = 1$):

$$u^\mu := \frac{dx^\mu}{d\tau} = \frac{dx^\mu}{dx^0} \frac{dx^0}{d\tau} = \gamma \left(1, u^i\right), \qquad \gamma := \frac{1}{\sqrt{1 - \|\mathbf{u}\|^2}}$$

   owing that the proper time τ is defined as follows:

$$(d\tau)^2 := \left(dx^0\right)^2 - \left(dx^1\right)^2 - \left(dx^2\right)^2 - \left(dx^3\right)^2 \quad \Longrightarrow \quad \frac{dx^0}{d\tau} = \gamma$$

with $\mathbf{u} := dx^i/dx^0 \mathbf{e}_i$ is a three-dimensional velocity. Form the four-velocity vector, we can derive the expression of the four-acceleration:

$$a^\mu := \frac{du^\mu}{d\tau} = \frac{d\gamma}{d\tau}\left(1, u^i\right) + \gamma\left(0, \frac{du^i}{d\tau}\right) = \frac{d\gamma}{dx^0}\frac{dx^0}{d\tau}\left(1, u^i\right) + \gamma\left(0, \frac{du^i}{dx^0}\frac{dx^0}{d\tau}\right)$$

By accounting for the derivative:

$$\frac{d}{dx^0}\left(\|\mathbf{u}\|^2\right) = 2\mathbf{u}\cdot\frac{d\mathbf{u}}{dx^0} = 2\mathbf{u}\cdot\mathbf{a}$$

we deduce the Lorentz invariant formulation of the first Newton's law:

$$a^\mu = \gamma^2\left[\gamma^2\mathbf{u}\cdot\mathbf{a}, \mathbf{a} + \gamma^2\mathbf{u}\otimes\mathbf{u}\,(\mathbf{a})\right] = 0 \tag{4.9}$$

which expresses the vanishing of the four-acceleration with respect to reference frame. Of course the law (4.9) reduces to (4.8) when the norm of the particle velocity $\mathbf{u}$ is small compared to the light speed.

3. The third case concerns the free fall of a particle with respect to an arbitrary reference. The covariant expression of the free fall equation projected onto a generalized coordinate system requires a vacuum spacetime with the Levi-Civita connection. The expression of the four-acceleration and by the way the covariant formulation of the Newton's first law are obtained:

$$a^\mu := \frac{du^\mu}{d\tau} + \hat{\overline{\Gamma}}^\mu_{\alpha\beta}u^\alpha u^\beta = 0 \tag{4.10}$$

with

$$\hat{\overline{\Gamma}}^\gamma_{\alpha\beta} := (1/2)\hat{g}^{\gamma\lambda}\left(\partial_\beta\hat{g}_{\alpha\lambda} + \partial_\alpha\hat{g}_{\lambda\beta} - \partial_\lambda\hat{g}_{\alpha\beta}\right)$$

The connection coefficients reduce to the symbols of Christoffel $\hat{\overline{\Gamma}}^\mu_{\alpha\beta}$ has zero torsion and zero curvature for a flat spacetime (free fall of the particle). Overline means Levi-Civita connection and hat the link with spacetime (see below). The acceleration (4.10) holds in any coordinate system (covariant) and in any reference frame (inertial or not).

Remark 4.3 The three relations (4.8)–(4.10) give the equations of a particle in free fall in Newton, special relativistic, and general relativistic mechanics respectively. Generally, an inertial frame may be defined by a frame of reference where free particles have zero acceleration. However in the general relativistic theory consider free fall means that the only force acting on the particle is the gravitation of the spacetime. In Newton's and special relativistic theories, free fall means that there are no gravitation at all. For nonuniform gravitation, inertial frames exist only locally.

4.2.1.2 Some Basic Recall

We consider the vacuous spacetime (or space) endowed with the metric $\hat{g}_{ij}$ which may depend or not of the coordinates x^μ. Associated to this metric can be defined connection coefficients $\overset{\wedge}{\overline{\Gamma}}{}^{\gamma}_{\alpha\beta}$, with zero torsion $\hat{\aleph}^{\gamma}_{\alpha\beta} \equiv 0$ but nonzero curvature $\hat{\Re}^{\lambda}_{\alpha\beta\mu} \neq 0$ e.g. Nakahara (1996). However, it is usual to start with the Minkowskian flat spacetime with the metric $\hat{g}_{\alpha\beta}$ with zero torsion, and zero curvature. Then we consider the actual spacetime metric $g_{\alpha\beta} := \hat{g}_{\alpha\beta} + h_{\alpha\beta}$, where $h_{\alpha\beta}(x^\mu)$ is a perturbation. With the metric and the curvature, Einstein built the theory of general relativistic gravitation. The spacetime may also have non symmetric connection but compatible with the metric. This gives the generic form of the coefficients $\hat{\Gamma}^{\gamma}_{\alpha\beta} :=$ $\overset{\wedge}{\overline{\Gamma}}{}^{\gamma}_{\alpha\beta} + \hat{\mathfrak{T}}^{\gamma}_{\alpha\beta}$ where we observe a contortion tensor $\hat{\mathfrak{T}}^{\gamma}_{\alpha\beta} \neq 0$. The Einstein–Cartan theory is the extension of the relativistic gravitation theory, allowing the spacetime to have nonzero torsion, it was suggested by Cartan in 1922. The application of the covariance theorem states that any Lagrangian function $\mathscr{L}(\hat{g}_{\alpha\beta}, \hat{\Gamma}^{\gamma}_{\alpha\beta}, \partial_\lambda \hat{\Gamma}^{\gamma}_{\alpha\beta})$ should be written as $\mathscr{L}(\hat{g}_{\alpha\beta}, \hat{\aleph}^{\gamma}_{\alpha\beta}, \hat{\Re}^{\lambda}_{\alpha\beta\mu})$ to be diffeomorphism invariant. Geometrical approach was applied both in the study of defects through continuous media e.g. Kröner (1981), Maugin (1993), Rakotomanana (2003), Wang (1967) and more generally in gravitation physics, e.g. Kleinert (2008), Lovelock (1971), Vitagliano et al. (2011). Previous papers interestingly considered the link between these two theories when dimension is reduced to three e.g. Katanaev and Volovich (1992), Verçyn (1990). In this section, we will omit if necessary the "hat" for the sake of the simplicity, and we will give some known illustrations.

We now consider a continuum matter endowed with the metric $g_{\alpha\beta}$ which may depend or not of the material coordinates x^α. The continuum evolves within a spacetime endowed with a metric $\hat{g}_{\alpha\beta} \equiv \delta_{\alpha\beta}$ for simplicity. Lagrangian of a continuum depends on strain ε and possibly on other arguments. In continuum mechanics, the metric is related with the Green-Lagrange strain by $g_{\alpha\beta} = \delta_{\alpha\beta} + 2\varepsilon_{\alpha\beta}$ where $\delta_{\alpha\beta}$ is the identity type $(0, 2)$ tensor e.g. Ruggiero and Tartaglia (2003). This relation expresses that the displacement field (multivalued or not) modifies the initial metric $\delta_{\alpha\beta}$ onto $g_{\alpha\beta}$ within matter. In the above relation, it is implicitly assumed that material coordinates within the continuum are used for describing the continuum transformation (in other words, we use Lagrangian description with deformed vector base e.g. Rakotomanana 2003). Consequently the first type of Lagrangian $\mathscr{L}$ we would like to analyze takes the form of (by abuse of notation) $\mathscr{L}(\varepsilon_{ij}, \partial_k \varepsilon_{ij}, \partial_l \partial_k \varepsilon_{ij}) \equiv \mathscr{L}(g_{ij}, \partial_k g_{ij}, \partial_l \partial_k g_{ij})$.[2] Most of strain gradient models use an affine connection which derives from metric. According to the Theorem 3.5 (Lovelock and Rund 1975), the form $\mathscr{L}(\varepsilon_{ij}, \partial_k \varepsilon_{ij})$ cannot exist. Such a finding

---

[2]Lagrangian density of type I as $\mathscr{L}(g_{ij}, \partial_k g_{ij}, \partial_l \partial_k g_{ij})$ or of type II as $\mathscr{L}(g_{ij}, \nabla_k g_{ij}, \nabla_l \nabla_k g_{ij})$ was considered in Manoff (1999) where three kinds of variational procedures, say the functional variation, the Lie variation, and the covariant variation, were used to derive the fields equations of Einstein's gravitation theory.

conforms to the results in Lovelock's theory of gravitation, where he showed
that the Lagrangian density concomitant to the metric and its first two derivatives
necessary takes the form of $\mathscr{L}\,(\mathbf{g}, \Re)$ (Lovelock 1971). By the way this class
of Lagrangian densities has the advantage to satisfy the consistency between the
Palatini formulation (introduction of an independent connection as arguments) and
the metric gravity formulation e.g. Exirifard and Sheikh-Jabbari (2008). Then the
possible forms are e.g. Agiasofitou and Lazar (2009): (a) $\mathscr{L}(\varepsilon_{ij})$ which is used in
classical elasticity theory; and (b) $\mathscr{L}(\varepsilon_{ij}, \partial_l \partial_k \varepsilon_{ij})$ and $\mathscr{L}(\varepsilon_{ij}, \partial_k \varepsilon_{ij}, \partial_l \partial_k \varepsilon_{ij})$ which
are used in strain gradient theory.

When the connection is Riemannian,[3] among the arguments of the material
Lagrangian function we necessarily consider second order derivative of the strain
as additional variable. We obtain continuum of grade three in terms of displacement
e.g. Agiasofitou and Lazar (2009). Let a continuum modeled by a metric-affine
manifold endowed with a connection compatible with the metric. (a) $\mathscr{L}(\mathbf{g}, \aleph = 0, \Re = 0)$ corresponds to an elastic strain energy function; (b) $\mathscr{L}(\mathbf{g}, \aleph)$ is associated
to an elastic continuum with dislocation. (c) $\mathscr{L}(\mathbf{g}, \aleph, \Re)$ is associated to an elastic
continuum with dislocation and disclination. The elasticity refers to the metric
as argument of the Lagrangian function. However the general form is far from
tractable. It is worth to introduce the Riemann curvature as $\Re_{\nu\alpha\beta\mu} := g_{\nu\gamma} \Re^{\gamma}_{\alpha\beta\mu}$
and also the Ricci curvature tensor $\Re_{\alpha\beta} := \Re^{\lambda}_{\lambda\alpha\beta}$. For compatible connection, say
$\nabla \mathbf{g} \equiv 0$, the Ricci curvature tensor is symmetric. Curvature of a three dimensional
manifold is uniquely determined by the Ricci tensor. The scalar curvature is
defined by the contraction $\Re := g^{\alpha\beta} \Re_{\alpha\beta}$. As for the metric-affine gravity theory
e.g. Vitagliano et al. (2011), the matter Lagrangian density takes the form of
$\mathscr{L}(g_{\alpha\beta}, \Re_{\alpha\beta})$, owing that metric and curvature are independent variables. This
defines models of second strain gradient continua, for which the torsion is equal
to zero.

4.2.1.3 Hierarchical Order of Continuum Structures

It is worth to present the hierarchical order of the geometric structures of spacetime
and gradient continua:

$$
\begin{matrix}
\text{Metric} \\
\text{affine} \quad \nabla\mathbf{g}\equiv 0 \\
\text{geometry}
\end{matrix}
\begin{matrix}
\text{Riemann/} \\
\text{Cartan} \\
\text{geometry}
\end{matrix}
\left\{
\begin{matrix}
\aleph \equiv 0 & \begin{matrix}\text{Riemann}\\\text{geometry}\end{matrix} & \Re \equiv 0 \\
\Re \equiv 0 & \begin{matrix}\text{Weitzenboeck}\\\text{geometry}\end{matrix} & \aleph \equiv 0
\end{matrix}
\right\}
\begin{matrix}
\text{Euclidean/} \\
\text{Minkowski} \\
\text{geometry}
\end{matrix}
$$

where the conditions given before a geometry are constraints to be applied to that
geometry. The metric compatibility condition is an essential property of a continuum

---

[3]The Euclidean connection derived from the metric tensor of a reference body was mostly the
connection used in continuum mechanics for over two centuries, e.g. Rakotomanana (2003).

Fig. 4.2 Somigliana dislocation. V is an added matter after cutting the continuum and separating the two opposite faces of the boundary ∂V with a small displacement field $\mathbf{b}(\mathbf{x})$, considered as discontinuity of vector field

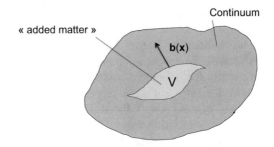

to survive as a continuum after non holonomic (plastic) deformation e.g. Verçyn (1990). From right to left, the geometric structure describes a strongly continuous body e.g. Marsden and Hughes (1983), a body with rotational and/or translational dislocations, and more general weakly continuous bodies e.g. Rakotomanana (1997). The analogy with spacetime holds. The non compatibility $\nabla_\gamma g_{\alpha\beta} \neq 0$ between the metric and the connection leads to another more extended version of dislocations, called Somigliana dislocations. The geometric background associated to Somigliana field of dislocation is the Weyl manifold where additional primal variables are the non-metricity tensor $Q^\gamma_{\alpha\beta} := \nabla^\gamma g_{\alpha\beta}$. A Somigliana dislocation is idealized in terms of a closed volume V of "added/subtracted matter" located within a continuum, as reported on Fig. 4.2. After cutting the body and separating two opposite faces, the operation is represented by a small displacement field $\mathbf{b}(\mathbf{x})$ (discontinuity of vector field), the empty space is then filled with "added matter" (or "subtracted matter" if the two faces penetrate each other after the cutting). The volume V is finally glued to obtain a new continuum with a Somigliana dislocation. Residual stress is generated by the presence of Somigliana dislocation e.g. Clayton et al. (2005). Volterra translational and rotational dislocations are particular case where the faces are linear and no matter is added or subtracted. Potential application of the Somigliana dislocations concept (added mass evolution) lies in the domain of living tissues adaption following non physiological stress in biomechanics of bone and ligaments e.g. Terrier et al. (2005). Indeed for these living matter, mass creation is the natural process for the growing.

4.2.2   Newton–Cartan Formalism for Classical Gravitation

Extension of the special non-relativistic spacetime to include gravitation is first due to Cartan (1986) and later in e.g. Havas (1964), and for continuum mechanics in e.g. Duval and Kunzle (1978). For classical continuum mechanics, metric and connection associated to gravity Lagrangian reduces to those of Newtonian spacetime, that is $\Gamma^k_{ij} = \overline{\Gamma}^k_{ij}$ with non zero curvature.

4.2.2.1 Classical Gravitation

In this subsection, to avoid unnecessary complication of notation, we denote Γ_{ij}^{k} without hat the spacetime connection coefficients. In both the weak field condition (for earth gravitation this means that $GM/(c^2R) \ll 1$) (G is the constant of gravitation, M earth mass, R earth radius, and c light speed) and the low speed motion i.e. $v/c \ll 1$ the difference between Newton gravitation and general relativistic gravitation may be neglected e.g. Shen and Moritz (1996). Connections in Newtonian gravitation are introduced as follows. We consider a particle in a gravitational potential $\mathcal{U}$ and write its linear momentum equation on the one hand, after worthily choosing a parameter $\lambda := a\,t + b$ to derive the second motion equation,

$$\frac{d^2x^i}{dt^2} = -\frac{\partial \mathcal{U}}{\partial x_i} \qquad \Longrightarrow \qquad \frac{d^2x^i}{d\lambda^2} + \frac{\partial \mathcal{U}}{\partial x_i}\left(\frac{dt}{d\lambda}\right)^2 = 0$$

On the other hand, let consider the usual geodesic equation:

$$\frac{d^2x^i}{d\lambda^2} + \Gamma_{jk}^i \frac{dx^j}{d\lambda}\frac{dx^k}{d\lambda} = 0$$

where Γ_{jk}^i are the connection coefficients of the spacetime with gravitational field. These two previous equations allow us to identify the connection, and then the torsion and curvature with physical environment. Let consider the geodesic for Riemannian metric manifold for analyzing the classical gravitation with a potential. As a remind we consider a variational approach. For a curved spacetime, the proper time is defined as $ds := d\tau$ with the coordinate system $(x^0 := ct, x^1, x^2, x^3)$. The propertime necessary for a particle to go from event A to event B along an arbitrary timelike curve is:

$$\tau_{AB} := \int_A^B d\tau = \int_A^B \sqrt{g_{\alpha\beta}(x^\mu)dx^\alpha dx^\beta} = \int_A^B \sqrt{g_{\alpha\beta}(x^\mu)\frac{dx^\alpha}{d\lambda}\frac{dx^\beta}{d\lambda}}d\lambda$$

where λ is an arbitrary affine (real) parameter. We can therefore introduce the Lagrangian function $\mathscr{L}(x^\mu, \dot{x}^\mu, \lambda) := \sqrt{g_{\alpha\beta}(x^\mu)\frac{dx^\alpha}{d\lambda}\frac{dx^\beta}{d\lambda}}$ in which dot means a derivative with respect to λ. Considering the action $\mathscr{S} := \int_A^B \mathscr{L}(x^\mu, \dot{x}^\mu, \lambda)d\lambda$ we obtain after some tedious calculus the following equation e.g. Kleinert (2008):

$$\frac{d^2x^\gamma}{d\lambda^2} + \overline{\Gamma}_{\mu\nu}^\gamma \frac{dx^\mu}{d\lambda}\frac{dx^\nu}{d\lambda} = 0, \qquad \overline{\Gamma}_{\mu\nu}^\gamma := \frac{1}{2}g^{\gamma\kappa}\left(\partial_\mu g_{\kappa\nu} + \partial_\nu g_{\mu\kappa} - \partial_\kappa g_{\mu\nu}\right)$$

$$(4.11)$$

This is the equation of motion for the particle moving on a timelike geodesic in the curved spacetime. It may be also considered as the demonstration that the Levi-Civita connection $\overline{\Gamma}_{\mu\nu}^\gamma$ is the appropriate connection with this approach. It is conventional to write the geodesic equation (4.11) by means of a differential operator $D/D\tau$:

$$\frac{Du^\gamma}{D\tau} \equiv 0, \quad \text{with} \quad \frac{Dv^\gamma}{D\tau} := \frac{dv^\gamma}{d\tau} + \overline{\Gamma}^\gamma_{\mu\nu} u^\mu v^\nu \tag{4.12}$$

for any vector field v^γ, and where τ is the proper time and $u^\mu := dx^\mu/d\tau$ the four-vector velocity. In classical gravitation, geodesic curves are metric geodesics meaning curves of extremal spacetime interval with respect to the metric, say $ds^2 := g_{\alpha\beta}dx^\alpha dx^\beta = g_{00}(dx^0)^2 - g_{ij}dx^i dx^j$. For classical mechanics, the time can be separated from the space. Owing that $x^0 := t$, this gives the only non zero coefficients, and curvatures

$$\Gamma^i_{00} := \frac{\partial \mathfrak{U}}{\partial x_i}, \quad \mathfrak{R}^i_{j00} = -\mathfrak{R}^i_{0j0} = \frac{\partial^2 \mathfrak{U}}{\partial x_i \partial x^j} \tag{4.13}$$

with all other components are vanishing. This confirms that for a free particle motion, if components $\mathfrak{R}^i_{j00}$ are equal to zero, then the curvature tensor vanishes everywhere e.g. Shen and Moritz (1996). We have two interpretations of the particle motion. The first is a description of the motion under the action of force field (as potential gradient) in a flat Minkowski spacetime. The second description considers a particle moving along a geodesic line in a Riemannian spacetime with curvature $\mathfrak{R}^i_{j00} \neq 0$. This illustrates the equivalence principle of Einstein. In the next chapter, we will consider the problem of the geodesic deviation defined by the acceleration of separation of two nearby material points with a gravitational field.

Remark 4.4 The explicit separation of the metric and the connection is worth in Newtonian gravitation theory. The curvature that bends the spacetime can not expressed in terms of metric tensor which is assumed uniform over the entire spacetime. The geometry of the space (hyperplane) for fixed x^0 is Euclidean. The curvature may rather be directly obtained from the affine connection derived from a scalar potential $\mathfrak{U}(x^\mu)$ e.g. Ehlers (1973), Havas (1964). This is one of the reasons why Einstein–Cartan gravitation theory should be considered as the natural extension of the Newtonian gravity, rather than the Einstein relativistic gravitation theory.

In a flat spacetime let assume the existence of an inertial frame characterized by zero affine connection $\Gamma^\gamma_{\alpha\beta}(\mathbf{x}) \equiv 0$ with $\alpha, \beta, \gamma = 0, 1, 2, 3$ where index 0 corresponds to time coordinate. Consider again the set of internal transformations defined by:

$$y^0 := y^0(x^0, x^1, x^2, x^3), \quad y^i := y^i(x^1, x^2, x^3)$$

where Latin indices hold for $(1, 2, 3)$, and Greek indices for $(0, 1, 2, 3)$. The connection coefficients $\tilde{\Gamma}^\gamma_{\alpha\beta}(y^\mu)$ associated to this new coordinate system $\{y^\mu\}$ are given by:

$$\tilde{\Gamma}^0_{\alpha\beta} = J^0_\lambda A^\lambda_{\alpha\beta}, \quad \tilde{\Gamma}^i_{0\beta} = 0, \quad \tilde{\Gamma}^i_{jk} = J^i_\ell A^\ell_{jk} \tag{4.14}$$

with the same notation as for the change of coordinate ($J_\lambda^\gamma := \partial y^\gamma / \partial x^\lambda$, and $A_{\alpha\beta}^\ell := \partial^2 x^\ell / \partial y^\alpha \partial x^\beta$). Then previous Eq. (4.14) shows that the vanishing of components $\tilde{\Gamma}_{0\beta}^i(x^\mu) \equiv 0$ represents a necessary and sufficient condition for the coordinate system (x^μ, $\mu = 0, 1, 2, 3$) to define an inertial frame of reference Krause (1976). Physical forces (either "external" or those due to acceleration of the relativistic frame of reference) are associated to the deviations from the affine geodesics. Particularly, if it happens that a frame is accelerated with respect to an inertial frame of reference (flat), there is no internal transformations able to eliminate all the components of the connection. Therefore, the difference between classical and relativistic mechanics lies upon the characteristics and properties of the affine connections linking microcosms e.g. Duval and Kunzle (1978). Newton calls gravitation what Einstein called curvature of spacetime. For relativistic gravity theory, gravitation is a property of the curvature of spacetime and is not an "external" forces.

Remark 4.5 Equation (4.13) when particularly $\Gamma_{00}^i \equiv 0$ may be related to Eq. (4.14) in classical mechanics (inertial frame in Galilean mechanics when gravitational fields is missing).

4.2.2.2   Newton–Cartan Spacetime

In the absence of gravitation, the spacetime of Newton physics may be described by a metric $g^{\alpha\beta} := \mathrm{diag}\{0, 1, 1, 1\}$, a vector $\tau^\alpha := (1, 0, 0, 0)$, and a symmetric affine connection $\Gamma_{\alpha\beta}^\gamma$, such that the metric is orthogonal to the vector τ^α. Metric compatibility of the connection $\Gamma_{\alpha\beta}^\gamma$ and of the vector τ^α, together with the vanishing of the curvature induce that there exists a family of coordinate system such that $\Gamma_{\alpha\beta}^\gamma \equiv 0$, identified with the so called reference frames e.g. Goenner (1974). Conversely to Galilean structure, in the formalism of Newton–Cartan, we have seen that the curvature cannot be equal to zero because of gravitation. The spacetime of Newton–Cartan physics is as usual described by a symmetric tensor $g^{\alpha\beta}$, a 1-form τ_α, and a symmetric affine connection $\Gamma_{\alpha\beta}^\gamma$ as for Galilean structure but satisfying the relationships e.g. Dixon (1975):

$$
\begin{cases}
g^{\alpha\beta} \tau_\beta = 0 \\
\nabla_\gamma g^{\alpha\beta} = 0 \quad \text{and} \quad \nabla_\alpha \tau_\beta = 0 \\
g^{\beta\sigma} \Re_{\alpha\sigma\gamma}^\delta = g^{\delta\sigma} \Re_{\gamma\sigma\alpha}^\beta
\end{cases}
\tag{4.15}
$$

The first line expresses the orthogonality condition of the space and time, the second the metric compatibility of the connection, and the third line means that gravitation forces exist (the curvature is not equal to zero). This structure has been developed by Duval et al. (1985) where they have used a extended five-dimensional spacetime together with Bargmann group of invariance which is an extension of the Galilean group. The Bargmann group plays in Newtonian mechanics the same role as Poincaré group invariance in special relativity. Finally, the set of Eqs. (4.26) and (4.15) constitutes the complete fields model resulting from a spatio-temporal distribution of matter density $\rho(x^\mu)$. Let now consider some discrete material points in motion within a Newton–Cartan spacetime. The classical equations of motions (see later (4.26)) keep the same shape under the Galilean group of transformations:

$$y^0 = x^0 + \xi^0, \qquad y^i = J_j^i x^j + v^i x^0 + \xi^i \qquad (4.16)$$

where $x^0 := t$, J_j^i is a constant group element of $SO(3)$ (isotropy of space), v^i a three-dimensional vector of the space (allowing us to define inertial frames from Galilean transformations), and ξ^α a four dimensional vector of the spacetime (homogeneity of space). The set of these transformations constitute a Lie group and provide the 10-parameter Galilean Lie group (Fig. 4.3). Now consider, as example, the motion of particles $\mathcal{P}^{(p)}$ of mass m_p evolving with respect to an inertial frame and defined by the Lagrangian function (Rosen 1972):

$$\mathscr{L}(\mathbf{x}^{(p)}, \dot{\mathbf{x}}^{(p)}, t) := \sum_p^N \frac{m_p}{2} \|\dot{\mathbf{x}}^{(p)}\|^2 - \mathcal{U}\left(\mathbf{x}^{(p)}, t\right), \qquad \mathcal{U} := \sum_{p<q}^N \frac{G m_p m_q}{\|\mathbf{x}^{(p)} - \mathbf{x}^{(q)}\|}$$

$$(4.17)$$

in which $\mathbf{x}^{(p)}$ denotes the Cartesian coordinates of the particle p, and G the gravitation constant. The first term of Lagrangian (4.17) represents the kinetic energy. The dot denotes derivative with respect to $x^0 := t$. Relations (4.13) and (4.17) allow us to calculate the field of connection of the spacetime $\Gamma_{\alpha\beta}^\gamma(x^\mu)$. The form of the Lagrangian function is invariant under the Galilean group of transformations but only in the particular case $v^i \equiv 0$. When $v^i \neq 0$ in the Galilean

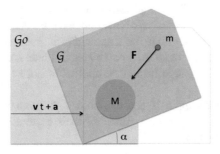

Fig. 4.3 Newton–Cartan spacetime. The Galilean group includes the uniform translation characterized by the constant velocity **v**, the constant displacement **a**, the shift of the time origin t_0, and the three angles α for changing the orientation of the spacetime then from $\mathscr{G}_0$ to $\mathscr{G}$, say $(v^i, \xi^i, \xi^0, J_j^i)$ according to Eq. (4.16). The invariance with respect to this group induces the gauge invariance. The presence of the body of mass M generates changes the Newton spacetime to Newton–Cartan spacetime with gravitation by creating a potential field $\mathcal{U}$ from which the force **F** on test mass m is derived. Newton–Cartan connection and curvature are thus calculated by means of Eq. (4.13), for which the covariance (directly) leads to a dependence only on curvature $\hat{\mathfrak{R}}_{\alpha\beta}$ after Eq. (4.13)

group of transformations (4.16) the Lagrangian takes the form of:

$$\mathscr{L}' = \mathscr{L} + \frac{d}{dt}\left[\sum_p^N m_p\left(\mathbf{v}\cdot\dot{\mathbf{x}}^{(p)} + (1/2)\|\mathbf{v}\|^2 t\right)\right] \tag{4.18}$$

(owing that $x^0 := t$). It shows that this Lagrangian function is not covariant. The Euler–Lagrange equations of this set of particles have exactly the same shape for these two forms of Lagrangian (particular case of Eq. (4.26)), say (*Newton's law of gravitation*):

$$\frac{d^2\mathbf{x}^{(p)}}{dt^2} + G\sum_{q\neq p}^N \frac{m_q\,(\mathbf{x}^{(p)} - \mathbf{x}^{(q)})}{\|\mathbf{x}^{(p)} - \mathbf{x}^{(q)}\|^3} = 0 \tag{4.19}$$

where the second term is the superposition of many vectors (the continuous version is also displayed on the right where $\mathbf{x}$ and $\mathbf{x}'$ are associated to the test mass point position and the mass distribution respectively, and $dm' := \rho(\mathbf{x}')\omega_n(\mathbf{x}')$ is the continuous element of mass which engenders the gravitation):

$$\mathbf{F}_p := -G\sum_{q\neq p}^N \frac{m_q\,(\mathbf{x}^{(p)} - \mathbf{x}^{(q)})}{\|\mathbf{x}^{(p)} - \mathbf{x}^{(q)}\|^3}, \qquad \mathbf{F} = -mG\int_{\mathscr{M}} \frac{\mathbf{x} - \mathbf{x}'}{\|\mathbf{x} - \mathbf{x}'\|^3}\rho(\mathbf{x}')\omega_n(\mathbf{x}')$$

This is the total gravitational force on the point mass m_p at $\mathbf{x}^{(p)}$ due to the set of many discrete points $\{m_q, \mathbf{x}^{(q)}\}$. In Newtonian mechanics, the mass m_p itself does not produce self-gravitation conversely to relativistic gravitation. This illustrates the difference between change of (space)-coordinates, and change of frame of reference. This later is related with the gauge invariance rather than with covariance. This example illustrates the two concepts of invariance we are dealing with in the present paper. We therefore can say that the Lagrangian is gauge-invariant (despite the additional time derivative) under a 10-parameter Galilean Lie group. Hereafter, we consider covariance and gauge invariance applied to both spacetime and material manifolds.

The most known example is the two-bodies problem with two particles of masses m_1 and m_2 subject to only forces of their mutual interaction (4.17). For this particular case, we have:

$$\mathscr{L} = \frac{m_1}{2}\|\mathbf{x}^{(1)}\|^2 + \frac{m_2}{2}\|\mathbf{x}^{(2)}\|^2 - \frac{Gm_1 m_2}{\|\mathbf{x}^{(1)} - \mathbf{x}^{(2)}\|} \tag{4.20}$$

Classically, introduction of the mass center and the relative position allows us to simplify the problem and define the Lagrangian as:

$$\mathscr{L} = \frac{M}{2}\|\dot{\mathbf{X}}\|^2 + \frac{m}{2}\|\dot{\mathbf{x}}\|^2 - \frac{GMm}{\|\mathbf{x}\|} \tag{4.21}$$

with the total mass, the reduced mass, the mass center position, and the relative position respectively:

$$M := m_1 + m_2, \ m := \frac{m_1 m_2}{m_1 + m_2}, \ \mathbf{X} := \frac{m_1 \mathbf{x}^{(1)} + m_2 \mathbf{x}^{(2)}}{m_1 + m_2}, \ \mathbf{x} := \mathbf{x}^{(2)} - \mathbf{x}^{(1)} \quad (4.22)$$

The motion of mass center is a straight line with a constant velocity (it may be easily proven by deriving the Euler–Lagrange equations and by observing that the potential does not depend on the variable $\mathbf{X}$. By choosing a spherical coordinate system (r, θ, φ) centered at the mass center, the Lagrangian holds:

$$\mathscr{L} = \frac{m}{2} \left(\dot{r}^2 + r^2 \dot{\theta}^2 + r^2 \sin^2 \theta \dot{\varphi}^2 \right) - \mathscr{U}(r), \quad \mathscr{U} := -G \frac{Mm}{r}$$

and by observing that the azimuthal angle φ is an ignorable coordinate say $\partial_\varphi \mathscr{L} = 0$, the Lagrangian takes the form of:

$$\frac{d}{dt} \left(\frac{\partial \mathscr{L}}{\partial \dot{\varphi}} \right) = 0 \implies \ell_0 := \frac{\partial \mathscr{L}}{\partial \dot{\varphi}} \implies \mathscr{L} = \frac{m}{2} \dot{r}^2 + \frac{\ell_0^2}{2mr^2} + G \frac{Mm}{r} \tag{4.23}$$

by assuming that the (invariant) angular momentum ℓ_0 is along the z axis at the time $t = 0$ $(\theta = \pi/2 \implies \dot{\theta} = 0)$. Indeed, since the angular momentum is always perpendicular to $\mathbf{x}$ and $\dot{\mathbf{x}}$, the motion of the fictitious point $\mathbf{x}$ is always in a plane perpendicular to the invariant angular momentum. Further simplification may be thus done by writing the conservation of the energy:

$$\mathscr{E}_0 := \frac{m}{2} \dot{r}^2 + \underbrace{\frac{\ell_0^2}{2mr^2} - G \frac{Mm}{r}}_{\text{effective potential}} = \text{Cst} \tag{4.24}$$

The tridimensional problem reduces to a one-dimensional one by considering the motion of the mass center with a reduced mass m, and subject to an effective potential in place of the Newtonian potential. Classical solutions of this problem allow us to analyze the various conic orbits either closed (elliptic) and open as parabolic and hyperbolic. The Kepler laws of planets are obtained accordingly e.g. Ryder (2009).

4.2.2.3 Newtonian Gravitation in Presence on Continuum Matter

We previously introduced the vacuous Newtonian spacetime by means of the contravariant metric $\hat{h}^{\alpha\beta}$, the covariant vector $\hat{\tau}_\alpha$, and the symmetric connection $\hat{\Gamma}^\gamma_{\alpha\beta}$. A zero curvature $\hat{\Re}^\gamma_{\alpha\beta\lambda} \equiv 0$ ensures a flat spacetime and the existence of inertial frames of references. In the presence of gravitation, covariant formulation of the Newtonian gravitation was first developed by Cartan to give the following relations e.g. Dixon (1975):

$$\hat{\mathfrak{R}}_{\alpha\beta} = 4\pi G\rho \; \hat{\tau}_\alpha \hat{\tau}_\beta, \qquad \hat{h}^{\gamma\sigma} \hat{\mathfrak{R}}^{\delta}_{\alpha\beta\sigma} = 0, \qquad \hat{h}^{\gamma\sigma} \hat{\mathfrak{R}}^{\delta}_{\alpha\sigma\lambda} = \hat{h}^{\delta\sigma} \hat{\mathfrak{R}}^{\beta}_{\lambda\sigma\alpha} \qquad (4.25)$$

with the Ricci tensor $\hat{\mathfrak{R}}_{\alpha\beta}$ of the curved spacetime due to the presence of matter with density $\rho(\mathbf{x})$. The raising of indices is not reversible since the metric is singular (and thus not invertible). The concept of the inertial frames of reference is cautious (inertial spacetime) and exists if only the curved spacetime is asymptotically flat. The set of Eq. (4.25) constitute a viable extension of the Newtonian theory of gravitation e.g. Dixon (1975). The uniqueness of this set of equations to describe Newtonian gravitation as a manifestation of the spacetime geometry, without restrictions on the mass density was shown by using the representation theory of general linear and orthogonal groups of transformations. We will give explicit examples of wave propagation within non homogeneous continuum in the last section of this paper. By calculating the Ricci curvature from contraction, we arrive to the classical Poisson equation for Newtonian gravitation $\mathfrak{R}_{00} := \Delta \mathcal{U} = 4\pi G\rho$ where ρ. Both Newtonian and relativistic gravity involve a privileged state of motion represented by affine connections in a four dimensional spacetime manifold e.g. Bernal and Sanchez (2003), Dixon (1975), Kadianakis (1996), Krause (1976). In sum, Newtonian gravity consists of two equations. The first comes from Newton's law of gravity as previous developed, whereas the second derives from the Newton's second law of motion (local Inertia Principle):

$$\Delta \mathcal{U} = 4\pi G\rho, \qquad \frac{d^2 x_i}{dt^2} = -\nabla_i \mathcal{U} \qquad (4.26)$$

where the Laplacian Δ and the gradient ∇ operators are obtained by using the Euclidean spacetime connection (torsionless and non curved manifold). Considering a spherical continuum matter, the gravitational potential $\mathcal{U} := (GM)/r$ is solution of Eq. (4.26a). The second equation (4.26b) describes how a material point (whose gravitational effects can be neglected) moves in the gravitational field provoked by the sphere. In the framework of classical mechanics limit, the geodesic equation and the Einstein's equations tend to the equations of Newton gravity, and to the Newton's second laws of motion (Fundamental Principle of Dynamics).

Remark 4.6 In a Newtonian–Cartan gravitation, let consider a body $\mathscr{B}$ of mass m immersed within an field of gravitation with total potential $\mathcal{U}_T$. The body $\mathscr{B}$ itself has itself a gravitation potential $\mathcal{U}_B$. Let define the "free fall" of the center of mass as the motion induced by the potential $\mathcal{U} := \mathcal{U}_T - \mathcal{U}_G$. Then the difference of the acceleration experienced by the center of mass and the "free fall" acceleration is bounded by the size of $\mathscr{B}$ and the variation of external field $\nabla\nabla\mathcal{U}$. Ehlers and Geroch have shown the extension of this theorem for relativistic gravitation in Ehlers and Geroch (2004).

For illustrating the fundamental solution of the gravitational field in Newtonian mechanics, let consider the gravitation field generated by a massive point $m > 0$

et let solve Eq. (4.26) assuming a spherical symmetric problem. In some sense, we give support for the shape of the potential energy in the Lagrangian (4.17). For a material point, the field equation (4.26) holds:

$$\Delta \mathcal{U} = 4\pi\, Gm\, \delta(\mathbf{x})$$

Integrating over a sphere $\mathscr{B}$ of center O and radius R, we have:

$$\int_{\mathscr{B}} 4\pi\, Gm\, \delta(\mathbf{x})dv = 4\pi\, Gm = \int_{\mathscr{B}} \nabla \cdot (\nabla \mathcal{U})\, dv = \int_{\partial\mathscr{B}} \nabla\mathcal{U} \cdot \mathbf{n}da$$

Since the problem is spherically symmetric, the potential takes the form of: $\mathcal{U} = U(r)$. Then we deduce:

$$4\pi\, Gm = U'(R)\mathcal{A} = 4\pi\, R^2 U'(R) \qquad \Longrightarrow \qquad U'(R) = G\frac{m}{R^2}$$

This is valid for any value of R, then, the Newton potential for gravitation of a material point of mass m, and the linear momentum equation of a mass M in the gravitational field of m, are given by:

$$\mathcal{U} = -G\frac{m}{r}, \qquad M\mathbf{a} := -M\nabla\mathcal{U} = -G\frac{Mm}{r^3}\mathbf{r} \tag{4.27}$$

where we have assumed that the potential vanishes for $r \to \infty$, and $\mathbf{r}$ denotes the vector position of mass M with respect to the origin where the attractive mass m is situated.

Remark 4.7 The second equation of (4.27) indeed expresses the idea of Newton's Universal Law of Gravitation. It states that any two objects of mass M and m exert a gravitational force of attraction on each other (here M is attracted by m). The direction of the force is along the line joining the two objects. The magnitude of the force $|\mathbf{F}\|$ is proportional to the product of the gravitational masses M an m of the objects, and inversely proportional to the square of the distance r between them.

4.2.3   Matter Within Torsionless Gravitation

Let us consider a torsionless but curved spacetime manifold $(\mathcal{M}, \mathbf{g}, \nabla)$ which is slightly more general than the Newton–Cartan spacetime. Such a model is considered as a "structureless continuum" if the torsion vanishes. The spacetime structure is summarized by the Riemannian spacetime structure with the metric $g^{\alpha\beta}(x^\mu)$, the torsion $\aleph^\gamma_{\alpha\beta}$, and the curvature $\mathfrak{R}^\delta_{\alpha\sigma\gamma}(x^\mu)$:

$$\nabla_\gamma g^{\alpha\beta} = 0, \quad \aleph^\gamma_{\alpha\beta} \equiv 0, \quad \mathfrak{R}^\delta_{\alpha\sigma\gamma} \neq 0 \tag{4.28}$$

The Lagrangian function $\mathscr{L}(\mathbf{g}, \nabla, \nabla\nabla)$ depends on the metric, the connection and the bi-connection in the general case. We consider first the corollary of covariance theorem by considering Levi-Civita connection. Under the same hypothesis as for the Theorem 3.6, we have the corollary:

Corollary 4.2 *On Riemannian manifold, there does not exist a covariant scalar field $\mathscr{L}$ (form-invariant) which depends on the metric $\mathbf{g}$ and the Levi-Civita connection $\overline{\nabla}$, say $\mathscr{L}(\mathbf{g}, \overline{\nabla})$.*

Proof According to the Theorems 2.1 and 3.6: $\mathscr{L}(\mathbf{g}, \overline{\nabla}) = \mathscr{L}(\mathbf{g}, \aleph = 0) = \mathscr{L}(\mathbf{g})$.

$\square$

Corollary 4.2 can be proven by using components formulation. Indeed, the coefficients of Levi-Civita connection are Christoffel symbols $\overline{\Gamma}^k_{ij}$ (2.41) which depend on the metric components and their first partial derivatives e.g. Nakahara (1996). According to Theorem 3.5: $\mathscr{L}(g_{ij}, \overline{\Gamma}^k_{ij}) = \mathscr{L}(g_{ij}, \partial_k g_{ij}) = \mathscr{L}(g_{ij})$. It may be related to the existence of local inertial frame of reference in general relativity—with Riemannian spacetime—where the coefficients of the connection may be set to zero in a normal coordinates system (strong equivalence principle e.g. Knox 2013). This also conforms to results obtained by Lovelock (1971) where the Lagrangian density $\mathscr{L}(g_{ij}, \partial_k g_{ij}, \partial_l \partial_k g_{ij})$ with a Levi-Civita connection $\overline{\Gamma}^k_{ij}$, operating on the manifold, leads to the modified gravity Lagrangian density $\mathscr{L}_G(g_{ij}, \overline{\Re}^k_{ijl})$ e.g. indirectly Cartan (1922), or directly Exirifard and Sheikh-Jabbari (2008). The relevant connection for Einstein relativistic gravitation is the Levi-Civita connection, with a non zero curvature describing the gravitational field. Moreover it constitutes the only Lorentz connection with zero torsion.

Remark 4.8 In continuum mechanics, further extension consists in using an affine connection which does not derive from a metric but compatible, and thus to introduce torsion and/or curvature to describe the dislocations and disclinations field e.g. Maugin (1993), Rakotomanana (2003). Although commonly adopted for modeling continuum, Levi-Civita connection $\overline{\nabla}$ is not worth in such a framework.

4.2.3.1 Einstein's Field Equation

Let consider a Lagrangian $\mathscr{L}(g_{\alpha\beta}, \overline{\Gamma}^\gamma_{\alpha\beta}, \partial_\lambda \overline{\Gamma}^\gamma_{\alpha\beta})$ corresponding to the Einstein–Hilbert action of classical relativistic gravitation e.g. Sotiriou and Faraoni (2010):

$$\mathscr{S}_G := (1/2\chi) \int \mathcal{R} \sqrt{\mathrm{Detg}}\, dx^0 \wedge \cdots \wedge dx^3 \tag{4.29}$$

where $\mathcal{R}$ is the scalar curvature. The constant factor is introduced to reproduce the classical mechanics of Newton when some matter is moving within this vacuum spacetime. For the variational formulation we introduce the metric variation $g_{\alpha\beta} \rightarrow g_{\alpha\beta} + \delta g_{\alpha\beta}$ (corresponding to the Eulerian variation of the metric at a fixed point of the spacetime). A straightforward calculus gives the variation of the connection and the Ricci curvature (we omit bar overline for connection and curvature for the sake of the notation simplicity):

$$\delta\Gamma^{\gamma}_{\alpha\beta} = (1/2)\, g^{\gamma\lambda}\left(\nabla_{\beta}\delta g_{\alpha\lambda} + \nabla_{\alpha}\delta g_{\lambda\beta} - \nabla_{\lambda}\delta g_{\alpha\beta}\right) \tag{4.30}$$

$$\delta\Re_{\alpha\beta} = \nabla_{\lambda}(\delta\Gamma^{\lambda}_{\beta\alpha}) - \nabla_{\beta}(\delta\Gamma^{\lambda}_{\lambda\alpha}) \tag{4.31}$$

where the covariant derivative is related to the unperturbed (metric compatible) connection. The equation of (4.31) is known as Palatini identity, showing that for torsionless continuum the variation of the Ricci tensor may be transferred to the boundary condition terms via the divergence theorem. Now we are concerned to derive the field equations. Notice $\chi := 8\pi G/c^4$. By writing the variation:

$$\delta\mathscr{L}_G = (1/2\chi)\left[\delta\left(\sqrt{\mathrm{Detg}}\right)g^{\alpha\beta}\Re_{\alpha\beta} + \sqrt{\mathrm{Detg}}\,\delta g^{\alpha\beta}\Re_{\alpha\beta} + \sqrt{\mathrm{Detg}}\,g^{\alpha\beta}\delta\Re_{\alpha\beta}\right]$$

where we have, thanks to the Palatini identity and the metric compatibility of ∇:

$$\sqrt{\mathrm{Detg}}\,g^{\alpha\beta}\,\delta\Re_{\alpha\beta} = \sqrt{\mathrm{Detg}}\left[\nabla_{\lambda}\left(g^{\alpha\beta}\delta\Gamma^{\lambda}_{\beta\alpha}\right) - \nabla_{\beta}\left(g^{\alpha\beta}\delta\Gamma^{\lambda}_{\lambda\alpha}\right)\right]$$

Integrating the divergence-like terms under brackets along the manifold boundary gives a zero term. Since $\delta\left(\sqrt{\mathrm{Detg}}\right) = -(1/2)\sqrt{\mathrm{Detg}}\,g_{\alpha\beta}\,\delta g^{\alpha\beta}$, the principle of least action $\delta\mathscr{L}_G \equiv 0$ for arbitrary variation of the metric gives the Einstein field equation of the general relativity e.g. Cartan (1922) (GR):

$$G_{\alpha\beta} := \Re_{\alpha\beta} - (1/2)\,\mathcal{R}\,g_{\alpha\beta} = 0 \tag{4.32}$$

which is the vacuum fields equation, the Euler–Lagrange equations associated to the Einstein–Hilbert action. Metric components $g_{\alpha\beta}(x^{\mu})$ are the unknown variables in this field equation. To this end, given a curvature field $\Re^{\gamma}_{\alpha\beta\lambda}(x^{\mu})$ on a metric manifold $\mathscr{M}$, determination of the 10 metric components $g_{\alpha\beta}(x^{\mu})$ needs integration of system of 20 second-order partial differential equations. For this to be possible, additional integrability must be satisfied for third-order derivatives of metric components assuming that they are $\mathcal{C}^3$. These conditions are the Bianchi identities e.g. Lovelock and Rund (1975), Nakahara (1996).

Remark 4.9 First, the derivation of the field equation of general relativity, is obtained accounting that the variation of the connection $\delta\Gamma^{\gamma}_{\alpha\beta}$ is shifted to the boundary by means of the divergence operator. Second, contrarily to the wave equation (2.13), the Einstein's gravitational equations are invariant under the group of (passive) diffeomorphisms (covariance), and not only for the group of Lorentz transformations. Under an arbitrary change of coordinate $\tilde{x}^{\alpha} = \tilde{x}^{\alpha}(x^{\mu})$, the same shape of equations is obtained to give exactly $\hat{G}_{\alpha\beta} = 0$.

4.2.3.2   Example: Schwarzschild Spherical Symmetric Spacetime

The most known example of spherical symmetric solution of vacuous spacetime
is the Schwarzschild spacetime. It is a good approximation of gravitation field produced by static spherically symmetric body at rest. The static condition imposes that
the metric components $\partial_0 \hat{g}_{\mu\nu} \equiv 0$ are independent of time coordinate x^0. The time
reversal independence $x^0 \rightarrow -x^0$ also induces that all off-diagonal terms vanish,
$\hat{g}_{0i} = \hat{g}_{i0} \equiv 0$. Without going in details, proofs are available in classical textbooks
e.g. Ryder (2009). The covariant components of the metric hold in the spherical
coordinate system (x^0, r, θ, ϕ): $\hat{g}_{\mu\nu} = \text{diag}\left\{e^{2f(r)}, -e^{2g(r)}, -r^2, -r^2 \sin^2 \theta\right\}$ in
which the two unknown functions $f(r)$ and $g(r)$ may be found by means of the
vacuum equations $\hat{\Re}_{\mu\nu} \equiv 0$. From the expression of the Christoffel symbols and
then the associated Ricci curvature, we obtain first $f(r) + g(r) = 0$, and by imposing
that the field of gravitation approaches the Minkowski space time at the infinity
$(r \rightarrow \infty)$, we get the Schwarzschild metric:

$$\hat{g}_{\mu\nu} = \text{diag}\left\{(1 - 2m/r), -(1 - 2m/r)^{-1}, -r^2, -r^2 \sin^2 \theta\right\} \tag{4.33}$$

with $2m := 2MG/c^2$ a constant called Schwarzschild radius. The details of the
proof will be reminded in the Chap. 6 to obtain the metric (6.97). This solution
holds for spacetime outside a body of (total) gravitating mass M. It conforms to
the Newtonian spacetime where the distribution of mass inside the gravitating body
does not play keyrole, only the total mass matters.

Remark 4.10 (Laplace "Black Hole", 1798) The Schwarzschild radius was
observed by Laplace in the context of Newtonian gravitation. Considering particle
of mass m moving at the light speed c, then with a kinetic energy $mc^2/2$, the
particle will not escape from the attraction of a massive body M with a Newtonian
potential mGM/r if potential is greater than the kinetic energy. This implies that
$R > 2GM/c^2$ e.g. Ryder (2009).

Remark 4.11 It was stated (Birkhoff theorem, 1923), see the book of Birkhoff
and Langer (1923) that, even for dynamic metric, the Schwarzschild spherical
metric is the only symmetric asymptotically flat spacetime solutions of Einstein's
vacuum equations. A consequence is that no gravitational waves can occur from
spherical pulsation of stars. Another consequence of Birkhoff's theorem is that
for a spherically symmetric thin shell (matter which is source of gravitation), the
interior spacetime must be Minkowskian with metric $g_{\mu\nu} := \text{diag}\{+1, -1, -1, -1\}$
meaning that the gravitational field must vanish inside a spherically symmetric shell.

The physics of black holes comes from the apparent singularity at $r = 2m$.
We observe that the signature of the metric is different inside $(-, +, -, -)$ and
outside $(+, -, -, -)$ the critical radius $r_c = 2m$. Eddington and Finkelstein
introduced (separately) a new coordinate system by defining a time coordinate
$d\bar{t} = dt + 2m \, dr/[c(r - 2m)]$, and suggested the metric components in the

Eddington-Finkelstein coordinate system $(\overline{x}^0 := c\overline{t}, r, \theta, \phi)$:

$$
\hat{g}_{\mu\nu} = \begin{bmatrix} (1 - 2m/r) & 4m/r & 0 & 0 \\ 4m/r & -(1 - 2m/r) & 0 & 0 \\ 0 & 0 & -r^2 & 0 \\ 0 & 0 & 0 & -r^2 \sin^2\theta \end{bmatrix} \tag{4.34}
$$

This ranges the radial coordinate solution for $0 < r < \infty$. Other coordinates system may be found in the literature e.g. Ryder (2009).

Remark 4.12 For either Schwarzschild or Eddington-Finkelstein metrics, the physical importance of this singularity radius may be neglected for some practical situation. For example, the mass of the Sun is $M \simeq 1.99 \times 10^{30}$[kg], giving the radius $m := r_S \simeq 1500$[m], showing that the Schwarzschild spherical surface of radius r_S is inside the Sun itself. Therefore, for regions outside the Sun at $r > 6.96 \times 10^8$[m], the vacuum solutions hold.

Remark 4.13 The metric (4.33), due to Hilbert, is not the original Schwarzschild solution which was calculated for a mass-point situated at the origin $r = 0$. The generalized Schwarzschild metric takes the form of:

$$
\hat{g}_{\mu\nu} = \text{diag} \left\{ (1 - 2m/R(r)), -(1 - 2m/R(r))^{-1}, -R(r)^2, -R(r)^2 \sin^2\theta \right\} \tag{4.35}
$$

where $R(r) := [r^3 + r_0^3]^{1/3}$ and r_0 is an integration constant obtained from boundary condition on the function $R(r)$. The Hilbert version and the Schwarzschild original solution are obtained for $r_0 = 0$ and $r := 2m$ respectively. Conversely to the Hilbert version, the original solution adapted for mass-point at the origin does not present singularity except for $r = 0$. The two versions are independent which means that there is no coordinate transform that can reduce one each other.

4.2.3.3 Example: Robertson-Walker Isotropic and Homogeneous Spacetime

The most known metric tensor underlying cosmological spacetime $\mathscr{C}$ is that of Robertson-Walker. The metric must account for the fundamental principle of cosmology: isotropy and homogeneity. The cosmological model are based on the assumptions:

- the spacetime $\mathscr{C}$ can be sliced into hypersurfaces (spaces) of constant time which are homogeneous and isotropic;
- there is a mean rest frame of all the galaxies that agrees with this definition of the simultaneity.

The spacetime $\mathscr{C}$ was developed to account for the expansion of the universe—the change of proper distance between galaxies—by means of the dependence of all components of the metric on the variable time. On the one hand, to ensure that all metric components increase at the same rate, the line element of the space may be written as:

$$d\ell^2 = a^2(x^0)g_{ij}dx^i dx^j$$

On the other hand, the possibility of slicing (first assumption) allows us to separate the time and the space to obtain the following shape of the line element:

$$ds^2 = \left(dx^0\right)^2 - a^2(x^0)g_{ij}dx^i dx^j$$

owing that the cross-components must vanish $g_{0i} \equiv 0, i = 1, 2, 3$ to satisfy the isotropy of the spacetime, or in other words to preclude the existence of privileged direction $\mathbf{e}_i$. The isotropy allows us also to assume that the metric has spherical symmetry about the origin of coordinates, which can be chosen at any point of $\mathscr{C}$. Accordingly the space line element should take the form of:

$$d\ell^2 := e^{2\lambda(r)}dr^2 + r^2 d\theta^2 + r^2 \sin^2\theta d\varphi^2 \tag{4.36}$$

where only the component g_{rr} may depend on the radius r by means of the scalar function $\lambda(r)$. To ensure the homogeneity of the spacetime it is necessary that the Ricci curvature of the three-dimensional space has the same value at every space point. A straightforward calculus gives the three non zero components of the Einstein tensor:

$$G_{rr} = \frac{1}{r^2}e^{2\lambda(r)}\left(1 - e^{-2\lambda(r)}\right), \quad G_{\theta\theta} = r\lambda'(r)e^{-2\lambda(r)}, \quad G_{\varphi\varphi} = \sin^2\theta\, G_{\theta\theta} \tag{4.37}$$

The homogeneity of the space means that the trace of Einstein tensor (space part) is constant:

$$\mathbf{Tr}\mathbf{G} := g^{ij}G_{ij} = \frac{1}{r^2}\left(1 - (re^{-2\lambda(r)})'\right) = G_0 \tag{4.38}$$

The integration of this differential equations gives the solution:

$$g_{rr} = e^{2\lambda(r)} = \frac{1}{1 - kr^2 - H/r} \tag{4.39}$$

where H is a constant of integration, and $k := G_0/3$ is the conventional curvature constant. Requirement of local flatness at the origin $r = 0$ induces $H = 0$. It can then be checked that the spacetime metric, denoted Robertson-Walker metric, obtained from this space metric is isotropic and homogeneous. The corresponding line element holds e.g. Ryder (2009):

$$ds^2 := \left(dx^0\right)^2 - a^2(x^0)\left(\frac{dr^2}{1-kr^2} + r^2d\theta^2 + r^2\sin^2\theta d\varphi^2\right) \tag{4.40}$$

where the curvature k may be positive, negative or zero. For $k = 0$, we obtain the flat Robertson-Walker spacetime. For $k > 0$, the model is called closed or spherical Robertson-Walker spacetime, and finally for $k < 0$, the spacetime is open or hyperbolic.

4.2.3.4   Cosmological Constant and (Anti)-de Sitter Spacetime

It is well known that the concept of dark energy, together with that of the dark matter, is not described by the Einstein's general relativity with the field equation (4.32). The accounting of the dark energy allows us to include the accelerated expansion of the universe, discovered by Hubble, by adding a cosmological constant Λ e.g. Padmanabhan (2003), Ryder (2009). The value of this cosmological constant is very low with respect to human length scale $\Lambda \simeq \leq 10^{-52}[\text{m}^{-2}]$. The Einstein–Hilbert action becomes

$$\mathscr{S}_{EHH} := (1/2\chi)\int (\mathcal{R} - 2\Lambda)\sqrt{\text{Detg}}\,dx^0 \wedge \cdots \wedge dx^3 \tag{4.41}$$

where the integration goes over the entire spacetime $\mathcal{M}$ with metric $\mathbf{g}$. As previously, the variation of the action $\mathscr{S}_{EHH}$ with respect to the metric leads to the equation:

$$\Delta\mathscr{S}_{EHH} := (1/2\chi)\int \left(\mathfrak{R}_{\alpha\beta} - \frac{\mathcal{R}}{2}\,g_{\alpha\beta} + \Lambda\,g_{\alpha\beta}\right)\sqrt{\text{Detg}}\Delta g^{\alpha\beta}dx^0 \wedge \cdots \wedge dx^3 = 0$$

where we dropped the boundary terms that are assumed to vanish for the sake of the simplicity. We deduce the associated field equation allowing us to take into account the universe expansion:

$$G_{\alpha\beta} + \Lambda\,g_{\alpha\beta} := \mathfrak{R}_{\alpha\beta} - (1/2)\,\mathcal{R}\,g_{\alpha\beta} + \Lambda\,g_{\alpha\beta} = 0 \tag{4.42}$$

The cosmological constant Λ is sometimes associated to "dark energy" of the universe and constitutes a new constant of the general relativity. From mathematical point of view, this constant may be positive or negative leading to de Sitter spacetime and to Anti-de Sitter spacetime respectively. The maximally spherical symmetric solutions of the field equation (4.42) gives the Schwarzschild-(Anti)-de Sitter metric:

$$ds^2 = \left(1 - \frac{2m}{r} - \frac{\Lambda}{3}r^2\right)(dx^0)^2 - \left(1 - \frac{2m}{r} - \frac{\Lambda}{3}r^2\right)^{-1}dr^2 - r^2d\theta^2 - r^2\sin^2\theta d\varphi^2 \tag{4.43}$$

where the mass m is located at the origin of the coordinate system, and we observe again that the metric is stationary. Let consider the (Anti)-de Sitter spacetime with $m = 0$. By defining the length $\ell := \sqrt{3\,|A|}$, we rewrite the metric (4.43) as:

$$ds^2 = \left(1 \pm \frac{r^2}{\ell^2}\right)(dx^0)^2 - \left(1 \pm \frac{r^2}{\ell^2}\right)^{-1} dr^2 - r^2 d\theta^2 - r^2 \sin^2\theta\, d\varphi^2 \qquad (4.44)$$

When the radius takes the particular value $r = \ell$ then there is degeneracy of the metric in the case of de Sitter metric, this value is denoted the horizon of the de Sitter spacetime. No degeneracy is observed for the Anti-de Sitter metric. We will investigate latter the influence of the torsion in the derivation of the spacetime metric. In the next following let consider some examples of matter within a curved (with gravitation) but spacetime without torsion.

Remark 4.14 Let us consider the action $\mathscr{S}$ including that of the spacetime $\mathscr{M}$ and that of some matter $\mathscr{B}$ evolving within the spacetime and write:

$$\mathscr{S}_T := \mathscr{S}_{ST} + \mathscr{S}_M = \frac{c^4}{16\pi G} \int_{\mathscr{M}} (\mathcal{R} - 2A)\,\omega_n + \int_{\mathscr{B}} \mathscr{L}_M \omega_n$$

win which we can assume that the domain of integration are the same $\mathscr{M} = \mathscr{B}$ without losing the generality of the purpose. Therefore we can group the cosmological constant with the matter to give:

$$\mathscr{S}_T := \mathscr{S}_{ST} + \mathscr{S}_M = \frac{c^4}{16\pi G} \int_{\mathscr{M}} \mathcal{R}\,\omega_n + \int_{\mathscr{M}} \left(\mathscr{L}_M - \frac{c^4}{8\pi G} A\right)\omega_n$$

Of course the variation of this action with respect to the spacetime metric exactly leads to the Einstein field equation with the cosmological constant term. The physical interpretation might be slightly modified to consider the Lagrangian $\mathscr{L}_M - c^4 A/8\pi G$ as a new Lagrangian including the cosmological term attributed as the Lagrangian of the "dark matter".

4.2.3.5 Dust Matter

The simplest example of matter would be the dust defined by the Lagrangian $\mathscr{S}_M := -\int \rho c \sqrt{u_\mu u^\mu}\,\omega_n$ e.g. Dirac (1974). This action is the analogous of the kinetic energy for non relativistic mechanics. We use temporarily the coordinates $(x^0 := t, x^1, x^2, x^3)$. When considering matter which is assumed reasonably to interact with gravity field through the energy-momentum tensor, this Lagrangian furnishes the equations of motion of the dust matter in the gravitation spacetime:

$$\mathfrak{R}_{\alpha\beta} - (1/2)\,(\mathcal{R} - 2A)\,g_{\alpha\beta} = (8\pi G/c^4)\,\rho\,u_\alpha u_\beta$$

where the presence of the stress-energy tensor in the right hand side of the equations is merely due to the variation of the action of the "dust" due to the variation of the spacetime metric. For dust matter evolving within a curved spacetime, the action finally takes the form of:

$$\mathscr{S} := \frac{c^4}{16\pi G} \int_{\mathscr{M}} \mathscr{R} \, \omega_n - \int_{\mathscr{M}} \rho c \sqrt{u^\alpha u_\alpha} \, \omega_n \tag{4.45}$$

where the integration is defined on the spacetime $\mathscr{M}$. The first term in Eq. (4.45) represents the spacetime gravitation whereas the second one defines the dust matter action, according to Dirac. We remind that $\omega_n := \sqrt{\text{Detg}} \, dx^0 \wedge dx^1 \wedge dx^2 \wedge dx^3$ and $dx^0 := dt$. To derive the field equation, it is worth to define the 4-momentum density as $p^\alpha := \rho u^\alpha \sqrt{\text{Detg}}$. This allows us to rewrite the action:

$$\mathscr{S} := \frac{c^4}{16\pi G} \int_{\mathscr{M}} \mathscr{R} \sqrt{\text{Detg}} \, dx^0 \wedge dx^1 \wedge dx^2 \wedge dx^3$$
$$- \int_{\mathscr{M}} c \sqrt{g^{\alpha\beta} p_\alpha p_\beta} \, dx^0 \wedge dx^1 \wedge dx^2 \wedge dx^3$$

(remind that $g^{\alpha\beta}$ are components of the spacetime metric tensor). The variation of the action with respect to the metric leads to the field equation of dust:

$$\mathfrak{R}_{\alpha\beta} - (1/2)\mathscr{R} \, g_{\alpha\beta} = (8\pi G/c^4) \, \rho \, u_\alpha u_\beta \tag{4.46}$$

where we remind the stress-energy tensor as:

$$T_{\alpha\beta} := -\frac{2}{\sqrt{\text{Detg}}} \frac{\partial}{\partial g_{\alpha\beta}} \left(\sqrt{\text{Detg}} \, \mathscr{L}_D \right) = \rho \, u_\alpha u_\beta$$

where $\mathscr{L}_D$ being the dust Lagrangian function. For more details on dust matter, the energy-momentum for dust is given by the $(2, 0)$ type tensor (contravariant components) $T^{\alpha\beta} := \rho_0 \, u^\alpha u^\beta$ where ρ_0 is the mass density in a rest frame. We respectively obtain the components:

$$\begin{cases} T^{00} = \rho_0 u^0 u^0 = \rho & \text{(energy density)} \\ T^{0i} = \rho_0 u^0 u^i = \rho_0 \frac{1}{c^2} \frac{dx^0}{d\tau} \frac{d^i}{d\tau} = \rho \frac{v^i}{c} & \text{(momentum density along i)} \\ T^{ij} = \rho_0 \frac{1}{c^2} \frac{dx^i}{d\tau} \frac{dx^j}{d\tau} = \rho \frac{v^i v^j}{c^2} & \text{(generalized stress tensor)} \end{cases} \tag{4.47}$$

where ρ is the mass density in a moving frame. For non-relativistic situation where $c \gg v^i$ the dominant component is the term $T^{00} = \rho \simeq \rho_0$, which is the matter density in space. A classical approach to obtain equations of motion for dust, we remind the basic steps as follows:

$$\delta \mathscr{S} = \delta \left(\frac{c^4}{16\pi G} \int_{\mathscr{M}} \mathscr{R} \, \omega_n - \int_{\mathscr{M}} \rho c \sqrt{u_\alpha u^\alpha} \, \omega_n \right)$$

which leads to the following variation, by dropping all terms of boundary conditions,

$$\delta \mathscr{S} = \int_{\mathscr{M}} \frac{c^4}{16\pi G} \left(\mathfrak{R}^{\alpha\beta} - \frac{\mathscr{R}}{2} g^{\alpha\beta} \right) \delta g_{\alpha\beta} \, \omega_n - \int_{\mathscr{M}} \frac{\rho}{2} u^\alpha u^\beta \, \delta g_{\alpha\beta} \, \omega_n + \text{B.C. terms}$$

where "B.C." holds for boundary conditions. We remind that $g^{\alpha\beta}$ are components of the spacetime metric in this previous equation. Now if we introduce the "variation" $\delta g_{\alpha\beta} = \nabla_\alpha \xi_\beta + \nabla_\beta \xi_\alpha$, we obtain the equations of motion:

$$\nabla_\beta \left[\left(\mathfrak{R}^{\alpha\beta} - (\mathscr{R}/2) g^{\alpha\beta} \right) - (8\pi G/c^4) \rho u^\alpha u^\beta \right] = 0 \tag{4.48}$$

In fact, the Lagrangian formulation based on the previous action is not the unique possibility to derive the dust motion within a relativistic gravitation spacetime. Such is the case for dust with viscous dissipation effects although is not always possible to include dissipative action in a Lagrangian function. It should be remarked that the four-dimensional vector u^α, according to Eq. (2.77), stands for velocity with respect to proper time (2.76) and not for displacement vector.

Remark 4.15 Without difficulty, the extension of the previous equations to matter within a (torsionless) relativistic gravitation takes the form of:

$$\nabla_\beta \left[\left(\mathfrak{R}^{\alpha\beta} - (\mathscr{R}/2) g^{\alpha\beta} \right) - \chi \, T^{\alpha\beta} \right] = 0 \tag{4.49}$$

where the first terms expresses the gravitation field (spacetime bending) whereas the second one gives the matter contribution (corresponding to stress for classical continuum mechanics).

In the framework of relativistic gravitation without torsion, we can remind the Bianchi identity to show that e.g. Ryder (2009):

$$\nabla_\beta \left[\left(\mathfrak{R}^{\alpha\beta} - (\mathscr{R}/2) g^{\alpha\beta} \right) \right] \equiv 0 \tag{4.50}$$

We deduce the well-known generic equations of motion for matter in a spacetime with gravitation: $\nabla_\beta \left(\chi \, T^{\alpha\beta} \right) = 0$. Provided that the χ is uniform, we arrive at the conservation laws for dust (and more precisely dust-fluid):

$$\nabla_\beta T^{\alpha\beta} = 0 \quad \Longrightarrow \quad \begin{cases} (\nabla_\alpha \rho) u^\alpha + \rho \nabla_\alpha u^\alpha = 0 \\ u^\beta \nabla_\beta u^\alpha = 0 \end{cases} \tag{4.51}$$

which are the classical continuity equation and the geodesic equation, respectively.

Remark 4.16 In the non relativistic approximation we expect to recover the classical conservation laws. Here for $\alpha = 0$ we obtain the mass conservation:

$$\nabla_0 T^{00} + \nabla_i T^{0i} = (1/c) \, \partial_t \rho + (1/c) \, \nabla_i (\rho v^i) = 0 \Longrightarrow \partial_t \rho + \mathrm{div}(\rho \mathbf{v}) = 0$$

(Euler description). For $\alpha = i$, we obtain the conservation of linear momentum:

$$\nabla_0 T^{i0} + \nabla_j T^{ij} = (1/c) \partial_t (\rho v^i / c)) + (1/c^2) \nabla_j (\rho v^i v^j)$$
$$= 0 \Longrightarrow \partial_t (\rho v^i) + \mathrm{div}(\rho v^i \mathbf{v}) = 0$$

The four equations are appropriate for dust matter on which no external forces act and furthermore the particles have non self interactions.

4.2.3.6   Perfect Barotropic Fluid

For perfect fluids, the same developments hold e.g. Taub (1954). For the particular case of barotropic fluid such as in Minazzoli and Karko (2012), the fluid action together with gravity is defined by Schutz (1970):

$$\mathscr{S} := \frac{c^4}{16\pi G} \int_{\mathscr{M}} \mathcal{R} \, \omega_n - \int_{\mathscr{B}} \rho \left(c^2 + \int \frac{P(\rho)}{\rho^2} d\rho \right) \omega_n \qquad (4.52)$$

where $\mathscr{M}$ is the spacetime and $\mathscr{B}$ the continuum matter. Starting from the definition of the stress-energy tensor, we first remind that for barotropic fluid the variation of the energy density ρ is given by $\delta\rho = (\rho/2) \left(g_{\alpha\beta} - u_\alpha u_\beta \right) \delta g^{\alpha\beta}$. Consider that the stress-energy tensor is derived according to the definition (Eq. (2.80)):

$$T_{\alpha\beta} = \frac{\partial \mathscr{L}_{fluid}}{\partial g^{\alpha\beta}} - \frac{\mathscr{L}_{fluid}}{2} g_{\alpha\beta}$$

We deduce the component form of the energy-momentum tensor:

$$T_{\alpha\beta} = \left\{ \rho \left[c^2 + \Pi(\rho) \right] + P(\rho) \right\} u_\alpha u_\beta + P(\rho) g_{\alpha\beta} \qquad (4.53)$$

which is usually considered in celestial relativistic mechanics. Another formulation may be adopted by defining the following e.g. Poplawski (2009):

$$\tilde{e} := \rho \left[c^2 + \Pi(\rho) \right], \qquad \tilde{p} := P(\rho) \qquad (4.54)$$

to give a classical form of the constitutive laws of relativistic perfect fluid:

$$T^{\alpha\beta} = (\tilde{e} + \tilde{p}) \, u^\alpha u^\beta + \tilde{p} \, g^{\alpha\beta} \qquad (4.55)$$

in which $\tilde{e}$ the relativistic rest energy density of the fluid, $\tilde{p}$ is the pressure within the fluid, u^β is the four-velocity of the fluid, and $g_{\alpha\beta}$ is the Minkowski metric tensor with signature $(+, -, -, -)$. In the scope of torsionless relativistic gravitation, the covariant derivative of Einstein tensor vanishes $\nabla_\beta G^{\alpha\beta} \equiv 0$ e.g. Lovelock and Rund (1975). This allows us to obtain the conservation laws, by applying the gauge invariance with $\delta g_{\alpha\beta} := \mathcal{L}_\xi g_{\alpha\beta} = \nabla_\beta \xi_\alpha + \nabla_\alpha \xi_\beta$ (see in the next section for extension of this Lie derivation in the scope of Einstein–Cartan gravitation). After integrating by parts, the conservation laws of barotropic perfect fluid hold e.g. Minazzoli and Karko (2012):

$$\nabla_\beta T^{\alpha\beta} = \nabla_\beta \left[(\tilde{e} + \tilde{p})\, u^\alpha u^\beta + \tilde{p}\, g^{\alpha\beta} \right] = 0 \qquad (4.56)$$

Distributing the derivative on all terms allows us to write:

$$\nabla_\beta \left(\frac{\tilde{e} + \tilde{p}}{n} u^\alpha \right) n\, u^\beta + g^{\alpha\beta}\, \nabla_\beta \tilde{p} = 0$$

owing that we suppose a metric compatible connection and the conservation of particles number $\nabla_\beta (nu^\beta) = 0$ (number of baryons). Let now multiply these equations with u^α. Owing that $u^\alpha u_\alpha = -1$, we obtain:

$$u^\beta \left[\nabla_\beta \tilde{p} - n\nabla_\beta \left(\frac{\tilde{e} + \tilde{p}}{n} \right) \right] = u^\beta \left(\nabla_\beta \tilde{e} - \frac{\tilde{e} + \tilde{p}}{n} \nabla_\beta n \right) = 0$$

At this stage, we remind the definition of "generalized" time derivative of a tensor field $\mathcal{A}$ in general relativity (in the comoving frame):

$$\frac{d\mathcal{A}}{d\tau} := \nabla_\beta \mathcal{A}\, u^\beta \qquad (4.57)$$

An alternative formulation of the previous equation is thus obtained (conservation of energy, without considering entropy,

$$\frac{d\tilde{e}}{d\tau} - \frac{\tilde{e} + \tilde{p}}{n} \frac{dn}{d\tau} = 0 \qquad (4.58)$$

Consider now the remaining other three components of the conservation laws:

$$n\, u^\beta\, \nabla_\beta \left(\frac{\tilde{e} + \tilde{p}}{n} u^\alpha \right) + g^{\alpha\beta}\, \nabla_\beta \tilde{p} = 0$$

For the sake of the simplicity, we work in the momentarily comoving reference frame (MCRF). Spatial components of the four-velocity $u^i = 0, i = 1, 2, 3$ in this comoving frame but its covariant derivative does not vanish $\nabla_\beta u^i \neq 0$. Owing again that the spatial components of the four-acceleration vector is defined as

$a^i := u^\beta \, \nabla_\beta u^i$, the relativistic momentum equation of the perfect barotropic fluid is deduced:

$$(\tilde{e} + \tilde{p}) \, u^\beta \, \nabla_\beta u^i + g^{i\beta} \, \nabla_\beta \tilde{p} = 0 \quad \Longrightarrow \quad (\tilde{e} + \tilde{p}) \, a_i + \nabla_i \tilde{p} = 0 \qquad (4.59)$$

In the particular case where $\|\mathbf{v}\| \ll c$, we recover the Euler non relativistic fluid dynamics equations e.g. Rosen (1972):

$$\rho \mathbf{a} + \nabla p = 0, \qquad \mathbf{a} := \dot{\mathbf{v}} + (\mathrm{grad}\mathbf{v}) \, \mathbf{v} \qquad (4.60)$$

An extension of variational formulation of relativistic initially proposed in Taub (1954) including Lagrange parameters takes the form of Poplawski (2010):

$$\mathscr{S} := \frac{c^4}{16\pi G} \int_{\mathscr{M}} \mathscr{R} \, \omega_n - \int_{\mathscr{B}} \rho \left[c^2 + \epsilon(\rho, \sigma) \right] \omega_n$$
$$- \int_{\mathscr{B}} \left[\frac{\mu_1}{2} \left(g_{\alpha\beta} u^\alpha u^\beta - 1 \right) + \mu_2 \frac{d\sigma}{d\lambda} \right] \omega_n \qquad (4.61)$$

where σ is the entropy density of the fluid at rest, and $\mu_i, i = 1, 2$ are Lagrange multipliers to ensure the geometrical (normalization of the four-velocity) and the entropy constraints. $\epsilon(\rho, \sigma)$ is the internal energy density per unit mass of the fluid. For this later action, the relativistic Euler's equation is given by (see Poplawski (2009) for details):

$$\left(\rho c^2 + \rho \epsilon + p \right) u^\beta \nabla_\beta u^\alpha - \left(g^{\alpha\beta} - u^\alpha u^\beta \right) \nabla_\beta p = 0 \qquad (4.62)$$

where the Lagrange multiplier is found to be $\mu_1 = \rho c^2 + \rho \epsilon + p$ (this relation is a strong compatibility requirement of the proposed action).

Remark 4.17 The Einstein–Hilbert action (4.29) was shown by mathematician D Hilbert to yield the Einstein field equations through a variational principle within a Riemannian manifold (with Levi-Civita connection $\overline{\Gamma}^\gamma_{\alpha\beta}$). The Ricci scalar curvature $\mathfrak{R}$ is the simplest scalar field that can be formed with the metric $g_{\alpha\beta}$ and its first, and second derivatives e.g. Cartan (1922).

4.2.4 Matter Within Curved Spacetime with Torsion

Riemann spacetime is extended to Riemann–Cartan spacetime by considering non zero torsion. It is known that the test of spacetime with torsion can be only conducted with particles or continuum with internal structure, such as intrinsic spin e.g. Hojman (1976), Papapetrou (1951), Puetzfeld and Obukhov (2008) or fluid with vortices e.g. Garcia de Andrade (2005), Garcia de Andrade (2004). It was shown that only intrinsic spin and not global angular rotation of matter couples with spacetime. Let us consider a curved spacetime manifold with torsion $(\mathcal{M}, \mathbf{g}, \nabla)$ with the metric $g^{\alpha\beta}(x^\mu)$, the torsion $\aleph^\gamma_{\alpha\beta}(x^\mu)$, and the curvature $\mathfrak{R}^\delta_{\alpha\sigma\gamma}(x^\mu)$:

$$\nabla_\gamma g^{\alpha\beta} = 0, \quad \aleph^\gamma_{\alpha\beta} \neq 0, \quad \mathfrak{R}^\delta_{\alpha\sigma\gamma} \neq 0 \tag{4.63}$$

Again, we omit the "hat" in this paragraph for the sake of the simplicity. In a Riemann–Cartan spacetime, geodesics defined by connection $\Gamma^\gamma_{\alpha\beta}(x^\mu)$ depend on its symmetric part $\overline{\Gamma}^\gamma_{\alpha\beta} + D^\gamma_{\alpha\beta}$ as sketched follows (we assume here a metric compatible connection). It is different from the geodesics defined by Levi-Civita connection $\overline{\Gamma}^\gamma_{\alpha\beta}$. Let first remind the geodesic on a connected manifold $\mathcal{M}$. For Riemann–Cartan manifold with non zero torsion, Eq. (4.11) is extended by considering an affine connection ∇ with coefficients $\Gamma^\gamma_{\mu\nu}$. A vector $\mathbf{v}$ is parallel transported along the curve $\mathbf{u} := \mathbf{x}(\lambda)$ if $\mathbf{v}$ satisfies $\nabla_\mathbf{u} \mathbf{v} \equiv 0$ e.g. Nakahara (1996). A curve of event of the spacetime $\mathcal{M}$ defined by its parametric equation $x^\mu(\lambda)$ is a geodesic if its tangent vector $u^\mu := dx^\mu/d\lambda$ is parallel along the curve, say $\nabla_\mathbf{u} \mathbf{u} \equiv 0$ say $\nabla_\mathbf{u} \mathbf{u} = \left(u^\mu \nabla_\mu u^\gamma\right) \mathbf{e}_\gamma = 0 = u^\mu \left(\partial_\mu u^\gamma + \Gamma^\gamma_{\mu\nu} u^\nu\right)$. Since λ is the parameter of the curve, then we have:

$$u^\mu := \frac{dx^\mu}{d\lambda}, \qquad u^\mu \frac{\partial}{\partial x^\mu} = \frac{d}{d\lambda} \implies \frac{d^2 x^\gamma}{d\lambda^2} + \Gamma^\gamma_{\mu\nu} \frac{dx^\mu}{d\lambda} \frac{dx^\nu}{d\lambda} = 0 \tag{4.64}$$

Here geodesics are called auto-parallel curves. Geodesics are extremal curves for (torsionless) Riemannian manifold whereas auto-parallel curves are extremal curves for Riemann–Cartan manifold. The definition of the autoparallel curves do not depend on metric tensor but only on connection. By extension, in the Riemann–Cartan spacetime (called $U(4)$), the decomposition $\Gamma^\gamma_{\alpha\beta} := \overline{\Gamma}^\gamma_{\alpha\beta} + D^\gamma_{\alpha\beta} + \Omega^\gamma_{\alpha\beta}$ (e.g. Rakotomanana 2005) allows us to have a slightly different conclusion stating that geodesic lines are changed to auto-parallel lines which are solutions of e.g. Acedo (2015), Hehl and von der Heyde (1973), Kleinert (2008):

$$\frac{d^2 x^\gamma}{d\lambda^2} + (\overline{\Gamma}^\gamma_{\alpha\beta} + D^\gamma_{\alpha\beta}) \frac{dx^\alpha}{d\lambda} \frac{dx^\beta}{d\lambda} = 0 \tag{4.65}$$

obtained from Eq. (4.64) where we observe that only the symmetric (but torsion dependent) part of the connection is involved. It should be observed that Eq. (4.65) is sound from a geometrical point of view but its direct relation with physics does not always exist e.g. Hehl et al. (2013). Its relation with the trajectories of particles moving within a Riemann–Cartan spacetime is not ensured at a first sight e.g. Mao et al. (2007). In the framework of general relativity, as initially introduced by Einstein, we have $D^\gamma_{\alpha\beta}(x^\mu) \equiv 0$. For spacetime with torsion, such is not the case, where the spacetime properties bend the trajectory of the particle since $D^\gamma_{\alpha\beta}(x^\mu) \neq 0$ acts as an external force e.g. Kleinert (2008). See Hehl et al. (2013) and references herein for a review on the chronological list of papers that investigated the background theory for measuring (by means of elementary particle spin) or for the impossibility to measure (by means of a coupling of orbital angular momentum of some planets) the torsion tensor $\aleph^\gamma_{\alpha\beta}$ e.g. Yasskin and Stoeger (1980). In addition to the dust, considered as a simple-pole test-particle (Papapetrou 1951), a rotating particle (pole-dipole) should be considered to analyze the overall motion of a small size body immersed within a curved spacetime with torsion $\mathscr{M}$ e.g. Papapetrou (1951), Hehl (1971), Shapiro (2002).

Remark 4.18 For body $\mathscr{B}$ with finite but relativistically small size, Ehlers and Geroch have shown that the center of mass of $\mathscr{B}$ moves along a geodesic in the framework of Einstein relativistic gravitation without torsion (Ehlers and Geroch 2004). The theorem they established hold outside regions of singularity and black holes. The fundamental concept of pole-dipole and more generally the multipole models in general relativity is mainly based on the expansion of the fields about a center of mass. The most known model is probably the spinning particle concept in general gravitation e.g. Papapetrou (1951), and the well-known equations of Mathisson-Papapetrou-Dixon constitute the basic theoretical model e.g. Leclerc (2005), Mathisson (1937), Papapetrou (1951). In the scope of continuum mechanics, the basic ingredients is the assumption of rigid local bases for Cosserat models of continuum and the section rigidity for Timoshenko beam models and Mindlin-Reissner plates in the domain of engineering structures e.g. Rakotomanana (2009). In his fundamental paper on the new mechanics and material systems, Mathisson proposed in 1937 the concept of *gravitational skeleton* which the keypoint for modelling the spinning particle in the motion of (slightly) extended body within gravitational field (Mathisson 1937). All along this paper, we restrict the modelling to expansion of the metric to obtain torsion and curvature, instead of introducing a spin tensor ad hoc, as geometrical arguments of the Lagrangian function.

Basically the dimensions of a test-particle, with or without internal rotation, are very small compared to a characteristic length which could be taken as the distance from the central body in the case of Schwarzschild gravitation field (4.33).

4.2.4.1 Equivalence Principle

The equivalence principle is reworded as follows. Given a sufficiently small region of the spacetime (Riemannian), it is possible to find a reference frame with respect to whose associated coordinates the metric reduces to Minkowskian metric $g_{\alpha\beta} = \mathrm{diag}\{c^2, -1, -1, -1\}$, and the connection coefficients and its derivatives do not appear in any field equations of matter, say $\overline{\Gamma}^\gamma_{\alpha\beta}(x^\mu) = 0$. In presence of torsion, terms $D^\gamma_{\alpha\beta}(x^\mu)$ are different of zero, then it is a priori rigorously impossible to find a inertial reference frame. This problem remains open in some sense and should involve the subtle coupling between spacetime and matter endowed with spin e.g. Hehl et al. (1976), Knox (2013).

4.2.4.2   Variational Formulation and Gauge Invariance

In the framework of Einstein gravitation, variational formulation may be found in e.g. Carter (1973), Taub (1954). Hereafter, we consider curved spacetime with torsion. Let $\mathscr{L}(g_{\alpha\beta}, \Gamma^{\gamma}_{\alpha\beta}, \partial_{\lambda}\Gamma^{\gamma}_{\alpha\beta})$ be a particular Lagrangian, which leads to the list of arguments $g_{\alpha\beta}, \aleph^{\gamma}_{\alpha\beta}, \mathfrak{R}^{\gamma}_{\lambda\alpha\beta}$. Then the geodesic trajectories in the spacetime should be replaced by the auto parallel trajectories whenever there is closure failure of Cartan's parallelogram in presence of torsion (Kleinert 1999). When $\aleph^{\gamma}_{\alpha\beta} \neq 0$, the standard variational procedure for the least action extrema must be modified. Due to torsion e.g. Rakotomanana (1997), the application of nonholonomic mapping principle (Fiziev and Kleinert 1995) has the consequence that even an action involving only metric as argument, a generalized stress due to torsion appears e.g. Kröner (1981), Rakotomanana (1997). This conforms to the differential geometry approach where torsion force is pointed out in the divergence operator on Riemann–Cartan manifold e.g. Rakotomanana (2003). Metric and affine connection are independent variables and accordingly curvature components $\mathfrak{R}^{\lambda}_{\alpha\beta\mu}$ do not a priori depend on the metric components $g_{\alpha\beta}$ (Hehl and Kerlick 1976). Consider the variation of the Lagrangian depending on three arguments $\mathscr{L}(g_{\alpha\beta}, \aleph^{\gamma}_{\alpha\beta}, \mathfrak{R}^{\gamma}_{\alpha\beta\lambda})\omega_n$ e.g. Duval and Kunzle (1978), Hojman (1976) which vanishes due to the invariance (also called Einstein invariance e.g. Kleinert 2008):

$$\mathscr{S} := \int \mathscr{L}\,\omega_n,$$

$$\delta\mathscr{L} = \left(\frac{\partial\mathscr{L}}{\partial g_{\alpha\beta}} - \frac{\mathscr{L}}{2}g^{\alpha\beta}\right)\delta g_{\alpha\beta} + \frac{\partial\mathscr{L}}{\partial\aleph^{\gamma}_{\alpha\beta}}\delta\aleph^{\gamma}_{\alpha\beta} + \frac{\partial\mathscr{L}}{\partial\mathfrak{R}^{\gamma}_{\lambda\alpha\beta}}\delta\mathfrak{R}^{\gamma}_{\lambda\alpha\beta} = 0 \quad (4.66)$$

where torsion and curvature are entirely calculated with connection coefficients $\Gamma^{\gamma}_{\alpha\beta}$, but not with metric $g_{\alpha\beta}$. ω_n denotes the volume-form of the spacetime. The variation is denoted δ although it is in fact a Lagrangian variation (Carter 1973). There are some hidden aspects in Eq. (4.66), the arbitrariness of the metric variation provides the field equation. At this step, we can not apply the arbitrariness of the torsion end curvature and have to go back to the connection to whom torsion and curvature are associated.

Remark 4.19 The action is an integral of a 4-form such as $\mathscr{L}\omega_n$, where the volume-form ω_n should satisfy among others a condition of compatibility with the connection ∇ e.g. Saa (1995). When the volume-form satisfies the relation (A.21) (appendix), ω_n is said compatible with the connection.

To go further, we now formulate the variation of primal independent variables as $\delta g_{\alpha\beta}$ and the connection $\delta\Gamma^{\gamma}_{\alpha\beta}$. For this purpose, we remind the Palatini relationships (Rakotomanana 1997):

$$\delta\aleph^{\gamma}_{\alpha\beta} = \delta\Gamma^{\gamma}_{\alpha\beta} - \delta\Gamma^{\gamma}_{\beta\alpha},$$

$$\delta\mathfrak{R}^{\lambda}_{\alpha\beta\mu} = \nabla_{\alpha}\left(\delta\Gamma^{\lambda}_{\beta\mu}\right) - \nabla_{\beta}\left(\delta\Gamma^{\lambda}_{\alpha\mu}\right) - \aleph^{\nu}_{\alpha\beta}\,\delta\Gamma^{\lambda}_{\nu\mu} \quad (4.67)$$

where the covariant derivatives use the connection with non zero torsion. The second equation (4.67) extends the Palatini identity when continuum has torsion. By analogy with generalized continuum mechanics, we define the *hyper-momenta* as in e.g. Obukhov et al. (1989) which are the constitutive laws of gradient continua:

$$\sigma^{\alpha\beta} := \frac{\partial \mathscr{L}}{\partial g_{\alpha\beta}} - \frac{\mathscr{L}}{2} g^{\alpha\beta}, \qquad \Sigma_\gamma^{\alpha\beta} := \frac{\partial \mathscr{L}}{\partial \aleph_{\alpha\beta}^\gamma}, \qquad \varXi_\gamma^{\lambda\alpha\beta} := \frac{\partial \mathscr{L}}{\partial \Re_{\lambda\alpha\beta}^\gamma} \qquad (4.68)$$

Dual variables $\sigma^{\alpha\beta}$, $\Sigma_\gamma^{\alpha\beta}$, and $\varXi_\gamma^{\lambda\alpha\beta}$ are also called *currents* in physics e.g. Forger and Römer (2004). Stress $\sigma^{\alpha\beta}$ in continuum mechanics (resp. energy-momentum tensor in relativistic gravitation theory) is the response to the local variation of the distances in the matter (resp. spacetime) e.g. Hehl and von der Heyde (1973). Moment stress (resp. spin-angular momentum) $\Sigma_\gamma^{\alpha\beta}$ is the response of change of discontinuity of scalar field within matter (resp. the torsion of spacetime). Curvature stress (resp. energy-momentum) is the response to variations of discontinuity of vector field (curvature field) e.g. Rakotomanana (1997). Introduction of these relations into the variation of the Lagrangian density gives e.g. Pons (2011):

$$\begin{aligned}
\delta\mathscr{L} &= \sigma^{\alpha\beta}\,\delta g_{\alpha\beta} \\
&\quad + \left[\Sigma_\gamma^{\alpha\beta} - \Sigma_\gamma^{\beta\alpha} - \varXi_\gamma^{\mu\nu\beta}\,\aleph_{\mu\nu}^\alpha - \nabla_\lambda \left(\varXi_\gamma^{\lambda\alpha\beta} - \varXi_\gamma^{\alpha\lambda\beta} \right) \right] \delta\Gamma_{\alpha\beta}^\gamma \qquad (4.69) \\
&\quad + \nabla_\lambda \left[\left(\varXi_\gamma^{\lambda\alpha\beta} - \varXi_\gamma^{\alpha\lambda\beta} \right) \delta\Gamma_{\alpha\beta}^\gamma \right]
\end{aligned}$$

We point out the worthiness of choosing appropriate variables from deriving the conservation laws. Despite the fact that metric and connection are independent variables, the independence of their variations $\delta g_{\alpha\beta}$, and $\delta\Gamma_{\alpha\beta}^\gamma$ in the variation (4.69) should be applied with great care. Of course, it is not acceptable to have $\sigma^{\alpha\beta} \equiv 0$ without additional explanation since at this stage the Lagrange function should be completed with additional term depending on the metric. The worth set of variables is a displacement variation δu_α and more generally an arbitrary non uniform translation ξ^α, and the connection $\delta\Gamma_{\alpha\beta}^\gamma$ as we will see later. For illustrating the second equation of motion, we will consider hereafter the case where a continuum evolves within a curved spacetime with a Hilbert–Einstein action, see a recent review in e.g. Sotiriou and Faraoni (2010).

Remark 4.20 At this step, the present method resembles to some developments conducted in e.g. Forger and Römer (2004). However the present development is based on a priori covariant Lagrangian function on a metric-affine manifold $\mathscr{M}$, which is not the case in this previous work.

Owing that metric and connection are independent variables, we respectively factorize with respect $\delta g_{\alpha\beta}$ and $\delta\Gamma_{\alpha\beta}^\gamma$ (both of them are tensor fields) e.g. Hehl et al. (1976), Sotiriou (2009). The last term of the (4.69) *rhs* represents a divergence term for the boundary conditions and could be dropped for the fields equations.

Scalar curvature includes only first and second-order derivatives of the metric $g_{\alpha\beta}$. Furthermore, terms in the first and second lines are at most first-order with respect to $\Gamma^\gamma_{\alpha\beta}$, with however the presence of third-order derivatives of the metric if the affine connection includes the symbols of Christoffel $\overline{\Gamma}^\gamma_{\alpha\beta}$. However, some specific conditions on the constitutive parameters allow us to reduce this order of derivatives to two at most e.g. Hammond (1990). The arbitrariness of the two basic variables $g_{\alpha\beta}$ (metric), and $\Gamma^\gamma_{\alpha\beta}$ (connection) allows us to deduce the field equations:

$$
\begin{cases}
\dfrac{\partial \mathscr{L}}{\partial g_{\alpha\beta}} - \dfrac{\mathscr{L}}{2} g^{\alpha\beta} = 0 \\[4mm]
\Sigma^{\alpha\beta}_\gamma - \Sigma^{\beta\alpha}_\gamma - \Xi^{\mu\nu\beta}_\gamma \, \aleph^\alpha_{\mu\nu} - \nabla_\lambda \left(\Xi^{\lambda\alpha\beta}_\gamma - \Xi^{\alpha\lambda\beta}_\gamma \right) = 0
\end{cases}
\tag{4.70}
$$

which extends the Einstein field equations to Einstein–Cartan field equations. A conceptual problem might pertain since the metric is a tensor whereas the connection is not. In the next section, we will consider both the metric and the contortion tensors as primal variables to overcome this apparently misconception modelling.

Remark 4.21 In the framework of the gradient continuum mechanics, there are alternatives of currents formulation e.g. Polizzotto (2013a) for quasi-static, and Polizzotto (2013b) for dynamic behaviors. In these papers, generalized stresses are defined accordingly following the approach of Mindlin (1964, 1965):

$$
\sigma^{\alpha\beta} := \frac{\partial \mathscr{L}_M}{\partial g_{\alpha\beta}} - \frac{\mathscr{L}_M}{2} g^{\alpha\beta}, \quad \Sigma^{\alpha\beta}_\gamma := \frac{\partial \mathscr{L}_M}{\partial \nabla^\gamma \varepsilon_{\alpha\beta}}, \quad \Xi^{\lambda\alpha\beta}_\gamma := \frac{\partial \mathscr{L}_M}{\partial \nabla_\lambda \nabla^\gamma \varepsilon_{\alpha\beta}} \tag{4.71}
$$

where $2\varepsilon_{\alpha\beta} := g_{\alpha\beta} - \hat{g}_{\alpha\beta}$. Nevertheless, physical meaning of Eq. (4.68) seems quite clear compared to that of Eq. (4.71). Torsion and curvature have precise physical or geometrical interpretation. Remark that for Levi-Civita connection $\overline{\nabla}$ which is metric compatible, covariant gradients of metric identically vanishes in principle e.g. Antonio and Rakotomanana (2011), Lovelock and Rund (1975). Care should be taken before considering the metric gradient as additional variables as we previously showed in the Corollary (3.1), where the explicit dependency of the Lagrangian on the covariant derivative of the metric could not be allowed since it identically vanishes. If such is not the case, then the geometric background is no more a Riemann–Cartan manifold but rather should be a Weyl manifold.

4.2.4.3   Conservation Laws

We can derive *conservation laws* when considering variation of metric and connection respectively. In the absence of external momenta loadings, the field equations for these dual variables hold from the variational Eq. (4.69) taking account of the arbitrariness of $\delta g_{\alpha\beta}$, and $\delta \Gamma^\gamma_{\alpha\beta}$ (Hehl and Kerlick 1976). It should be stressed however that the two obtained equations would not be at "at the same level". The first is related to a tensor variable and the second to a non tensorial variable. The first equation may be usually re-derived by reminding the relation between the Lagrangian variation and the Eulerian variation $\Delta g_{\alpha\beta} = \delta g_{\alpha\beta} + \mathcal{L}_\xi g_{\alpha\beta}$ where we recognize the Live derivative of $g_{\alpha\beta}$ along the displacement ξ : $\mathcal{L}_\xi g_{\alpha\beta} := \nabla_\alpha \xi_\beta + \nabla_\alpha \xi_\beta$ e.g. Carter and Quintana (1977) for a metric compatible Levi-Civita connection. By shifting the boundary terms the two conservation laws take the slightly re-arranged form of:

$$\begin{cases} \nabla_\beta \sigma^{\alpha\beta} = 0 \\ \nabla_\lambda \left(\Xi_\gamma^{\lambda\alpha\beta} - \Xi_\gamma^{\alpha\lambda\beta} \right) = \Sigma_\gamma^{\alpha\beta} - \Sigma_\gamma^{\beta\alpha} - \Xi_\gamma^{\mu\nu\beta} \aleph_{\mu\nu}^{\alpha} \end{cases} \qquad (4.72)$$

where the first equation, deduced from the arbitrariness of δu_α, means that the energy-momentum tensor is divergence free. The second equation, deduced from the arbitrariness of $\delta \Gamma_{\alpha\beta}^\gamma$, suggests that the hypermomentum associated to the torsion acts as an external source forces.

Remark 4.22 The derivation of constitutive laws was conducted by varying arbitrarily the two variables δu_α (displacement), and $\delta \Gamma_{\alpha\beta}^\gamma$ (connection). In the case of relativistic gravitation the Lagrangian $\mathcal{L}$ admits the scalar curvature $\mathcal{R}$ as argument. Then the stress-energy tensor, as defined in the relation (4.71), may be rewritten as:

$$\sigma^{\alpha\beta} = \sigma_m^{\alpha\beta} + \Xi^{\alpha\beta\lambda\mu} G_{\lambda\mu}, \qquad G_{\lambda\mu} := \mathfrak{R}_{\lambda\mu} - (1/2)\mathcal{R}g_{\lambda\mu}$$

where $\sigma_m^{\alpha\beta}$ is the part of the Lagrangian due the matter. In the next paragraph, we will show that the Bianchi identities induce the vanishing of the gravitational part of the stress-energy tensor, where $G_{\lambda\mu}$ is called Einstein tensor. For instance, the gravitational stress vanishes in the particular case where $\Xi^{\alpha\beta\lambda\mu}$ takes the form of $\Xi^{\alpha\beta\lambda\mu} := \Lambda_1 g^{\alpha\beta} g^{\lambda\mu} + \Lambda_2 \left(g^{\alpha\lambda} g^{\beta\mu} + g^{\alpha\mu} g^{\beta\lambda} \right)$ and where the connection is metric compatible.

4.2.4.4 Bianchi Identities

For torsionless manifold (model of either spacetime or matter continuum) the Riemann–Cartan curvature has some symmetry properties. Two of them are the Bianchi identities, they express some relations between the curvature components and the covariant derivatives of curvature components. Let consider the particular case of Einstein gravitation theory where the torsion vanishes everywhere. The second Bianchi identity takes the form of:

$$\sum_{(\alpha\beta\gamma)} \nabla_\gamma \mathfrak{R}_{\alpha\beta\sigma}^\kappa = 0 \quad \Longrightarrow \quad \delta_\alpha^\kappa \sum_{(\alpha\beta\gamma)} \nabla_\gamma \mathfrak{R}_{\alpha\beta\sigma}^\kappa = 0$$

Since the associated connection is assumed metric compatible, we can introduce the metric within the covariant derivative. Owing the symmetry properties of the curvature and contracting again by multiplication with $g^{\beta\sigma}$, we obtain, after applying the circular permutation on $(\alpha\beta\gamma)$, $\nabla_\gamma \mathcal{R} - \nabla^\sigma \mathfrak{R}_{\gamma\sigma} - \nabla_\alpha \mathfrak{R}_\gamma^\alpha = 0$. Again by using the metric compatibility of the connection we deduce:

$$\nabla_\gamma G^{\alpha\gamma} = 0, \qquad \text{with} \qquad G^{\alpha\gamma} := \mathfrak{R}^{\alpha\gamma} - (1/2)\mathcal{R}g^{\alpha\gamma} \tag{4.73}$$

in which we recognize the Einstein tensor $G^{\alpha\gamma}$. We check that the Einstein tensor is divergence free (as a direct consequence of the second Bianchi identity) and this is true only for Riemannian manifold. Furthermore, for term in front of $\delta g_{\alpha\beta} := \nabla_\alpha \delta u_\beta + \nabla_\alpha \delta u_\beta$ (it is usually assumed but it is indeed not complete variation in a metric-affine manifold, as we will see hereafter when the torsion does not vanish). Classically the divergence-free property is then deduced from the second Bianchi identity. It is a very particular situation. We can observe that there is no explicit extension of the Einstein tensor in the Einstein–Cartan spacetime. The first Bianchi identity, sometimes called fundamental identity follows mainly from the commutativity of the second partial derivatives of the metric components $g_{\alpha\beta}(x^\mu)$ (integrability condition). The second Bianchi identity is deduced from the single-valuedness of the affine connection $\Gamma^\gamma_{\alpha\beta}(x^\mu)$ that is the commutativity of the second derivatives of its coefficients e.g. Kleinert (2008) (integrability condition). The Einstein tensor $G^{\alpha\beta}$ is not divergence-free when the torsion is not equal to zero. As extension to manifolds with torsion, and for the sake of the self-consistency, let remind the two Bianchi identities e.g. Rakotomanana (2003). They extend the Bianchi identities formulae of Riemannian geometry to Einstein–Cartan manifold where $(\alpha\beta\gamma)$ (curved manifold with torsion) is a circular permutation. In a more convenient form, the first and the second Bianchi identities hold:

$$\begin{cases} \displaystyle\sum_{(\alpha\beta\gamma)} \mathfrak{R}^\sigma_{\alpha\beta\gamma} + \nabla_\alpha \aleph^\sigma_{\beta\gamma} + \aleph^\sigma_{\mu\gamma} \aleph^\mu_{\alpha\beta} = 0 \\[2mm] \displaystyle\sum_{(\alpha\beta\gamma)} \nabla_\gamma \mathfrak{R}^\kappa_{\alpha\beta\sigma} + \aleph^\mu_{\alpha\beta}\mathfrak{R}^\kappa_{\mu\gamma\sigma} \qquad = 0 \end{cases} \tag{4.74}$$

In the Einstein–Cartan theory of gravitation, the stress-energy tensor may thus be not divergence-free.

Remark 4.23 Recent studies show that, depending on how the coupling between gravitational and matter parts of Lagrangian acts, some gravitation theories are not viable e.g. Sotiriou (2008). Shortcomings are mainly due to facts that differentiation order of the matter field may be higher than that of the spacetime metric (not to be confused with that of material metric). This motivated the development of Palatini variational principle. However, the conservation laws deduced from the gauge $\delta\Gamma^\gamma_{\alpha\beta}$), and conservation laws corresponding to linear momentum highlight that application of Palatini variational principle has also these shortcomings because no hypermomentum associated to torsion was introduced in those previous studies.

We observe that for the gauge invariance, consistency problems also appear for strain gradient continuum. Derivative operator using affine connection with non zero torsion and curvature should then be used on the gradient continuum. This extends both the notion of objective time derivatives to higher gradient spacetime physics e.g. Rakotomanana (2003), and the variation procedure accounting for the non zero torsion e.g. Kleinert (2008). Further studies should be done for exploring the consequences of field equations relating discontinuity of scalar field, and discontinuity of vector fields, and in a broad sense the interaction between derivatives of strain for high gradient spacetime. Particularly, the presence of torsion suggests to fundamentally use the concept of auto parallel trajectories rather than geodesic ones e.g. Kleinert (1999). In practise, it modifies the nature of wave propagation within non homogeneous continuum e.g. Antonio et al. (2011), Futhazar et al. (2014). Rigorous definitions of the general variations $\delta g_{\alpha\beta}$, $\delta\aleph^{\gamma}_{\alpha\beta}$, and $\delta\Re^{\gamma}_{\alpha\beta\lambda}$ is the key point to obtain conservation laws and the additional Eq. (4.68) defining the covariant hypermomenta.

In order to explore the features of $f(\Re)$-theory of relativistic gravitation, which extends the Einstein–Hilbert action, some studies analyze the entanglement of the interrelations between metric and connection (with torsion) to point out the role of torsion when the curvature is calculated with the help of connection with torsion e.g. Dadhich and Pons (2012), Hehl et al. (1976), Sotiriou (2009). By considering a particular case of projective variation, they showed that the $f(\Re)$-theory of relativistic gravitation does not support the existence of independent connection which carries additional dynamical degrees of freedom. This states that the spacetime manifold is necessarily pseudo-Riemannian. In the presence of matter, however this result is only valid when the matter action is not coupled with spacetime connection. This result thus does not apply when we a priori split the Lagrangian into two contributions $\mathscr{L} := \mathscr{L}_G(\hat{g}_{\alpha\beta}, \hat{\Re}^{\gamma}_{\alpha\beta\lambda}) + \mathscr{L}_M(\hat{g}_{\alpha\beta}, g_{\alpha\beta}, \aleph^{\gamma}_{\alpha\beta}, \Re^{\gamma}_{\alpha\beta\lambda})$ e.g. Hehl et al. (1976). The presence of the spacetime metric $\hat{g}_{\alpha\beta}$ is mandatory for the matter Lagrangian (Lehmkuhl 2011). The choice of such a Lagrangian suggests that there can be no torsion of the spacetime outside the eventually spinning/strain gradient defected matter distribution contained in $\mathscr{L}_M$. We will see hereafter that the presence of torsion should be considered in principle when the connection is coupled with matter in $\mathscr{L}_M$ e.g. Sotiriou (2009). We focus on the spacetime behavior without matter.

4.2.4.5 Fields Equations in Curved Spacetime with Torsion

Let $(\mathscr{M}, \hat{g}_{\alpha\beta}, \hat{\Gamma}^{\gamma}_{\alpha\beta})$ be a spacetime where the connection may not be a priori metric compatible. We now consider the Einstein–Hilbert action for relativistic gravitation by considering three formalisms for the same action:

$$\mathscr{S}_{EH} := \int_{\mathscr{M}} \hat{g}^{\beta\lambda}\hat{\Re}^{\alpha}_{\alpha\beta\lambda}\hat{\omega}_n, \qquad \hat{\omega}_n := \sqrt{\mathrm{Det}\hat{g}}\, dx^0 \wedge dx^1 \wedge dx^2 \wedge dx^3 \qquad (4.75)$$

First, the *metric formalism* (classic relativistic gravitation) is based on the Einstein gravitation, where the connection is reduced to the Levi-Civita torsionless connection. From Eq. (4.69), the variation with respect to metric $\delta \hat{g}_{\alpha\beta}$ leads to the field equations:

$$\hat{\mathcal{R}}_{\alpha\beta} - (1/2)\, \hat{\mathcal{R}}\, \hat{g}_{\alpha\beta} = 0 \tag{4.76}$$

which leads to the classical field equations of relativistic gravitation. No continuum matter is considered in this section and the spacetime $\mathcal{M}$ is assumed unbounded.

Remark 4.24 First, the unknowns of the field equations (4.76) are the components of the spacetime metric $\hat{g}_{\alpha\beta}(x^\mu)$. Then at a second step, we calculate the associated Levi-Civita connection $\hat{\Gamma}^\gamma_{\alpha\beta}(x^\mu)$.

Second, the *Einstein–Palatini formalism* is based on the same shape as Einstein–Hilbert action: $\mathscr{S}_{EP} := \int \hat{\mathcal{R}}[\hat{\Gamma}^\gamma_{\alpha\beta}]\omega_n$ where the connection is independent on the metric $\hat{g}_{\alpha\beta}$. The independent variation of the metric and the connection leads to, thanks to Eqs. (4.69), and (4.72):

$$\begin{cases} \hat{\mathcal{R}}_{\alpha\beta} - (1/2)\, \hat{\mathcal{R}}\, \hat{g}_{\alpha\beta} = 0 \\ \hat{\nabla}_\lambda \left(\Xi^{\lambda\alpha\beta}_\gamma - \Xi^{\alpha\lambda\beta}_\gamma \right) = 0 \end{cases} \tag{4.77}$$

In Einstein relativistic gravitation, only the first row of Eq. (4.77) holds since the angular momentum equation is trivial, it is satisfied by the symmetry of the energy-momentum tensor (dual of the metric variable). In the spacetime of Riemann–Cartan, a second row appears in (4.77), which emerges as second independent equation of the gravity, considered as the conservation laws of angular momentum e.g. Mao et al. (2007). For calculating $\Xi^{\lambda\alpha\beta}_\gamma$, the directional derivative is used:

$$\lim_{h \to 0} \frac{1}{h} \left[\mathscr{L}(\hat{\mathfrak{R}}^\gamma_{\lambda\alpha\beta} + h\delta\hat{\mathfrak{R}}^\gamma_{\lambda\alpha\beta}) - \mathscr{L}(\hat{\mathfrak{R}}^\gamma_{\lambda\alpha\beta}) \right] = \frac{\partial \mathscr{L}}{\partial \hat{\mathfrak{R}}^\gamma_{\lambda\alpha\beta}} \delta\hat{\mathfrak{R}}^\gamma_{\lambda\alpha\beta} \tag{4.78}$$

The trick is also to introduce the skew symmetry of the curvature with respect to the two lower indices $\hat{\mathfrak{R}}^\gamma_{\lambda\alpha\beta} = -\hat{\mathfrak{R}}^\gamma_{\alpha\lambda\beta}$. Then, a straightforward calculus gives the derivatives:

$$\Xi^{\lambda\alpha\beta}_\gamma = \frac{1}{2} \left(\hat{g}^{\alpha\beta} \delta^\lambda_\gamma - \hat{g}^{\lambda\beta} \delta^\alpha_\gamma \right), \qquad \Xi^{\alpha\lambda\beta}_\gamma = \frac{1}{2} \left(\hat{g}^{\lambda\beta} \delta^\alpha_\gamma - \hat{g}^{\alpha\beta} \delta^\lambda_\gamma \right)$$

The field equations associated to the Einstein–Palatini action then become:

$$\begin{cases} \hat{\mathcal{R}}_{\alpha\beta} - (1/2)\, \hat{\mathcal{R}}\, \hat{g}_{\alpha\beta} = 0 \\ \left(\delta^\alpha_\mu \delta^\lambda_\gamma - \delta^\lambda_\mu \delta^\alpha_\gamma \right) \hat{\nabla}_\lambda \hat{g}^{\mu\beta} = 0 \end{cases} \tag{4.79}$$

where we introduced $\hat{g}^{\alpha\beta}\delta^\lambda_\gamma - \hat{g}^{\lambda\beta}\delta^\alpha_\gamma = \left(\delta^\alpha_\mu\delta^\lambda_\gamma - \delta^\lambda_\mu\delta^\alpha_\gamma\right)\hat{g}^{\mu\beta}$, which is a skew-symmetric projection of the metric.

Remark 4.25 The unknowns of this coupled system of equations (4.79) are the metric $\hat{g}_{\alpha\beta}(x^\mu)$ of the spacetime and the coefficients of the connection $\hat{\Gamma}^\gamma_{\alpha\beta}(x^\mu)$. Both of them determine the dynamics of the spacetime. We observe that the system (4.79) constitutes a system of first-order coupled partial differential equations in $\hat{g}_{\alpha\beta}(x^\mu)$ and $\hat{\Gamma}^\gamma_{\alpha\beta}(x^\mu)$.

The second equation in (4.79) is a constraint equation for the unknown connection coefficients $\hat{\Gamma}^\gamma_{\alpha\beta}$ which are a priori arbitrary e.g. Sotiriou and Faraoni (2010). The second equation (4.79) states that only a projection of the metric gradient vanishes. If the metric is assumed to be a priori compatible then this is in conformity to the study of Sotiriou and coworkers e.g. Sotiriou and Faraoni (2010) stating that when the gravitation Lagrangian is chosen as $f(\hat{\mathcal{R}}) := \hat{\mathcal{R}}$, the Einstein–Palatini leads to metric formalism if the torsion tensor $\hat{\aleph}^\gamma_{\alpha\beta} \equiv 0$ is assumed to vanish. In such a case, $\hat{\Gamma}^\gamma_{\alpha\beta}$ reduces to the Levi-Civita connection. This may be related to the fundamental theorem of Ricci e.g. Nakahara (1996). But in such a way the compatibility is not a conclusion, it is an assumption. Third, the *metric-affine formalism* is based on the action: $\mathscr{S}_{MA} := \int \hat{\mathcal{R}}[\hat{\Gamma}^\gamma_{\alpha\beta}]\omega_n$ with non vanishing torsion. In the scope of Einstein–Cartan relativistic gravitation, we keep the Einstein–Hilbert action as the spacetime action, but we do not introduce the concept of matter spin at this stage. Conversely to classical approach, we then assume in the present work that the Lagrangian density for the gravitational field in the Einstein–Cartan theory is proportional to the scalar Ricci curvature. Remind nevertheless that the curvature in such a case is calculated explicitly by means of connection $\Gamma^\gamma_{\alpha\beta}$. The spacetime action implicitly involves metric and torsion where the spacetime Lagrangian holds $\mathscr{L} := \hat{\mathcal{R}} = \hat{g}^{\alpha\beta}\hat{\mathcal{R}}_{\alpha\beta} = \hat{g}^{\alpha\beta}\hat{\mathcal{R}}^\lambda_{\lambda\alpha\beta}$. The curvature is calculated with the connection which is independent on the metric $\hat{g}_{\alpha\beta}$. The arbitrariness of the variation of the metric $\delta\hat{g}_{\alpha\beta}$ and the connection $\delta\hat{\Gamma}^\gamma_\alpha$ leads to the field equations, see Eq. (4.69):

$$\begin{cases} \hat{\mathcal{R}}_{\alpha\beta} - (1/2)\,\hat{\mathcal{R}}\,\hat{g}_{\alpha\beta} = 0 \\ \left(\Sigma^{\alpha\beta}_\gamma - \Sigma^{\beta\alpha}_\gamma - \Xi^{\mu\nu\beta}_\gamma\,\hat{\aleph}^\alpha_{\mu\nu}\right) + \hat{\nabla}_\lambda\left(\Xi^{\lambda\alpha\beta}_\gamma - \Xi^{\alpha\lambda\beta}_\gamma\right) = 0 \end{cases} \tag{4.80}$$

The same calculus of derivatives as previously leads to the field equations for metric-affine formalism:

$$\begin{cases} \hat{\mathcal{R}}_{\alpha\beta} - (1/2)\,\hat{\mathcal{R}}\,\hat{g}_{\alpha\beta} = 0 \\ 2\left(\delta^\alpha_\mu\delta^\lambda_\gamma - \delta^\lambda_\mu\delta^\alpha_\gamma\right)\hat{\nabla}_\lambda\hat{g}^{\mu\beta} + \hat{\aleph}^\alpha_{\gamma\mu}\,\hat{g}^{\mu\beta} = 0 \end{cases} \tag{4.81}$$

Equation (4.81) are derived for modelling an ideal Riemann–Cartan vacuum spacetime, which may be questionable. This is slightly different from e.g. Hehl et al. (1974) where they considered both the intrinsic spin of microscopic elementary particles and the spacetime torsion, or randomly distributed macroscopic particles but with extremely high density where the use of U_4 theory still remains necessary. In the absence of particles either microscopic or macroscopic, for our purpose, the unknowns of the coupled system of equations (4.81) are the metric $\hat{g}_{\alpha\beta}(x^\mu)$ and the torsion $\hat{\aleph}^\gamma_{\alpha\beta}(x^\mu)$ of the spacetime. As first guess, the solving of Eq. (4.81) in the case of spherical symmetry would be a first challenge. Conversely to e.g. Maier (2014) which is not concerned with the torsion source and then assumed the existence of a matter Lagrangian to preclude the necessity of contortion tensor, the two equations (4.81) are expected to provide, at least in principle, both the metric and the torsion (algebraically deduced from the covariant derivative of the metric). In the presence of torsion source, the right-hand-side of Eq. (4.81) is not equal to zero.

Remark 4.26 Metric compatibility might be pictured as the condition for the non zero torsion and/or curved spacetime to be a set of "glued" Minkowskian microcosms e.g. Hehl et al. (1976), the non metricity behaves a like a source of torsion field.

Equations (4.81) lead to different conclusions than that of the Einstein–Palatini formalism whenever the spacetime is with non vanishing torsion. This conforms to the fact that the non zero torsion of spacetime cannot appear without matter spin since the torsion is directly obtained by algebraic calculus in the second equation (4.81), and the spacetime torsion $\hat{\aleph}^\gamma_{\alpha\beta} \equiv 0$ vanishes if the connection is a priori assumed metric compatible e.g. Capoziello et al. (2009), Sotiriou and Faraoni (2010). Without a priori metric compatibility assumption, we observe that only a projection of the non metricity vanishes, the connection may then be generally incompatible with metric. For a metric-affine spacetime $\mathcal{M}$ where the connection is not metric compatible, we obtain an algebraic equation for the torsion tensor in terms of the non metricity. The torsion of the spacetime $\mathcal{M}$ could not be highlighted with a non rotating test-particle, but a test-particle (pole-dipole) with spin e.g. Hehl (1971), Papapetrou (1951), or continuum medium as fluid with vortex e.g. Garcia de Andrade (2004), or other physical phenomenon as quantum effects e.g. Hammond (2002) are suggested to allow us (at least theoretically) to measure the torsion field of spacetime $\mathcal{M}$.

Remark 4.27 In view of the results obtained with three previous models (Einstein–Hilbert, Einstein–Palatini, and Einstein–Cartan), the metric compatibility (leading to a Riemann–Cartan spacetime) induces that the torsion vanishes. The inverse is not true. For more general theory, it is worth to consider the general form of Lagrangian (4.82) (hereafter) if we would like to highlight the specific role of the torsion of the spacetime.

4.2.5 Lagrangian for Coupled Spacetime and Matter

Let now consider a strain gradient continuum $(\mathcal{B}, g_{\alpha\beta}, \Gamma^\gamma_{\alpha\beta})$ evolving within a metric-affine spacetime $(\mathcal{M}, \hat{g}_{\alpha\beta}, \hat{\Gamma}^\gamma_{\alpha\beta})$. For testing the non zero torsion and curved spacetime, it is mandatory to consider structured matter as gradient and continuum with torsion e.g. Puetzfeld and Obukhov (2008), particle with intrinsic spin e.g. Hehl (1971), or fluid with vortices e.g. Garcia de Andrade (2005). One of the fundamental problem lies upon the definition of the deformation of such a generalized continuum with respect to the spacetime. We start by deriving the concept of *generalized strain* accounting for the loss of affine equivalence between the matter and the spacetime. Some authors e.g. Baldacci et al. (1979), Kleinert (1987) already suggested to investigate the dislocation dynamics in stressed metallic bodies with the help of relativistic gravitation theory by introducing micro-universe, which is in some extend the concept of geometric microcosm in e.g. Gonseth (1926), Malyshev (2000). In the same way, we consider the common mathematical background between strain gradient continuum and relativistic gravity theories. Coupling of matter and spacetime is not a trivial problem e.g. Lehmkuhl (2011), Sotiriou (2008). The simplest illustration consists in an elastic simple material $(\mathcal{B}, g_{\alpha\beta})$ evolving within a Newton–Cartan spacetime $(\mathcal{N}, \hat{g}_{\alpha\beta} \equiv [\delta_{\alpha\beta}, \hat{\tau}_\alpha \equiv 1], \hat{\Gamma}^\gamma_{\alpha\beta})$ and submitted to a conservative force (Maugin 1978) deriving from a potential $\mathcal{U}_{\text{ext}}(x^\alpha)$ with (the spacetime connection coefficients are obtained from the external potential (4.13)):

$$\mathscr{L} = \mathscr{L}(\varepsilon_{\alpha\beta}, \hat{\mathfrak{R}}^\gamma_{\alpha\beta\lambda}), \quad \varepsilon_{\alpha\beta} := (1/2)\left(g_{\alpha\beta} - \hat{g}_{\alpha\beta}\right), \quad \hat{\mathfrak{R}}^\beta_{\alpha 00} := \partial^\beta \partial_\alpha \, \mathcal{U}_{\text{ext}}$$

where the dependence on the strain tensor ε may be included in a potential strain energy $\mathcal{U}_{\text{int}}\left(\varepsilon_{\alpha\beta}\right)$. It was argued that the presence of torsion tensor as arguments of the Lagrangian density was reasonable at microscopic level e.g. Hehl (1985), Yasskin and Stoeger (1980). First, the reason supporting the inclusion of the three arguments (metric, torsion, and curvature) is to relate the curvature to mass and the torsion to spin at small scale level, both of them may be source of gravitation e.g. Aldrovandi and Pereira (2007). Second, for 3D strain gradient continuum, there are three independent invariants in torsion, and three independent invariants in curvature which can be introduced as "fine tuning" variables. These lead to propose a general form of Lagrangian density in terms of torsion and curvature including some important particular cases e.g. Katanaev and Volovich (1992), Kleinert (2008), Kobelev (2010), Maluf et al. (2002):

$$\mathscr{S} := \int_{\mathscr{M}} \mathscr{L}\left(\hat{g}_{\alpha\beta}, \hat{\aleph}^\gamma_{\alpha\beta}, \hat{\mathcal{R}}^\gamma_{\alpha\beta\lambda}; g_{\alpha\beta}, \aleph^\gamma_{\alpha\beta}, \mathfrak{R}^\gamma_{\alpha\beta\lambda}\right) \omega_n \tag{4.82}$$

where the domain of integration is, for example, over a volume of the spacetime swept out by the worldlines of all particles of the continuum matter e.g. Taub (1954). In a general manner, Lagrangian (4.82) includes different models e.g. Lompay (2014), Maier (2014), Sotiriou and Faraoni (2010) (in the last reference the analysis of viability of each of these models with their field equations is conducted in details), Vitagliano et al. (2011). Two possibilities to account for spacetime-matter coupling already exist (minimal coupling Lehmkuhl 2011): (a) the tetrads formulation $g_{\alpha\beta} := F^i_\alpha \, \hat{g}_{ij} \, F^j_\beta$, and (b) the strain method $\varepsilon_{\alpha\beta} := (1/2)(g_{\alpha\beta} - \hat{g}_{\alpha\beta})$ where the presence of spacetime metric is implicitly assumed in the matter Lagrangian e.g. Lehmkuhl (2011). They are not exclusive. When considering the motion of a continuum matter $\mathscr{B}$ within a spacetime $\mathscr{M}$ with respectively their own metric and (Levi-Civita) connection, Taub (1954) and later Carter (1973) analyzed the perturbation of a continuum motion due of both the infinitesimal displacement of the medium and the infinitesimal variation of the spacetime metric tensor for deriving the associated variational principle, for perfect fluids and slightly extended for elastic continuum matter respectively. Carter has shown that it is unnecessary to make distinction between the Lagrangian and the Eulerian variations of the action $\mathscr{S} := \int_{\mathscr{M}} \mathscr{L}\omega_n$, if the action $\mathscr{L}$ or the local infinitesimal translation (Poincaré gauge) vanishes on the boundary of the volume of integration $\partial \mathscr{M}$. For curved continuum with torsion evolving within a curved spacetime with torsion, further studies are needed. Let proceed to a more complete definition of generalized transformations e.g. Hammond (1990), Lovelock (1971), Utiyama (1956), Vitagliano et al. (2011) which include change of shape and change of topology e.g. Verçyn (1990) of manifolds. The basic idea is to highlight the change of local topology within continuum matter during the transformations.

4.2.5.1 Example of Nonminimal Coupling Application

The simultaneous presence of spacetime metric and matter metric, matter torsion and curvature, and spacetime torsion and curvature means a coupling of the spacetime and the matter. It was shown in e.g. Puetzfeld and Obukhov (2008) that under minimal coupling theory of gravitation the detection of spacetime torsion was not possible for bodies without intrinsic spin, or microstructural architecture. To go further, and as an illustration of nonminimal coupling between curved spacetime with torsion, and a matter with microstructure, the following Lagrangian, which is an example of (4.82) was proposed by Puetzfeld and Obukhov (2013b) : $\mathscr{L} := \mathscr{F}(\hat{g}_{\alpha\beta}, \hat{\aleph}^{\gamma}_{\alpha\beta}, \hat{\mathscr{R}}^{\gamma}_{\alpha\beta\lambda}) \, \mathscr{L}_M(\hat{g}_{\alpha\beta}, \Psi^A, \nabla_{\alpha}\Psi^A)$ where $\mathscr{F}$ is a scalar function whose arguments are metric, torsion and curvature components of the spacetime. General formulation of conservations laws deduced from this Lagrangian by means of Noether's theorem were obtained in Puetzfeld and Obukhov (2013b), where they developed the equations for pole-dipole bodies under gravitation. This paper based on nonminimal coupling opens the way to derive theory able to detect torsion of spacetime.

4.2.5.2 Strain Tensor in Relativity

One main difficulty in defining the relativistic strain in deformable solid is the missing of a natural undeformed state for solid continuum. This difficulty does not appear for fluid mechanics in relativistic mechanics. For solid-like continuum and for the concept of body's strain in the framework of relativistic gravitation, we start by defining the basic unknown as a mapping from the four-dimensional spacetime onto the three-dimensional body (or material space) where $\mu = 0, 1, 2, 3$ and $i = 1, 2, 3$:

$$\varphi : x^\mu \in \mathcal{M} \rightarrow X^i := \varphi^i \left(x^\mu \right) \in \mathcal{B} \tag{4.83}$$

with the following conditions: (a) The linear mapping $\partial_\mu \varphi^i$ should has maximal rank; (b) the null space of the linear mapping $\partial_\mu \varphi^i$ is a time-like with respect to the spacetime metric $\hat{g}_{\alpha\beta}$. In other words, there exists a unique time-like four-vector field $u^\mu \in T_\mathbf{x}\mathcal{M}$ such that:

$$u^\mu \, \partial_\mu \varphi^i = 0, \qquad \hat{g}_{\mu\nu} \, u^\mu u^\nu = 1, \qquad u^0 > 0 \tag{4.84}$$

The mapping φ is called configuration of the body. Solving the above equations allows us to express the velocity field of the matter u^μ as function of the components $\partial_\mu \varphi^i$, and the spacetime metric $\hat{g}_{\mu\nu}$. We remind the use the coordinate system $(x^0 := ct, x^1, x^2, x^3)$ in this section and in most part of this paper. The relation $\hat{g}_{\alpha\beta}u^\alpha u^\beta = c^2$ can be replaced by $\hat{g}_{\alpha\beta}u^\alpha u^\beta = 1$ obviously in the adopted coordinate system. The following relation holds e.g. Kijowski and Magli (1992):

$$\overline{\nabla}_\gamma \left(\hat{g}_{\alpha\beta} u^\alpha u^\beta \right) = 2 u_\alpha \overline{\nabla} u^\alpha = 0 \tag{4.85}$$

This constitutes the geodesic equation of a timelike curve: $\hat{g}_{\mu\nu} \, u^\mu u^\nu = 1$ with $u^\mu := dx^\mu/d\tau$.

Definition 4.1 The body-metric (or matter-metric) in $\mathcal{B}$ corresponding to the configuration φ of the body evolving within a spacetime $\mathcal{M}$ endowed with the metric $\hat{g}_{\alpha\beta}$ is defined as:

$$g^{ij} := \partial_\mu \varphi^i \, \hat{g}^{\mu\nu} \, \partial_\nu \varphi^j \tag{4.86}$$

This metric although depending on x^0 is a definite positive symmetric associated to the matter. It measures the space distance between microcosms of matter at each time x^0. This is similar approach compared to the tetrads method, however $\partial_\mu \varphi^i$ is not truly analogous to the classical deformation gradient but corresponds to its inverse e.g. Kijowski and Magli (1992). The line element in the body $\mathcal{B}$ holds: $dS^2 := g_{ij}dX^i dX^j$. Since the matter-metric $g_{ij}(x^\mu)$ depends on the x^μ (four spacetime coordinates) and the configuration φ is not a diffeomorphism, a supplemental (necessary and sufficient) condition is required.

Definition 4.2 The orthogonal projection operator $\mathscr{P}^{\alpha\beta} := \hat{g}^{\alpha\beta} - u^\alpha u^\beta$ is a projection from spacetime $\mathcal{M}$ onto the spacelike hypersurface orthogonal at event x^μ.

Each tangent vector to the manifold $\mathcal{M}$ can be decomposed into a component tangent to the velocity u^μ and a component orthogonal to it. The supplemental condition applies on the space projection operator which, first, satisfies some trivial properties:

$$\mathscr{P}^{\alpha\beta} u_\beta = 0, \qquad \mathscr{P}^{\alpha\beta} \hat{g}_{\alpha\beta} = 3 \tag{4.87}$$

Then, the Lie derivative of the metric projection orthogonal to four-vector $\mathbf{u}$ in the direction of $\mathbf{u}$ itself should vanish: $\mathcal{L}_{\mathbf{u}} \mathscr{P}^{\alpha\beta} \equiv 0$. We have defined the contravariant components of the matter metric $g^{ij}(x^\mu)$ in the relations (4.86). An alternative way to calculate its covariant components consists in calculating the line element of the projection of a four-dimensional vector $\mathscr{M}$ to the three-dimensional body $\mathscr{B}$:

$$ds^2 = \mathscr{P}_{\mu\nu} dx^\mu dx^\nu \equiv g_{ij} dX^i dX^j, \qquad dx^\mu = F_i^\mu dX^i$$

where $F_i^\mu(X^k, \tau)$ is the direct deformation gradient (not invertible) from $\mathscr{B}$ to $\mathscr{M}$, and τ is merely a time parameter and not argument. We deduce the metric components and by the way we can define the extended version of the Green-Lagrange strain tensor in relativistic continuum mechanics, e.g. Kijowski and Magli (1992), Maugin (1978):

$$g_{ij} = F_i^\mu \mathscr{P}_{\mu\nu} F_j^\nu, \qquad \varepsilon_{ij} := (1/2)\left(F_i^\mu \mathscr{P}_{\mu\nu} F_j^\nu - \hat{g}_{ij} \right) \tag{4.88}$$

The contravariant components (4.86) are defined by means of the transformation φ whereas the covariant components are obtained with the help of the projection operator and the gradient of deformation. Covariant components reduce to classical Green-Lagrange strain tensor for classical continuum mechanics e.g. Marsden and Hughes (1983).

Remark 4.28 On the one hand, starting from the definition of the gradient of deformation $dx^\mu := F_i^\mu dX^i$ and the transformation φ from $\mathscr{M}$ to $\mathscr{B}$, say $dX^i = \partial_\nu \varphi^i dx^\mu$, we obtain the relation: $F^\mu \partial^i \varphi_\nu = \delta_\nu^\mu$, since $\delta_\nu^\mu = g^{\mu\alpha} g_{\alpha\nu}$, where Greek indices run for $(0, 1, 2, 3)$, and the gradient of deformation is not invertible. On the other hand, we can write $dX^i = \partial_\mu \varphi^i dx^\mu$, and $dx^\mu = F_j^\mu dX^j$, we can deduce:

$$\partial_\mu \varphi^i F_j^\mu = \mathscr{P}_j^i, \qquad \mathscr{P}_\nu^\mu := g^{\mu\alpha} \mathscr{P}_{\alpha\nu} \tag{4.89}$$

For flat Minkowskian spacetime $\mathscr{M}$ and with the coordinate system (x^0, x^1, x^2, x^3), the mixed components of the projector reduce to:

$$\mathscr{P}_\nu^\mu := g^{\mu\alpha} \mathscr{P}_{\alpha\nu} = \begin{bmatrix} 0 & 0 & 0 & 0 \\ 0 & 1 & 0 & 0 \\ 0 & 0 & 1 & 0 \\ 0 & 0 & 0 & 1 \end{bmatrix} \tag{4.90}$$

where spacelike projection reduces to identity tensor δ^i_j if only $3D$ space and matter are considered.

Now, let consider a spacetime configuration of a material and let define the pull-back of the three-dimensional space metric of $\mathscr{B}$ (Riemann manifold) as follows:

$$h_{\mu\nu} := \partial_\mu \varphi^i \, g_{ij} \, \partial^j_\nu \qquad (4.91)$$

where $g_{ij}(x^0, x^i), i = 1, 2, 3$ is a metric of the Riemann configuration of the material body $\mathscr{B}$ at the event x^μ. We have directly the following properties:

$$h_{\mu\nu} = h_{\nu\mu}, \qquad h_{\mu\nu} u^\nu = \partial_\mu \varphi^i \, g_{ij} \, \partial_\nu \varphi^j u^\nu = 0$$

showing that the symmetric spacetime metric $h_{\mu\nu}$ is orthogonal to the velocity field u^ν deduced from Eq. (4.84). Let define the tensor

$$\mathscr{K}^\mu_\nu := \hat{g}^{\mu\sigma} h_{\sigma\nu} + u^\mu u_\nu \qquad (4.92)$$

The velocity vector is an eigenvector of the symmetric tensor $\mathscr{K}$ with eigenvalue 1:

$$\mathscr{K}^\mu_\nu u^\nu = \hat{g}^{\mu\sigma} \underbrace{h_{\sigma\nu} u^\nu}_{=0} + u^\mu \underbrace{u_\nu u^\nu}_{=1} = u^\mu$$

Several strain tensors may be designed by means of this tensor $\mathscr{K}$ which is analogous to the Cauchy-Green strain tensor and the related generalized strain measures e.g. Curnier and Rakotomanana (1991):

$$E^\mu_\nu := \frac{1}{2} \left(\mathscr{K}^\mu_\nu - \delta^\mu_\nu \right), \qquad E^\mu_\nu := \frac{1}{2} \ln \left(\mathscr{K}^\mu_\nu \right) \qquad (4.93)$$

The two previous definitions of relativistic strain tensor were respectively introduced by Maugin e.g. Maugin (1993), and Kijowski and Magli e.g. Kijowski and Magli (1992). Tensor $\mathbf{E}$ contains all information about the local state of strain within matter, and satisfies the relation $E^\mu_\nu u^\nu = 0$.

4.2.5.3 Metric Strain Tensor

The difference between metric of spacetime and that of the matter is defined by the strain tensor that is similar to the Green-Lagrange strain in relativistic theory (4.88):

$$2\varepsilon_{\alpha\beta} := g_{\alpha\beta} - \hat{g}_{\alpha\beta} \qquad (4.94)$$

defining modification of shape of the continuum with respect to spacetime. In this definition, it is implicitly assumed that the continuum matter is initially endowed by the spacetime metric $\hat{g}_{\alpha\beta}$. For the sake of the simplicity, this metric is always

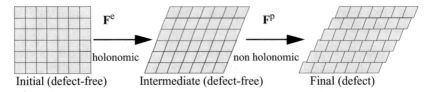

Initial (defect-free)	Intermediate (defect-free)	Final (defect)

Fig. 4.4 Change of local topology and stress relaxation. When an external force is applied, an internal strain is produced in the continuum. For the sake of the simplicity, it is assumed that an internal elastic stress accompanies this strain during a finite interval of time. If we keep the deformation much longer as the relaxation time of the material, the bonding structure of microcosms change and the internal strain diminishes (rightmost state) or eventually disappears. In fact a new type of strain appears and corresponds to plastic deformation/or more generally a change of local topology within the matter e.g. Bilby et al. (1955), Le and Stumpf (1996)

assumed to be Euclidean/(flat) Minkowskian. The definition (4.94) is appropriate for elastic deformation of matter when no internal change of microstructure changes. For understanding how material evolves in time, let consider a material consisting of many microcosms (represented by rectangular box on the below Fig. 4.4) bonded each other. Let assume that their are at their equilibrium state in the absence of strains (on the leftmost state).

In Fig. 4.4, the central and the rightmost figures have the shame shape, mathematically the same induced metric but with different bonding structure between microcosms. More in details, two points x^μ and $x^\mu + dx^\mu$ are separated by the (squared) distance $ds^2 = g_{\alpha\beta}(\mathbf{x}) dx^\alpha dx^\beta$. If we remove virtually the strain field in the region surrounding these two points, then these two points will take other position $\tilde{x}^\mu$, and $\tilde{x}^\mu + d\tilde{x}^\mu$ respectively. This corresponds to a map $\varphi : \mathbf{x} \to \tilde{\mathbf{x}}(\mathbf{x})$. We can define therefore a metric $\tilde{g}_{\alpha\beta}(\mathbf{x})$ measuring the virtual distance between the transformed points (commonly it is called the pullback of the metric $g_{\alpha\beta}$ for the map, $\tilde{\mathbf{g}} := \varphi^* \mathbf{g}$ e.g. Marsden and Hughes 1983):

$$d\tilde{s}^2 := g_{\alpha\beta}(\tilde{\mathbf{x}}) d\tilde{x}^\alpha d\tilde{x}^\beta = g_{\alpha\beta}(\tilde{\mathbf{x}}) \frac{\partial \tilde{x}^\alpha}{\partial x^\mu} \frac{\partial \tilde{x}^\beta}{\partial x^\nu} dx^\mu dx^\nu := \tilde{g}_{\mu\nu}(\mathbf{x}) dx^\mu dx^\nu$$

An alternative (Green-Lagrange) strain tensor may be also defined as the half-difference of metric:

$$2\varepsilon_{\alpha\beta}(\mathbf{x}) := g_{\alpha\beta}(\mathbf{x}) - \tilde{g}_{\alpha\beta}(\mathbf{x}) \tag{4.95}$$

which is different than the definition in Eq. (4.94). The strain (4.95) might be related to the concept of locally relaxed configuration in the framework of continua elastoplasticity e.g. Rakotomanana (2003).

Remark 4.29 The definitions of strain (4.94) and (4.95) may be considered as a relativistic strain tensor compared to the spacetime and to a non defected configuration of the continuum, respectively. In the framework of general relativity where the spacetime is no more flat, the spacetime metric is field dependent on the coordinates (x^μ). When dealing with perturbation analysis of a continuum motion, say $\mathscr{B}$ in the spacetime $\mathscr{M}$, it is well sound that perturbations of the motion are induced both by the displacements of $\mathscr{B}$ and by the non uniformity of the spacetime $\mathscr{M}$ metric field e.g. Carter (1973), Carter and Quintana (1977). Accordingly, the accounting for the two contributions should be done when deriving the change of the metric field for a variation principle.

4.2.5.4   Physical Interpretation on Material Manifold and Defects

For a n-dimensional continuum ($n \leq 3$), the Ricci curvature may play the role of irreversible plastic deformation e.g. Kobelev (2010). In this work, the emergence of torsion and curvature is treated as the transition from elastic, defect-free state, with Euclidean geometry to plastic, defect nucleation, endowed with Riemann–Cartan geometry (Kleman and Friedel 2008). In some sense, it is analogous to the nucleation of dislocations and disclinations, the microcracking within idealized virgin continuum e.g. Rakotomanana (1997), Ramaniraka and Rakotomanana (2000). The accounting of the torsion, and curvature fields on the metric-affine continuum may be done for different length scales as illustrated by Fig. 4.5 ranging from crystalline defects to continuous distributions of microcracking. Design of strain energy density of gradient continuum may be recast into the mathematical background of relativistic gravitation with these new insights. Previous original energy function in Mindlin (1964, 1965), and all particular cases developed after, would be better reshaped with a metric-affine manifold as geometric basis. In any case, there is a plethora of new constants and certainly with little guiding physics

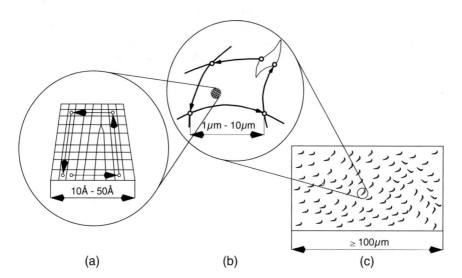

(a)   (b)   (c)

Fig. 4.5 Various length scales of a model: (**a**) crystalline—with a Burgers vector, (**b**) microscopic with cracking—showing a non closed Cartan path, and (**c**) macroscopic defected continuum $\simeq 100\,\mu m$

principle to help the significance of each new constant. For instance, Katanaev and Volovich reduced the number of independent parameters of previous Lagrangian to 2. They have adapted the Lagrangian method of metric-affine gravity to three-dimensional elasticity with defects e.g. Verçyn (1990). To this end, they find solutions of the associated equilibrium and analyze the set of coupling constants that may lead to physically acceptable solutions: the requirement is to impose the model to describe solutions with only dislocations, with only disclinations, and the case with neither dislocations nor disclinations Katanaev and Volovich (1992).

4.2.5.5 Torsion Change

The existence of both the displacement of the continuum $\mathscr{B}$ and the non uniformity of the spacetime $\mathscr{M}$ metric has also its consequences for the connection which is point-dependent. The difference between the matter connection and the Levi-Civita connection of the matter is the *matter contortion*, owing that both the spacetime and the matter may have their own connection e.g. Koivisto (2011),

$$\mathcal{T}^{\gamma}_{\alpha\beta} := \Gamma^{\gamma}_{\alpha\beta} - \overline{\Gamma}^{\gamma}_{\alpha\beta} \tag{4.96}$$

where $\overline{\Gamma}^{\gamma}_{\alpha\beta}$ are the Christoffel symbols associated to the metric $g_{\alpha\beta}$. The contortion tensor describes the deviation of the matter geometry from the Riemannian geometry one, whose connection reduces to the Christoffel symbols (Fig. 4.6).

In the framework of microphysics and relativistic gravitation, Hehl et al. have already observed that the intrinsic spin of microparticles is rather associated to contortion but not to torsion e.g. Hehl et al. (1976). From mathematical point of view, it should be reminded that torsion is a more basic variable since contortion includes both the metric and the torsion tensors. When the connection is assumed metric compatible, contortion tensor $\mathcal{T}^{\gamma}_{\alpha\beta}$ includes symmetric and skew-symmetric parts, see Eq. (2.44) where $\Omega^{\gamma}_{\alpha\beta} := (1/2)\,\aleph^{\gamma}_{\alpha\beta}$,

$$\mathcal{T}^{\mu}_{\alpha\beta} := \Omega^{\gamma}_{\alpha\beta} + g^{\gamma\lambda}g_{\alpha\mu}\,\Omega^{\mu}_{\lambda\beta} + g^{\gamma\lambda}g_{\mu\beta}\,\Omega^{\mu}_{\lambda\alpha},$$

Fig. 4.6 Stress-strain curve during plastic deformation of continuum. During the loading of the matter, beyond the stress yield σ_Y, there is a nucleation of dislocations within the body and then nucleation of discontinuities (torsion)

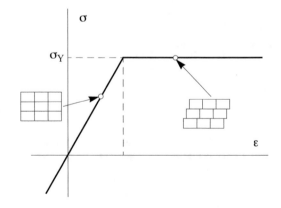

Matter contortion in Eq. (4.96) includes both the metric and the torsion. As alternative (by analogy to the metric), if we choose the variable $\tilde{\mathfrak{T}}^{\gamma}_{\alpha\beta} := \Gamma^{\gamma}_{\alpha\beta} - \hat{\Gamma}^{\gamma}_{\alpha\beta}$ (two independent connections for spacetime and matter e.g. Koivisto 2011), the interpretation is modified:

$$\tilde{\mathfrak{T}}^{\gamma}_{\alpha\beta} = \mathfrak{T}^{\gamma}_{\alpha\beta} + (1/2)\, g^{\gamma\sigma}(\partial_{\alpha} g_{\sigma\beta} + \partial_{\beta} g_{\alpha\sigma} - \partial_{\sigma} g_{\alpha\beta})$$

$$- (1/2)\, \hat{g}^{\gamma\sigma}(\partial_{\alpha} \hat{g}_{\sigma\beta} + \partial_{\beta} \hat{g}_{\alpha\sigma} - \partial_{\sigma} \hat{g}_{\alpha\beta}) \tag{4.97}$$

Nevertheless, the choice of the contortion $\mathfrak{T}^{\gamma}_{\alpha\beta}$ seems worth because it captures the change of the continuum matter topology (locally) from holonomic to non holonomic deformation, rather than $\hat{\mathfrak{T}}^{\gamma}_{\alpha\beta}$ which is rather the difference of local topology between the matter and the spacetime.

Remark 4.30 In the framework of continuum physics, it is worth to consider description at different length scales, say "macroscopic deformation" (associated to the change of metric) and "mesoscopic deformation" (change of other geometric variables). For instance, the first length scale constitutes the basis of classical continuum mechanics e.g. Marsden and Hughes (1983), whereas the second one is used for internal architecture of matter e.g. Rakotomanana (2003). A real situation mixed both of them. A practical approach would assume that at the macroscopic length scale, the deformation of the continuum is a assumed a diffeomorphism. In a such a particular way, the Levi-Civita connection coefficients $\overline{\Gamma}^{\gamma}_{\alpha\beta}$ associated to the metric $g_{\alpha\beta}$ do engender neither torsion nor curvature, then only contortion $\mathfrak{T}^{\gamma}_{\alpha\beta}$ matters in Eq. (4.96).

4.2.5.6 Curvature Change

Let first define the general context on curvature. For relativistic gravitation, the symbols of Christoffel $\overline{\Gamma}^{\gamma}_{\alpha\beta} := (1/2)g^{\gamma\lambda}(\partial_{\alpha} g_{\lambda\beta} + \partial_{\beta} g_{\alpha\lambda} - \partial_{\lambda} g_{\alpha\beta})$ were often considered as possible additional arguments of the Lagrangian density, in addition to metric $g_{\alpha\beta}$, and other kinematical variables related to the geometry of spacetime e.g. Forger and Römer (2004).

However, on a Riemannian continuum, Lovelock and Rund have shown that if one assumes a dependence as $\mathscr{L}(g_{\alpha\beta}, \partial_{\gamma} g_{\alpha\beta})$, the Lagrangian takes necessarily the form of $\mathscr{L}(g_{\alpha\beta})$ to satisfy covariance. This is a major result on the covariance condition. The first part of the present paper was inspired in part from results in Lovelock and Rund (1975). Lovelock (1971) then extended its investigation to Lagrangian density $\mathscr{L}(g_{\alpha\beta}, \partial_{\gamma} g_{\alpha\beta}, \partial_{\gamma}\partial_{\mu} g_{\alpha\beta})$ Forger and Römer (2004), Lovelock (1971) and applied the covariance condition. Earlier, Cartan (1922) faced the problem of finding a symmetric second order contravariant tensor (called Einstein tensor) $T^{\mu\nu}(g_{\alpha\beta}, \partial_{\gamma} g_{\alpha\beta}, \partial_{\gamma}\partial_{\mu} g_{\alpha\beta})$ which is divergence free and linear with respect to second derivatives $\partial_{\gamma}\partial_{\mu} g_{\alpha\beta}$ of the metric, assuming a Minkowskian spacetime background. See Anderson (1978) for more recent results of divergence-free second order contravariant tensors. To satisfy the covariance condition, any Lagrangian density depending on the metric and its first and second derivatives is necessarily function of only the metric and the curvature (see Lovelock (1971) for general form of Lagrangian) $\mathscr{L}(g_{\alpha\beta}, \overline{\mathfrak{R}}^{\lambda}_{\alpha\beta\mu})$, where curvature components are directly obtained from derivatives of metric components (Lovelock and Rund 1975):

$$\overline{\mathfrak{R}}^{\lambda}_{\alpha\beta\mu} = (1/2)g^{\lambda\sigma}\left(\partial_{\mu}\partial_{\alpha}g_{\sigma\beta} - \partial_{\mu}\partial_{\beta}g_{\sigma\alpha} + \partial_{\sigma}\partial_{\beta}g_{\mu\alpha} - \partial_{\sigma}\partial_{\alpha}g_{\mu\beta}\right)$$
$$+ (1/4)g^{\lambda\sigma}\left(\partial_{\alpha}g_{\sigma\gamma} + \partial_{\gamma}g_{\alpha\sigma} - \partial_{\sigma}g_{\alpha\gamma}\right)g^{\gamma\kappa}\left(\partial_{\beta}g_{\kappa\mu} + \partial_{\mu}g_{\beta\kappa} - \partial_{\kappa}g_{\beta\mu}\right)$$
$$- (1/4)g^{\lambda\sigma}\left(\partial_{\beta}g_{\sigma\gamma} + \partial_{\gamma}g_{\beta\sigma} - \partial_{\sigma}g_{\beta\gamma}\right)g^{\gamma\kappa}\left(\partial_{\alpha}g_{\kappa\mu} + \partial_{\mu}g_{\alpha\kappa} - \partial_{\kappa}g_{\alpha\mu}\right)$$

$$(4.98)$$

The change of curvature is induced by the presence of the matter contortion (4.96). The non uniform metric $g_{\alpha\beta}(x^{\mu})$ induces connection $\overline{\Gamma}^{\gamma}_{\alpha\beta}(x^{\mu})$ and then nonzero curvature $\overline{\mathfrak{R}}^{\gamma}_{\alpha\beta\lambda}(x^{\mu})$ (except in elasticity for strongly continuum mechanics e.g. Marsden and Hughes 1983). Indeed, another still open question is the possible coupling between the spacetime curvature $\hat{\mathfrak{R}}$ and the matter curvature $\mathfrak{R}$, whenever curvature is zero for neither spacetime nor matter. By choosing a metric connection on the manifold, we have:

$$\mathfrak{R}^{\gamma}_{\alpha\beta\lambda} = \partial_{\alpha}(\overline{\Gamma}^{\gamma}_{\beta\lambda} + \mathfrak{T}^{\gamma}_{\beta\lambda}) - \partial_{\beta}(\overline{\Gamma}^{\gamma}_{\alpha\lambda} + \mathfrak{T}^{\gamma}_{\alpha\lambda})$$
$$- (\overline{\Gamma}^{\mu}_{\alpha\lambda} + \mathfrak{T}^{\mu}_{\alpha\lambda})(\overline{\Gamma}^{\gamma}_{\beta\mu} + \mathfrak{T}^{\gamma}_{\beta\mu}) + (\overline{\Gamma}^{\mu}_{\beta\lambda} + \mathfrak{T}^{\mu}_{\beta\lambda})(\overline{\Gamma}^{\gamma}_{\alpha\mu} + \mathfrak{T}^{\gamma}_{\alpha\mu})$$

We arrive to the expression of the curvature strain which is defined as the difference between the matter and the curvature calculated with Levi-Civita connection:

$$\mathcal{K}^{\gamma}_{\alpha\beta\lambda} := \mathfrak{R}^{\gamma}_{\alpha\beta\lambda} - \overline{\mathfrak{R}}^{\gamma}_{\alpha\beta\lambda} = \overline{\nabla}_{\alpha}\mathfrak{T}^{\gamma}_{\beta\lambda} - \overline{\nabla}_{\beta}\mathfrak{T}^{\gamma}_{\alpha\lambda} - (\mathfrak{T}^{\gamma}_{\beta\mu}\mathfrak{T}^{\mu}_{\alpha\lambda} - \mathfrak{T}^{\gamma}_{\alpha\mu}\mathfrak{T}^{\mu}_{\beta\lambda}) \qquad (4.99)$$

where another interest appears when we calculate the scalar curvature e.g. Sotiriou et al. (2011). Now, by analogy with torsion, we can define the curvature "strain" with respect to the spacetime curvature by writing $\tilde{\mathcal{K}}^{\gamma}_{\alpha\beta\lambda} := \mathfrak{R}^{\gamma}_{\alpha\beta\lambda} - \hat{\mathfrak{R}}^{\gamma}_{\alpha\beta\lambda} = \mathcal{K}^{\gamma}_{\alpha\beta\lambda} + (\overline{\mathfrak{R}}^{\gamma}_{\alpha\beta\lambda} - \hat{\mathfrak{R}}^{\gamma}_{\alpha\beta\lambda})$ where we remind the components from Eq. (4.98):

$$\overline{\mathfrak{R}}^{\lambda}_{\alpha\beta\mu} = (1/2)g^{\lambda\sigma}\left(\partial_{\mu}\partial_{\alpha}g_{\sigma\beta} - \partial_{\mu}\partial_{\beta}g_{\sigma\alpha} + \partial_{\sigma}\partial_{\beta}g_{\mu\alpha} - \partial_{\sigma}\partial_{\alpha}g_{\mu\beta}\right)$$
$$+ (1/4)g^{\lambda\sigma}\left(\partial_{\alpha}g_{\sigma\gamma} + \partial_{\gamma}g_{\alpha\sigma} - \partial_{\sigma}g_{\alpha\gamma}\right)g^{\gamma\kappa}\left(\partial_{\beta}g_{\kappa\mu} + \partial_{\mu}g_{\beta\kappa} - \partial_{\kappa}g_{\beta\mu}\right)$$
$$- (1/4)g^{\lambda\sigma}\left(\partial_{\beta}g_{\sigma\gamma} + \partial_{\gamma}g_{\beta\sigma} - \partial_{\sigma}g_{\beta\gamma}\right)g^{\gamma\kappa}\left(\partial_{\alpha}g_{\kappa\mu} + \partial_{\mu}g_{\alpha\kappa} - \partial_{\kappa}g_{\alpha\mu}\right)$$

and with the hat:

$$\hat{\mathfrak{R}}^{\lambda}_{\alpha\beta\mu} = (1/2)\hat{g}^{\lambda\sigma}\left(\partial_{\mu}\partial_{\alpha}\hat{g}_{\sigma\beta} - \partial_{\mu}\partial_{\beta}\hat{g}_{\sigma\alpha} + \partial_{\sigma}\partial_{\beta}\hat{g}_{\mu\alpha} - \partial_{\sigma}\partial_{\alpha}\hat{g}_{\mu\beta}\right)$$
$$+ (1/4)\hat{g}^{\lambda\sigma}\left(\partial_{\alpha}\hat{g}_{\sigma\gamma} + \partial_{\gamma}\hat{g}_{\alpha\sigma} - \partial_{\sigma}\hat{g}_{\alpha\gamma}\right)\hat{g}^{\gamma\kappa}\left(\partial_{\beta}\hat{g}_{\kappa\mu} + \partial_{\mu}\hat{g}_{\beta\kappa} - \partial_{\kappa}\hat{g}_{\beta\mu}\right)$$
$$- (1/4)\hat{g}^{\lambda\sigma}\left(\partial_{\beta}\hat{g}_{\sigma\gamma} + \partial_{\gamma}\hat{g}_{\beta\sigma} - \partial_{\sigma}\hat{g}_{\beta\gamma}\right)\hat{g}^{\gamma\kappa}\left(\partial_{\alpha}\hat{g}_{\kappa\mu} + \partial_{\mu}\hat{g}_{\alpha\kappa} - \partial_{\kappa}\hat{g}_{\alpha\mu}\right)$$

For a flat spacetime as Minkowski $\mathcal{M}$ and Galilean $\mathcal{G}$ spacetime (without "external forces of gravitation"), this later vanishes $\hat{\mathfrak{R}}^\gamma_{\alpha\beta\lambda} \equiv 0$ as well as the torsion $\hat{\aleph}^\gamma_{\alpha\beta} \equiv 0$. In such a case, we exactly have the curvature expression $\overline{\mathfrak{R}}^\lambda_{\alpha\beta\mu}$ by changing the metric into the strain $2\,\varepsilon_{\alpha\beta}$ instead of $g_{\alpha\beta}$ (both first and second derivatives occur in the overlined curvature). The same remark as for the torsion holds for the curvature tensor.

Definition 4.3 Consider a curved spacetime with torsion, modeled with four dimensional metric-affine manifold $\mathcal{M}$ endowed with metric $\hat{g}_{\alpha\beta}$, torsion $\hat{\aleph}^\gamma_{\alpha\beta}$, and curvature $\hat{\mathfrak{R}}^\gamma_{\alpha\beta\lambda}$. A second gradient continuum $\mathcal{B}$ is a continuum whose matter Lagrangian is defined by $\mathscr{L}_M := \mathscr{L}_M(\varepsilon_{\alpha\beta}, \mathfrak{T}^\gamma_{\alpha\beta}, \mathcal{K}^\gamma_{\alpha\beta\lambda})$, where $\varepsilon_{\alpha\beta}$ (Eq. (4.94)), $\mathfrak{T}^\gamma_{\alpha\beta}$ (Eq. (4.96)), and $\mathcal{K}^\gamma_{\alpha\beta\lambda}$ (Eq. (4.99)) are respectively the strain, the torsion, and the bending of the continuum matter evolving in the spacetime.

Remark 4.31 An extension of the Einstein–Hilbert action was defined in e.g. Tamanini (2012) by considering two independent connections $\Gamma^\gamma_{\alpha\beta}$ and $\hat{\Gamma}^\gamma_{\alpha\beta}$ (for instance associated to spacetime and to matter respectively) in addition to the metric $g_{\alpha\beta}$ as arguments of the Lagrangian density. Recent and exhaustive review of arguments list for Lagrangian function may be found in e.g. Clifton et al. (2012).

4.2.5.7 Relativistic Volume-Forms and Matter Current

For studying the motion of a continuum within a relativistic spacetime, it is also necessary to consider two volume-forms. Let introduce a volume-form in the three-dimensional material space $\mathcal{B}$:

$$\omega_3 := \sqrt{\mathrm{Detg}}\, dX^1 \wedge dX^2 \wedge dX^3 \qquad (4.100)$$

The pull-back of ω_3 to the four-dimensional spacetime is a 3-form field in the four-dimensional spacetime $\mathcal{M}$. It means that ω_3 is a vector density. It is usual to define the matter current as the vector density:

$$\mathbf{J} := \rho_0\, \omega_3 \qquad (4.101)$$

where $\rho_0(X^i)$ is assumed as the density of the matter. Determination of the components of $\mathbf{J}$ in a coordinate system x^μ is done by substituting the material space coordinates $X^i := \varphi^i(x^\mu)$ by their values in function of x^μ:

$$\mathbf{J} = \rho_0\sqrt{\mathrm{Detg}}\left(\partial_\nu\varphi^1\partial_\mu\varphi^2\partial_\sigma\varphi^3\right) dx^\nu \wedge dx^\mu \wedge dx^\sigma$$

in which Greek indices run from 0 to 3, and where we can define components projected onto each of the four 4-form fields:

$$\left\{dx^0 \wedge dx^1 \wedge dx^2, dx^1 \wedge dx^2 \wedge dx^3, dx^2 \wedge dx^3 \wedge dx^0, dx^3 \wedge dx^0 \wedge dx^1\right\}$$

The associated components may thus be written in a condensed formulation:

$$J^\mu = \rho_0 \sqrt{\text{Detg}}\ \epsilon^{\mu\nu\rho\sigma} \left(\partial_\nu \varphi^1 \partial_\rho \varphi^2 \partial_\sigma \varphi^3 \right) \tag{4.102}$$

where $\epsilon^{\mu\nu\rho\sigma} = 1$ is for a cyclic circular permutation (0123), -1 for anti-cyclic permutation and 0 whenever two indices are equal. The exterior derivative of the 3-form ω_3 vanishes, say $d\omega_3 \equiv 0$ since it gives a 4-form in the three dimensional space $\mathscr{B}$. Therefore, the exterior derivative of its pull-back is also equal to zero: $d\mathbf{J} = d\,(\varphi^*\omega_3) = \varphi^*\,(d\omega_3) \equiv 0$. It is equivalent to write down a free-divergence field (which is an extension of the mass conservation):

$$d\mathbf{J} = \mathbf{DivJ}\ \omega_{\mathscr{M}} = 0 \qquad \Longrightarrow \qquad \mathbf{DivJ} = 0 \tag{4.103}$$

Following the method of e.g. Kijowski and Magli (1992), we observe that multiplying the matter current by the deformation gradient gives $\partial_\mu \varphi^a\, J^\mu \equiv 0$ since $\epsilon^{\mu\nu\rho\sigma} \left(\partial_\nu \varphi^1 \partial_\rho \varphi^2 \partial_\sigma \varphi^3 \right) \partial_\mu \varphi^a$ is the determinant of a matrix with two identical columns, since $a = 1, 2,$ or 3. Both $\partial_\mu \varphi^i\, u^\mu$ and $\partial_\mu \varphi^i\, J^\mu$ are equal to zero, then the matter current J^μ is parallel to the velocity field u^μ, say:

$$J^\mu = \rho \sqrt{\text{Detg}}\, u^\mu, \qquad \rho := \sqrt{\frac{J^\mu J_\mu}{\text{Detg}}} \tag{4.104}$$

where the matter current is given by relation (4.102). The term $\rho(\partial_\mu \varphi^i)$ is called actual rest frame density e.g. Kijowski and Magli (1992). Since $u^0 = 1$, the component $J^0 := \rho\sqrt{\text{Detg}} = \rho_0 \sqrt{\text{Detg}}\ \epsilon^{0\nu\rho\sigma} (\partial_\nu \varphi^1 \partial_\rho \varphi^2 \partial_\sigma \varphi^3)$ leads to:

$$J^0 = \rho_0 \sqrt{\text{Detg}}\ \text{Det} \left(\partial_\mu \varphi^i \right)$$

4.2.5.8 Lagrange for Coupled Gravity and Matter

We observe that: (a) curvature $\mathfrak{R}^\gamma_{\alpha\beta\lambda}$ is a second gradient variable in terms of strain $\varepsilon_{\alpha\beta}$, and then a third gradient one for displacement u_α; and (b) curvature is not linear with respect to contortion tensor $\mathcal{T}^\gamma_{\alpha\beta}$. The previous development shows us that the generalized deformation of the matter continuum with respect to the ambient spacetime (relativistic or classical) includes three contributions: (1) the metric strain $\varepsilon_{\alpha\beta}(x^\mu)$, (2) the loss of affine equivalence of matter during deformation (linear change of the connection) $\mathcal{T}^\gamma_{\alpha\beta}(x^\mu)$ in Eq. (4.97), and (3) the bending of the matter in the course of time (second order change of the connection) $\mathcal{K}^\gamma_{\alpha\beta\lambda}(x^\mu)$ in Eq. (4.99). Thanks to the expression of change of metric, torsion, and curvature, we obtain the following action by introducing the "generalized deformations" in the Lagrangian (4.82):

$$\mathscr{S} := \int \mathscr{L}\left(\hat{g}_{\alpha\beta}, \hat{\aleph}^{\gamma}_{\alpha\beta}, \hat{\mathscr{R}}^{\gamma}_{\alpha\beta\lambda}; \varepsilon_{\alpha\beta}, \mathscr{T}^{\gamma}_{\alpha\beta}, \mathscr{K}^{\gamma}_{\alpha\beta\lambda}\right) \omega_n \qquad (4.105)$$

which includes the spacetime gravitation (relativistic or not), and the part of Lagrangian due to the transformations of the matter. The presence of torsion of the continuum $\mathscr{B}$ and that of spacetime $\mathscr{M}$ is suggested by the seek of coupling of the Riemann–Cartan geometry of both of them. For instance, minimal coupling of electromagnetic field and gravitation may be done by means of the torsion of a spacetime e.g. Prasanna (1975a), Smalley and Krisch (1992). The presence of the spacetime torsion $\hat{\aleph}^{\gamma}_{\alpha\beta}$ may be illustrated by the model in e.g. Poplawski (2010) where the torsion is shown to be a candidate to be an electromagnetic potential by considering an Einstein–Cartan spacetime, the curl of the torsion trace constitutes the electromagnetic field tensor whereas the curvature represents the gravity.

Remark 4.32 Function (4.105) is a different formulation of the general Lagrangian suggested in e.g. Obukhov and Puetzfeld (2014), Puetzfeld and Obukhov (2013a) where they propose the contortion as primal variable but they keep the curvature as an argument of the Lagrangian instead of the change of the curvature. Two classes of models with non minimal coupling have been studied in Obukhov and Puetzfeld (2014), say,

$$\mathscr{L} = \mathscr{L}_G\left(\hat{g}_{\alpha\beta}, \hat{\aleph}^{\gamma}_{\alpha\beta}, \hat{\mathfrak{R}}^{\gamma}_{\alpha\beta\lambda}\right) \mathscr{L}_M\left(g_{\alpha\beta}, \aleph^{\gamma}_{\alpha\beta}, \mathfrak{R}^{\gamma}_{\alpha\beta\lambda}\right)$$

$$\mathscr{L} = \mathscr{L}_G\left(\hat{g}_{\alpha\beta}, \hat{\aleph}^{\gamma}_{\alpha\beta}, \hat{\mathfrak{R}}^{\gamma}_{\alpha\beta\lambda}\right) + \mathscr{L}_M\left(\hat{g}_{\alpha\beta}, g_{\alpha\beta}, \aleph^{\gamma}_{\alpha\beta}, \mathfrak{R}^{\gamma}_{\alpha\beta\lambda}\right)$$

The first one can be considered as multiplicative model, whereas the second one is a quite classical additive model. The present model (4.105) is slightly different.

In the case of classical Galilean spacetime, the Lagrangian density (4.105) extends the Mindlin's model in Mindlin (1964) for first strain gradient linear elasticity, and in Mindlin (1965) for the second strain gradient models. This also modified version of the Lagrangian proposed by Hehl et al. (2013) in the framework of Poincaré Gauge Theory where these authors propose a standard theory of gravitation with torsion, when a curved continuum matter with torsion is evolving within a curved spacetime with torsion. For inertial terms, add $\rho c\sqrt{u^{\alpha}u_{\alpha}}$ where $u^{\alpha} := \hat{g}^{\alpha\beta}u_{\beta}$ is the contravariant components of the four-vector velocity. The inertial terms in such a case is obtained by applying the condition of slow velocity $c^2 >> v^2$ and the four-vector velocity reduces to $u^{\alpha} \rightarrow (1, v^1, v^2, v^3)$. For that purpose, the (three dimensional) generalized deformation of the generalized continuum is defined by a symmetric strain $\varepsilon_{\alpha\beta}$, a skew-symmetric torsion deformation $\tilde{\tau}_{\alpha\beta}$, and a bending deformation $\tilde{\kappa}_{\alpha\beta}$, calculated by means of the space vector v^i (more precisely of the displacement associated to the velocity v^i). The last two tensors are obtained from the torsion, and the Ricci curvature. Both of them have only six components for three dimensional manifolds e.g. Nakahara (1996).

4.2.5.9 Remarks on Teleparallel Gravity (TPG)

Now, let go back to the vacuum spacetime with respect to torsion and to curvature fields. From Eq. (4.99) we may calculate the curvature $g^{\beta\lambda}\mathfrak{R}^{\alpha}_{\alpha\beta\lambda}$, and obtain the relationship:

$$\mathcal{R} = \overline{\mathcal{R}} + \mathcal{T} + \overline{\nabla}_{\alpha}(g^{\beta\lambda}\,\mathcal{T}^{\alpha}_{\lambda\beta}) - \overline{\nabla}_{\beta}(g^{\beta\lambda}\,\mathcal{T}^{\alpha}_{\lambda\alpha}) \tag{4.106}$$

where $\overline{\mathcal{R}}$ is the curvature obtained from the Levi-Civita connection. The quadratic contortion scalar function

$$\mathcal{T} := g^{\beta\lambda}(-\mathcal{T}^{\alpha}_{\beta\mu}\mathcal{T}^{\mu}_{\alpha\lambda} + \mathcal{T}^{\alpha}_{\alpha\mu}\mathcal{T}^{\mu}_{\beta\lambda}) \tag{4.107}$$

is called the quadratic torsion Lagrangian and is utilized in the tele parallel gravity e.g. Sotiriou et al. (2011). By the way, the two last terms are divergence terms with respect to the Levi-Civita connection. This means that for Weitzenböck connection with a zero curvature $\mathcal{R} \equiv 0$, the action of the Einstein relativistic gravitation (function of the curvature $\overline{\mathcal{R}}$) differs from the action of the tele parallel gravity (function of $\mathcal{T}$) by a boundary term (integral of a divergence term). This explains why the Einstein–Hilbert Lagrangian function has an equivalent Lagrangian function in terms of torsion (tele parallel gravitation) e.g. Cho (1976a). In this paper, Cho showed that if the gauge principle is applied for the group of translations T_4, the obtained gauge theory is unique and reduces to the Einstein classical theory of gravitation. The relation (4.99) relates the curvature and the contortion (and thus the torsion), and then it was showed that the use of torsion $\mathcal{T}$ rather than curvature $\mathcal{R}$ in the Lagrangian formulation gives an alternative geometric interpretation of the Einstein's theory with the help of translational gauge formalism (Cho 1976a). Such a approach allows us to develop generalized continuum models which showed that edge end screw Volterra dislocations are associated to gauge invariance in the three dimension material manifolds e.g. Malyshev (2000). In this paper, Malyshev has interestingly shown that the Einstein–Hilbert Lagrangian is suitable to study Volterra dislocations in elastic material. By the way, the use of Weitzenböck connection together with tele parallel gravitation was shown to be an alternative way of modelling Volterra dislocations. We recover the following theorem.

Theorem 4.2 *Consider a metric-affine manifold $(\mathscr{B}, \mathbf{g}, \nabla)$. Let assume that the curvature associated to the metric identically vanishes, say $\overline{\mathfrak{R}}^{\gamma}_{\alpha\beta\lambda} \equiv 0$. Then, any Lagrangian density $\mathscr{L}(\mathfrak{R}^{\gamma}_{\alpha\beta\lambda})$ function of the scalar curvature may be expressed as a Lagrangian density function of the contortion $\mathscr{L}(\mathcal{T}^{\gamma}_{\alpha\beta})$ (and its Levi-Civita covariant derivative).*

Proof It suffices to apply the relation (4.99) and to transfer the divergence-like terms at the manifold boundary $\partial\mathcal{B}$. See also proof in e.g. Sotiriou et al. (2011).   □

The most known Lagrangian density and related action built upon this theorem is the tele parallel gravity given by: $\mathscr{S} := (1/2\chi) \int_{\mathscr{B}} \mathfrak{T}\, \omega_n$, directly drawn from the relation (4.99), which is quadratic in torsion terms (but linear in curvature). This Lagrangian is obviously covariant, but Li et al. have shown its non invariance under the (active) Lorentz diffeomorphism (Li et al. 2001). This also true for its extension $f(\mathfrak{T})$ e.g. Li et al. (2001).

Remark 4.33 In some particular cases, curvature tensor may be defined without making use of (independent) connection but entirely based on the derivatives of the metric tensor $g_{\alpha\beta}$ (for Levi-Civita connection). Conversely, torsion has no metrical interpretation, it is an intrinsic property of the connection $\Gamma^{\gamma}_{\alpha\beta}$.

A class of Lagrangian density was proposed by using the Weitzenböck spacetime with torsion but zero curvature e.g. Hayashi (1979), Maluf et al. (2002) to obtain the so called "new general relativity". Relations between curvature based gravitation and torsion based gravitation, mainly boosted by the works of Kibble, Utiyama and others were developed by imposing a second order equations, rather than fourth order equations ones Kibble (1961), Utiyama (1956). This allowed to consider the torsion as gauge field of conformal transformation e.g. Hammond (1990). A Lagrangian density in the framework of tele parallel gravity was suggested, with the constant $\chi := 8\pi G/c^4$,

$$\mathscr{L}_G := \mathbb{T}^{\alpha\beta,\alpha'\beta'}_{\gamma,\gamma'}\, \mathfrak{T}^{\gamma}_{\alpha\beta}\mathfrak{T}^{\gamma'}_{\alpha'\beta'}, \qquad \mathbb{T}^{\alpha\beta,\alpha'\beta'}_{\gamma,\gamma'} = \frac{\mathrm{Det}(F^i_{\alpha})}{8\chi} g_{\gamma\gamma'} \left(g^{\alpha\alpha'}g^{\beta\beta'} - g^{\alpha\beta'}g^{\alpha'\beta} \right)$$

There is a long last debate on the consistency of tele parallel gravity compared to the theory of gravity developed with the curvature tensor as source of gravity. We do not go into detail of this interesting debate, however we may explore new method to be applied for relativistic gravitation and strain gradient continuum. Recently, some authors have suggested to recover the theory of general relativity from continuum mechanics e.g. Boehmer and Downes (2014) by starting to consider a Lagrangian function of the type:

$$\mathscr{L} := \frac{1}{2} \sum_{m=1}^{m=15} c_m \delta^{(\alpha\beta,\alpha'\beta')}_{m(\gamma,\gamma')} \mathfrak{T}^{\gamma}_{\alpha\beta}\mathfrak{T}^{\gamma'}_{\alpha'\beta'}$$

where c_m are undetermined coefficients and $\delta^{(\alpha\beta,\alpha'\beta')}_{m(\gamma,\gamma')}$ are the 15 possible combinations of the covariant $\hat{g}_{\kappa\nu}$ or contravariant $\hat{g}^{\kappa\nu}$ metric components of a flat Minkowskian spacetime $\mathscr{M}$. Replacing the curvature with the torsion-scalar variable $\mathfrak{T}$ or a scalar function of it constitutes the basis for the derivation of the teleparralel gravity equivalent with the curvature-based gravitation either for with or without spacetime e.g. Hayashi (1979), Ferraro and Fiorini (2011). In this later reference, search of spherically symmetric vacuous spacetime is conducted in presence of torsion tensor. This form is obtained by imposing isotropy of the spacetime, and by adding the Lorentz invariance of the Lagrangian with respect to all transformations defined by Eq. (2.24), they arrived to a function analogous to by retaining only one element $m = 1$ among the 15:

$$\mathscr{L} = c_1 \, \hat{g}_{\gamma\gamma'} \left(\hat{g}^{\alpha\alpha'} \hat{g}^{\beta\beta'} - \hat{g}^{\alpha\beta'} \hat{g}^{\alpha'\beta} \right) \mathfrak{T}^\gamma_{\alpha\beta} \mathfrak{T}^{\gamma'}_{\alpha'\beta'} \tag{4.108}$$

It should be stressed that the derivation of this (con)-torsion-based Lagrangian which is merely an expression of the Lagrangian in tele parallel gravitation, and by the way equivalent to the Hilbert–Einstein approach up to a surface term (see Eq. (4.99)), was deduced from the symmetry of the skew-symmetry of the torsion tensor with respect to covariant indices, the isotropy of the spacetime, and the invariance of the Lagrangian with respect to Lorentz group. We remind however that the paper (Boehmer and Downes 2014) rather directly used contortion tensor rather the torsion tensor.

4.3 Gauge Invariance on a Riemann–Cartan Continuum

The Principle of Least Action may be used to express the laws of continuum mechanics and relativistic gravitation. Symmetry transformations are changes in the coordinates $\tilde{x}^\mu(x^\alpha)$ that leave the action $\mathscr{S} := \int_{\mathscr{M}} \mathscr{L}(\mathbf{g}, \aleph, \mathfrak{R}) \, \omega_n$ invariant. Continuous symmetries (Lie group) are usually exploited to generate conservation laws (the most known method comes from the Noether's Theorem) e.g. Lazar and Anastassiadis (2008). Remind that the accounting of the Lagrangian variation of any tensor, for instance $\Delta g_{\alpha\beta}$, includes the Eulerian variation and the variation due to the "Lie derivative variation" according to $\Delta g_{\alpha\beta} = \delta g_{\alpha\beta} + \mathcal{L}_\xi g_{\alpha\beta}$, Carter (1973), Manoff (1999). The same principle holds for torsion and curvature. The goal of this last section is to generalize the translational invariance by considering both spatially-varying shifts $\xi(\mathbf{x})$ e.g. Utiyama (1956) and coordinate transformations $\tilde{\mathbf{x}}(\mathbf{x})$ that leave the action $\mathscr{S}$ invariant e.g. Lazar and Anastassiadis (2008).

Let consider the link between Minkowskian spacetime and global invariance. Among all physical processes involving low energies, the gravitational field induces weak forces compared to other fundamental forces acting in physics, it is called a weak interaction. We have seen that in the absence of gravitation, the spacetime reduces to four-dimensional Minkowskian spacetime $\mathscr{M}$ with metric $\hat{g}_{\mu\nu} :=$ diag$\{+1, -1, -1, -1\}$ uniform all over spacetime. The group of transformations: $x^\mu \rightarrow y^\mu = x^\mu + \Lambda^\mu_\nu \, x^\nu + \epsilon^\mu$, where $\Lambda^{\nu\mu}$ (such that $\Lambda^{\nu\mu} + \Lambda^{\mu\nu} = 0$), and ϵ^μ constitute 10 constant parameters and define the isometry group of the spacetime, called global Poincaré group of transformations (Lorentz rotations and translations) e.g. McKellar (1981). In the following, we localize the Poincaré's group of transformations in order to develop the local gauge symmetry of Lagrangian function

$\mathscr{L}$. More precisely, the localization consists in defining a non uniform and arbitrary vector field $\xi^\mu(x^\nu)$ on the manifold $\mathscr{M}$.

4.3.1 Lie Derivative and Gauge Invariance

Starting from the causality principle, Lorentz transformations are not the only symmetries of the Minkowski spacetime. Translation in space and translation in time constitute also symmetries. In the sixties, Zeeman (1964) has shown that the causality group is generated by the orthochronous Poincaré group together with dilations. Later, Williams (1973) has shown that the complete Poincaré group is the group that preserves the time-like vectors of the Minkowski spacetime, and even further the Poincaré group follows from the principle of the invariance of the light velocity.

4.3.1.1 Lie Derivative of a Vector, and 1-Form

Now we consider translations (space and time) as groups with smooth continuous parameters denoted ξ^μ, called *Lie groups*. For the Lorentz transformations, there are six transformations that generate the entire Lorentz group $\mathscr{O}(1, 3)$, three of these generators are spatial rotations, and the other three generators are time-space operators, called boosts. As physical interpretation, boosts are related to the change of observer's reference frame from another's. Mathematically, Lorentz boosts and spacetime translation that not necessary allow to bring you back to where you started, the Poincaré group is not compact. Let report on Fig. 4.7 the elements for defining the Lie derivative. On a manifold $\mathscr{M}$, the mapping $\psi(s, \mathbf{x}) := \psi_s(\mathbf{x})$ $(s \in \mathbb{R})$ transforms the point $\mathbf{x} \rightarrow \mathbf{y} = \psi_\epsilon(\mathbf{x})$ for an infinitesimal value of the parameter $s = \epsilon$. Conversely we have $\mathbf{x} = \psi_{-\epsilon}(\mathbf{y})$. The tangent space $\mathscr{T}_\mathbf{x}\mathscr{M}$ at $\mathbf{x}$ is transformed by the linear tangent mapping $d\psi_\epsilon$ at $\mathbf{x}$ to the tangent space $\mathscr{T}_\mathbf{y}\mathscr{M}$ at the point $\mathbf{y}$. Then the vector $\mathbf{u}$ is transformed as $\mathbf{v} = d\psi_\epsilon[\mathbf{u}]$, owing that the two tangent spaces are not the same. For comparing $\mathbf{u}$ and $\mathbf{v}$, we must pull back the vector $\mathbf{v}$ to the vector $\mathbf{v}' \in \mathscr{T}_\mathbf{x}\mathscr{M}$, calculated as $\mathbf{v}' := d\psi_{-\epsilon}[\mathbf{v}]$ where the linear tangent mapping is defined at $\mathbf{y}$. The Lie derivative of the vector $\mathbf{u}$ along the vector field ξ generator of the mapping ψ_s is defined as:

$$\mathcal{L}_\xi \mathbf{u} := \lim_{\epsilon \to 0} \frac{1}{\epsilon} \{d\psi_{-\epsilon}[\mathbf{u}(\psi_\epsilon(\mathbf{x}))] - \mathbf{u}(\mathbf{x})\} \qquad (4.109)$$

Fig. 4.7 Lie derivative of a vector field. Calculus of the difference $\mathbf{v}' - \mathbf{u}$

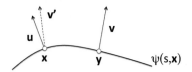

where the generator vector of the mapping ψ_ϵ is defined as:

$$\xi := \frac{d\psi(s, \mathbf{x})}{ds} \tag{4.110}$$

The derivation of various formulae of Lie derivative is given in appendix. The necessity to introduce Lie derivative is due to the fact that we cannot merely take the difference between components of the vector fields $\mathbf{u}$ and $\mathbf{v}$ to calculate the derivative since they are not on at the same tangent space, it is necessary to pull back $\mathbf{v}$ to $\mathbf{v}'$.

Now, consider a metric-affine manifold $\mathcal{M}$ ($3D$ material continuum or $4D$ space-time continuum), and spacetime local translation. Say an (active) diffeomorphism which is a smooth invertible map $\varphi : \mathcal{M} \rightarrow \mathcal{M}$. Let θ a scalar field on $\mathcal{M}$, and a new scalar field $\tilde{\theta}$ defined as $\tilde{\theta}(\mathbf{x}) := \theta[\varphi(\mathbf{x})]$, meaning that the initial scalar field is pushed forward to a new manifold, using nevertheless the same coordinate values. For instance the transformed 1-form under an active diffeomorphism (particular case of Poincaré's transformations e.g. McKellar (1981) of the form $\tilde{\mathbf{x}} := \mathbf{x} - \xi \ d\lambda$ (the "small" term $d\lambda \in \mathbb{R}$ is introduced here to mean an infinitesimal diffeomorphism) holds:

$$\tilde{u}_\mu(\mathbf{x}) = J_\mu^\alpha \, u_\alpha \left[\mathbf{x} + \xi(\mathbf{x}) \, d\lambda\right] = u_\mu(\mathbf{x}) + \left[\xi^\alpha(\mathbf{x}) \, \partial_\alpha u_\mu(\mathbf{x}) + u_\alpha(\mathbf{x}) \, \partial_\mu \xi^\alpha(\mathbf{x})\right] d\lambda$$

since $J_\mu^\alpha(\mathbf{x}) = \delta_\mu^\alpha + \partial_\mu \xi^\alpha(\mathbf{x}) \, d\lambda$, and where we might drop the dependence on $\mathbf{x}$ for the simplicity of notation. With the help of formulae and definition in the appendix, we obtain the Lie derivative of scalar, covariant and contravariant vectors:

$$\delta\phi(\mathbf{x}) := \mathcal{L}_\xi \phi(\mathbf{x}) = \xi^\alpha(\mathbf{x}) \partial_\alpha \phi(\mathbf{x}) \tag{4.111}$$

$$\delta\omega_\mu(\mathbf{x}) := \mathcal{L}_\xi \omega_\mu(\mathbf{x}) = \xi^\alpha(\mathbf{x}) \, \partial_\alpha \omega_\mu(\mathbf{x}) + \omega_\alpha(\mathbf{x}) \, \partial_\mu \xi^\alpha(\mathbf{x}) \tag{4.112}$$

$$\delta u^\mu(\mathbf{x}) := \mathcal{L}_\xi u^\mu(\mathbf{x}) = \xi^\alpha(\mathbf{x}) \, \partial_\alpha u^\mu(\mathbf{x}) - u^\alpha(\mathbf{x}) \, \partial_\alpha \xi^\mu(\mathbf{x}) \tag{4.113}$$

The relations (4.111)–(4.113) can be re-written by means of covariant derivatives as follows:

$$\mathcal{L}_\xi \phi(\mathbf{x}) := \xi^\alpha(\mathbf{x}) \nabla_\alpha \phi(\mathbf{x}) \tag{4.114}$$

$$\mathcal{L}_\xi \omega_\mu(\mathbf{x}) := \xi^\alpha(\mathbf{x}) \, \nabla_\alpha \omega_\mu(\mathbf{x}) + \omega_\alpha(\mathbf{x}) \, \nabla_\mu \xi^\alpha(\mathbf{x}) + \xi^\alpha(\mathbf{x}) \, \aleph_{\alpha\mu}^\rho(\mathbf{x}) \, \omega_\rho(\mathbf{x}) \tag{4.115}$$

$$\mathcal{L}_\xi u^\mu(\mathbf{x}) := \xi^\alpha(\mathbf{x}) \, \nabla_\alpha u^\mu(\mathbf{x}) - u^\alpha(\mathbf{x}) \, \nabla_\alpha \xi^\mu(\mathbf{x}) - \xi^\alpha(\mathbf{x}) \, \aleph_{\alpha\rho}^\mu(\mathbf{x}) \, u^\rho(\mathbf{x}) \tag{4.116}$$

where ∇ and $\aleph$ are connection and its torsion respectively. This is a rewriting of the variations/Lie derivatives which have been a priori defined independently of any connection on the manifold $\mathcal{B}$.

4.3.1.2 Lie Derivative of Metric, Torsion, and Curvature

For a 2-covariant tensor field on $\mathcal{M}$, we accordingly obtain (by neglecting second order terms in $d\lambda$):

$$\tilde{g}_{\mu\nu}(\mathbf{x}) = J_\mu^\alpha(\mathbf{x}) J_\nu^\beta(\mathbf{x}) g_{\alpha\beta}(\mathbf{x} + \xi\, d\lambda)$$

$$= g_{\mu\nu}(\mathbf{x}) + \left[\xi^\alpha(\mathbf{x})\partial_\alpha g_{\mu\nu}(\mathbf{x}) + g_{\alpha\nu}(\mathbf{x})\partial_\mu\xi^\alpha(\mathbf{x}) + g_{\mu\alpha}(\mathbf{x})\partial_\nu\xi^\alpha(\mathbf{x})\right] d\lambda \quad (4.117)$$

$$= g_{\mu\nu}(\mathbf{x}) + \left[\xi^\alpha(\mathbf{x})\nabla_\alpha g_{\mu\nu}(\mathbf{x}) + g_{\alpha\nu}(\mathbf{x})\nabla_\mu\xi^\alpha(\mathbf{x}) + g_{\mu\alpha}(\mathbf{x})\nabla_\nu\xi^\alpha(\mathbf{x})\right] d\lambda$$

$$+ \xi^\alpha \left(\aleph_{\alpha\nu}^\gamma g_{\mu\gamma} + \aleph_{\alpha\mu}^\gamma g_{\gamma\nu}\right) d\lambda = g_{\mu\nu}(\mathbf{x}) + \mathcal{L}_{\xi(\mathbf{x})} g_{\mu\nu}(\mathbf{x}) d\lambda \quad (4.118)$$

where the Lie derivative of the metric holds for compatible connection:

$$\mathcal{L}_\xi g_{\mu\nu} := g_{\alpha\nu}\nabla_\mu\xi^\alpha + g_{\mu\alpha}\nabla_\nu\xi^\alpha + \xi^\alpha\left(\aleph_{\alpha\nu}^\gamma g_{\mu\gamma} + \aleph_{\alpha\mu}^\gamma g_{\gamma\nu}\right) \quad (4.119)$$

It shows that the infinitesimal active transformation may be interpreted as generated by a Lie derivative, including the torsion field effects e.g. Nakahara (1996). It should be pointed out that $\mathcal{L}_{\xi(\mathbf{x})} g_{\mu\nu}(\mathbf{x}) \neq \xi^\alpha(\mathbf{x})\nabla_\alpha g_{\mu\nu}(\mathbf{x}) = \nabla_{\xi(\mathbf{x})} g_{\mu\nu}(\mathbf{x})$. First, originally the Lie derivative needs no connection, second the vector $\xi(\mathbf{x})$ are not differentiated in the covariant derivative, and third the torsion has influence apart for the Lie derivative. We remind that a metric $g_{\alpha\beta}(\mathbf{x})$ on a metric-affine manifold $\mathcal{M}$ is compatible with the affine connection ∇_γ if and only if $\nabla_\gamma g_{\alpha\beta} \equiv 0$ on $\mathcal{M}$. For metric compatible and torsion free connection, the Lie derivative (4.119) reduces to e.g. Marsden and Hughes (1983):

$$\mathcal{L}_{\xi(\mathbf{x})} g_{\mu\nu}(\mathbf{x}) = \nabla_\mu \xi_\nu(\mathbf{x}) + \nabla_\nu \xi_\mu(\mathbf{x}) \quad (4.120)$$

The active diffeomorphism invariance of the metric induces the vanishing of this tensor expressing the so called Killing's equation. Its solutions are called the Killing vector fields.

Remark 4.34 For the two conditions as torsion free and metric compatibility, the Lie derivative $\mathcal{L}_{\xi(\mathbf{x})} g_{\mu\nu}(\mathbf{x})$ is equivalent to the variation $\delta g_{\alpha\beta}$ along the vector field $\xi^\gamma(\mathbf{x})$. This is not the case when one of these two conditions are not satisfied.

If the previous transformation is considered as a coordinate system change, i.e. passive diffeomorphism $x^\mu = \tilde{x}^\mu + \xi^\mu(x^\alpha) d\lambda$ (re-labelling of coordinates), the transformation rule for 2-covariant tensor holds (by retaining first order terms in $d\lambda$):

$$\tilde{g}_{\mu\nu}(\tilde{x}^\gamma) = J_\mu^\alpha(x^\gamma) J_\nu^\beta(x^\gamma) g_{\alpha\beta}(x^\gamma)$$

$$= g_{\mu\nu}(x^\gamma) + \left[g_{\alpha\nu}(x^\gamma)\partial_\mu\xi^\alpha(x^\gamma) + g_{\mu\alpha}(x^\gamma)\partial_\nu\xi^\alpha(x^\gamma)\right] d\lambda \quad (4.121)$$

where $J_\mu^\alpha(x^\gamma) = \delta_\mu^\alpha + \partial_\mu\xi^\alpha(x^\gamma) d\lambda$ e.g. Kleinert (2000). We observe that the right-hand-side of Eqs. (4.117) and (4.121) looks like each other whenever the metric is compatible with the connection. With respect to a (local) normal coordinate system, the two equations are exactly the same, with however the difference that the first expresses the change of the 2-covariant field by the active diffeomorphism whereas the second expresses the metric 2-covariant tensor in terms of the new coordinates. The passive diffeomorphisms group (covariance group) characterizes the mathematical formulation of theory. The active diffeomorphisms group (symmetry group) characterizes the relativity principle which is a physical fact (Ehlers 1973). For short, two types of groups of transformations may be used for relativistic gravitation, and also for strain gradient continuum theory. Hereafter, we consider a diffeomorphism invariance in the sense of active transformation.

Remark 4.35 Definition of Lie derivative usually starts with the transformation $J_\mu^\alpha(\mathbf{x}) := \delta_\mu^\alpha + \partial_\mu \xi^\alpha(\mathbf{x}) \, d\lambda$, where λ is merely introduced to express that the field ξ has infinitesimal amplitude. When considering a metric-affine continuum (gravitation fields or generalized continuum models), it is worth to consider the infinitesimal vector field on the manifold, and thus to use the covariant derivative instead of the partial one. The infinitesimal transformation becomes:

$$J_\mu^\alpha(\mathbf{x}) := \delta_\mu^\alpha + \nabla_\mu \xi^\alpha(\mathbf{x}) d\lambda = \delta_\mu^\alpha + \partial_\mu \xi^\alpha(\mathbf{x}) d\lambda - \overline{\Gamma}_{\mu\gamma}^\alpha \xi^\gamma d\lambda - \mathcal{T}_{\mu\gamma}^\alpha \xi^\gamma d\lambda \quad (4.122)$$

where the two first terms are exactly the same as previously, whereas the contortion tensor $\mathcal{K}_{\mu\gamma}^\alpha$ constitutes an additional gauge field. For instance in Banerjee and Roy (2011), the skew symmetric part of the contortion tensor is replaced by a rotation angle. The set of all gauge parameters allows us to highlight what they called Hamiltonian gauge symmetries. However, its was shown that the Poincaré (see hereafter) and the Hamiltonian gauge symmetries are equivalent, modulo the trivial gauge transformations e.g. Banerjee and Roy (2011).

The Lie derivative along a vector field is a way of differentiating tensor fields on the continuum e.g. Petrov and Lompay (2013). It is an advantageous method to derive the Poincaré gauge theory (based on the transformations generated by the local translation $\xi^\mu(x^\alpha)$) because it is defined only from manifold structure of $\mathcal{B}$, without any reference to metric $g_{\alpha\beta}$ or to extra structure such as a connection $\Gamma_{\alpha\beta}^\lambda$. The metric, the torsion, and the curvature are independently varied for the general case by using non uniform translational vector $\xi^\mu(x^\nu)$ e.g. Lazar and Anastassiadis (2008), limited to infinitesimal mapping. Whenever this vector represents a (virtual) motion of a symmetry group, then its is referred to as a symmetry generator. The most usual case is the Killing vector solutions. Let start with the definition of Lie derivative from Eq. (A.23) in the appendix and applied to $(0, 2)$ tensor $\mathbf{g}$, $(1, 2)$ tensor $\aleph$, and $(1, 3)$ tensor $\mathfrak{R}$ e.g. Lovelock and Rund (1975):

$$\mathcal{L}_\xi \, g_{\alpha\beta} = \xi^\gamma \, \partial_\gamma g_{\alpha\beta} + g_{\gamma\beta} \, \partial_\alpha \xi^\gamma + g_{\alpha\gamma} \, \partial_\beta \xi^\gamma \quad (4.123)$$

$$\mathcal{L}_\xi \, \aleph_{\alpha\beta}^\lambda = \xi^\gamma \, \partial_\gamma \aleph_{\alpha\beta}^\lambda - \aleph_{\alpha\beta}^\gamma \, \partial_\gamma \xi^\lambda + \aleph_{\gamma\beta}^\lambda \, \partial_\alpha \xi^\gamma + \aleph_{\alpha\gamma}^\lambda \, \partial_\beta \xi^\gamma \quad (4.124)$$

$$\mathcal{L}_\xi \, \mathfrak{R}_{\alpha\beta\mu}^\lambda = \xi^\gamma \partial_\gamma \, \mathfrak{R}_{\alpha\beta\mu}^\lambda - \mathfrak{R}_{\alpha\beta\mu}^\gamma \, \partial_\gamma \xi^\lambda + \mathfrak{R}_{\gamma\beta\mu}^\lambda \, \partial_\alpha \xi^\gamma + \mathfrak{R}_{\alpha\gamma\mu}^\lambda \, \partial_\beta \xi^\gamma + \mathfrak{R}_{\alpha\beta\gamma}^\lambda \, \partial_\mu \xi^\gamma$$
$$(4.125)$$

In addition to the infinitesimal transformations of the spacetime coordinates δx^α, these "Lie variation" of metric, torsion, and curvature define the changes of gravity fields. The four arbitrary functions $\xi^\alpha(x^\mu)$ define an arbitrary local diffeomorphism. For an affinely connected manifold endowed with a metric, the knowledge of the torsion and curvature tensors are sufficient to determine locally both the metric, and the affine connection. This permits to consider the variation up to "second derivatives". Moreover, the local transformations $\xi^\mu(x^\alpha)$ which extend the concept of Poincaré's local group of transformations permits to define the change of these three variables (arguments of the Lagrangian function) following a generalized coordinate transformations.

Remark 4.36 Lie derivative of tensor is also a tensor. For the proof, it is sufficient to rewrite it in terms of covariant derivatives (2.33) and torsion tensor. We obtain the relations for the metric:

$$\mathcal{L}_\xi\, g_{\alpha\beta} = \xi^\gamma\, \nabla_\gamma g_{\alpha\beta} + g_{\gamma\beta}\, \nabla_\alpha\xi^\gamma + g_{\alpha\gamma}\, \nabla_\beta\xi^\gamma + \xi^\gamma\left(g_{\alpha\nu}\,\aleph^\nu_{\gamma\beta} + g_{\nu\beta}\,\aleph^\nu_{\gamma\alpha}\right),$$
(4.126)

for torsion:

$$\mathcal{L}_\xi\, \aleph^\lambda_{\alpha\beta} = \xi^\gamma\, \nabla_\gamma\aleph^\lambda_{\alpha\beta} - \aleph^\gamma_{\alpha\beta}\, \nabla_\gamma\xi^\lambda + \aleph^\lambda_{\gamma\beta}\, \nabla_\alpha\xi^\gamma + \aleph^\lambda_{\alpha\gamma}\, \nabla_\beta\xi^\gamma$$
$$- \xi^\gamma\, \aleph^\lambda_{\gamma\nu}\, \aleph^\nu_{\alpha\beta} + \xi^\gamma\, \aleph^\nu_{\gamma\alpha}\, \aleph^\lambda_{\nu\beta} + \xi^\gamma\, \aleph^\nu_{\gamma\beta}\, \aleph^\lambda_{\alpha\nu},$$
(4.127)

and for curvature:

$$\mathcal{L}_\xi\, \Re^\lambda_{\alpha\beta\mu} = \xi^\gamma\nabla_\gamma\Re^\lambda_{\alpha\beta\mu} - \Re^\gamma_{\alpha\beta\mu}\nabla_\gamma\xi^\lambda + \Re^\lambda_{\gamma\beta\mu}\nabla_\alpha\xi^\gamma + \Re^\lambda_{\alpha\gamma\mu}\nabla_\beta\xi^\gamma + \Re^\lambda_{\alpha\beta\gamma}\nabla_\mu\xi^\gamma$$
$$- \xi^\gamma\Re^\nu_{\alpha\beta\mu}\aleph^\lambda_{\gamma\nu} + \xi^\gamma\Re^\lambda_{\nu\beta\mu}\aleph^\nu_{\gamma\alpha} + \xi^\gamma\Re^\lambda_{\alpha\nu\mu}\aleph^\nu_{\gamma\beta} + \xi^\gamma\Re^\lambda_{\alpha\beta\nu}\aleph^\nu_{\gamma\mu}$$
(4.128)

Relations (4.126)–(4.128) are the expressions of Lie derivative in Riemann–Cartan and metric-affine manifolds. However, the connection ∇ appearing in these relationships include the torsion tensor and cannot utilized in practise to solve the conservation laws in continuum physics, and theory of gravitation. These expressions should be modified to separate the Levi-Civita connection and the contortion tensor $\mathcal{T}^\gamma_{\alpha\beta}$ which become an unknown variable.

4.3.2 Poincaré's Group of Transformations

The invariance of the velocity of light, which is strongly related to the causality principle, allows us to deduce that the Poincaré group of transformations is the invariance group of relativistic gravitation e.g. Williams (1973), Zeeman (1964) (for this later reference, the theorem is focused to Lorentz group $\mathcal{O}(1, 3)$ which is a subgroup of Poincaré group). From the metric tensor, we can define the group of Poincaré in classical and relativistic mechanics e.g. Ali et al. (2009), Capoziello and de Laurentis (2009), Kibble (1961), McKellar (1981). Gauge transformations were originally introduced by Weyl (1918) to relate gravitation and electromagnetic fields. Later, the work of Yang and Mills (1954), generalized by Utiyama (1956), treated the non commutative group of internal symmetries for relativistic gravitation and spins. With the help of fundamental results on manifolds with torsion Cartan (1986), Kibble (1961) used the inhomogeneous Lorentz group to derive the gauge invariance for relativistic gravity. In the same way, we consider the local group of transformations: $y^\mu = A_\rho^\mu(x^\alpha) \ x^\rho + \xi^\mu(x^\alpha)$. The local spacetime translation $\xi^\mu(x^\alpha)$ is added to the transformations $A_\rho^\mu(x^\alpha)$. In fact, the next step considers a modification of this local Poincaré transformations by only retaining the transformation $\xi^\mu(x^\alpha)$ that can include the so-called Poincaré's local transformations e.g. Capoziello and de Laurentis (2011), which could be interpreted as a general coordinate transformations.[4] This allows us to obtain the Poincaré group, which is also called full inhomogeneous Lorentz group, a subgroup of this later restricted by the condition (cf. isometry definition (2.23)):

$$\hat{g}_{\mu\nu} A_\rho^\mu J_\sigma^\nu \equiv \hat{g}_{\rho\sigma} \tag{4.129}$$

where $\hat{g}_{\mu\nu}$ are the components of a nonsingular symmetric tensor with signature -2. It can be observed that such a metric $\hat{\mathbf{g}}$ has numerically the same components in all allowed coordinate systems, according to the tensor components transformation rules. The Minkowskian metric is a typical example of such a tensor. Indeed, in a (flat) spacetime, the Minkowskian metric $\hat{g}_{\alpha\beta}$, and its inverse $\hat{g}^{\alpha\beta}$ have (diagonal) matrix components as in e.g. Havas (1964) (in some sense we introduce here the line element $ds^2 = c^2[dt^2 - (1/c^2)(dx^2+dy^2+dz^2) := c^2 \ \hat{g}^{\alpha\beta} dx_\alpha dx_\beta$ for determining the inverse of the metric and then the metric itself):

$$\hat{g}^{\alpha\beta} := \text{diag}\left(1, -c^{-2}, -c^{-2}, -c^{-2}\right), \qquad \hat{g}_{\alpha\beta} := \text{diag}\left(1, -c^2, -c^2, -c^2\right) \tag{4.130}$$

where c is the light speed. The determinant of A_ρ^μ is equal to ± 1. As only six of the components of the transformations A_ρ^μ are independent, the Poincaré's group contains ten independent components. The group may be reduced to the inhomogeneous Galilean group if the following additional conditions are accounted for:

$$A_0^0 = \pm 1, \qquad A_i^0 = 0, \qquad A_k^i \ (A^T)_j^k \equiv \delta_j^i \tag{4.131}$$

where Latin indices vary from 1 to 3. The upper index T holds for the tensor transpose. This later equation expresses that A_k^i reduces to the group of orthogonal transformations of the three-dimensional space. For the Galilean group, we can introduce two separated space like and time like metrics by considering the

---

[4]As shown by Kibble (1961), the Lagrangian density of the form $\mathscr{L}(x^\mu, \Phi, \partial_\alpha \Phi, \partial_\beta \partial_\alpha \Phi)$ should be replaced by a Lagrangian of the form $\mathscr{L}(\Phi, \nabla_\alpha \Phi, \nabla_\beta \nabla_\alpha \Phi)\sqrt{\text{Detg}}$ which is invariant under a general coordinates transformations, which are nothing else than Poincaré's local transformations.

following limits (the second definition implicitly assumes the unity $c = 1$) e.g. Havas (1964):

$$\hat{h}_{\alpha\beta} := \lim_{c\to\infty} \frac{\hat{g}_{\alpha\beta}(c)}{c^2} = \mathrm{diag}\,(0, -1, -1, -1)\,, \tag{4.132}$$

$$\hat{\tau}_\alpha \hat{\tau}_\beta = \hat{\tau}_{\alpha\beta} := \lim_{c\to\infty} \hat{g}_{\alpha\beta}(c) = \mathrm{diag}\,(1, 0, 0, 0) \tag{4.133}$$

We obtain the metric $\hat{g}_{\alpha\beta} := \hat{h}_{\alpha\beta} + \hat{\tau}_{\alpha\beta}$. The (non tensorial) conditions (Eq. 4.131) are equivalent to:

$$\begin{cases} \hat{\tau}_{\mu\nu}\, A^\mu_\rho\, A^\nu_\sigma = \hat{\tau}_{\rho\sigma} \\ \hat{h}^{\mu\nu}\, A^\rho_\mu\, A^\sigma_\nu = \hat{h}^{\rho\sigma} \end{cases} \tag{4.134}$$

We note that both the two tensors $\hat{\tau}$ and $\hat{\mathbf{h}}$ taken separately are singular. They satisfy the condition $\hat{\tau}_{\mu\rho}\, \hat{h}^{\rho\nu} = 0$ (orthogonality condition) e.g. Bain (2004). These conditions again restrict the number of independent parameters to 10 for the Galilean group.

Remark 4.37 Wigner first realized that the true symmetry group for particle physics is not the homogeneous Lorentz group $\mathcal{SO}^+(1, 3)$. Rather, the underlying symmetry group for particle physics must consist of translations in spacetime in addition to the Lorentz generators which are called boosts and rotations. The extension to inhomogeneous Lorentz group leads to the Poincaré group (cf. the reprint of a 1939 paper Wigner 1939). This should be related to the work of Williams in e.g. Williams (1973) which shows that the Poincaré group in relativistic theory follows from the invariance of the light velocity, and by the way from the causality principle (Zeeman 1964).

4.3.3   Poincaré's Gauge Invariance and Conservation Laws

Covariance requires that Lagrangian $\mathscr{L}$ displays the same functional form in terms of transformed arguments as it does in terms of the original arguments (metric, torsion, and curvature tensors).[5] Under a change of coordinate system (passive diffeomorphism) the Lagrangian is a scalar (when multiplied by the volume form) whose shape and value are left unchanged. Say $\mathscr{L}(g_{\alpha\beta}, \aleph^\gamma_{\alpha\beta}, \Re^\lambda_{\alpha\beta\mu})$, for either spacetime or strain gradient matter. We now consider the symmetry group by considering the coordinate transformations following infinitesimal local translations that generalize the uniform translations of Newtonian mechanics to (active) diffeomorphisms, which include *local translations*, and *local rotations*. It

---

[5]The dependence on coordinates x^μ is dropped according to results in e.g. Kibble (1961).

constitutes in some sense the Poincaré's group of transformations e.g. McKellar (1981) for 4 dimensional spacetime. In addition to the 3 rotation and 3 boost generators, the Poincaré's group includes 4 translation generators. There are thus 10 generators in the Poincaré's group.

It should be noticed that, as for metric variables (Carter 1973) and by analogy to the variations in Manoff (1999), there are also three set of variations: the Lagrangian (comoving) variations $(\Delta g_{\alpha\beta}, \Delta \aleph^{\gamma}_{\alpha\beta}, \Delta \mathfrak{R}^{\lambda}_{\alpha\beta\mu})$ which is the sum of the Eulearian (fixed point) variations $(\delta g_{\alpha\beta}, \delta \aleph^{\gamma}_{\alpha\beta}, \delta \mathfrak{R}^{\lambda}_{\alpha\beta\mu})$ and the Lie derivative variations $(\mathcal{L}_{\xi} g_{\alpha\beta}, \mathcal{L}_{\xi} \aleph^{\gamma}_{\alpha\beta}, \mathcal{L}_{\xi} \mathfrak{R}^{\lambda}_{\alpha\beta\mu})$. Now consider the Lie derivatives along the arbitrary vector field $\xi^{\mu}(x^{\alpha})$ which are particular class of variations. Lie derivatives are closely related to the gauge invariance and conservation laws due to the presence of this arbitrary vector field. The basic principle is to express the Lie derivatives of metric, torsion, and curvature along a an arbitrary (non uniform) vector field and to deduce the conservation laws after imposing the gauge invariance. When the volume-form $\omega_n := \sqrt{\mathrm{Det}\mathbf{g}} \, dx^0 \wedge \cdots \wedge dx^n$ includes the determinant of the metric, the above Lagrangian function $\mathcal{L}$ becomes a scalar (tensor) and not a density.

The Lagrangian is modified when the trajectory is shifted following an active diffeomorphism, such as generalized (local) translation $\xi(x^{\nu})$ on the manifold. For some shifting, the action may be left e.g. Petrov and Lompay (2013), and then these (virtual) translations define the symmetries of the problem. In this section, we consider the conservation laws that govern the continuum when the trajectory is (infinitesimally) shifted while the action is left unchanged. We need to find trajectory that makes the action stationary (Principle of Least Action) and the resulting necessary (and hopefully sufficient) condition will be conservation laws e.g. Capoziello and de Laurentis (2009). Global translation invariance implies the conservation laws for linear momentum, and a global rotation to the conservation laws for angular momentum. We now apply the (active diffeomorphism invariance) to extend the equation of Killing, and its solutions, to the Killing vector in the framework of Riemann–Cartan manifold. For that purpose, we suggest to analyze the change of the action integral under an active diffeomorphism:

$$\Delta \mathscr{S} := \mathscr{S}\left[x^{\mu}(\tau) + \xi^{\mu}[x^{\nu}(\tau)] \, d\lambda\right] - \mathscr{S}\left[x^{\mu}(\tau)\right] \tag{4.135}$$

where $d\lambda$ is a small parameter and τ time parameter. Neglecting terms of higher order than $\mathcal{O}(d\lambda)$, this difference gives:

$$\Delta \mathscr{S} = d\lambda \int_{x^0=\tau_i}^{x^0=\tau_f} \int_B \left(\sigma^{\alpha\beta} \mathcal{L}_{\xi} g_{\alpha\beta} + \Sigma^{\alpha\beta}_{\gamma} \mathcal{L}_{\xi} \aleph^{\gamma}_{\alpha\beta} + \Xi^{\alpha\beta\mu}_{\lambda} \mathcal{L}_{\xi} \mathfrak{R}^{\lambda}_{\alpha\beta\mu} \right) \omega_n \tag{4.136}$$

4.3.3.1 Lagrangian of the Form $\mathscr{L}(g_{\alpha\beta})$

For the sake of the simplicity, we consider Lagrangian function depending only on $g_{\alpha\beta}$ as explicit primal variables. For Riemann–Cartan manifold, it may model some class of elastic continua with possible change of internal structure since the torsion $\aleph^{\gamma}_{\alpha\beta}$ is implicitly an argument as intrinsically a part of the connection $\Gamma^{\gamma}_{\alpha\beta}$.

Theorem 4.3 *Let a metric-affine continuum* $(\mathscr{B}, \mathbf{g}, \nabla)$ *with a Lagrangian function* $\mathscr{L}(g_{\alpha\beta})$, *depending only on the metric tensor. The connection is assumed metric compatible. If the Lagrangian is gauge invariant under the vector field* $\xi^{\mu}(x^{\alpha})$ *then the following conservation law holds:*

$$\nabla_{\alpha}\sigma^{\alpha}_{\gamma} + \sigma^{\alpha}_{\lambda}\aleph^{\lambda}_{\alpha\gamma} = 0 \tag{4.137}$$

where tensorial function $\sigma^{\alpha\beta}(g_{\lambda\mu})$ *defines the constitutive laws of the continuum.*

Proof After a straightforward calculus, and by using the formula for the covariant derivative of $(1,0)$ vector $\xi^{\mu}(x^{\nu})$, the Lie derivative for metric is given by Eq. (4.126), where we observe in the last term the influence of torsion field. Let now consider a continuum with the Lagrangian $\mathscr{L}(g_{\alpha\beta})$ depending only on the metric. The Lagrangian is gauge invariant if for any vector field $\xi^{\mu}(x^{\nu})$, the action $\mathscr{S}$ remains unchanged. Then, we can define the constitutive laws (relating "stress" in function of "strain") and write:

$$\sigma^{\alpha\beta}\mathcal{L}_{\xi}g_{\alpha\beta} = \sigma^{\alpha\beta}\left[\xi^{\gamma}\nabla_{\gamma}g_{\alpha\beta} + g_{\gamma\beta}\nabla_{\alpha}\xi^{\gamma} + g_{\alpha\gamma}\nabla_{\beta}\xi^{\gamma} + \xi^{\gamma}\left(g_{\alpha\nu}\aleph^{\nu}_{\gamma\beta} + g_{\nu\beta}\aleph^{\nu}_{\alpha\gamma}\right)\right]$$

where the "stress" σ is defined as:

$$\sigma^{\alpha\beta} := \frac{\partial\mathscr{L}_{M}}{\partial g_{\alpha\beta}} - \frac{\mathscr{L}_{M}}{2}g^{\alpha\beta}$$

Factorize the vector ξ^{β}. By accounting for the metric compatibility of the connection, and by eliminating the divergence term which may be shifted at the boundary (and worthily chosen to vanish at the boundary $\partial\mathscr{B}$), we arrive at the conservation laws: $2\nabla_{\alpha}\sigma^{\alpha}_{\gamma} - \sigma^{\alpha}_{\lambda}\left(\aleph^{\lambda}_{\alpha\gamma} + \aleph^{\lambda}_{\alpha\gamma}\right) = 0$ leading to $\nabla_{\alpha}\sigma^{\alpha}_{\gamma} + \sigma^{\alpha}_{\lambda}\aleph^{\lambda}_{\alpha\gamma} = 0$, showing by the way that independently on the values of the torsion field, the conservation laws holds as Div $\sigma = 0$ in coordinate-free form e.g. Marsden and Hughes (1983), and here in a covariant form. $\qquad\square$

Remark 4.38 It should be pointed out that the considered connection is the continuum connection (for instance the material connection in the large deformation of elastic continuum with a distribution of dislocations e.g. Le and Stumpf 1996), then the following coefficients of connection are used: $\Gamma^{\gamma}_{\alpha\beta} = \overline{\Gamma}^{\gamma}_{\alpha\beta} + D^{\gamma}_{\alpha\beta} + \Omega^{\gamma}_{\alpha\beta}$ with the Levi-Civita connection $\overline{\Gamma}^{\gamma}_{\alpha\beta}$, the anholonomy $\Omega^{\gamma}_{\alpha\beta} := (1/2)\aleph^{\gamma}_{\alpha\beta}$, and $D^{\gamma}_{\alpha\beta} := g^{\gamma\lambda}g_{\alpha\mu}\Omega^{\mu}_{\lambda\beta} + g^{\gamma\lambda}g_{\mu\beta}\Omega^{\mu}_{\lambda\alpha}$. They define the contortion tensor $\mathcal{T}^{\gamma}_{\alpha\beta} := \Omega^{\gamma}_{\alpha\beta} + D^{\gamma}_{\alpha\beta}$.

Continuum $\mathscr{B}$ is a metric-affine manifold but Lagrangian is assumed to depend only on metric. The contortion tensor is a source of space fading of waves within strain gradient elastic matter e.g. Antonio et al. (2011), Futhazar et al. (2014).

Remark 4.39 Although the action involves only the metric tensor, we notice the presence of torsion "force" in the conservation laws. It will always be the case whenever the continuum is modeled by a Riemann–Cartan manifold with non vanishing torsion and curvature.

4.3.3.2 Lagrangian of the Form $\mathscr{L}(\aleph^{\lambda}_{\alpha\beta})$

Consider a Lagrangian function depending explicitly on and only on torsion tensor as primal variables. It models some spacetime continua in the framework of tele parallel gravitation.

Theorem 4.4 *Let a Riemann–Cartan continuum $(\mathscr{B}, \mathbf{g}, \nabla)$ defined by a Lagrangian $\mathscr{L}(\aleph^{\lambda}_{\alpha\beta})$, depending on the torsion tensor. The connection is assumed metric compatible. If the Lagrangian is gauge invariant then the conservation law holds:*

$$\nabla_\lambda \left(\Sigma^{\lambda\beta}_\alpha \aleph^\alpha_{\gamma\beta} \right) - \Sigma^{\alpha\beta}_\lambda \left(\aleph^\nu_{\gamma\beta} \aleph^\lambda_{\alpha\nu} \right) = 0 \tag{4.138}$$

where hypermomenta functions $\Sigma^{\alpha\beta}_\lambda(\aleph^\lambda_{\alpha\beta})$ define the continuum constitutive laws.

Proof Let write the Lie derivative of torsion in terms of covariant derivatives from Eq. (4.127). Lie derivative of torsion is coupled neither to metric nor to curvature. Consider a continuum with the Lagrangian $\mathscr{L}(\aleph^{\lambda}_{\alpha\beta})$ depending only on torsion. The Lagrangian is gauge invariant if for any vector gauge, the action $\mathscr{S}$ remains unchanged, meaning that $\Sigma^{\alpha\beta}_\lambda \mathcal{L}_\xi \aleph^\lambda_{\alpha\beta} = 0$ for any vector field $\xi^\mu(x^\nu)$. A straightforward calculus leads to the conservation laws after factorizing ξ^γ:

$$\nabla_\lambda (\Sigma^{\alpha\beta}_\gamma \aleph^\lambda_{\alpha\beta} \underbrace{- \Sigma^{\lambda\beta}_\alpha \aleph^\alpha_{\gamma\beta} - \Sigma^{\alpha\lambda}_\beta \aleph^\beta_{\alpha\gamma}}_{= 2\, \Sigma^{\alpha\lambda}_\beta \aleph^\beta_{\gamma\alpha}})$$

$$+ \Sigma^{\alpha\beta}_\lambda (\nabla_\gamma \aleph^\lambda_{\alpha\beta} - \aleph^\lambda_{\gamma\nu} \aleph^\nu_{\alpha\beta} + \underbrace{\aleph^\nu_{\gamma\alpha} \aleph^\lambda_{\nu\beta} + \aleph^\nu_{\gamma\beta} \aleph^\lambda_{\alpha\nu}}_{= 2\, \aleph^\nu_{\gamma\beta} \aleph^\lambda_{\alpha\nu}}) = 0$$

where divergence terms have been eliminated by means of boundary conditions. If the Lagrangian does not depend on the curvature then from the relation (4.72), we deduce that $\Sigma^{\alpha\beta}_\gamma = \Sigma^{\beta\alpha}_\gamma$. This permits to obtain relation (4.138). □

A Lagrangian depending only on torsion may be introduced to describe edge and screw dislocations, they called quadratic gauge translational Lagrangian. Inspiring from the tele parallel formulation of the Einstein–Hilbert Lagrangian, the incompatibility of deformation resulting from edge and screw dislocation fields is summarized by Malyshev into the Lagrangian $\mathscr{L} := (-2\kappa)(\partial_\alpha \phi_\beta^i - \partial_\beta \phi_\alpha^i)(\partial^\alpha \phi_i^\beta - \partial^\beta \phi_i^\alpha)$, where the coefficient κ is a "coupling" constant (Malyshev 2000). He introduced gauge fields including vector δu^i and tetrad $\delta \phi_\beta^i$. Conversely, we have considered here one vector $\xi^i(x^\mu) := \delta u^i$ as gauge field resulting to a only one conservation laws. Einstein–Hilbert Lagrangian density $\mathscr{L}_{HE}$ would be a better model for capturing both edge and screw dislocations, thanks that this Lagrangian is expressed in terms of matter contortion tensor $\mathfrak{T}_{\alpha\beta}^\gamma$ as in Eq. (4.97) e.g. Malyshev (2000).

4.3.3.3 Lagrangian of the Form $\mathscr{L}(\mathfrak{R}_{\alpha\beta\mu}^\lambda)$

Let consider a Lagrangian depending on and only on the curvature tensor as primal variables. It models spacetime continua in the framework of curvature-based gravitation. For the curvature, the Lie derivative holds from Eq. (4.128), where the second line vanishes for torsionless manifold. When the dependence is on the Ricci curvature, we may write $\mathscr{L}(\mathfrak{R}_{\alpha\beta})$. This is particularly convenient for three-dimensional manifold, for which the Ricci curvature entirely defines the Riemann–Cartan curvature tensor. Let consider the Einstein–Hilbert Lagrangian density where the metric and the connection (then the curvature) are independent: $\mathscr{L} = (1/2\chi)g^{\alpha\beta}\mathfrak{R}_{\alpha\beta}$. The Lie derivative of the Lagrangian along vector field ξ gives (do not confuse the notation for Lie derivative and the Lagrangian): $\Delta\mathscr{L} = (1/2\chi)\,\mathfrak{R}_{\alpha\beta}\,\mathcal{L}_\xi g^{\alpha\beta} + (1/2\chi)\,g^{\alpha\beta}\,\mathcal{L}_\xi\mathfrak{R}_{\alpha\beta}$ with:

$$\mathcal{L}_\xi \mathfrak{R}_{\alpha\beta} = \xi^\gamma \,\nabla_\gamma \mathfrak{R}_{\alpha\beta} + \mathfrak{R}_{\gamma\beta}\,\nabla_\alpha\xi^\gamma + \mathfrak{R}_{\alpha\gamma}\,\nabla_\beta\xi^\gamma + \xi^\gamma\left(\mathfrak{R}_{\alpha\nu}\,\aleph_{\gamma\beta}^\nu + \mathfrak{R}_{\nu\beta}\,\aleph_{\alpha\gamma}^\nu\right)$$

$$\mathcal{L}_\xi g^{\alpha\beta} = -g^{\alpha\mu}\,g^{\nu\beta}\,\mathcal{L}_\xi g_{\mu\nu}$$

$$= -\left(g^{\alpha\mu}\nabla_\mu\xi^\beta + g^{\mu\beta}\nabla_\mu\xi^\alpha + \xi^\gamma g^{\mu\beta}\,\aleph_{\gamma\mu}^\alpha + \xi^\gamma g^{\alpha\mu}\,\aleph_{\mu\gamma}^\beta\right)$$

when accounting for metric compatibility of the connection. Introducing these two previous equations in the variation of the Lagrangian gives:

$$\Delta\mathscr{L} = (1/\chi)\xi^\beta\,\nabla_\alpha\mathfrak{R}_\beta^\alpha - (1/\chi)\nabla_\alpha\left(\mathfrak{R}_\beta^\alpha\xi^\beta\right) - \underbrace{\left(\mathfrak{R}_\alpha^\mu\aleph_{\beta\mu}^\alpha + \mathfrak{R}_\alpha^\mu\aleph_{\mu\beta}^\alpha\right)}_{\equiv 0}\xi^\beta$$

$$+ (1/\chi)\xi^\beta\,\nabla_\beta\mathfrak{R} - (1/\chi)\xi^\beta\,\nabla_\alpha\mathfrak{R}_\beta^\alpha + (1/\chi)\nabla_\alpha\left(\mathfrak{R}_\beta^\alpha\,\xi^\beta\right)$$

$$+ \underbrace{\left(\aleph_{\gamma\beta}^\nu\mathfrak{R}_\nu^\gamma + \aleph_{\beta\gamma}^\nu\mathfrak{R}_\nu^\gamma\right)}_{\equiv 0}\xi^\beta$$

$$= (1/\chi)\,\xi^\beta\,\nabla_\beta\mathfrak{R}$$

Consider now a non uniform vector field $\xi^\beta(\mathbf{x}) \neq 0$ (gauge fields), the invariance with respect to this active diffeomorphism allows us to deduce the conservation laws, without transferring the divergence terms to the boundary conditions, $\nabla_\beta \mathcal{R} = 0$. This result means that curvature remains constant on the continuum $\mathscr{B}$ for an arbitrary continuous symmetry generated by the vector field ξ (local translation). Nevertheless this equation can be completed by including the volume-form change as: $\mathcal{L}_\xi (\mathcal{R}\omega_n) = (\mathcal{L}_\xi \mathcal{R}) \omega_n + \mathcal{R}\mathcal{L}_\xi \omega_n$. Lie derivative can be re-written as follows for a particular volume-form:

$$\mathcal{L}_\xi (\mathcal{R}\omega_n) = \left[\mathcal{L}_\xi \left(\mathfrak{R}_{\alpha\beta} g^{\alpha\beta} \right) \sqrt{\mathbf{Detg}} + \mathcal{R}\mathcal{L}_\xi \sqrt{\mathbf{Detg}} \right] dx^0 \wedge \cdots \wedge dx^3$$

$$= \left(G_{\alpha\beta} \mathcal{L}_\xi g^{\alpha\beta} + g^{\alpha\beta} \mathcal{L}_\xi \mathfrak{R}_{\alpha\beta} \right) \omega_n$$

where the Lie derivative of the Ricci tensor may be transferred to the manifold boundary $\partial\mathcal{M}$, and thus can be skipped. Indeed we recover the classical vacuum equations of gravitation in general relativity. However, gauge invariance allows us to obtain the influence of scalar curvature on Lagrangian, for either gradient continuum or relativistic gravitation. We can express the "vacuum equations" as:

$$\mathcal{L}_\xi ((1/2\chi)\mathcal{R}\omega_n) = \left((1/2\chi)G^{\alpha\beta} \mathcal{L}_\xi g_{\alpha\beta} \right) \omega_n, \quad G^{\alpha\beta} := \mathfrak{R}^{\alpha\beta} - (\mathfrak{R}/2) g^{\alpha\beta} \tag{4.139}$$

Introducing the expression of the metric Lie derivative (4.119) we obtain:

$$G^{\alpha\beta} \mathcal{L}_\xi g_{\alpha\beta} = G^{\alpha\beta} \left[g_{\mu\beta}\nabla_\alpha \xi^\mu + g_{\alpha\mu}\nabla_\beta \xi^\mu + \xi^\mu \left(\aleph^\gamma_{\mu\beta} g_{\alpha\gamma} + \aleph^\gamma_{\mu\alpha} g_{\gamma\beta} \right) \right]$$

By transferring the divergence term at the manifold boundary, and owing that $G^{\alpha\beta} = G^{\beta\alpha}$ we deduce the local equations for the vacuum spacetime with torsion (or equivalently a continuum with non zero torsion and non zero curvature but for which Lagrangian density $\mathscr{L}$ depends only on curvature):

$$\nabla_\alpha G^\alpha_\beta = \aleph^\gamma_{\beta\alpha} G^\alpha_\gamma \tag{4.140}$$

Relationship of (4.140) with Bianchi identities (4.73) should be clarified in the future.

Remark 4.40 In the scope of Einstein gravitation, the connection $\overline{\Gamma}^\gamma_{\alpha\beta}$ is deduced from the metric tensor $g_{\alpha\beta}$. The second Bianchi identity thus induces that the Einstein tensor $\overline{G}_{\alpha\beta} := \overline{\mathfrak{R}}_{\alpha\beta} - (\overline{\mathcal{R}}/2)g_{\alpha\beta}$ is divergence-free. We observe that such is not the case for the Einstein–Cartan gravitation (4.140). It is worth to remind that divergence-free property of the Einstein tensor may directly be deduced from the second Bianchi identity extended to metric-affine manifold e.g. Rakotomanana (2003). In the same way, we may also check that the second Bianchi identity (4.73) allows us to obtain the previous relationship assessing that the divergence of the analogous Einstein tensor is not divergence free.

Remark 4.41 We now focus on the difference between Einstein relativistic gravitation *vs* Einstein–Cartan relativistic gravitation that may be sketched in the following. Consider a continuum matter $\mathscr{B}$ evolving within a spacetime (either Einstein spacetime or Einstein–Cartan spacetime). We consider that the action of this continuum is calculated by a Lagrangian including the gravitation and the matter as:

$$\mathscr{S} = \int_{\mathscr{B}} \left(-\frac{1}{2\chi} \mathcal{R} + \mathscr{L}_M \right) \omega_n \qquad (4.141)$$

where $\mathscr{L}_M$ stands for the matter contribution whereas $\mathcal{R}$ is the spacetime curvature (we limit to Einstein–Hilbert action for capturing the gravitational effects). For Einstein relativistic gravitation within a curved Riemannian spacetime e.g. Nakahara (1996), we have after a standard variational procedure, assuming worth boundary conditions, to the field equations:

$$\frac{1}{2\chi} \overline{G}^{\alpha\beta} = \overline{T}^{\alpha\beta} \qquad (4.142)$$

where $\overline{T}^{\alpha\beta}$ is the energy-momentum tensor (2.80) and with:

$$\begin{cases} \overline{\Gamma}^{\gamma}_{\alpha\beta} := (1/2)g^{\gamma\lambda}\left(\partial_\alpha g_{\lambda\beta} + \partial_\beta g_{\alpha\lambda} - \partial_\lambda g_{\alpha\beta}\right) \\ \overline{\aleph}^{\gamma}_{\alpha\beta} = 0 \\ \overline{\mathfrak{R}}^{\lambda}_{\alpha\beta\mu} := (\partial_\alpha \overline{\Gamma}^{\lambda}_{\beta\mu} + \overline{\Gamma}^{\nu}_{\beta\mu}\overline{\Gamma}^{\lambda}_{\alpha\nu}) - (\partial_\beta \overline{\Gamma}^{\lambda}_{\alpha\mu} + \overline{\Gamma}^{\nu}_{\alpha\mu}\overline{\Gamma}^{\lambda}_{\beta\nu}) \neq 0 \end{cases} \qquad (4.143)$$

and then, we deduce:

$$\begin{cases} \overline{G}^{\alpha\beta} := \overline{\mathfrak{R}}^{\alpha\beta} - (\overline{\mathfrak{R}}/2)\, g^{\alpha\beta} & \text{(Einstein tensor)} \\ \overline{\nabla}_\alpha \overline{G}^{\alpha}_{\beta} = 0 & \text{(Bianchi second identity)} \\ \overline{\nabla}_\alpha \overline{T}^{\alpha\beta} = 0 & \text{(Conservation laws)} \end{cases} \qquad (4.144)$$

The conservation laws result from the second identity of Bianchi e.g. Rakotomanana (2003) (in this reference, Bianchi identities are derived in the framework of Einstein–Cartan manifolds). For Einstein–Cartan relativistic gravitation, the Levi-Civita connection is replaced by an affine connection:

$$\begin{cases} \Gamma^{\gamma}_{\alpha\beta} := \mathbf{u}^{\gamma}\left(\nabla_{\mathbf{u}_\alpha}\mathbf{u}_\beta\right) \\ \aleph^{\gamma}_{\alpha\beta} = \Gamma^{\gamma}_{\alpha\beta} - \Gamma^{\gamma}_{\beta\alpha} \\ \mathfrak{R}^{\lambda}_{\alpha\beta\mu} := (\partial_\alpha \Gamma^{\lambda}_{\beta\mu} + \Gamma^{\nu}_{\beta\mu}\Gamma^{\lambda}_{\alpha\nu}) - (\partial_\beta \Gamma^{\lambda}_{\alpha\mu} + \Gamma^{\nu}_{\alpha\mu}\Gamma^{\lambda}_{\beta\nu}) - \aleph^{\nu}_{0\alpha\beta}\Gamma^{\lambda}_{\nu\mu} \neq 0 \end{cases} \qquad (4.145)$$

showing that the torsion is now an additional independent geometric variable of the continuum motion.

We propose in the next subsection a general framework to model the coupling of generalized continua with a curved spacetime with torsion. Equation (4.144) should be re-derived in deep. The first point is the inclusion of the torsion as independent variable. The second point is to highlight that the covariant derivative associated to the connection implicitly contains the torsion which is an unknown variable. It is not useful as such in practise. It is necessary to split the connection with unique decomposition $\Gamma_{\alpha\beta}^{\gamma} = \overline{\Gamma}_{\alpha\beta}^{\gamma} + \mathcal{T}_{\alpha\beta}^{\gamma}$ where $\mathcal{T}_{\alpha\beta}^{\gamma}$ is the contortion tensor (see Eq. (2.44)).

4.3.4 Conservation Laws in a Curved Spacetime with Torsion

Let consider a generalized continuum $(\mathscr{B}, \mathbf{g}, \nabla)$ evolving within a spacetime $(\mathscr{M}, \hat{\mathbf{g}}, \hat{\nabla})$, both of them are modeled with Riemann–Cartan manifold. We adopt the definition of generalized transformations where metric strain $\varepsilon_{\alpha\beta}$, changes of topology $\mathcal{T}_{\alpha\beta}^{\gamma}$ and $\mathcal{K}_{\alpha\beta\lambda}^{\gamma}$ are the primal variables.

4.3.4.1 Lie Derivatives on Metric Compatible Manifolds

We develop in this paragraph some basic expressions of the Lie derivatives for vector, 1-form, metric, torsion and curvature in terms of the Levi-Civita connection $\overline{\nabla}$ and the contortion tensor $\mathcal{T}$.

Theorem 4.5 *Let $(\mathscr{B}, \mathbf{g}, \nabla)$ a Riemann–Cartan manifold, where the connection ∇ is metric compatible, and say $\overline{\nabla}$ the Levi-Civita connection associated to the metric $\mathbf{g}$. Then, we have respectively the following Lie derivative of vector u^{μ}, 1-form ω_{μ}, metric $g_{\alpha\beta}$, and the torsion $\aleph_{\alpha\beta}^{\gamma}$:*

$$\mathcal{L}_{\xi} u^{\mu} = \xi^{\alpha} \overline{\nabla}_{\alpha} u^{\mu} - u^{\alpha} \overline{\nabla}_{\alpha} \xi^{\mu} \tag{4.146}$$

$$\mathcal{L}_{\xi} \omega_{\mu} = \xi^{\alpha} \overline{\nabla}_{\alpha} \omega_{\mu} + \omega_{\alpha} \overline{\nabla}_{\mu} \xi^{\alpha} \tag{4.147}$$

$$\mathcal{L}_{\xi} g_{\alpha\beta} = g_{\alpha\gamma} \overline{\nabla}_{\beta} \xi^{\gamma} + g_{\gamma\beta} \overline{\nabla}_{\alpha} \xi^{\gamma} + g_{\alpha\gamma} \mathcal{T}_{\rho\beta}^{\gamma} \xi^{\rho} + g_{\gamma\beta} \mathcal{T}_{\rho\alpha}^{\gamma} \xi^{\rho} \tag{4.148}$$

$$\mathcal{L}_{\xi} \aleph_{\alpha\beta}^{\lambda} = \xi^{\gamma} \overline{\nabla}_{\gamma} \aleph_{\alpha\beta}^{\lambda} - \aleph_{\alpha\beta}^{\gamma} \overline{\nabla}_{\gamma} \xi^{\lambda} + \aleph_{\alpha\gamma}^{\lambda} \overline{\nabla}_{\beta} \xi^{\gamma} - \aleph_{\beta\gamma}^{\lambda} \overline{\nabla}_{\alpha} \xi^{\gamma} \tag{4.149}$$

Proof Starting with the relation (4.114) and owing that: $\nabla_{\alpha} u^{\mu} = \overline{\nabla}_{\alpha} u^{\mu} + \mathcal{T}_{\alpha\rho}^{\mu} u^{\rho}$ and $\nabla_{\alpha} \xi^{\mu} = \overline{\nabla}_{\alpha} \xi^{\mu} + \mathcal{T}_{\alpha\rho}^{\mu} \xi^{\rho}$, we deduce the first relation by accounting for the decomposition of the contortion $\mathcal{T}_{\alpha\rho}^{\mu} = D_{\alpha\rho}^{\mu} + \Omega_{\alpha\rho}^{\mu}$ into a symmetric part and a skew-symmetric part. By using the same method, starting from relation (4.114) and owing that:

$$\nabla_{\alpha} \omega_{\mu} = \overline{\nabla}_{\alpha} \omega_{\mu} - \mathcal{T}_{\alpha\mu}^{\rho} \omega_{\rho}, \qquad \nabla_{\mu} \xi^{\alpha} = \overline{\nabla}_{\mu} \xi^{\alpha} + \mathcal{T}_{\mu\rho}^{\alpha} \xi^{\rho}$$

we obtain the second relation. We start with Eq. (4.126). By accounting for the metric compatibility of the connection ∇ and using the following relations:

$$\mathfrak{T}^{\gamma}_{\beta\rho} = \Omega^{\gamma}_{\beta\rho} + D^{\gamma}_{\beta\rho} = -\frac{1}{2}\aleph^{\gamma}_{\rho\beta} + D^{\gamma}_{\rho\beta} \quad\Longrightarrow\quad \mathfrak{T}^{\gamma}_{\beta\rho} + \aleph^{\gamma}_{\rho\beta} = \frac{1}{2}\aleph^{\gamma}_{\rho\beta} + D^{\gamma}_{\rho\beta} = \mathfrak{T}^{\gamma}_{\rho\beta}$$

The same relation holds for $\mathfrak{T}^{\gamma}_{\alpha\rho} + \aleph^{\gamma}_{\rho\alpha} = \mathfrak{T}^{\gamma}_{\rho\alpha}$. This allows us to deduce the third relation. The Lie derivative of the torsion is obtained with analogous method. First from the original definition of the Lie derivative we introduce the relations:

$$\begin{cases} \nabla_{\gamma}\aleph^{\lambda}_{\alpha\beta} = \partial_{\gamma}\aleph^{\lambda}_{\alpha\beta} + \Gamma^{\lambda}_{\gamma\rho}\aleph^{\rho}_{\alpha\beta} - \Gamma^{\rho}_{\gamma\alpha}\aleph^{\lambda}_{\rho\beta} - \Gamma^{\rho}_{\gamma\beta}\aleph^{\lambda}_{\alpha\rho} \\ \nabla_{\alpha}\xi^{\gamma} = \partial_{\alpha}\xi^{\gamma} + \Gamma^{\gamma}_{\alpha\rho}\xi^{\rho} \\ \nabla_{\beta}\xi^{\gamma} = \partial_{\beta}\xi^{\gamma} + \Gamma^{\gamma}_{\beta\rho}\xi^{\rho} \end{cases}$$

Introducing these relations into the definition of the Lie derivative of $\aleph^{\gamma}_{\alpha\beta}$ leads to Eq. (4.127). Finally, introducing the contortion tensor into the torsion covariant derivative, we have:

$$\nabla_{\gamma}\aleph^{\lambda}_{\alpha\beta} = \overline{\nabla}_{\gamma}\aleph^{\lambda}_{\alpha\beta} + \mathfrak{T}^{\lambda}_{\gamma\rho}\aleph^{\rho}_{\alpha\beta} - \mathfrak{T}^{\rho}_{\gamma\alpha}\aleph^{\lambda}_{\rho\beta} - \mathfrak{T}^{\rho}_{\gamma\beta}\aleph^{\lambda}_{\alpha\rho}$$

which allows us to obtain the Lie derivative of the torsion $\aleph^{\gamma}_{\alpha\beta}$ in terms of the Levi-Civita connection. $\qquad\square$

Remark 4.42 The relation $\mathcal{L}_{\xi}u^{\mu} = \xi^{\alpha}\overline{\nabla}_{\alpha}u^{\mu} - u^{\alpha}\overline{\nabla}_{\alpha}\xi^{\mu}$ conforms to the intrinsic formulation of the Lie derivative of a contravariant vector $\mathcal{L}_{\xi}\mathbf{u} := \nabla_{\xi}\mathbf{u} - \nabla_{\mathbf{u}}\xi - \aleph(\xi, \mathbf{u})$ where $\aleph$ is the torsion operator for any affine connection with torsion e.g. Manoff (2001b).

4.3.4.2 Lagrangian of the Form $\mathscr{L}(g_{\alpha\beta}, \mathfrak{T}^{\gamma}_{\alpha\beta})$

Let now combine the two aspects, the first is the requirement of gauge invariance for the Lagrangian, and second the dependence of the Lagrangian not on the torsion but on the metric and the contortion tensor. First of all, the covariance theorem implies that a Lagrangian of the form $\mathscr{L}(g_{\alpha\beta}, \Gamma^{\gamma}_{\alpha\beta})$ should be formulated as $\mathscr{L}(g_{\alpha\beta}, \aleph^{\gamma}_{\alpha\beta})$. Secondly, if we want the Lagrangian to depend on the change of internal topology, the dependence becomes $\mathscr{L}(g_{\alpha\beta}, \mathfrak{T}^{\gamma}_{\alpha\beta})$. This is not exactly the same as the Lagrangian proposed in Maier (2014) where the solutions of vacuum spacetime in non-Riemannian gravitation is investigated. Owing that the contortion tensor includes a skew-symmetric and a symmetric terms (2.44) for metric compatible connection:

$$\Omega^{\gamma}_{\alpha\beta} = (1/2)(\Gamma^{\gamma}_{\alpha\beta} - \Gamma^{\gamma}_{\beta\alpha}), \qquad D^{\gamma}_{\alpha\beta} = g^{\gamma\lambda}g_{\alpha\mu}\,\Omega^{\mu}_{\lambda\beta} + g^{\gamma\lambda}g_{\mu\beta}\,\Omega^{\mu}_{\lambda\alpha}$$

we can assume the form $\mathscr{L}(g_\alpha, \mathfrak{J}^\gamma_{\alpha\beta})$. In this paragraph we define the two momenta by their constitutive laws, slightly modified for the second one because the matter dependence is assumed on contortion instead on the torsion,

$$\sigma^{\alpha\beta} := \frac{\partial \mathscr{L}}{\partial g_{\alpha\beta}} - \frac{\mathscr{L}}{2} g^{\alpha\beta}, \qquad \Sigma^{\alpha\beta}_\gamma := \frac{\partial \mathscr{L}}{\partial \mathfrak{J}^\gamma_{\alpha\beta}} \qquad (4.150)$$

where $\Sigma^{\alpha\beta}_\gamma$ is called the spin angular momentum tensor of matter (Hehl et al. 1976). The Eulerian (fixed point) variation of the corresponding action takes the form of:

$$\delta(\mathscr{L} \omega_n) = \left(\sigma^{\alpha\beta} \, \delta g_{\alpha\beta} + \Sigma^{\alpha\beta}_\gamma \, \delta \mathfrak{J}^\gamma_{\alpha\beta} \right) \omega_n.$$

After a straightforward calculus, we obtain the two contributions of the variation leading to the field equations, the first factor of the metric $\delta g_{\alpha\beta}$ (symmetric), and the second factor of the anholonomy $\delta \Omega^\gamma_{\alpha\beta}$ (skew-symmetric with respect the two lower indices):

$$\begin{cases} \sigma^{\alpha\beta} - \Sigma^{\kappa\nu}_\gamma g^{\gamma\alpha} \left(g_{\kappa\mu} \Omega^\mu_{\lambda\nu} + g_{\mu\nu} \Omega^\mu_{\lambda\kappa} \right) g^{\lambda\beta} + g^{\gamma\lambda} \left(\Sigma^{\alpha\kappa}_\gamma \Omega^\beta_{\lambda\kappa} + \Sigma^{\kappa\beta}_\gamma \Omega^\alpha_{\lambda\kappa} \right) = 0 \\ \Sigma^{\alpha\beta}_\gamma + g_{\gamma\lambda} \left(g^{\alpha\mu} \Sigma^{\lambda\beta}_\mu - g^{\beta\mu} \Sigma^{\alpha\lambda}_\mu \right) = 0 \end{cases}$$
$$(4.151)$$

in which it is reminded that the hypermomentum $\Sigma^{\alpha\beta}_\gamma$ is neither symmetric nor skew-symmetric with respect to the two upper indices. Therefore, the second equation (4.151) should be used very cautiously and even modified. It should be rewritten as follows:

$$(1/2) \left(\Sigma^{\alpha\beta}_\gamma - \Sigma^{\beta\alpha}_\gamma \right) + g_{\gamma\lambda} \left(g^{\alpha\mu} \Sigma^{\lambda\beta}_\mu - g^{\beta\mu} \Sigma^{\alpha\lambda}_\mu \right) = 0 \qquad (4.152)$$

This means that only the skew-symmetric part of $\Sigma^{\alpha\beta}_\gamma$ with respect the two indices (α, β) only should be taken into account in the relation. For establishing the above conservation laws, it was necessary to calculate the variation of $g^{\gamma\lambda}$ as follows:

$$\delta \left(g^{\gamma\lambda} g_{\lambda\mu} \right) = 0 \qquad \Longrightarrow \qquad \delta g^{\gamma\lambda} = -g^{\gamma\kappa} \delta g_{\kappa\nu} g^{\nu\lambda} \qquad (4.153)$$

This Eq. (4.152) is the *angular momentum equation* for this matter model. We observe that the conservation laws deduced from relations (4.151) are very different of those models where the metric and the torsion are considered as the primal variables. We do not develop details of the expected conservations laws since we will proceed to that later. The keypoint is here to consider the change of connection with respect to the Levi-Civita connection (contortion) as the primal variable modeling the internal change of topology. It seems that this keypoint is a physical assumption rather than a mathematical consequence of covariance or any other invariance principle.

4.3.4.3 Alternative Formulation for $\mathcal{L}(g_{\alpha\beta}, \mathcal{T}^\gamma_{\alpha\beta})$

Consider the same system as in the previous paragraph and focus is on the conservation laws. Alternative and simpler derivation of conservation laws may be conducted by considering the Lie derivative of metric and contortion (the proof is done by a straightforward calculus):

$$\begin{cases} \mathcal{L}_\xi g_{\alpha\beta} = \overline{\nabla}_\alpha \xi_\beta + \overline{\nabla}_\beta \xi_\alpha + g_{\alpha\gamma}\mathcal{T}^\gamma_{\rho\beta}\xi^\rho + g_{\gamma\beta}\mathcal{T}^\gamma_{\rho\alpha}\xi^\rho \\ \mathcal{L}_\xi \mathcal{T}^\gamma_{\alpha\beta} = \xi^\rho \overline{\nabla}_\rho \mathcal{T}^\gamma_{\alpha\beta} - \mathcal{T}^\rho_{\alpha\beta}\overline{\nabla}_\rho \xi^\gamma + \mathcal{T}^\gamma_{\rho\beta}\overline{\nabla}_\alpha \xi^\rho + \mathcal{T}^\gamma_{\alpha\rho}\overline{\nabla}_\beta \xi^\rho \end{cases}$$

Theorem 4.6 *Let a continuum matter $(\mathcal{B}, g_{\alpha\beta}, \Gamma^\gamma_\alpha)$ modeled with Riemann–Cartan manifold, in motion within a Riemann–Cartan spacetime $(\mathcal{M}, \hat{g}_{\alpha\beta}, \hat{\Gamma}^\gamma_\alpha)$. The Lagrangian of the continuum is assumed to depend on the metric and on the contortion tensor $\mathcal{L}(g_{\alpha\beta}, \mathcal{T}^\gamma_{\alpha\beta})$. The conservation laws hold for continuum matter with arbitrary contortion field and for which Lagrangian depends on the contortion explicitly*

$$\overline{\nabla}_\alpha \left(g_{\rho\beta}\sigma^{\alpha\beta} + \Sigma^{\alpha\beta}_\gamma \mathcal{T}^\gamma_{\rho\beta} \right) + \overline{\nabla}_\beta \left(g_{\alpha\rho}\sigma^{\alpha\beta} + \Sigma^{\alpha\beta}_\gamma \mathcal{T}^\gamma_{\alpha\rho} \right) =$$

$$\sigma^{\alpha\beta}\left(g_{\alpha\gamma}\mathcal{T}^\gamma_{\rho\beta} + g_{\gamma\beta}\mathcal{T}^\gamma_{\rho\alpha} \right) + \Sigma^{\alpha\beta}_\gamma \overline{\nabla}_\rho \mathcal{T}^\gamma_{\alpha\beta} + \Sigma^{\alpha\beta}_\rho \overline{\nabla}_\gamma \mathcal{T}^\gamma_{\alpha\beta} \qquad (4.154)$$

Proof Starting from the variational formula: $\delta\mathcal{L} = \sigma^{\alpha\beta}\mathcal{L}_\xi g_{\alpha\beta} + \Sigma^{\alpha\beta}_\gamma \mathcal{L}_\xi \mathcal{T}^\gamma_{\alpha\beta}$, we introduce the previous formulae of Lie derivatives of $\mathcal{L}_\xi g_{\alpha\beta}$ and $\mathcal{L}_\xi \mathcal{T}^\gamma_{\alpha\beta}$. By shifting to the boundary $\partial\mathcal{B}$ all the divergence terms $\overline{\nabla}_\alpha(\cdots)$ and after factorization by ξ^ρ, we obtain:

$$\delta\mathcal{L} = \left[\sigma^{\alpha\beta}\left(g_{\alpha\gamma}\mathcal{T}^\gamma_{\rho\beta} + g_{\gamma\beta}\mathcal{T}^\gamma_{\rho\alpha} \right) + \Sigma^{\alpha\beta}_\gamma \overline{\nabla}_\rho \mathcal{T}^\gamma_{\alpha\beta} + \Sigma^{\alpha\beta}_\rho \overline{\nabla}_\gamma \mathcal{T}^\gamma_{\alpha\beta} \right]\xi^\rho$$

$$- \left[\overline{\nabla}_\alpha \left(g_{\rho\beta}\sigma^{\alpha\beta} + \Sigma^{\alpha\beta}_\gamma \mathcal{T}^\gamma_{\rho\beta} \right) + \overline{\nabla}_\beta \left(g_{\alpha\rho}\sigma^{\alpha\beta} + \Sigma^{\alpha\beta}_\gamma \mathcal{T}^\gamma_{\alpha\rho} \right) \right]\xi^\rho$$

$$+ \text{ boundary terms on } \partial\mathcal{B}$$

This allows us to obtain the conservation laws:

$$\overline{\nabla}_\alpha \left(g_{\rho\beta}\sigma^{\alpha\beta} + \Sigma^{\alpha\beta}_\gamma \mathcal{T}^\gamma_{\rho\beta} \right) + \overline{\nabla}_\beta \left(g_{\alpha\rho}\sigma^{\alpha\beta} + \Sigma^{\alpha\beta}_\gamma \mathcal{T}^\gamma_{\alpha\rho} \right)$$

$$= \sigma^{\alpha\beta}\left(g_{\alpha\gamma}\mathcal{T}^\gamma_{\rho\beta} + g_{\gamma\beta}\mathcal{T}^\gamma_{\rho\alpha} \right) + \Sigma^{\alpha\beta}_\gamma \overline{\nabla}_\rho \mathcal{T}^\gamma_{\alpha\beta} + \Sigma^{\alpha\beta}_\rho \overline{\nabla}_\gamma \mathcal{T}^\gamma_{\alpha\beta} \qquad \square$$

We observe that some additional terms are present in the divergence operator, $\sigma^{\alpha\beta}$ becomes $g_{\rho\beta}\sigma^{\alpha\beta} + \Sigma^{\alpha\beta}_\gamma \mathcal{T}^\gamma_{\rho\beta}$. It is pointed out that the divergence operator is associated to the Levi-Civita connection.

Corollary 4.3 *We deduce from (4.154) the following conservation laws:*

1. for classical continuum matter $\mathscr{B}$ with zero contortion $\mathfrak{T}^{\gamma}_{\alpha\beta} \equiv 0$

$$\overline{\nabla}_{\alpha}\sigma^{\alpha}_{\rho} = 0 \tag{4.155}$$

2. for continuum matter $\mathscr{B}$ with uniform contortion and with Lagrangian not depending explicitly on contortion, say $\Sigma^{\alpha\beta}_{\gamma} \equiv 0$,

$$\overline{\nabla}_{\alpha}\sigma^{\alpha}_{\rho} = \sigma^{\beta}_{\gamma}\mathfrak{T}^{\gamma}_{\rho\beta} \tag{4.156}$$

The conservation laws Eq. (4.156) are easier to handle than (4.137) because the divergence operator does not contain the torsion implicitly. This advantage is particularly obvious when deriving equations in Cartesian coordinate system. The two conservation Eqs. (4.155) and (4.156)—(which can be related to Eq. (4.137))—govern the classical Einstein gravitation and continuum mechanics e.g. Marsden and Hughes (1983), and the generalized continuum mechanics with torsion e.g. Rakotomanana (1997), respectively. These conservation laws are similar to those of Yasskin and Stoeger in e.g. Yasskin and Stoeger (1980) (equation 19 in this reference) and even extend them to the coupling of non zero torsion and curved matter and spacetime. Equation (4.155) reduces to the Einsteinian relativistic gravitation equation of fields, which can be obtained by means of Bianchi (4.73) identities too.

Remark 4.43 First, it is worth to remind that the contortion $\mathfrak{T}^{\gamma}_{\alpha\beta}$ is a tensor and therefore its intrinsic properties does not depend on the choice of the coordinate system x^{μ}. The conservation laws (4.156) are covariant in which the contortion tensor behaves as primal variables and unknowns. Implicitly, the metric tensor $g_{\alpha\beta}$ captures a holonomic deformation whereas the evolution of the contortion tensor $\mathfrak{T}^{\gamma}_{\alpha\beta}$ captures the non holonomic part of the deformation. In principle, an additional conservation laws should be derived due to the increase of unknowns number. However, modelling the evolution of the contortion tensor as internal variables might be based on thermomechanics of dissipating continuum e.g. Rakotomanana (2003).

4.3.4.4 Lagrangian of the Form $\mathscr{L}(\varepsilon_{\alpha\beta}, \mathfrak{T}^{\gamma}_{\alpha\beta}, \mathcal{K}^{\gamma}_{\alpha\beta\lambda})$

Let consider a continuum body $(\mathscr{B}, \mathbf{g}, \nabla)$ in motion in the spacetime $(\mathscr{M}, \hat{\mathbf{g}}, \hat{\nabla})$ (which may be assumed Minkowskian for the sake of the simplicity) where the Lagrangian depends on the strain, the contortion and the change of curvature according to the relations (4.95), (4.97), and (4.99):

$$\begin{cases} 2\varepsilon_{\alpha\beta} := g_{\alpha\beta} - \hat{g}_{\alpha\beta} \\ \mathfrak{T}^{\gamma}_{\alpha\beta} := \Gamma^{\gamma}_{\alpha\beta} - \overline{\Gamma}^{\gamma}_{\alpha\beta}\left[g_{\lambda\mu}, \partial_{\nu}g_{\lambda\mu}\right] \\ \mathcal{K}^{\gamma}_{\alpha\beta\lambda} := \overline{\nabla}_{\alpha}\mathfrak{T}^{\gamma}_{\lambda\beta} - \overline{\nabla}_{\beta}\mathfrak{T}^{\gamma}_{\lambda\alpha} + \left(\mathfrak{T}^{\gamma}_{\beta\mu}\mathfrak{T}^{\mu}_{\alpha\lambda} - \mathfrak{T}^{\gamma}_{\alpha\mu}\mathfrak{T}^{\mu}_{\beta\lambda}\right) \end{cases}$$

The first relation expresses the strain of the matter $\mathscr{B}$ with respect to the Minkowski spacetime $\mathscr{M}$. In fact, we can replace the strain $\varepsilon_{\alpha\beta}$ by the metric $g_{\alpha\beta}$ without loss of the generality of our purpose. The second and third relations are merely the change of the connection from its Levi-Civita part to the actual connection which is curved with non zero torsion. In the remaining part of this paragraph we consider $g_{\alpha\beta}$ instead of $\varepsilon_{\alpha\beta}$. Introducing these relations within the Lagrangian arguments, we suggest the Lagrangian function with a slightly modified list of arguments:

$$\mathscr{L} := \mathscr{L}\left(g_{\alpha\beta}, \mathfrak{T}^{\gamma}_{\alpha\beta}, \overline{\nabla}_{\lambda}\mathfrak{T}^{\gamma}_{\alpha\beta}\right) \tag{4.157}$$

where the dependence is on strain with respect to the ambient spacetime, the change of topology and its gradient with respect to the Levi-Civita covariant derivative. Covariance of (4.157) could be considered as ensured since it is a consequence of the fact that all arguments of this Lagrangian are covariant objects. Nevertheless, the covariance of such form of Lagrangian has to be carefully checked. This is similar to the problem investigated on the development of conservation laws in a unified framework by e.g. Obukhov and Puetzfeld (2014). The difference is that we consider here the change of curvature with respect to Riemann curvature rather than the curvature itself. Remind the following definition of the hypermomenta by means of their constitutive laws (they are slightly different compared to the previous definition in (4.68)):

$$\sigma^{\alpha\beta} := \frac{\partial\mathscr{L}}{\partial g_{\alpha\beta}}, \qquad \Sigma^{\alpha\beta}_{\gamma} := \frac{\partial\mathscr{L}}{\partial\mathfrak{T}^{\gamma}_{\alpha\beta}}, \qquad \Xi^{\lambda\alpha\beta}_{\gamma} := \frac{\partial\mathscr{L}}{\partial\overline{\nabla}_{\lambda}\mathfrak{T}^{\gamma}_{\alpha\beta}} \tag{4.158}$$

As for the Lagrangian depending on the metric and the contortion only, we obtain the following theorem for derivation of fields equations.

Theorem 4.7 *Let a Riemann–Cartan continuum matter* $(\mathscr{B}, g_{\alpha\beta}, \Gamma^{\gamma}_{\alpha})$ *in motion within a Minkowski spacetime* $(\mathscr{M}, \hat{g}_{\alpha\beta}, \hat{\Gamma}^{\gamma}_{\alpha})$. *The Lagrangian of the continuum is assumed to depend on the metric and on the contortion tensor and its metric covariant derivative* $\mathscr{L}(g_{\alpha\beta}, \mathfrak{T}^{\gamma}_{\alpha\beta}, \overline{\nabla}_{\lambda}\mathfrak{T}^{\gamma}_{\alpha\beta})$. *The conservation laws hold for continuum matter with arbitrary contortion field and for which Lagrangian depends on the contortion explicitly:*

$$\overline{\nabla}_{\alpha}\left(g_{\rho\beta}\sigma^{\alpha\beta} + \Sigma^{\alpha\beta}_{\gamma}\mathfrak{T}^{\gamma}_{\rho\beta} + \Xi^{\lambda\alpha\beta}_{\gamma}\overline{\nabla}_{\lambda}\mathfrak{T}^{\gamma}_{\rho\beta}\right)$$

$$+ \overline{\nabla}_{\beta}\left(g_{\alpha\rho}\sigma^{\alpha\beta} + \Sigma^{\alpha\beta}_{\gamma}\mathfrak{T}^{\gamma}_{\alpha\rho} + \Xi^{\lambda\alpha\beta}_{\gamma}\overline{\nabla}_{\lambda}\mathfrak{T}^{\gamma}_{\alpha\rho}\right)$$

$$= \sigma^{\alpha\beta}\left(g_{\alpha\gamma}\mathfrak{T}^{\gamma}_{\rho\beta} + g_{\gamma\beta}\mathfrak{T}^{\gamma}_{\rho\alpha}\right) + \Sigma^{\alpha\beta}_{\gamma}\overline{\nabla}_{\rho}\mathfrak{T}^{\gamma}_{\alpha\beta} + \Sigma^{\alpha\beta}_{\rho}\overline{\nabla}_{\gamma}\mathfrak{T}^{\gamma}_{\alpha\beta}$$

$$+ \Xi^{\lambda\alpha\beta}_{\gamma}\overline{\nabla}_{\rho}\overline{\nabla}_{\lambda}\mathfrak{T}^{\gamma}_{\alpha\beta} + \overline{\nabla}_{\gamma}\left(\Xi^{\lambda\alpha\beta}_{\rho}\overline{\nabla}_{\lambda}\mathfrak{T}^{\gamma}_{\alpha\beta}\right) - \overline{\nabla}_{\lambda}\left(\Xi^{\lambda\alpha\beta}_{\gamma}\overline{\nabla}_{\rho}\mathfrak{T}^{\gamma}_{\alpha\beta}\right) \tag{4.159}$$

where hypermomenta are given by constitutive relations (4.158).

Proof The proof is based on the same method as previously by writing first the variational equation:

$$\delta\mathcal{L} = \sigma^{\alpha\beta}\mathcal{L}_\xi g_{\alpha\beta} + \Sigma_\gamma^{\alpha\beta}\mathcal{L}_\xi \mathcal{T}_{\alpha\beta}^\gamma + \Xi_\gamma^{\lambda\alpha\beta}\mathcal{L}_\xi \overline{\nabla}_\lambda \mathcal{T}_{\alpha\beta}^\gamma$$

where the Lie derivatives $\mathcal{L}_\xi g_{\alpha\beta}$ and $\mathcal{L}_\xi \mathcal{T}_{\alpha\beta}^\gamma$ have the same form as previously. It is easily shown that the Lie derivative of the covariant derivative takes the form of:

$$\mathcal{L}_\xi \overline{\nabla}_\lambda \mathcal{T}_{\alpha\beta}^\gamma = \xi^\rho \overline{\nabla}_\rho \overline{\nabla}_\lambda \mathcal{T}_{\alpha\beta}^\gamma$$
$$- \overline{\nabla}_\lambda \mathcal{T}_{\alpha\beta}^\rho \overline{\nabla}_\rho \xi^\gamma + \overline{\nabla}_\rho \mathcal{T}_{\alpha\beta}^\gamma \overline{\nabla}_\lambda \xi^\rho + \overline{\nabla}_\lambda \mathcal{T}_{\alpha\beta}^\gamma \overline{\nabla}_\alpha \xi^\rho + \overline{\nabla}_\lambda \mathcal{T}_{\alpha\beta}^\gamma \overline{\nabla}_\beta \xi^\rho$$

Introducing these relations into the first variational equation allows us to obtain after straightforward but tedious calculus the conservation laws (4.159).   □

Remark 4.44 The conservation laws (4.159) are similar to those derived in the 18th chapter on covariant conservation law of the book (Kleinert 2008). They extend the previous conservation laws to continuum whose Lagrangian depend not only on metric and contortion tensors but also on the Levi-Civita covariant derivative of the contortion tensor. The dependence on this latter argument follows from the dependence on the curvature of the Lagrangian.

4.3.4.5 Field Equations

In some sense, Eq. (4.159) may be exploited to derive the equation of conservation of a Riemann–Cartan U_4 theory of gravitation. If assuming the Einstein–Hilbert action for the Riemann–Cartan spacetime, the field equations have been derived previously (4.81), where the hat is omitted since we are dealing with the matter metric $g_{\alpha\beta}$,

$$\begin{cases} \mathcal{R}_{\alpha\beta} - (1/2)\,\mathcal{R}\,g_{\alpha\beta} = 0 \\ 2\left(\delta_\mu^\alpha \delta_\gamma^\lambda - \delta_\mu^\lambda \delta_\gamma^\alpha\right)\nabla_\lambda g^{\mu\beta} + \aleph_{\gamma\mu}^\alpha\, g^{\mu\beta} = 0 \end{cases} \tag{4.160}$$

Conversely to the work of Hehl et al., the main assumption here is to start with a vacuous Riemann–Cartan spacetime (indeed analogous to curved continuous matter with torsion) and this explains the missing of right-hand side term in the first equation (4.160) e.g. Hehl et al. (1974), in which the equations derived were devoted to microscopic particles with canonical spin.

Remark 4.45 As for the Lagrangian depending only on the metric and contortion, we observe again that some additional terms are present in the divergence operator, $\sigma^{\alpha\beta}$ becomes $g_{\rho\beta}\sigma^{\alpha\beta} + \Sigma_\gamma^{\alpha\beta}\mathcal{T}_{\rho\beta}^\gamma + \Xi_\gamma^{\lambda\alpha\beta}\overline{\nabla}_\lambda \mathcal{T}_{\rho\beta}^\gamma$ e.g. Yasskin and Stoeger (1980). Published works on gravitation with torsion stated that the torsion, and by the way the contortion, couples to the microscopic spin but not to the rotation e.g. Hehl et al. (2013), Yasskin and Stoeger (1980) and this is expressed in the missing of torsion in the conservation laws at the macroscopic level. This not in agreement with the results of some publications e.g. Mao et al. (2007) which supports that Gravity Probe B can be used to experimentally highlight the existence of torsion. However, some investigations are still needed when we look at Eq. (4.159) where we observe that the contortion explicitly appears in the conservation laws even in the missing of hypermomentum $\Sigma_\gamma^{\alpha\beta}$. This still remains a long debate as far as we know.

Remark 4.46 As for conservation laws developed in Obukhov and Puetzfeld (2014), Eq. (4.159) may be used as starting point to investigate either relativistic gravitation within Riemann framework, or Riemann–Cartan gravity in the case of compatible connection. We rewrite conservation laws where covariant derivative is defined with the Levi-Civita connection $\overline{\nabla}$ for practical reason: to avoid torsion tensor in the derivative operator. Of course, we recover the particular case when the contortion vanishes.

Conservation laws (4.151) and (4.159) involve the contortion field and its metric-covariant derivatives on the manifold $\mathscr{B}$ immersed within a spacetime $\mathscr{M}$, both of them are curved with non zero torsion. We have seen that the motion of a particle test within a non zero curved spacetime is along the autoparallels that are different from the geodesics ones e.g. Hehl et al. (1976), Kleinert (2000), Papapetrou (1951). But the equation of motion does not allow us to determine the skew-symmetric part of the contortion tensor $\mathfrak{T}$. What should be mentioned is that the presence of the contortion adds 24 unknowns to the continuum primal variables (ten for the metric). At least theoretically, it is necessary to propose a "gedanken experiment" background to measure the torsion field e.g. Hehl (1971), Garcia de Andrade (2004). This is a main motivation to develop the model of Riemann–Cartan continuum moving within a non zero torsion curved spacetime $\mathscr{M}$. Some attempts have been also made to suggest theory supporting the existence, and even the necessity to adopt a Riemann–Cartan manifold to mimic the rotational flows of fluid with vorticity e.g. Garcia de Andrade (2004, 2005), Rakotomanana (2003). The trace of the Cartan torsion is expressed in terms of the background frequency of the flow, and the theory supporting this suggestion is based on the gauge invariance including torsion tensor.

Chapter 5
Topics in Continuum Mechanics and Gravitation

5.1 Introduction

Modelling spacetime and more generally an arbitrary continuum requires the definition of the background geometry adapted for capturing all subtleties of the medium. As for the evolution of the spacetime concept from Newtonian to Riemann and even Riemann–Cartan manifolds, mathematical models of continuum also has been extended independently on the speed of bodies. Namely transformation of strain gradient continuum do not necessarily involve large speed of particles constituting the body. Extension of non relativistic spacetime to include gravitation was first due to Cartan to obtain Newton–Cartan gravitation. Without reference to gravitation, the Cartan geometric background was worthily used to account for dislocations and disclinations within otherway virgin matter. This chapter explores some aspects of continuum mechanics and gravitation within this context, namely in the domain of wave propagation.

Two types of waves are mostly considered and assumed to capture signals from faraway regions (in the sense of spacetime distance) to detect the merging of black holes, the presence of astronomic planets, and any other astrophysical phenomena. First, electromagnetic waves, including light wave propagation, whose theory is based on the Maxwell equations, are actually the most mastered and intensively exploited to support for many years signal science and technology. Second, gravitational waves although predicted by Einstein relativistic gravitation almost 100 years ago, are now on the way of experimental validation. Conversely to electromagnetic waves that are considered to be dispersed and faded with environment, gravitational waves which are spacetime waves are thought to propagate with neither dispersion nor fading. This concept may nevertheless evolve in the future if we work in the framework of Einstein–Cartan curved spacetime with torsion rather staying in the Einstein curved spacetime framework.

We consider in this chapter some applications in the domain of strain gradient continuum, gravitational waves, and will sketch some aspects of electromagnetism

© Springer International Publishing AG, part of Springer Nature 2018 177
L. R. Rakotomanana, *Covariance and Gauge Invariance in Continuum Physics*,
Progress in Mathematical Physics 73, https://doi.org/10.1007/978-3-319-91782-5_5

in the next chapter. The link between these examples is related to the developments of the curvature tensor, and the eventual presence of torsion in the spacetime and the matter continuum. The derivation of the deviation equation was a starting point for the detection of gravitational waves in physics. Introducing an affine connection with torsion may render the derivation more complex since there is no more geodesic but autoparallel curves in the spacetime, and the principle of equivalence should be revised. We first in this section remind the basics of gravitational waves within a Einstein gravitation theory. Then we derive the deviation equation for both geodesics and autoparallels.

5.2   Continuum Mechanics in a Newton Spacetime

In this section we consider a background spacetime without gravitation, and more generally without external applied forces. The spacetime of Newton mechanics is thus described by a Cartesian metric $g^{\alpha\beta} := \mathrm{diag}\{0, 1, 1, 1\}$, a vector $\tau^{\alpha} := (1, 0, 0, 0)$, and a and symmetric affine connection $\Gamma^{\gamma}_{\alpha\beta}$, such that the metric is orthogonal to the vector τ^{α} e.g. Goenner (1974). Metric compatibility of $\Gamma^{\gamma}_{\alpha\beta}$ and τ^{α}, ensure the existence of a family of coordinate system such that $\Gamma^{\gamma}_{\alpha\beta} \equiv 0$ (referential frames). We now investigate some types of waves in this spacetime. The most basic of them and mostly observed in everyday life is the elastic waves as during vibrations of beams, seismic perturbations to name but a few. In such a case, the waves are the motion of matter within a flat spacetime endowed with the metric $\hat{g}_{\alpha\beta}$ where the space is separated from the time.

5.2.1   Classical Continuum in Newtonian Spacetime

In addition to the flat spacetime, we also consider a classical continuum characterized by a zero torsion and zero curvature for the continuum matter e.g. Marsden and Hughes (1983). Accordingly, we can define a global coordinate system on the continuum manifold $\mathscr{B}$, and a metric compatible flat connection $\hat{\nabla}$.

5.2.1.1   Continuum Transformations and Integrability

By limiting our purpose to elastic deformation, both the torsion and the curvature of not only the spacetime but also the continuum matter are zero in this subsection, say the two conditions $\aleph_r = \aleph_c \equiv 0$ and $\Re_r = \Re_c \equiv 0$. Indeed these two conditions are merely requirements that the metric and the connection are single-valued tensor fields on the continuum and smooth enough to be twice differentiable e.g. Maugin (1993), Rakotomanana (2003). For the sake of the simplicity, we choose a Cartesian coordinate system in an inertial Galilean reference frame. The tetrads restricted to infinitesimal transformations and the metric tensor are given by the relationships:

$$F^i_\lambda = \delta^i_\alpha + \partial_\lambda u^i, \quad g_{\lambda\kappa} = \delta_{\lambda\kappa} + (\partial_\lambda u_\kappa + \partial_\kappa u_\lambda), \quad \Gamma^\gamma_{\lambda\kappa} = \partial_\lambda \partial_\kappa u^\gamma$$

where the use of Latin and Greek indexes may be done arbitrarily without difficulties since we are adopting a Cartesian coordinate system for both reference and current configurations. Postulating that the metric and the connection are single-valued and sufficiently smooth, the infinitesimal integrability conditions (Schwarz conditions) take the form of:

$$\begin{cases} \left(\partial_\alpha \partial_\beta - \partial_\beta \partial_\alpha\right) g_{\lambda\kappa} = 0 \\ \left(\partial_\alpha \partial_\beta - \partial_\beta \partial_\alpha\right) \Gamma^\gamma_{\lambda\kappa} = 0 \end{cases} \tag{5.1}$$

First, from the expression of the connection in terms of displacement field, we observe that $\aleph^\gamma_{\lambda\kappa} := \partial_\lambda \partial_\kappa u^\gamma - \partial_\kappa \partial_\lambda u^\gamma = 0$ because the displacement mapping is single-valued and twice differentiable. Second, from the first row of Eq. (5.1), we deduce the vanishing of the Riemann curvature tensor. This may be also checked by calculating the Riemann–Cartan curvature:

$$\begin{aligned} \Re^\gamma_{\alpha\beta\lambda} &:= \partial_\alpha \Gamma^\gamma_{\beta\lambda} - \partial_\beta \Gamma^\gamma_{\alpha\lambda} + \Gamma^\gamma_{\alpha\nu} \Gamma^\nu_{\beta\lambda} - \Gamma^\gamma_{\beta\nu} \Gamma^\nu_{\alpha\lambda} \\ &= \left(\partial_\alpha \partial_\beta - \partial_\beta \partial_\alpha\right) \partial_\lambda u^\gamma + \text{second order terms} \\ &\simeq 0 \end{aligned}$$

which vanishes at first order when assuming a (single-valued and) smooth displacement gradient. Proof for finite transformations and nonlinear case may be found in e.g. Rakotomanana (2003).

5.2.1.2 Constitutive Laws and Conservation Laws

The basic theoretical model for analyzing the wave propagation is derived from the Navier equation in linear elasticity with homogeneous continuum matter. For deriving the conservation laws, we apply the Poincaré's invariance. We remind here the basic steps for driving the conservation laws for a small strain elastic continuum. The action associated to an arbitrary but admissible potential energy $\mathscr{U}$ takes the form of:

$$\mathscr{S} := \int_{t_0}^{t_1} \int_{\mathscr{B}} \left[\frac{\rho}{2} \hat{g}_{ij} \partial_t u^i \partial_t u^j - \mathscr{U}\left(\varepsilon_{ij}\right)\right] dvdt \tag{5.2}$$

where the only argument of the potential energy $\mathscr{U}$ is the small strain tensor $\varepsilon_{ij} := (1/2)\left(\hat{\nabla}_i u_j + \hat{\nabla}_j u_i\right)$. The Lie derivative variation, say $\delta := \mathcal{L}_\xi$ of the action is easily written after integrating by parts (on time for the first row, and on space for the second row):

$$\delta \mathscr{S} = \left[\int_{\mathscr{B}} \rho \xi_i \partial_t u^i \, dv \right]_{t_0}^{t_1} - \int_{t_0}^{t_1} \int_{\mathscr{B}} \rho \xi_i \partial_{tt} u^i \, dv dt$$

$$- \int_{t_0}^{t_1} \int_{\mathscr{B}} \hat{\nabla}_j \left(\sigma^{ij} \xi_i \right) dv dt + \int_{t_0}^{t_1} \int_{\mathscr{B}} \xi_i \hat{\nabla}_j \sigma^{ij} \, dv dt$$

where we have defined the constitutive laws relating the stress tensor and the strain tensor as:

$$\sigma^{ij} := \frac{\partial \mathscr{U}}{\partial \varepsilon_{ij}}, \qquad \text{and} \qquad \sigma^{ij} (1/2) \left(\hat{\nabla}_i \xi_j + \hat{\nabla}_j \xi_i \right) = \sigma^{ij} \hat{\nabla}_j \xi_i \qquad (5.3)$$

since the symmetry of the stress tensor is here obviously deduced from that of the strain tensor. We also remind the Lie derivative variation of the strain tensor as $\mathcal{L}_\xi \varepsilon_{ij} = \hat{\nabla}_i \xi_j + \hat{\nabla}_j \xi_i$. From the Haar lemma and worth choice of boundary conditions (in time and in space), we can localize and obtain the usual conservation laws in elasticity theory e.g. Marsden and Hughes (1983):

$$\rho \partial_{tt} u^i = \hat{\nabla}_j \sigma^{ij} \qquad (5.4)$$

where $\hat{\nabla}$ is an Euclidean flat connection (without torsion and without curvature too) e.g. Flügge (1972). Closure of this linear momentum equation is obtained by defining the constitutive equations relating stress and strain $\sigma^{ij} (\varepsilon_{kl})$ or equivalently the shape of the energy function $\mathscr{U} (\varepsilon_{kl})$.

5.2.1.3   Navier Equation

We now consider elastic wave propagation within linear, isotropic, and homogeneous continuum. Development of the theory is ruled by the derivation of exact linear momentum equations and by the way boundary conditions if any. The method includes three steps: definition of the Lagrangian and namely the strain energy potential, formulation of the linear momentum equation by accounting for the stress-strain law. The linear elastic deformation of a continuum in the framework of non relative small strains may be captured by the Lagrangian:

$$\mathscr{L} := (\rho/2) \hat{g}_{ij} \partial_t u^i \partial_t u^j - \left[(\lambda/2) \mathrm{Tr}^2 \varepsilon + \mu \mathrm{Tr} \left(\varepsilon^2 \right) \right] \qquad (5.5)$$

in which $\hat{g}_{\alpha\beta}$ denotes the metric tensor of the three dimension space, $u^\alpha (t, x^\gamma)$ the displacement field of the continuum. The two constants λ and μ are the Lamé coefficients of elasticity of the continuum and ρ the matter density. Constant μ is the shear modulus of the material. The metric induced by the deformation of the continuum $\mathscr{B}$ may be written as $g_{\alpha\beta} \simeq \hat{g}_{\alpha\beta} + 2\varepsilon_{\alpha\beta}$ with $2\varepsilon_{\alpha\beta} := \mathcal{L}_{\mathbf{u}}\hat{g}_{\alpha\beta} = \overline{\nabla}_{\alpha}u_{\beta} + \overline{\nabla}_{\beta}u_{\alpha}$. Choosing the Lagrangian (5.18), we obtain the Navier equation. As a remind the resulting elastic wave equation in linear isotropic homogeneous elasticity takes then the form of e.g. Rakotomanana (2009):

$$(\lambda + \mu)\overline{\nabla}(\overline{\nabla} \cdot \mathbf{u}) + \mu\overline{\Delta}\mathbf{u} = \rho\mathbf{a} \tag{5.6}$$

where $\mathbf{a}$ denotes the acceleration. The nabla operator $\overline{\nabla}$ and Laplacian operator $\overline{\Delta}$ are derived by means of the Levi-Civita connection $\overline{\Gamma}^{\gamma}_{\alpha\beta}$ of the flat spacetime. Classically, application of the divergence $\overline{\nabla}\cdot$ and rotational $\overline{\nabla}\times$ operators allows us to point out the two waves included in this model. Practically, we can decompose the displacement field into a gradient and a rotational field $\mathbf{u} := \overline{\nabla}\Phi + \overline{\nabla} \times \mathbf{H}$, called Helmholtz decomposition,

$$\overline{\nabla}\left[(\lambda + 2\mu)\,\overline{\Delta}\Phi - \rho\,\partial_t^2\Phi\right] + \overline{\nabla} \times \left\{\mu\,\overline{\Delta}\mathbf{H} - \rho\,\partial_t^2\mathbf{H}\right\} = 0$$

The volume and shear wave propagation equations which are sufficient conditions of this previous equation are deduced accordingly:

$$\begin{cases} (\lambda + 2\mu)\,\overline{\Delta}\Phi - \rho\,\partial_t^2\Phi = 0 \\ \mu\,\overline{\Delta}\mathbf{H} - \rho\,\partial_t^2\mathbf{H} = 0 \end{cases} \implies \begin{cases} \partial_t^2\Phi - c_L^2\,\overline{\Delta}\Phi = 0 \\ \partial_t^2\mathbf{H} - c_T^2\,\overline{\Delta}\mathbf{H} = 0 \end{cases} \tag{5.7}$$

where two elastic waves co-exist and whose celebrities are obtained $c_L^2 := (\lambda + 2\mu)/\rho$, and $c_T^2 := \mu/\rho$. Celebrities are different. A change in volume which is a dilatational perturbation will propagate at the velocity c_L whereas a distortional wave will propagate at the velocity c_T.

5.2.1.4 Four-Dimensional Formulation

Conversely to acoustic waves and electromagnetic waves in vacuum spacetime, elastic waves include two celebrities (for isotropic continuum). A coordinate system may then be defined as $x^{\mu} := (x^0 := c_L t, x^1, x^2, x^3)$, and the wave equations (5.7) may be written respectively as:

$$\hat{g}^{\alpha\beta}\,\hat{\nabla}_{\alpha}\hat{\nabla}_{\beta}A^{\mu} = 0, \qquad A^{\mu} := \left(\Phi, \frac{c_L^2}{c_T^2}H^1, \frac{c_L^2}{c_T^2}H^2, \frac{c_L^2}{c_T^2}H^3\right) \tag{5.8}$$

where $\hat{g}^{\alpha\beta} = \text{diag}\{+1, -1, -1, -1\}$ is the Minkowskian metric of the spacetime, and A^{μ} a kind of four potential. It is similar to an electromagnetic potential we will see in the next chapter. The simplest solution of the four-dimensional wave equation (5.8) is obtained by assuming a plane wave as:

$$A^\mu = \Re e \left\{ \hat{A}^\mu \exp \left(i \kappa_\alpha x^\alpha \right) \right\} \tag{5.9}$$

where $\hat{A}^\mu$ is the wave (complex) amplitude, and κ_α is a null four-vector such that $\kappa_\alpha \kappa^\alpha = \hat{g}^{\alpha\beta} \kappa_\alpha \kappa_\beta = \kappa_0 \kappa^0 - \kappa_i \kappa^i \equiv 0$. The link with the three dimensional spatial solution and temporal solution is then given by the relationships (dispersion equation):

$$\begin{cases} \dfrac{\omega}{c_L} := \sqrt{\kappa_0 \kappa^0} = \sqrt{\kappa_i \kappa^i} \\[2mm] \mathbf{k} := \left(\kappa^1, \kappa^2, \kappa^3 \right) \end{cases} \tag{5.10}$$

where ω and $\mathbf{k}$ are the usual frequency and the wave number vector respectively. The longitudinal wave celerity is introduced since the time coordinate was defined as $x^0 := c_L t$. An alternative description would be based on another definition $x^0 := c_T t$ and the definition of the vector A^μ would be changed accordingly.

Remark 5.1 The first line of Eq. (5.10) provides the relation between the frequency and the wave number and constitutes nothing more than the dispersion equation. The nullity of the four-vector, sometimes also called four-frequency, $\kappa_\alpha \kappa^\alpha \equiv 0$ means that κ_α is a lightlike vector. It should be pointed out that considering κ_α as a lightlike vector gives the dispersion equation. There is a strong similarity with the special relativity theory, and also with the electromagnetism theory.

5.2.1.5   Example of Spherical Waves

For illustrating Eq. (5.6) let us consider the radial vibration of a continuum medium with spherical symmetry about the origin O of the space. The displacement is assumed to be $\mathbf{u} := u(t, r) \, \mathbf{e}_r$ where we use a spherical coordinate system (r, θ, φ), φ denoting the azimuthal coordinate. The motion equation (5.6) that does not vary with θ or φ reduces to a radial motion and is governed by:

$$(\lambda + 2\mu) \left[\partial_{rr} u + (2/r) \partial_r u - 2u/r^2 \right] = \rho \partial_{tt} u$$

Searching for solutions with the form of $u(r, t) := U(r) T(t)$, the function $U(r)$ should satisfy the radial equation:

$$r^2 U''(r) + 2r U(r) + \left(k^2 r^2 - 2 \right) U(r) = 0, \quad \text{with} \quad k^2 := \rho \omega^2 / (\lambda + 2\mu) \tag{5.11}$$

owing that the time solution takes the form of $T(t) = A \sin(\omega t) + B \cos(\omega t)$, in which ω is the frequency of the wave. The space part of the spherical wave equation (5.11) is a spherical Bessel equation of order $\ell = 1$. It is worth to define a function $V(r) := U(r) \sqrt{kr}$ which allows us to obtain the equation:

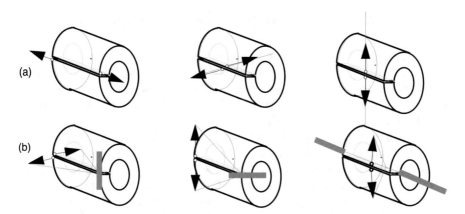

Fig. 5.1 First line (**a**): Volterra process for the two types of dislocations (relative displacement of two contacting surfaces): (a1) screw dislocation; (a2) edge dislocation; and (a3) edge dislocation which separates two surfaces; Second line (**b**): Volterra process of two types of disclinations (relative rotation of two contacting surfaces): (b1) and (b2) twist disclinations, and (b3) wedge disclination

$$r^2 V''(r) + r V'(r) + \left(k^2 r^2 - (\ell + 1/2)^2\right) = 0$$

which is Bessel equation of order $\ell + 1/2$. By accounting for the change of variable, the solutions of the initial equation are thus ($\ell = 1$):

$$j_\ell(r) = \sqrt{\frac{\pi}{2kr}} J_{\ell+1/2}(kr), \qquad n_\ell(r) = \sqrt{\frac{\pi}{2kr}} Y_{\ell+1/2}(kr), \tag{5.12}$$

which are called spherical Bessel functions of first kind and second kind respectively. It is noticed that spherical Bessel functions are simpler compared to Bessel functions. They are related to trigonometric functions as follows:

$$j_1(r) = \frac{\sin(kr)}{k^2 r^2} - \frac{\cos(kr)}{kr}, \qquad n_1(r) = -\frac{\cos(kr)}{k^2 r^2} - \frac{\sin(kr)}{kr}, \tag{5.13}$$

The general solution holds:

$$U(r) = C\left[\frac{\sin(kr)}{k^2 r^2} - \frac{\cos(kr)}{kr}\right] + D\left[-\frac{\cos(kr)}{k^2 r^2} - \frac{\sin(kr)}{kr}\right] \tag{5.14}$$

in which the constants C and D may be calculated by means of the boundary conditions. For plain sphere, the finiteness value of the displacement amplitude at the origin allows us to eliminate the second term when $r \to 0$, then $D = 0$. By the way, we notice that $j_1(r) \simeq kr/3 + \mathcal{O}(r)$.

5.2.2 Continuum with Torsion in a Newtonian Spacetime

As previously, we consider a simple material model, and follows the three steps as before in order to derive the extended linear momentum equation. Extension of Navier equations is necessary because we deal no more with Riemann continuum but rather a Riemann–Cartan continuum. Various theoretical approaches may be used to classify the dislocations (line defects). We consider one of most classical of them which is the Volterra process as sketched on Fig. 5.1. The first line of the figure reports the translational dislocations corresponding to the relative displacement of contacting surfaces along axial, radial, and azimuthal directions respectively. The second line displays the relative rotations (disclinations) of two contacting surfaces about azimuthal, radial, and axial directions respectively. The first modes of continuous fields of dislocations and disclinations with non-Riemannian geometry approach draw back to the fifties e.g. Bilby et al. (1955). Since then, numerous models have been proposed e.g. Maugin (1993), Noll (1967) to name but a few. A common feature is that the main ingredients to model continuous distributions of line defects are the torsion and curvature fields associated to an affine connection on a Riemann–Cartan manifold e.g. Rakotomanana (1997). Particularly, the torsion tensor measures not only the density of dislocation within a material but is also able to capture the continuous distribution of non smoothness of scalar field over the manifold (Rakotomanana 2003). The torsion distribution is then of great interest if we want to compare properties of different materials or to characterize the material response under waves. Indeed, the use of elastic waves to determine the amount of torsion field presents some interest. In a dislocated continuum approach, the concept of Bürgers vector **b** is a common variable allowing the generation of dislocation. Relationships between **b** and the torsion holds:

$$b^\gamma := \oint_{\mathscr{C}} \aleph^\gamma_{\alpha\beta} \, dx^\alpha \wedge dx^\beta$$

where the integration is done along a loop $\mathscr{C}$ surrounding a dislocation. The Bürgers vector is not topologically invariant. This vector violates the covariance $\mathbf{x} \to \mathbf{x}'$:

$$b^{\gamma'} = \oint_{\mathscr{C}} \aleph^{\gamma'}_{\alpha'\beta'} \, dx^{\alpha'} \wedge dx^{\beta'} = \oint_{\mathscr{C}} \frac{\partial x^{\gamma'}}{\partial x^\gamma} \aleph^\gamma_{\alpha\beta} \, dx^\alpha \wedge dx^\beta \neq \frac{\partial x^{\gamma'}}{\partial x^\gamma} b^\gamma$$

Therefore, the torsion field $\aleph^\gamma_{\alpha\beta}$ may be considered as more fundamental variable than the vector b^γ. Practically, despite its nice physical interpretation, the Bürgers vector cannot be used as a argument of the Lagrangian $\mathscr{L}$. The next example is thus motivated by the derivation of a model for measuring torsion tensor in a continuum matter with internal architecture undergoing small strain and small displacement. We consider a particular continuum with torsion $(\mathscr{B}, g_{\alpha\beta}, \Gamma^\alpha_{\alpha\beta})$ in motion within a Newtonian spacetime $(\mathscr{N}, \hat{g}_{\alpha\beta} \equiv \delta_{\alpha\beta}, \hat{\tau}_\alpha := 1, \hat{\Gamma}^\gamma_{\alpha\beta} \equiv 0)$.

5.2.2.1 Lagrangian Function

Consider the motion of a continuum matter $\mathscr{U}$ in a Newtonian spacetime. We assume that the mathematical model of $\mathscr{B}$ is entirely described by the action, where we have dropped any external action but the gravitation,

$$\mathscr{S} = -\int_{\mathscr{B}} \mathscr{L}_M \left(\varepsilon_{\alpha\beta}, \mathfrak{T}^{\gamma}_{\alpha\beta} \right) \omega_n - \int_{\mathscr{B}} \rho c \sqrt{u^{\alpha} u_{\alpha}} \, \omega_n \tag{5.15}$$

where ρ is the density (mass per unit volume), $v_{\alpha}(t, x^{\mu})$ denotes the four-vector velocity field on $\mathscr{B}$, $\varepsilon_{\alpha\beta} := (1/2) \left(\partial_{\alpha} u_{\beta} + \partial_{\beta} u_{\alpha} \right)$ denotes the Cauchy strain tensor with $u_{\alpha}(t, x^{\mu})$ the displacement field, and $\kappa_{\alpha\beta} := \mathfrak{R}^{\lambda}_{\lambda\alpha\beta}$ the $(3D)$-Ricci curvature tensor on $\mathscr{B}$. As for linearized gravitation waves, we can introduce the strain as $g_{\alpha\beta} = \hat{g}_{\alpha\beta} + 2\varepsilon_{\alpha\beta}$ with $|\varepsilon_{\alpha\beta}| \ll 1$. When the velocity of a material point of the continuum is relatively low $v^i \ll c$, the four-velocity (2.77) reduces to:

$$v^{\mu} := \frac{dx^{\mu}}{d\tau} = \frac{1}{\sqrt{1 - (v^2/c^2)}} \left(1, v^i \right) \quad \Longrightarrow \quad v^{\mu} \simeq \left(1, v^i \right) \simeq \left(1, \frac{\partial u^i}{\partial t} \right) \tag{5.16}$$

where u^i is the displacement components of the point. Then, assuming a low velocity situation and small perturbation for the displacement field (we neglect the nonlinear part of the strain $\mathrm{Tr}(\nabla^{\mathsf{T}} \mathbf{u} \nabla \mathbf{u})$), the Lagrangian reduces to the Newtonian limit to give ($\alpha, \beta, \gamma, \nu = 1, 3$):

$$\mathscr{L} := (\rho/2) \, \partial_t u_{\alpha} \partial_t u^{\alpha} - (1/2) \, \mathbb{E}^{\alpha\beta\gamma\nu} \varepsilon_{\alpha\beta} \varepsilon_{\gamma\nu}, \quad \varepsilon_{\alpha\beta} := (1/2) \left(\nabla_{\alpha} u_{\beta} + \nabla_{\beta} u_{\alpha} \right) \tag{5.17}$$

where the relativistic Green-Lagrange strain (4.88) is linearized to obtain the strain tensor in (5.17). The dependence on the contortion tensor implicitly appears in the strain tensor since the covariant derivative involves the connection coefficients $\Gamma^{\gamma}_{\alpha\beta} = \overline{\Gamma}^{\gamma}_{\alpha\beta} + \mathfrak{T}^{\gamma}_{\alpha\beta}$. We remark that the gravitational part of the Lagrangian (kinetic energy) is merely given by $\mathscr{L}_G := (\rho/2) \, \partial_t u_{\alpha} \partial_t u^{\alpha}$ which is an approximation of the dust matter inertial terms $-\rho c \sqrt{v^{\alpha} v_{\alpha}} \, \omega_n$ (v_{α} stands for four-velocity vector here) in Eq. (2.85).

5.2.2.2 Conservation Laws

Say a continuum $\mathscr{B}$ with non zero torsion (which may be also called first gradient continuum, even if the kinematic variable is rather the torsion than the covariant derivative of the strain). The continuum is evolving within a Newton space-time. A common method is to split the Lagrangian into two parts: the kinetic energy due to the motion within the spacetime, and the potential energy associated to matter. Let consider an extension of the Lagrangian (5.17) as:

$$\mathscr{L} = (\rho/2)\,\partial_0 u_\alpha \partial_0 u^\alpha - \mathscr{L}_M\left(\varepsilon_{\alpha\beta}\right) \tag{5.18}$$

The variation of the associated action leads to:

$$\delta\mathscr{L} = \frac{\partial\mathscr{L}}{\partial(\partial_0 u_\alpha)}\delta(\partial_0 u_\alpha) + \frac{\partial\mathscr{L}}{\partial\varepsilon_{\alpha\beta}}\delta\varepsilon_{\alpha\beta} = \rho\partial_0 u^\alpha \delta(\partial_0 u_\alpha) - \frac{1}{2}\sigma^{\alpha\beta}\delta\left(\nabla_\alpha u_\beta + \nabla_\beta u_\alpha\right)$$

where we remind the definition of the stress $\sigma^{\alpha\beta}$ as the derivative of the matter Lagrangian $\mathscr{L}_M$ with respect to the stress $\varepsilon_{\alpha\beta}$. This variation may be written as follows:

$$\delta\mathscr{L} = \partial_0(\rho\partial_0 u^\alpha \delta u_\alpha) - \partial_0\left(\rho\partial_0 u^\alpha\right)\delta u_\alpha - \nabla_\alpha\left(\sigma^{\alpha\beta}\delta u_\beta\right) + \left(\nabla_\alpha\sigma^{\alpha\beta}\right)\delta u_\beta$$

The first term and the third term may be pushed to the spacetime boundary and included in a divergence term at the boundary. Therefore, the remaining terms hold as the conservation laws:

$$\nabla_\beta\sigma^{\beta\alpha} - \partial_0\left(\rho\partial_0 u^\alpha\right) = 0 \tag{5.19}$$

which resembles to the classical conservation laws for continuum mechanics e.g. Marsden and Hughes (1983). In the material part of the Lagrangian, the strain may be obtained by using either a Riemann–Cartan connection $\Gamma^\gamma_{\alpha\beta}$ of the continuum (non zero torsion, and non zero curvature), or the Levi-Civita connection $\overline{\Gamma}^\gamma_{\alpha\beta}$ of the Cartesian ambient space. In all cases, we worthily assume a metric compatibility of the connection. See e.g. Sharma and Ganti (2005) for the importance of the angular gauge-field in the definition of the strain for strain gradient elasticity. For isotropic, and uniform density (which does not mean homogeneous matter), the tangent stiffness holds (usual Hooke's law) $\mathbb{E}^{\alpha\beta\lambda\nu}$ as constant.

5.2.2.3 Extended Navier Equation in a Riemann–Cartan Continuum

We consider a Riemann–Cartan continuum $\mathscr{B}$ with the Lagrangian (5.18) where the strain tensor $\varepsilon_{\alpha\beta}$ is calculated as the symmetrized part of $\nabla_\alpha u_\beta$ for material strain or that of $\overline{\nabla}_\alpha u_\beta$ for the so-called spatial strain. The assumed Lagrangian explicitly depends neither on the curvature nor on the torsion. By applying the Poincaré gauge invariance and then (5.19), we first derive the divergence of the stress tensor:

$$\nabla_\beta\sigma^{\alpha\beta} = (1/2)\nabla_\beta\left[\mathbb{E}^{\alpha\beta\mu\nu}\left(\nabla_\mu u_\nu + \nabla_\nu u_\mu\right)\right] = \mathbb{E}^{\alpha\beta\mu\nu}\nabla_\beta\nabla_\mu u_\nu \tag{5.20}$$

owing the symmetry $\mathbb{E}^{\alpha\beta\mu\nu} = \mathbb{E}^{\alpha\beta\nu\mu}$ and the metric compatibility of the connection. Let us remind some properties of the torsion tensor (2.35) and the curvature (2.38) tensor in the framework of Riemann–Cartan geometry, for any scalar field Φ and vector field $\mathbf{u}$ on the manifold $\mathscr{B}$:

$$\begin{cases} -\aleph^{\gamma}_{\beta\mu}\nabla_{\gamma}\Phi = \nabla_{\beta}\nabla_{\mu}\Phi - \nabla_{\mu}\nabla_{\beta}\Phi \\ \Re^{\nu}_{\beta\mu\gamma}u^{\gamma} = \nabla_{\beta}\nabla_{\mu}u^{\nu} - \nabla_{\mu}\nabla_{\beta}u^{\nu} + \aleph^{\gamma}_{\beta\mu}\nabla_{\gamma}u^{\nu} \end{cases} \qquad (5.21)$$

For the sake of the simplicity, we consider the case of linear isotropic elasticity for which the tangent tensor takes the form of:

$$\mathbb{E}^{\alpha\beta\mu\nu} := \lambda\,\hat{g}^{\alpha\beta}\hat{g}^{\mu\nu} + \mu\left(\hat{g}^{\alpha\mu}\hat{g}^{\beta\nu} + \hat{g}^{\alpha\nu}\hat{g}^{\beta\mu}\right) \qquad (5.22)$$

Introduction of this isotropic tangent tensor (5.20) and the relation (5.21) for commutation of covariant derivatives allows us to extend the classical Navier equation (5.6) to take the form of Futhazar et al. (2014):

$$(\lambda + \mu)\,\nabla^{\alpha}\nabla_{\beta}u^{\beta} + \mu\nabla^{\beta}\nabla_{\beta}u^{\alpha} + \mu\hat{g}^{\alpha\mu}\left(\Re_{\mu\nu}\,u^{\nu} - \aleph^{\gamma}_{\beta\mu}\nabla_{\gamma}u^{\beta}\right) = \rho\partial^{2}_{t}u^{\alpha} \qquad (5.23)$$

where $\Re_{\mu\nu} := \Re^{\gamma}_{\gamma\mu\nu}$ is the Ricci curvature which entirely defines the curvature tensor in a three dimensional space. We observe that the Navier equation (5.19) for defected matter includes the influence of torsion and curvature Kleman and Friedel (2008). Some aspects of the influence of torsion and curvature on the linear momentum conservation has been investigated in e.g. Maugin (1993) in a rather general purpose. The form of the obtained equation suggested similarity with the gravitation-spin theory. We will develop in the next chapter the influence of the Riemann–Cartan geometry on other waves as electromagnetic fields. Of course we recover the classical elastic wave equation for non curved and zero torsion continuum. The first two terms in Eq. (5.23) are analogous to the classical Navier equation although the connection here includes the contortion tensor. The third term is the contribution of torsion and curvature and they are only related to the shear modulus μ although their interaction appears for all three components of the displacement. In the last two terms of the left hand side of the extended Navier equation, the curvature part $\mu g^{\alpha\mu}\Re_{\mu\nu}\,u^{\nu}$ induces an auto-oscillation of the continuum whereas the torsion part $\mu g^{\alpha\mu}\aleph^{\gamma}_{\beta\mu}\nabla_{\gamma}u^{\beta}$ engenders a space fading of the wave. An example for dislocated continuum is reported in the next paragraph.

Remark 5.2 Conversely to the classical wave propagation equation (5.6) it is difficult to separate the dilatational wave Φ and the shear wave $\mathbf{H}$ since they are highly coupled in Eq. (5.23).

5.2.2.4   Example of a $\aleph^{1}_{23} \neq 0$

In a recent work, we considered a uniform distribution of screw dislocation characterized by one component of the torsion $\aleph_{0} := (1/2)\aleph^{1}_{23} \neq 0$, all other components are zero. We work in a Cartesian basis for the sake of the simplicity. It is however stressed that the Bianchi identities for Einstein–Cartan manifolds e.g. Rakotomanana (2003) (and reported in relation (4.74)) should be satisfied for the model, such is the case in this example. A uniform distribution of defects is now introduced in the conservation laws. Details of field equations deduced from the Lagrangian (5.17) may be found in Futhazar et al. (2014). It may be observed that the use of Riemann–Cartan was also developed for fluid flows with vortex e.g. Garcia de Andrade (2004), Rakotomanana (2003) where the contortion tensor is added to extend the classical wave operator in compressible fluid to account for the rotational flow with vortex. For the strain is calculated by means of the entire connection, the propagation equation (5.23) reduces to the conservation laws where the divergence operator also includes the torsion of the connection ∇,

$$
\begin{aligned}
\partial_{tt} u_1 &= (c_L^2 - c_T^2)\partial_{1j} u_j + c_T^2 \partial_{jj} u_1 + c_T^2 \aleph_0 (\partial_3 u_2 - \partial_2 u_3) \\
\partial_{tt} u_2 + (c_T \aleph_0)^2 u_2 &= (c_L^2 - c_T^2)\partial_{2j} u_j + c_T^2 \partial_{jj} u_2 + c_T^2 \aleph_0 (\partial_3 u_1 + 2\partial_1 u_3) \\
\partial_{tt} u_3 + (c_T \aleph_0)^2 u_3 &= (c_L^2 - c_T^2)\partial_{3j} u_j + c_T^2 \partial_{jj} u_3 - c_T^2 \aleph_0 (\partial_2 u_1 + 2\partial_1 u_2)
\end{aligned}
\tag{5.24}
$$

For this model, additional body forces (configurational forces including wave propagation, diffusion, and breathing) appear in the wave propagation equations, they correspond to famous Peach-Koehler forces in dislocation theory. Analysis of the solution have been investigated elsewhere (Futhazar et al. 2014), showing drastically different behavior for this models particularly for the wave attenuation and polarization. It is astonishing that use of strain (with torsion) allows us to highlight a superimposed self-vibration phenomenon due to the uniform array of dislocations (see e.g. Barra et al. (2009) for some experimental results), with a specific frequency $\omega_0 := \sqrt{\mu \aleph_0 / \rho}$ along the directions 2, and 3. Physical interpretation of self-vibration of dislocations is the following: an incident wave—coming from any other region of the solid—hits the dislocation, causing it to oscillate in response. It induces scattering of the incident wave.

We sum up some results of previous (Futhazar et al. 2014) to highlight the influence of the choice of the connection. Let consider plane wave solution with gradient continuum of the form $\mathbf{u} := \mathbf{u}_0 e^{i(\mathbf{k}\cdot\mathbf{x}-\omega t)}$ where $\mathbf{u}_0$ is the polarization vector, and $\mathbf{k} := k\,\mathbf{n}$ is the wave vector such that $\mathbf{n} = \cos\phi\,\mathbf{e}_1 + \sin\phi\cos\theta\,\mathbf{e}_2 + \sin\phi\sin\theta\,\mathbf{e}_3$ is the unit propagation director. Introduction of the plane wave solutions allows us to determine the dispersion patterns. The dispersion equation thus takes the form of:

$$
\mathscr{P}_6(k) = \text{Det}\left(-k^2 \mathbb{M} + ik\mathbb{D} + \mathbb{B} + \omega^2 \mathbb{I}\right) = 0
\tag{5.25}
$$

with $c_l^2 := (\lambda + 2\mu)/\rho$ and $c_t^2 := \mu/\rho$. The matrices $\mathbb{M}$, $\mathbb{D}$, and $\mathbb{B}$ are explicitly given by:

$$\mathbb{M} = c_t^2 \, \mathbb{I} + \left(c_l^2 - c_t^2 \right) \mathbf{n} \otimes \mathbf{n},$$

$$\mathbb{D} = c_t^2 \aleph_0 \begin{pmatrix} 0 & -\sin\phi\sin\theta & \sin\phi\cos\theta \\ \sin\phi\sin\theta & 0 & 2\cos\phi \\ -\sin\phi\cos\theta & -2\cos\phi & 0 \end{pmatrix},$$

$$\mathbb{B} = -c_t^2 \aleph_0^2 \begin{pmatrix} 0 & 0 & 0 \\ 0 & 1 & 0 \\ 0 & 0 & 1 \end{pmatrix}$$

They represent the elastic wave, the diffusion, and the breathing modes respectively.

Remark 5.3 It is interesting to observe that, independently on the classical elastic wave and wave diffusion, a theoretical periodic motion appears in time in addition to space motion. Then we notice that continuum particles move rigidly and periodically and return back to their original configuration, namely all of them move on an ellipse.

Some illustrations of dispersion curves are given in Figs. 5.2 and 5.3, for two different directions for illustrating the anisotropy of wave propagation. Dispersion equation (cubic polynomial functions of k^2 with real constants) shows one quasi-longitudinal wave and two quasi-transversal waves. For the two particular polarization directions $\phi = 0$ and $\phi = \pi/6$, the imaginary part of the wavenumber vanish meaning a non attenuation of the wave. For other directions, there is attenuation. For illustrating, real parts of wavenumbers calculated with the material strain are displayed in Fig. 5.3 for two directions of propagation ϕ.

It was by the way shown that the material strain (use of ∇) involves the disclination effects through the Ricci tensor's contribution. In view of this simple application, spatial strain and material strain models have different behavior. However, it seems worth to include contortion in strain (see e.g. Sharma and Ganti (2005) for strain gradient continuum), that allows one to capture a known physical self-vibration of dislocations. This also conforms to the idea that strain is physically due to *relative motion* of matter but not a relative movement of corresponding portion of space (remind that displacement may be multivalued in microfractured media).

Remark 5.4 The example of elastic continuum following the Hooke's law involves more terms than usual model of elasticity since the strain tensor itself includes additional contribution which very interestingly looks like general relativistic effect, see Maugin (1978) and references herein for detail. This again highlights the analogy between relativistic gravitation and dislocations/disclinations theory e.g. Baldacci et al. (1979), Malyshev (2000) (in this reference, Malyshev showed that the Einstein–Hilbert Lagrangian $\mathscr{L}_{EH}$ is suitable for capturing the Volterra dislocations). The additional terms for strain may be used for an initially (non holonomic) strained continuum.

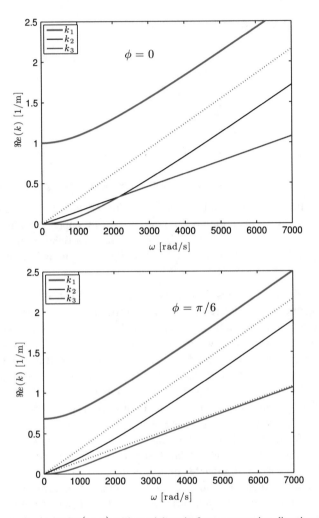

Fig. 5.2 Dispersion curves $\Re\big(k(\omega)\big)$ with spatial strain for a propagation direction $\phi = 0$, $\pi/6$. Roots are labeled arbitrarily k_i, $i = 1, 2, 3$. Dotted lines are the dispersion curves in a perfect medium: $k_l = \omega/c_l$ and $k_t = \omega/c_t$. Simulation is performed for $\aleph_0 = 1\ m^{-1}$ and typical steel values: $\rho_0 = 7500\,\mathrm{kg\,m^{-3}}$, Young modulus $E = 210\,\mathrm{GPa}$ and Poisson ratio $\nu = 1/3$

In practise, care should thus be done for physical interpretation and operating usage of experimental *wave attenuation* results during wave non-destructive testing e.g. Barra et al. (2009), where its was shown that it is experimentally possible to measure dislocation density within aluminium. In the general framework of three-dimensional continuum mechanics, the linearized version of this elastic model holds with:

$$\mathscr{L} := -\rho c \sqrt{u^\alpha u_\alpha} - (1/2)\mathbb{E}^{\alpha\beta\lambda\mu}\varepsilon_{\alpha\beta}\varepsilon_{\lambda\mu}, \quad \varepsilon_{\alpha\beta} \simeq (1/2)\big(\nabla_\alpha u_\beta + \nabla_\beta u_\alpha\big)$$

Fig. 5.3 Dispersion curves $\Re\big(k(\omega)\big)$ with material strain at $\phi = 0$, $\pi/6$. The vertical line localizes ϖ. Simulation is performed for $\aleph_0 = 1\,\mathrm{m}^{-1}$ and typical steel values: $\rho_0 = 7500\,\mathrm{kg\,m}^{-3}$, Young modulus $E = 210\,\mathrm{GPa}$ and Poisson ratio $\nu = 1/3$

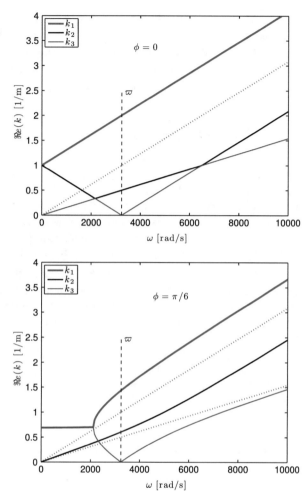

where connection includes the contortion tensor $\Gamma^{\gamma}_{\alpha\beta} = \overline{\Gamma}^{\gamma}_{\alpha\beta} + D^{\gamma}_{\alpha\beta} + \Omega^{\gamma}_{\alpha\beta}$. Again we have $u^{\alpha} := \hat{g}^{\alpha\beta} u_{\beta}$. Using the result on the tele parallel gravity, we recall that $\mathcal{R} = \overline{\mathcal{R}} + \mathcal{T} + \overline{\nabla}_{\alpha}(g^{\beta\lambda}\,\mathcal{T}^{\alpha}_{\lambda\beta}) - \overline{\nabla}_{\beta}(g^{\beta\lambda}\,\mathcal{T}^{\alpha}_{\lambda\alpha})$ where $\overline{\mathcal{R}}$ with $\mathcal{T} := g^{\beta\lambda}(\mathcal{T}^{\alpha}_{\mu\beta}\mathcal{T}^{\mu}_{\lambda\alpha} - \mathcal{T}^{\alpha}_{\mu\alpha}\mathcal{T}^{\mu}_{\lambda\beta})$ is the quadratic torsion utilized in the tele parallel gravity e.g. Sotiriou et al. (2011). Similar developments were conducted by Malyshev (2000) to demonstrate that the use of gauge theory allows us to describe both screw and edge dislocations in the framework of linear elasticity. And more generally, he showed that an appropriate Lagrangian function on a Riemann–Cartan manifold enabled to model continuous matter with defects. For that purpose, the equivalence between Einstein–Hilbert density and the quadratic torsion is worthily used. Transferring divergence terms at boundaries, and assuming that $\overline{\mathcal{R}} \equiv 0$ (the continuum is assumed to evolve in an Galilean spacetime), the equation shows that the curvature, via the contortion tensor, acts as an external source of force, whereas the torsion acts within not only the derivation of stress but also for the calculus of the strain within matter $\sigma_\beta^\alpha \left[\overline{\nabla}_\mu u_\nu, \mathcal{T}_{\mu\nu}^\gamma \right]$. By the way, a more complete model would include the evolution laws of the contortion tensor $\dot{\mathcal{T}}_{\alpha\beta}^\gamma (\mathcal{T}_{\mu\kappa}^\nu, \overline{\nabla}_\mu u_\nu, \cdots)$ which are variables independent of displacement fields e.g. Rakotomanana (1997).

5.2.3 Curved Continuum in a Newtonian Spacetime

Let consider a second gradient continuum $\mathcal{B}$ with a Lagrangian depending on metric and its partial derivatives $\mathcal{L}(g_{\alpha\beta}, \partial_\lambda g_{\alpha\beta}, \partial_\mu \partial_\lambda g_{\alpha\beta})$ (geometric background is a torsionless Riemannian manifold) in motion within a Newtonian spacetime.

We work with a Cartesian coordinate system. The covariance theorem implies that the Lagrangian should depend only on the metric (and by the way on the strain), and on the curvature to fulfill the passive diffeomorphism invariance $\mathcal{L}(\varepsilon_{\alpha\beta}, \mathfrak{R}_{\alpha\beta})$ where $\mathfrak{R}_{\alpha\beta} := \overline{\mathfrak{R}}_{\lambda\alpha\beta}^\lambda + \mathcal{K}_{\lambda\alpha\beta}^\lambda$ defines the change of the Ricci curvature tensor on $\mathcal{B}$. As we can see, the Ricci curvature evolution during transformation includes two parts: the variation of the curvature due to metric and then the symbols of Christoffel, and the variation of the curvature due to the change of the contortion tensor. For short, we notice the change of metric the "macroscopic deformation" and the change of contortion the "mesoscopic deformation". Accordingly, at least two extreme situations may occur for characterizing the curvature $\mathfrak{R}_{\alpha\beta}$. The general case would be in between, nevertheless for the sake of the clarity, we shortly analyze the two extreme cases.

5.2.3.1 Holonomic and Nonholonomic Transformations

The first extreme situation is to consider the "macroscopic" deformation to be holonomic and by the way the metric $g_{\alpha\beta}$ is smooth then the corresponding curvature $\overline{\mathfrak{R}}$, assumed to be zero initially, remains unchanged during the macroscopic deformation. The non holonomic part of the deformation is then only captured by the curvature $\kappa_{\alpha\beta} \neq 0$ which may be deduced entirely and conceptually after Eq. (4.99) by means of the contortion tensor and its first covariant derivative (by using of Levi-Civita connection which remains unchanged during the deformation). The Ricci curvature $\kappa_{\alpha\beta}$ is neither symmetric nor skew-symmetric in a general case, then it is common to introduce a 1-form θ_α, $\alpha = 0, 1, 2, 3$ such that $\kappa_{\alpha\beta} := \overline{\nabla}_\beta \theta_\alpha$ which is analogous to the electromagnetic potential A_α. θ_α and u_α are actually the unknown.[1] The Lagrangian of the continuum takes the form of: $\mathcal{L}_M = \mathcal{L}_M \left(\varepsilon_{\alpha\beta}, \theta_\alpha, \kappa_{\alpha\beta} \right)$ where the contortion does not explicitly appear as the unknown variable. For

---

[1]From the physics point of view on dimension, θ_α measured an angle [rad] and $\kappa_{\alpha\beta}$ behaves like an inverse of a curvature radius [m^{-1}].

constitutive laws, the Lagrangian of the continuum and the spacetime holds:

$$\mathscr{L} := (\rho/2)\partial_t u_\alpha \partial_t u^\alpha + (\rho/2)\mathbb{I}_{\alpha\beta}\partial_t \theta^\alpha \partial_t \theta^\beta - \mathscr{L}_M\left(\varepsilon_{\alpha\beta}, \theta_\alpha, \kappa_{\alpha\beta}\right) \tag{5.26}$$

where the unknowns are the displacement u_α and the potential θ_α. The tensor $\mathbb{I}_{\alpha\beta}$ is to be defined. See Appendix A.4 for an engineering example of (5.26) for Timoshenko beam in structural mechanics. First, application of Lagrangian variation Δ allows us to deduce the covariant constitutive laws as:

$$\sigma^{\alpha\beta} := \frac{\partial \mathscr{L}_M}{\partial \varepsilon_{\alpha\beta}}, \quad \Xi^{\alpha\beta} := \frac{\partial \mathscr{L}_M}{\partial \kappa_{\alpha\beta}}, \quad \Lambda^\alpha := \frac{\partial \mathscr{L}_M}{\partial \theta_\alpha}. \tag{5.27}$$

where the symmetric tensor $\sigma^{\alpha\beta}$ defines the reaction of the continuum to the macroscopic deformation as for classical continuum mechanics e.g. Marsden and Hughes (1983).

Remark 5.5 Constitutive equations (5.27) are usually obtained from the conservation of energy of an infinitesimal portion of the continuum body $\mathscr{B}$ by analogy to the usual case of hyperelasticity e.g. Marsden and Hughes (1983). $\sigma^{\alpha\beta}$ is symmetric whereas $\Xi^{\alpha\beta}$ may be not.

Second, application of Lie-derivative variation $\mathcal{L}_{\delta\mathbf{u},\delta\theta}$ gives the conservation laws:

$$\begin{aligned}
\delta\mathscr{S} = &\int_{t_1}^{t_2}\int_{\mathscr{B}} \rho\left[\partial_t\left(\partial_t u_\alpha \delta u^\alpha\right) - \delta u_\alpha \partial_{tt} u^\alpha\right.\\
&\left.+ \mathbb{I}_{\alpha\beta}\left(\partial_t\left(\partial_t\theta^\alpha \delta\theta^\beta\right) - \delta\theta^\beta \partial_{tt}\theta^\alpha\right)\right] dt \wedge \omega_{\mathscr{B}}\\
&- \int_{t_1}^{t_2}\int_{\mathscr{B}}\left[\sigma^{\alpha\beta}\overline{\nabla}_\beta \delta u_\alpha + \left(\Lambda^\alpha - \overline{\nabla}_\beta \Xi^{\alpha\beta}\right)\delta\theta_\alpha\right] dt \wedge \omega_{\mathscr{B}} + \text{B.C. terms}
\end{aligned}$$

in which we may introduce the constitutive laws (5.27). It is worth to define $\omega_{\mathscr{B}}$ as a three dimensional volume-form of the continuum body $\mathscr{B}$ since space is decoupled from the time-form dt in a Newtonian spacetime. Assuming kinematic compatibility of the variations at the boundary $\partial\mathscr{B}$, and thanks to the gauge invariance of

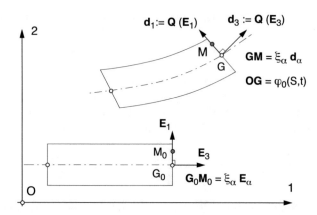

Fig. 5.4 Timoshenko beam. The initial configuration of the beam is defined by the straight line $(G_0, \mathbf{E}_3)$ with the triad $(\mathbf{E}_1, \mathbf{E}_2, \mathbf{E}_3)$ at each point G_0. The deformed configuration is defined by the curve $\{G\}$ the set of material points initially at positions G_0, and the deformed triad $(\mathbf{d}_1, \mathbf{d}_2, \mathbf{d}_3)$ such that $\mathbf{d}_i := \mathbf{Q}(\mathbf{d}_i)$. The displacement is defined by $\mathbf{u}_G := \mathbf{G}_0\mathbf{G}$ and triad rotation by $\mathbf{Q}(\theta_1, \theta_2, \theta_3)$ where θ_i are parameters of the rotation tensor

the action, we deduce the local conservation laws due to the arbitrariness of the variations (boundary terms on $\partial\mathcal{M}$ are dropped for convenience):

$$\begin{cases} \partial_t\left(\rho\partial_t u^\alpha\right) = \overline{\nabla}_\beta\sigma^{\alpha\beta} \\ \partial_t\left(\rho\mathbb{I}^{\alpha\beta}\,\partial_t\theta_\beta\right) = \Lambda^\alpha - \overline{\nabla}_\beta\,\Xi^{\alpha\beta} \end{cases} \tag{5.28}$$

where we have considered the Lie derivative $\delta\varepsilon_{\alpha\beta} = (1/2)(\overline{\nabla}_\alpha\delta u_\beta + \overline{\nabla}_\beta\delta u_\alpha)$, and $\delta\kappa_{\alpha\beta} = (1/2)(\overline{\nabla}_\alpha\delta\theta_\beta + \overline{\nabla}_\beta\delta\theta_\alpha)$. The first row expresses of the system (5.28) the analogous of linear momentum conservation whereas the second row one the angular momentum equation. Remind the dependence of the constitutive dual variables $\sigma^{\alpha\beta}(\varepsilon_{\lambda\mu}, \theta_\lambda, \kappa_{\lambda\mu})$, $\Xi^{\alpha\beta}(\varepsilon_{\lambda\mu}, \theta_\lambda, \kappa_{\lambda\mu})$, and $\Lambda^\alpha(\varepsilon_{\lambda\mu}, \theta_\lambda, \kappa_{\lambda\mu})$. Causality in the sense of Einstein is ensured for this conservation laws (5.28) because the derivative orders are the same for space and time. Both of them have second-order derivatives (Fig. 5.4).

5.2.3.2   Example of Curved Continuum: Mindlin-Reissner Elastic Plate

A particular example of two-dimensional continuum (5.26) is the Mindlin-Reissner plate model initially in the plane Ox^1x^2. This two-dimensional continuum $\mathscr{B}$ is in motion within a three dimensional space $\mathscr{M}$. During its transformation, this $2D$-continuum undergoes a change of curvature initially set to zero. For completeness, the displacement field in $\mathscr{B}$ is submitted to the Navier constraint (a transversal plane section remains plane in the course of time):

$$\begin{cases} u_1(x^1, x^2, x^3) = \bar{u}_1(x^1, x^2) + x^3\, \theta_1(x^1, x^2) \\ u_2(x^1, x^2, x^3) = \bar{u}_2(x^1, x^2) + x^3\, \theta_2(x^1, x^2) \\ u_3(x^1, x^2, x^3) = w(x^1, x^2) \end{cases} \tag{5.29}$$

where $\bar{u}_\alpha, \alpha = 1, 2$ is the in-plane displacement field and w is the out-of-plane displacement field one. The angles $\theta_\alpha, \alpha = 1, 2$ denote the rotation of transversal section. The resulting strain tensor allows us to define three strain variables e.g. Flügge (1972):

$$\begin{cases} \varepsilon_{\alpha\beta} := (1/2)\left(\overline{\nabla}_\alpha \bar{u}_\beta + \overline{\nabla}_\beta \bar{u}_\alpha\right) \\ \kappa_{\alpha\beta} := (1/2)\left(\overline{\nabla}_\alpha \theta_\beta + \overline{\nabla}_\beta \theta_\alpha\right) \\ \gamma_\alpha := \theta_\alpha + \overline{\nabla}_\alpha w \end{cases} \tag{5.30}$$

denoted in-plane strain, curvature strain, and transversal shear strain respectively. By means of these arguments, the Lagrangian function of this continuum model is defined as follows:

$$\mathscr{L} := (\bar{\rho}/2)\partial_t u_\alpha \partial_t u^\alpha + (\bar{\rho}/2)\partial_t w \partial_t w + (\bar{\rho}/2)\mathbb{I}_{\alpha\beta}\partial_t \theta^\alpha \partial_t \theta^\beta - \mathscr{L}_M\left(\varepsilon_{\alpha\beta}, \kappa_{\alpha\beta}, \gamma_\alpha\right)$$

where the continuum matter contribution to the Lagrangian holds, for instance,

$$\mathscr{L}_M\left(\varepsilon_{\alpha\beta}, \kappa_{\alpha\beta}\right) := \frac{1}{2}\left(\mathbb{E}^{\alpha\beta\lambda\mu}\varepsilon_{\alpha\beta}\varepsilon_{\lambda\mu} + \mathbb{K}^{\alpha\beta\lambda\mu}\kappa_{\alpha\beta}\kappa_{\lambda\mu} + \mathbb{C}^{\alpha\beta}\gamma_\alpha\gamma_\beta\right) \tag{5.31}$$

where the potential θ_α denotes the angle of rotation of a cross section, in addition to the slope due to the gradient of the displacement u_α. Such a model is called Mindlin-Reissner thin shell model. The elastic stiffness $\mathbb{E}^{\alpha\beta\lambda\mu}$, $\mathbb{K}^{\alpha\beta\lambda\mu}$, and $\mathbb{C}^{\alpha\beta}$ are material properties. Due to the fact that this two dimensional model evolves within a three dimensional space, the conservation laws (5.28) can be derived to include this three dimension effects since we have introduced the out-of-plane displacement field e.g. Rakotomanana (2009). Now, let us apply the variational procedure by means of Poincaré's invariance with the Lie derivative variations $\mathcal{L}_{\delta\mathbf{u}}\varepsilon_{\alpha\beta}$, $\mathcal{L}_{\delta\theta}\kappa_{\alpha\beta}$, and $\mathcal{L}_{\delta\theta, \delta w}\gamma_\alpha$. By worthily choosing the boundary conditions at $\partial\mathscr{B}$, the local conservations laws are deduced as:

$$\begin{cases} \overline{\nabla}_\beta N^{\alpha\beta} = \bar{\rho}\, \partial_{tt}\bar{u}^\alpha \\ \overline{\nabla}_\beta M^{\alpha\beta} - Q^\alpha = \bar{\rho}\, \mathbb{I}^{\alpha\beta}\partial_{tt}\theta_\beta \\ \overline{\nabla}_\beta Q^\alpha = \bar{\rho}\, \partial_{tt}w \end{cases} \tag{5.32}$$

in which we have defined the hypermomentum as usually:

$$N^{\alpha\beta} := \frac{\partial\mathscr{L}_M}{\partial\varepsilon_{\alpha\beta}}, \qquad M^{\alpha\beta} := \frac{\partial\mathscr{L}_M}{\partial\kappa_{\alpha\beta}}, \qquad Q^\alpha := \frac{\partial\mathscr{L}_M}{\partial\gamma_\alpha} \tag{5.33}$$

called in-plane tension, bending moment, and shear transversal force respectively.

Remark 5.6 The Mindlin-Reissner plate model is a typical second strain continuum model when it is considered as a $2D$-continuum. This point is essential because, if the Mindlin-Reissner plate is considered as a $3D$-continuum, it could not be considered as a strain gradient continuum.

As an illustration, let us consider a Mindlin-Reissner plate with the spacetime coordinates $x^\mu := (t, r, \theta, z)$ (cylindrical coordinates for the space) associated with the local vector base $(\mathbf{e}_r, \mathbf{e}_\theta, \mathbf{e}_z)$ (orthonormal). The conservation laws (5.32) take the explicit form for the linear momentum equation along the radial and the azimuthal directions:

$$
\begin{cases}
\dfrac{\partial N_{rr}}{\partial r} + \dfrac{1}{r}\dfrac{\partial N_{r\theta}}{\partial \theta} + \dfrac{1}{r}(N_{rr} - N_{\theta\theta}) = \overline{\rho}\,\dfrac{\partial^2 \overline{u}_r}{\partial t^2} \\[4mm]
\dfrac{\partial N_{\theta r}}{\partial r} + \dfrac{1}{r}\dfrac{\partial N_{\theta\theta}}{\partial \theta} + \dfrac{1}{r}(N_{\theta r} + N_{r\theta}) = \overline{\rho}\,\dfrac{\partial^2 \overline{u}_\theta}{\partial t^2}
\end{cases}
$$

The second row of (5.32) gives the angular momentum equation:

$$
\begin{cases}
\dfrac{\partial M_{rr}}{\partial r} + \dfrac{1}{r}\dfrac{\partial M_{r\theta}}{\partial \theta} + \dfrac{1}{r}(M_{rr} - M_{\theta\theta}) - Q_r = \overline{\rho}\,\dfrac{h^3}{12}\dfrac{\partial^2 \theta_r}{\partial t^2} \\[4mm]
\dfrac{\partial M_{\theta r}}{\partial r} + \dfrac{1}{r}\dfrac{\partial M_{\theta\theta}}{\partial \theta} + \dfrac{1}{r}(M_{\theta r} + M_{r\theta}) - Q_\theta = \overline{\rho}\,\dfrac{h^3}{12}\dfrac{\partial^2 \theta_\theta}{\partial t^2}
\end{cases}
$$

where h is the thickness of the plate. The last row of (5.32) expresses the linear momentum equation along Oz:

$$
\frac{\partial Q_r}{\partial r} + \frac{1}{r}\left(\frac{\partial Q_\theta}{\partial \theta} + Q_r\right) = \overline{\rho}\,\frac{\partial^2 \overline{w}}{\partial t^2}
$$

For the particular case of homogeneous, isotropic, and linear elastic continuum, the constitutive laws take the form of ($\sigma^{ij} = \mathbb{E}^{ijkl}\varepsilon_{kl}$):

$$
\begin{cases}
N_{11} = \mathcal{D}_t\,(\overline{\varepsilon}_{11} + \nu\,\overline{\varepsilon}_{22}) \\
N_{22} = \mathcal{D}_t\,(\nu\,\overline{\varepsilon}_{11} + \overline{\varepsilon}_{22}) \\
N_{12} = \mathcal{D}_t\,(1-\nu)\,\overline{\varepsilon}_{12}
\end{cases}
,\quad
\begin{cases}
M_{11} = \mathcal{D}\,(\overline{\kappa}_{11} + \nu\overline{\kappa}_{22}) \\
M_{22} = \mathcal{D}\,(\nu\,\overline{\kappa}_{11} + \overline{\kappa}_{22}) \\
M_{12} = \mathcal{D}\,(1-\nu)\,\overline{\kappa}_{12}
\end{cases}
,\quad
\begin{cases}
Q_1 = k_F\mathbb{G}h\,\gamma_1 \\
Q_2 = k_F\mathbb{G}h\,\gamma_2
\end{cases}
$$

where: $\mathcal{D}_t := \mathbb{E}h$, and $\mathcal{D} := \mathbb{E}h^3/12$ are the membrane stiffness and the bending stiffness of the plate. The material parameters $\mathbb{E}$, $\mathbb{G}$, and ν are the Young's modulus, the shear modulus and the Poisson's ratio of the elastic material respectively. $k_F = 5/6$ a specific coefficient to account for the non-uniformity of the shear stress on a transverse section of the plate. For illustration, let consider a wave within the plate with the condition $\overline{u}^\alpha \equiv 0$, $\alpha = 1, 2$ (the case $\overline{u}^\alpha \neq 0$, $\alpha = 1, 2$ corresponds to the classic membrane wave and is independent of the rest of Eq. (5.32)). Searching for monochromatic wave of the type:

$$w(\mathbf{x}, t) = \exp[j(\mathbf{k} \cdot \mathbf{x} + \omega t)], \qquad \theta(\mathbf{x}, t) = \exp[j(\mathbf{k} \cdot \mathbf{x} + \omega t)]$$

we arrive to the dispersion equation of the Mindlin-Reissner (thick) plate model:

$$\begin{cases} \left(\rho h \omega^2 - k_F G h |\mathbf{k}|^2\right) \left\{\rho \frac{h^3}{12} \omega^2 - \left[k_F G h + \mathcal{D}\left(k_1^2 + \frac{1-\nu}{2} k_2^2\right)\right]\right\} + (k_F G h)^2 k_1^2 = 0 \\ \\ \rho \frac{h^3}{12} \omega^2 - \frac{\mathcal{D}}{2}(1-\nu)\|\mathbf{k}\|^2 - k_F G h = 0 \end{cases}$$

$$(5.34)$$

Remark 5.7 It is worth to notice that for wavenumber ($m = 0, n = 0$), the fundamental frequency does not vanish according to the send row of the system (5.34):

$$\omega_{c00} = \frac{1}{h}\sqrt{12 k_F \frac{G}{\rho}}, \qquad \omega_{c00} h \simeq \sqrt{10}\, c_T, \qquad c_T^2 := \frac{G}{\rho}$$

where c_T denotes the shear wave celerity. ω_{c00} is a cut-off frequency.

5.2.3.3 Example of Gradient Continuum: Timoshenko Beam

Let consider a one-dimensional continuum $\mathcal{B}$ with assumed shape of displacement (rigidity of the section). The continuum $\mathcal{B}$ evolves within a three dimension space endowed with an Euclidean metric. A Cartesian coordinate system ($x^0 :=$ $t, , x^1, x^2, x^3$) is used to derive the Lagrangian and the conservation laws. The Timoshenko beam model takes into account shear deformation and rotational inertia effects. We only consider infinitesimal displacements and infinitesimal rotations for the sake of the simplicity. Such a model is suitable for describing the behavior of short beams, and for dynamics the situation of beams subject to high-frequency excitation when the wavelength approaches the thickness of the beam. Details of the model could be found in e.g. Rakotomanana (2009). This Cosserat-Timoshenko beam model is one of the most known example of strain gradient continuum. For beam undergoing small deformation, for either strain and rotation, where rotation of section is projected onto the base as $\theta := \theta_i\, \mathbf{E}_i$, the Lagrangian scalar holds, where $\Phi := \{u_{G1}, u_{G2}, u_{G3}, \theta_1, \theta_2, \theta_3\}$:

$$\mathcal{L} = (\overline{\rho}/2)\left[(\partial_0 u_{G1})^2 + (\partial_0 u_{G2})^2 + (\partial_0 u_{G3})^2\right] dx^0 \wedge dx^3$$
$$+ (\overline{\rho}/2A)\left[\mathbb{I}_1(\partial_0\theta_1)^2 + \mathbb{I}_2(\partial_0\theta_2)^2 + \mathbb{J}(\partial_0\theta_3)^2\right] dx^0 \wedge dx^3$$

$$- (1/2) \left[k_1 GA \, (\partial_3 u_{G1} - \theta_2)^2 + k_2 GA \, (\partial_3 u_{G2} + \theta_1)^2 + EA \, (\partial_3 u_{G3})^2 \right] dx^0 \wedge dx^3$$

$$- (1/2) \left[E\mathbb{I}_1 \, (\partial_3 \theta_1)^2 + E\mathbb{I}_2 \, (\partial_3 \theta_2)^2 + G\mathbb{J} \, (\partial_3 \theta_3)^2 \right] dx^0 \wedge dx^3$$

where the transversal displacement $\mathbf{u}_G \, (S, t)$ of the mean fiber, and the section rotation $\theta \, (S, t)$ are the primal variables (dependent variables). $\overline{\rho}$ and A are linear density and section area of the beam. They are projected onto the local base of the mean fiber $(\mathbf{E}_1, \mathbf{E}_2, \mathbf{E}_3)$, at each point x^3. Direction $\mathbf{E}_3$ is along the beam; $\mathbb{I}_\alpha$ is inertia moment of the beam section about $\mathbf{E}_\alpha$ and $\mathbb{J}$, and its polar inertia about the direction $\mathbf{E}_3$ at the geometric center of the section. E and G are respectively the shear modulus, and the Young's modulus. Coefficients k_α are correction factors due to the uniformity of shear stress along the direction $\mathbf{E}_\alpha$ in a section. We have assumed that the vectors $(\mathbf{E}_1, \mathbf{E}_2)$ are along the principal directions. Application of the variational principle allows us to derive $\delta \int_{t_1}^{t_2} \mathscr{L} dx^0 = 0$ to obtain the local equations for linear momentum conservation:

$$\partial_3 \left[k_1 GA \, (\partial_3 u_{G1} - \theta_2) \right] = \overline{\rho} \, \partial_{00}^2 u_{G1}$$

$$\partial_3 \left[k_2 GA \, (\partial_3 u_{G2} + \theta_1) \right] = \overline{\rho} \, \partial_{00}^2 u_{G2}$$

$$\partial_3 \left(EA \, \partial_3 u_{G3} \right) = \overline{\rho} \, \partial_{00}^2 u_{G3}$$

and for angular momentum conservation:

$$\partial_3 \left(E\mathbb{I}_1 \, \partial_3 \theta_1 \right) - k_2 GA \, (\partial_3 u_{G2} + \theta_1) = (\overline{\rho}/A) \, \mathbb{I}_1 \partial_{00}^2 \theta_1$$

$$\partial_3 \left(E\mathbb{I}_2 \, \partial_3 \theta_2 \right) + k_1 GA \, (\partial_3 u_{G1} - \theta_2) = (\overline{\rho}/A) \, \mathbb{I}_2 \partial_{00}^2 \theta_2$$

$$\partial_3 \left(G\mathbb{J} \, \partial_3 \theta_3 \right) = (\overline{\rho}/A) \, \mathbb{I}_3 \partial_{00}^2 \theta_3$$

where the directions of derivation are along δu_{Gi}, and $\delta \theta_i$. It is worth to mention that the model of Timoshenko beam belongs to the class of Cosserat medium, and accordingly is implicitly a particular strain gradient continuum e.g. Bideau et al. (2011). This is a particular illustration of Cosserat (strain gradient) continuum, with practical application in engineering design. The curvature $\kappa_{ij} := \partial_i \theta_j$ has a dimension of gradient of strain $[\mathrm{m}^{-1}]$.

Remark 5.8 The design of the continuum matter Lagrangian is analogous to the strain energy density of the gradient elastic continuum e.g. Askes and Aifantis (2011). The quadratic energy density proposed by Mindlin in e.g. Mindlin (1964, 1965) includes macroscopic displacement, macroscopic strain (the same as metric), the microscopic deformation, relative deformation (difference between macroscopic strain and microscopic deformation), and gradient of microscopic deformation. The isotropic case involves 903 independent coefficients. At least from the dimensional point of view, the model (5.27) is analogous since our mesoscopic variable $\kappa_{\lambda\mu}$ is first derivatives of angle (called microscopic variables in the review (Askes and Aifantis 2011), or in the original references Mindlin 1964, 1965) are curvature components then derivatives of angles and then may be considered as second gradient of the displacement. The difference comes from the fact that we add the angle θ_μ as additional variable rather that the gradient of the metric. This approach is quite usual in structural mechanics as for Timoshenko beams or Mindlin-Reissner plates and shells e.g. Rakotomanana (2009). This helps for ensure causality of models and avoids higher spatial order derivatives compared to time derivatives.

For illustration, let consider the transversal vibration of a Timoshenko bean in a plane where the displacement reduces to $\mathbf{u}_G = \bar{u}_{G1}\mathbf{e}_1$ and the rotation of the section to $\theta = \theta_2\mathbf{e}_2$. Then the conservation laws include two equations:

$$\begin{cases} \partial_3 \left[k_1 GA \left(\partial_3 u_{G1} - \theta_2\right)\right] = \bar{\rho}\, \partial_{00}^2 u_{G1} \\ \partial_3 \left(E\mathbb{I}_2\, \partial_3\theta_2\right) + k_1 GA \left(\partial_3 u_{G1} - \theta_2\right) = (\bar{\rho}/A)\,\mathbb{I}_2\partial_{00}^2\theta_2 \end{cases} \tag{5.35}$$

Searching for monochromatic waves of the form $\bar{u}_{G1} := U_0 \exp i\,(\omega t + kx)$ and $\theta_2 := \Theta_0 \exp i\,(\omega t + kx)$ allows us to obtain the dispersion equation in a matrix form:

$$\begin{pmatrix} \bar{\rho}\,\omega^2 - k_F GA\, k^2 & -ik_F GA\, k \\ ik_F GA\, k & \bar{\rho}\frac{\mathbb{I}}{A}\,\omega^2 - E\mathbb{I}\, k^2 - k_F GA \end{pmatrix} \begin{pmatrix} U_0 \\ \Theta_0 \end{pmatrix} = \begin{pmatrix} 0 \\ 0 \end{pmatrix} \tag{5.36}$$

Non zero solutions exist if and only if the determinant of the system is equal to zero. We deduce the dispersion equation of the Timoshenko beam (gradient continuum):

$$\frac{E\mathbb{I}}{\bar{\rho}}\, k^4 - \left(\frac{\mathbb{I}}{A} + \frac{E\mathbb{I}}{k_F GA}\right)\omega^2\, k^2 - \omega^2 + \frac{\bar{\rho}\mathbb{I}}{k_F GA^2}\,\omega^4 = 0 \tag{5.37}$$

As for Mindlin-Reissner plate model, a cut-off frequency also exists for Timoshenko beam. A swift analysis shows that this cut-off frequency is calculated as, accounting for both the material property and the shape of the beam section,

$$\omega^2 = \omega_c^2 := \frac{k_F GA^2}{\bar{\rho}\,\mathbb{I}} \tag{5.38}$$

It may be formulated as follows:

$$\omega_c^2 = k_F\, \frac{A}{\mathbb{I}}\, \frac{GA}{\bar{\rho}} = \frac{k_F}{\varrho^2}\, c_T^2$$

in which $\varrho := \sqrt{\mathbb{I}/A}$ et $c_T^2 := G/\rho$ are the gyration radius of the section and the square of the shear wave celerity (three dimensional continuum). An associated wave number may be defined as $k_c^2 := k_F/\varrho^2$. For example, for steel $c_T \simeq 3160$ [m/s]. For a rectangular shape section with b and height h, we have $\mathbb{I} = bh^3/12$ et $\varrho^2 = h^2/12$. Owing that $k_F \simeq 5/6$, we find $\omega_c^2 = 10\, c_T^2/h^2$:

$$\begin{aligned} h &= 1\,[\text{cm}]\,, & f_c &:= \omega_c/2\pi \simeq 159{,}040\,[\text{Hz}] \\ h &= 1\,[\text{mm}]\,, & f_c &:= \omega_c/2\pi \simeq 1{,}590{,}400\,[\text{Hz}] \end{aligned}$$

assessing that the gyration radius ϱ of the section plays a keyrole to highlight if waves are of below (low) or beyond (high) the cut-off frequency.

5.2.4 Causal Model of Curved Continuum

The goal of this subsection is to discuss about the various possibilities to define causal models of curved continuum evolving within a Newton–Cartan spacetime. For that purpose, let consider a model of continuum with a Lagrangian function depending on the metric, and the curvature $\mathscr{L}_M(\hat{g}_{\alpha\beta}, \hat{\Re}^\gamma_{\alpha\beta\mu}, g_{\alpha\beta}, \Re^\gamma_{\alpha\beta\mu})$, where the spacetime metric is necessary argument to ensure a minimal coupling e.g. Anderson (1981), and where the spacetime curvature may be introduced in the presence of gravitation.

5.2.4.1 Inertial Terms Within a Newton–Cartan Spacetime

The Lagrangian function for the spacetime may be written as $\mathscr{L}_G(\hat{g}_{\alpha\beta}, \hat{\Re}^\gamma_{\alpha\beta\mu})$. For the sake of simplicity, we first focus on the curvature terms to investigate the inertial terms. The continuum evolves within a Newton–Cartan spacetime meaning a classical spacetime with gravitation. For simplifying the notation we skip the hat of the curvature hereafter.

Next step would be the derivation of variational form in the framework of Newton–Cartan spacetime to verify if an additional inertial (rotational) term as in e.g. Mindlin (1965), Polyzos and Fotiadis (2012) (review of lattice approach by means of Taylor expansion and continuum approach is given in this later reference) may be recovered in a geometric way. Adding higher (rotational) inertia term in the kinetic energy term is a well-known method to include higher derivatives with respect to time and space as in e.g. Bideau et al. (2011) for Timoshenko beam, and in e.g. Polizzotto (2012) for gradient elastic material models. Let go back to the four-dimensional spacetime and have a look at the curvature field $\kappa_{\alpha\beta}$ where the Greek indices range from 0 to 3. Say the component

$$\kappa_{00} = \delta^{\lambda\mu}\left(\partial_0\partial_\lambda\varepsilon_{\mu 0} - \partial_0\partial_0\varepsilon_{\mu\lambda} + \partial_\mu\partial_0\varepsilon_{0\lambda} - \partial_\mu\partial_\lambda\varepsilon_{00}\right)$$

$$= -\partial_0\partial_0(\varepsilon_{00} - \delta^{ij}\varepsilon_{ij}) - \partial_0\partial_0\varepsilon_{00} + \delta^{ij}\partial_i\partial_j\varepsilon_{00}$$

$$= -\partial_t\partial_t(\varepsilon_{00} - \mathrm{div}\mathbf{u}) - \underbrace{(\partial_t\partial_t\varepsilon_{00} - \Delta\varepsilon_{00})}_{\text{gravitational waves}}$$

if we assume that components $\varepsilon_{0i}, i = 1, 2, 3$ vanish. This corresponds to the transverse-traceless small perturbation in general relativity and propagation of gravitational waves. The term $\mathrm{div}\mathbf{u}$ correspond to a three dimensional spatial divergence. For systematic search for Lagrangian arguments, let report all the components of the curvature fields. In the following Greek indices ranges from 0 to 3, whereas Latin indices from 1 to 3. From the definition of the curvature tensor, we obtain the linearized curvature as follows (in a Cartesian coordinate system):

$$\mathfrak{R}^{\gamma}_{\alpha\beta\mu} := \hat{g}^{\gamma\sigma}\left(\partial_\mu\partial_\alpha\varepsilon_{\sigma\beta} - \partial_\mu\partial_\beta\varepsilon_{\sigma\alpha} + \partial_\sigma\partial_\beta\varepsilon_{\mu\alpha} - \partial_\sigma\partial_\alpha\varepsilon_{\mu\beta}\right) + \mathcal{O}\left(\varepsilon\right) \qquad (5.39)$$

where the inverse of the metric takes the form of: $\hat{g}^{\alpha\sigma} = \mathrm{diag}\{1, -c_\ell^{-2}, -c_\ell^{-2}, -c_\ell^{-2}\}$, where the introduction of the parameter c_ℓ is related to the physical requirement (for instance $c_\ell := \sqrt{\mathbb{E}/\rho}$ for elastic material) that any signal (wave) within the continuum matter cannot propagate with an infinite celerity (as for the special relativity) e.g. Baldacci et al. (1979). This is a basic assumption for propagation of wave within a gradient material. Let calculate the components ($j = 1, 2, 3$):

$$\mathfrak{R}^{0}_{\alpha\beta\mu} = \left(\partial_\mu\partial_\alpha\varepsilon_{0\beta} - \partial_\mu\partial_\beta\varepsilon_{0\alpha} + \partial_0\partial_\beta\varepsilon_{\mu\alpha} - \partial_0\partial_\alpha\varepsilon_{\mu\beta}\right)$$

$$\mathfrak{R}^{j}_{\alpha\beta\mu} = -(1/c_\ell^2)\left(\partial_\mu\partial_\alpha\varepsilon_{j\beta} - \partial_\mu\partial_\beta\varepsilon_{j\alpha} + \partial_j\partial_\beta\varepsilon_{\mu\alpha} - \partial_j\partial_\alpha\varepsilon_{\mu\beta}\right)$$

Say for $\mu = 0$ or for $\alpha = 0$:

$$\mathfrak{R}^{0}_{\alpha\beta 0} = \left(\partial_0\partial_\alpha\varepsilon_{0\beta} - \partial_0\partial_\beta\varepsilon_{0\alpha} + \partial_0\partial_\beta\varepsilon_{0\alpha} - \partial_0\partial_\alpha\varepsilon_{0\beta}\right) = 0$$

$$\mathfrak{R}^{0}_{0\beta\mu} = \left(\partial_\mu\partial_0\varepsilon_{0\beta} - \partial_\mu\partial_\beta\varepsilon_{00} + \partial_0\partial_\beta\varepsilon_{\mu 0} - \partial_0\partial_0\varepsilon_{\mu\beta}\right)$$

This last component is symmetric with respect to the indices β, and μ. For the other components we have ($i \neq 0, l \neq 0$):

$$\mathfrak{R}^{0}_{\alpha\beta l} = \left(\partial_l\partial_\alpha\varepsilon_{0\beta} - \partial_l\partial_\beta\varepsilon_{0\alpha} + \partial_0\partial_\beta\varepsilon_{l\alpha} - \partial_0\partial_\alpha\varepsilon_{l\beta}\right)$$

$$\mathfrak{R}^{0}_{i\beta\mu} = \left(\partial_\mu\partial_i\varepsilon_{0\beta} - \partial_\mu\partial_\beta\varepsilon_{0i} + \partial_0\partial_\beta\varepsilon_{\mu i} - \partial_0\partial_\alpha\varepsilon_{\mu i}\right)$$

We again calculate the components for $\gamma = k$:

$$\mathfrak{R}^{k}_{\alpha\beta 0} = -(1/c_\ell^2)\left(\partial_0\partial_\alpha\varepsilon_{k\beta} - \partial_0\partial_\beta\varepsilon_{k\alpha} + \partial_k\partial_\beta\varepsilon_{0\alpha} - \partial_k\partial_\alpha\varepsilon_{0\beta}\right)$$

$$\mathfrak{R}^{k}_{\alpha\beta l} = -(1/c_\ell^2)\left(\partial_l\partial_\alpha\varepsilon_{k\beta} - \partial_l\partial_\beta\varepsilon_{k\alpha} + \partial_k\partial_\beta\varepsilon_{l\alpha} - \partial_k\partial_\alpha\varepsilon_{l\beta}\right)$$

For mixed spacetime components we have ($\gamma = k$):

$$\mathfrak{R}^{k}_{i\beta 0} = -(1/c_\ell^2)\left(\partial_0\partial_i\varepsilon_{k\beta} - \partial_0\partial_\beta\varepsilon_{ki} + \partial_k\partial_\beta\varepsilon_{0i} - \partial_k\partial_i\varepsilon_{0\beta}\right)$$

$$\mathfrak{R}^{k}_{i\beta l} = -(1/c_\ell^2)\left(\partial_l\partial_i\varepsilon_{k\beta} - \partial_l\partial_\beta\varepsilon_{ki} + \partial_k\partial_\beta\varepsilon_{li} - \partial_k\partial_i\varepsilon_{l\beta}\right)$$

For mixed spacetime components we have ($\alpha = 0$, and $\gamma = k$):

$$\mathfrak{R}^{k}_{0\beta 0} = -(1/c_\ell^2)\left(\partial_0\partial_0\varepsilon_{k\beta} - \partial_0\partial_\beta\varepsilon_{k0} + \partial_k\partial_\beta\varepsilon_{00} - \partial_k\partial_0\varepsilon_{0\beta}\right)$$

$$\mathfrak{R}^{k}_{0\beta l} = -(1/c_\ell^2)\left(\partial_l\partial_0\varepsilon_{k\beta} - \partial_l\partial_\beta\varepsilon_{k0} + \partial_k\partial_\beta\varepsilon_{l0} - \partial_k\partial_0\varepsilon_{l\beta}\right)$$

For "pure" spatial components, this gives:

$$\mathfrak{R}^k_{ij0} = -(1/c_\ell^2)\left(\partial_0\partial_i\varepsilon_{kj} - \partial_0\partial_j\varepsilon_{ki} + \partial_k\partial_j\varepsilon_{0i} - \partial_k\partial_i\varepsilon_{0j}\right)$$

$$\mathfrak{R}^k_{ijl} = -(1/c_\ell^2)\left(\partial_l\partial_i\varepsilon_{kj} - \partial_l\partial_j\varepsilon_{ki} + \partial_k\partial_j\varepsilon_{li} - \partial_k\partial_i\varepsilon_{lj}\right)$$

For these curvature components, differentiation $\partial_0 := (c_\ell)^{-1}\partial_t$ since $x^0 := c_\ell t$. These results suggest that the time derivative and the gradient of strain have influence on the spacetime curvature of the gradient continuum. The "pure" spatial components show that the Lagrangian depend on the second strain derivative. It is observed that the order of time and space derivatives are the same, which may guarantee the causality of the model e.g. Metrikine (2006), Polyzos and Fotiadis (2012). It is nevertheless easier for tractation if we utilize the Riemann curvature instead of Riemann–Cartan one $\mathfrak{R}_{\nu\alpha\beta\mu} := g_{\nu\gamma}\,\mathfrak{R}^\gamma_{\alpha\beta\mu}$. This permits to eliminate the scale c_ℓ to obtain directly the linearized components of the Riemann curvature:

$$\mathfrak{R}_{\nu\alpha\beta\mu} = \partial_\mu\partial_\alpha\varepsilon_{\nu\beta} - \partial_\mu\partial_\beta\varepsilon_{\nu\alpha} + \partial_\nu\partial_\beta\varepsilon_{\mu\alpha} - \partial_\nu\partial_\alpha\varepsilon_{\mu\beta} \tag{5.40}$$

Remark 5.9 For the sake of the completeness, the previous development suggests that the Lagrangian should include mixed temporal and spatial derivatives of the displacement as arguments. This is clearly stated by the presence of curvature components where the derivatives ∂_0 and ∂_k are present in most of terms.

5.2.4.2 Curved Continuum and Relativistic Gravitation

Inspiring from the free fall in general relativity, it is worth to define the local coordinates of the observer's proper reference frame according to Fermi coordinates. Ni and Zimmermann extended this concept to expand the metric to obtain second-order terms of the metric e.g. Ni and Zimmermann (1978). This method may be used here to obtain the Taylor expansion about the normal coordinates. The idea is to build a frame such that a "point particle" is free-falling along the geodesics using Fermi normal coordinates. Let consider a particle with spatial origin at $x^i = 0$ and a time coordinate equal to the proper time $x^0 = \tau = ct$. Consider a local (inertial) frame attached to this point. The line element corresponding to flat Minkowskian metric holds: $ds^2 := (dx^0)^2 - (dx^1)^2 - (dx^2)^2 - (dx^3)^2$. The second-order expansion relative to distance with respect to the origin writes:

$$ds^2 := (dx^0)^2 - (dx^1)^2 - (dx^2)^2 - (dx^3)^2 + \mathcal{O}\left(\|\mathbf{x}\|^2/\mathcal{R}^2\right)$$

where $\mathcal{R}$ is the curvature radius $\mathcal{R}^{-2} := \|\mathfrak{R}_{\nu\alpha\beta\mu}\|$. The lightcone structure of the special relativity theory is till kept in general relativity. However, the orientation of the lightcone and its open angle can vary across the spacetime, as sketched on Fig. 5.5. On this figure, the light cone is different at each point but the light speed c

Fig. 5.5 At each point of the continuum matter or spacetime, local Minkowski spacetime is attached (here points are represented by its neighborhood x^μ, y^μ, and z^μ)

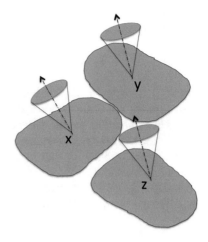

remains the same at each point. See Eq. (5.41) for the metric Taylor expansion. In other words, at various positions in the gravitational field, clocks do not flow with the same speed: some run faster, and others run slower that their neighborhood. This is so the case within a continuum immersed under gravitation: there is no unique time pertinent for the whole spacetime. Ni and Zimmermann explicitly obtained the following Taylor expansion (with our convention for the indices of the curvature) (Ni and Zimmermann 1978):

$$ds^2 = \left(1 - \Re_{ji00}\, x^i x^j\right) (dx^0)^2$$
$$+ (4/3)\, \Re_{i0jk}\, x^j x^k\, dx^0 dx^i - \left[\delta_{ij} - (\Re_{kijl}/3)\, x^k x^l\right] dx^i dx^j \quad (5.41)$$

Three points should be pointed out in this line element. The first one is the additional presence of the component $\Re_{ji00}\, x^i x^j$ in the "time" part of the line. In terms of strain this gives from Eq. (5.40):

$$\Re_{ji00} = \partial_0 \partial_i \varepsilon_{j0} - \partial_0 \partial_0 \varepsilon_{ji} + \partial_j \partial_0 \varepsilon_{0i} - \partial_j \partial_i \varepsilon_{00}$$

in which the mixed space and time derivatives appear together with second derivatives with respect to time of the strain and the Hessian of the potential ε_{00}. This clearly shows that the kinetic energy should include these terms in dynamics. The second point is the presence of the spatial components of the curvature e.g. Ni and Zimmermann (1978):

$$\Re_{kijl} = \partial_l \partial_i \varepsilon_{kj} - \partial_l \partial_j \varepsilon_{ki} + \partial_k \partial_j \varepsilon_{li} - \partial_k \partial_i \varepsilon_{lj} \quad (5.42)$$

leading the second strain gradient continuum models. Finally the last and most interesting point is the coupling between space and matter due to the terms:

$$\mathfrak{R}_{i0jk} = \partial_k \partial_0 \varepsilon_{ij} - \partial_k \partial_j \varepsilon_{i0} + \partial_i \partial_j \varepsilon_{k0} - \partial_i \partial_0 \varepsilon_{kj} \tag{5.43}$$

explicitly showing a dependence of the Lagrangian with respect to the first derivatives of the strain with respect to spatial coordinates but also in presence of time derivatives. The second-order expansion of the metric was mainly used in the framework of linearized gravitation and gravitational waves detection, but may be worthily exploited to give a strong mathematical background for the strain gradient continuum model. This description is correct as long as the we can disregard the "correction $\|\mathbf{x}\|^2 / \mathcal{R}^2$". For non zero curvature any vector field on the continuum manifold has distributed discontinuity of vector fields e.g. Rakotomanana (1997).

5.2.4.3 Examples of Strain Gradient Continuum

For modelling strain gradient continuum, the problem of inertial terms conceptually remains a problem. For instance, the Timoshenko beam model (which is implicitly a strain gradient continuum) includes the rotational inertial terms (e.g. Rakotomanana 2009) to ensure a causal beam model. For three-dimensional continua, adding quadratic terms in gradient velocity was already formulated by Mindlin in 1964 for his model of high gradient continuum (Mindlin 1964). The Lagrangian function takes the form of e.g. Polizzotto (2013a,b):

$$\mathscr{L} := \underbrace{(1/2)\rho\mathbf{v} \cdot \mathbf{v} + (1/2)\rho\ell_G^2 \nabla\mathbf{v} : \nabla\mathbf{v}}_{kinetic\ energy} - \underbrace{(1/2)\varepsilon : \mathbb{E} : \varepsilon - (1/2)\ell_M^2 \nabla\varepsilon : \mathbb{E} : \nabla\varepsilon}_{potential\ energy}$$

where ρ is the density, ℓ_G and ℓ_M two internal length scale parameters (gravity and matter), and $\mathbb{E}$ the usual moduli tensor of elasticity. Firstly, it may observed that the gradient of velocity includes two distinct parts $\nabla\mathbf{v} = \Omega + \mathbf{D}$ where Ω is the uniquely defined skew symmetric part whereas $\mathbf{D}$ is the symmetric strain rate tensor. It should be stressed that in an Euclidean space, and for small strain assumption, the component of the strain rate tensor reduce to $\partial_0 \varepsilon_{ij}$. The second term of the kinetic energy includes a kind of inertia of rotation as $\mathscr{K}_{rot} := (1/2)\,\rho\ell_G^2\,\omega \cdot \omega$, where ω is the axial rotation vector isomorph to the skew-symmetric tensor Ω (only for three-dimensional space).

The connection utilized in these theories was the Levi-Civita connection e.g. Polizzotto (2012), which may present some drawbacks when considering the covariance of the model e.g. Antonio and Rakotomanana (2011). Indeed, the covariant derivative of the metric vanishes due to metric compatibility. In view of the covariance requirement of the Lagrangian function, it seems more rigorous to a priori consider a curved connection with torsion $\nabla := \overline{\nabla} + \mathcal{K}$. Indeed, the strain gradient $\overline{\nabla}^\gamma \varepsilon_{\alpha\beta}$ in the last term of the Lagrangian would be replaced by the torsion tensor $\aleph_{\alpha\beta}^{\gamma}$ which expresses the covariant argument of the Lagrangian. In such a case, there is no need to explicitly introduce a matter scale length ℓ_M. Similarly, an open question would be the replacement of the velocity gradient by the time derivative of the torsion for the kinetic energy.

In a series of papers, Polizzotto assumes that the kinetic energy is rather function of the velocity, gradient of velocity, and the second gradient of the velocity $\mathcal{T}(\mathbf{v}, \nabla\mathbf{v}, \nabla\nabla\mathbf{v})$, whereas the potential energy admits as argument the strain and the first and second gradients of the strain $\mathcal{U}(\varepsilon, \nabla\varepsilon, \nabla\nabla\varepsilon)$ (Polizzotto 2013a,b):

$$\mathcal{L} := \mathcal{T}(\mathbf{v}, \nabla\mathbf{v}, \nabla\nabla\mathbf{v}) - \mathcal{U}(\varepsilon, \nabla\varepsilon, \nabla\nabla\varepsilon)$$

where the connection again Levi-Civita connection $\nabla := \overline{\nabla}$. Explicitly, this Lagrangian may be written as follows by assuming a Cartesian frame for the sake of the simplicity:

$$\mathcal{L} := \mathcal{T}(\partial_0 u^i, \partial_j \partial_0 u^i, \partial_k \partial_j \partial_0 u^i) - \mathcal{U}(\partial_k \partial_j u^i, \partial_l \partial_k \partial_j u^i, \partial_m \partial_l \partial_k \partial_j u^i) \qquad (5.44)$$

in which we explicitly highlight the dependence with respect to the displacement and its derivatives. We recognize the formal dependence of the Lagrangian with respect to mixed temporal and spatial derivatives of the displacement as sketched in Eq. (5.40). In sum, the assumption that the Lagrangian density should depend on the all set of curvature components may be a better approach since this method a priori allows us to obtain directly a covariant Lagrangian. The expression (5.44) should additionally satisfy the covariance requirement. Remind that $\mathcal{L}(\mathfrak{R}_{\alpha\beta\lambda}^{\gamma})$ is already covariant.

5.2.4.4 General Remark on the Strain Gradient Continuum

The previous model constitutes an improvement of Polizzotto's previous paper (Polizzotto 2012), owing the lack of covariance of first gradient model $\mathcal{V}(\varepsilon, \nabla\varepsilon)$. Some particular forms of energy are found in e.g. Polizzotto (2013b) in dynamic situation. We should however draw attention to the covariance condition of the Lagrangian $\mathcal{L}$ to be admissible and then acceptable (with regards of the diffeomorphism-invariance) list of arguments:

$$\mathcal{L}(\varepsilon_{\alpha\beta}, \partial_\gamma \varepsilon_{\alpha\beta}, \partial_\lambda \partial_\gamma \varepsilon_{\alpha\beta}) \rightarrow \mathcal{L}(\varepsilon, \nabla\varepsilon, \nabla\nabla\varepsilon) \rightarrow \mathcal{L}(\mathbf{g}, \nabla, \nabla\nabla) \rightarrow \mathcal{L}(\mathbf{g}, \aleph, \mathfrak{R})$$
$$(5.45)$$

as has been shown in the first part of the present paper devoted to covariance of Lagrangian function. We remind the first step which is the MCP (Minimal Coupling Procedure) an usual method in relativistic gravitation theory. It should be stressed that the introduction of the connection as arguments e.g. Krause (1976), rather than the gradient of strain and its gradient, of the Lagrangian function is coherent with the principle of equivalence applied to both the spacetime or the strain gradient continuum. Similarly, at every point in an gravitational field, or correspondingly at every point of the gradient continuum, we can choose locally an inertial frame or correspondingly a field of connection, in which the physics laws take the same form, or for continuum mechanics correspondingly, the same Lagrangian functions have the same form. The role of the connection goes beyond the covariance of three-dimensional continuum and includes the aspects of gauge invariance by introducing the local Poincaré group of transformations.

5.3 Gravitational Waves

In this subsection, we derive the basic equations due to linear perturbation of the Minkowskian metric as for linear gravity phenomenon. Gravity is the consequence of how massive object deforms the spacetime. Near any massive body, the spacetime becomes curved following the change of the spacetime metric. The deformation does not stay only near the massive body. The field equations of Einstein suggested that the deformation can propagate throughout the entire spacetime. The main difference compared to seismic waves is that gravitational waves can travel in empty space at the light speed. This is typical example where the gauge invariance is useful for deriving the wave equations of relativistic gravitation. The method is based on linear perturbation of the metric, the Ricci curvature tensor, and the Einstein tensor.

5.3.1 Basic Equations

As an example from the strain (4.88), consider a weak field gravitation where metric is close to Minkowski metric tensor $g_{\alpha\beta} \simeq \hat{g}_{\alpha\beta} + 2\varepsilon_{\alpha\beta}$, with $\|\varepsilon_{\alpha\beta}\| << 1$.[2] It is also usual to assume that at large distance from sources, the spacetime becomes Minkowskian e.g. Dixon (1975). For classical relativistic gravitation, the Einstein–Hilbert action is proposed by using the Ricci and scalar curvature deduced from Eq. (4.98) as $\overline{\mathfrak{R}}_{\alpha\beta} := \overline{\mathfrak{R}}^{\lambda}_{\lambda\alpha\beta}$ and $\overline{\mathcal{R}} := g^{\alpha\beta}\overline{\mathfrak{R}}_{\alpha\beta}$, and its variation hold:

$$\mathscr{S}_G := \frac{1}{2\chi} \int_{\mathscr{M}} \overline{\mathcal{R}} \, \omega_n, \qquad \delta\mathscr{S}_G = \frac{1}{2\chi} \int_{\mathscr{M}} \left[\overline{\mathfrak{R}}_{\alpha\beta} - (1/2)\overline{\mathcal{R}} \, g_{\alpha\beta}\right] \delta g^{\alpha\beta} \, \omega_n$$

---

[2] Application of the Lagrangian formalism in general relativity may induce some difficulties, because physical quantities in classical or special relativity framework require fixed geometric background (Newtonian or Minkowskian spacetime). Indeed, for general relativity the spacetime geometry is itself a dynamical object. Separation of the metric into two parts that may be respectively assigned to inertia and gravity is an affair of taste e.g. Shen and Moritz (1996).

In the second equation, the vanishing of the variation of the Einstein–Hilbert action implies the famous Einstein field equation for gravitation:

$$\overline{\mathfrak{R}}_{\alpha\beta} - (1/2)\overline{\mathcal{R}}\, g_{\alpha\beta} = 0 \tag{5.46}$$

where it should be stressed that the partial differential equations of the metric components $g_{\alpha\beta}$ are of second order. Remind that they are in principle expected to be of fourth order as being Euler–Lagrange equations associated to Lagrange function of type $\mathscr{L}(g_{\alpha\beta}, \partial_\gamma g_{\alpha\beta}, \partial_\lambda \partial_\gamma g_{\alpha\beta})$. Theory of special gravitation allows us to obtain the conservation laws associated to the linearized part of Hilbert–Einstein Lagrangian $\mathscr{L}(g_{\alpha\beta}, \partial_\gamma \partial_\lambda g_{\alpha\beta}) := (1/2\chi)\,\overline{\mathcal{R}}$ with:

$$\overline{\mathfrak{R}}^\lambda_{\alpha\beta\mu} = \hat{g}^{\lambda\sigma}\left(\partial_\mu\partial_\alpha\varepsilon_{\sigma\beta} - \partial_\mu\partial_\beta\varepsilon_{\sigma\alpha} + \partial_\sigma\partial_\beta\varepsilon_{\mu\alpha} - \partial_\sigma\partial_\alpha\varepsilon_{\mu\beta}\right) \tag{5.47}$$

Equation (5.46) governs the dynamics of vacuum spacetime in relativistic gravitation. where the unknowns are the metric components (4.98). In the presence of moving bodies within the spacetime, the problem in relativistic gravitation is that we have to solve, at the same time, the gravitation fields induced by the bodies and the motion of the bodies e.g. Papapetrou (1951).

Theorem 5.1 *Let $\mathcal{M}$ be a Minkowski spacetime with $\eta_{\alpha\beta} := \operatorname{diag}\{+1, -1, -1, -1\}$ its metric. Assume that the metric is changed to $g_{\beta\mu}(x^\lambda)$ due a small perturbation such that $g_{\alpha\beta} = \eta_{\alpha\beta} + 2\varepsilon_{\alpha\beta}$, with $\varepsilon_{\alpha\beta} << 1$ following the deformation of the spacetime $\mathcal{M}$ to a Riemannian manifold (spacetime). Then the scalar curvature of the deformed spacetime $\mathcal{B}$ is given by:*

$$\mathcal{R} = 2\left[\partial^\alpha\partial^\nu\varepsilon_{\alpha\nu} - \eta^{\alpha\nu}\partial_\alpha\partial_\nu\,(\mathrm{Tr}\varepsilon)\right] \tag{5.48}$$

Proof Let remind the components of the Cartan curvature tensor:

$$\mathfrak{R}^\lambda_{\alpha\beta\mu} = (\partial_\alpha\Gamma^\lambda_{\beta\mu} + \Gamma^\sigma_{\beta\mu}\Gamma^\lambda_{\alpha\sigma}) - (\partial_\beta\Gamma^\lambda_{\alpha\mu} + \Gamma^\sigma_{\alpha\mu}\Gamma^\lambda_{\beta\sigma}) \tag{5.49}$$

and the Ricci curvature tensor as:

$$\mathfrak{R}_{\beta\mu} = \mathfrak{R}^\alpha_{\alpha\beta\mu} = (\partial_\alpha\Gamma^\alpha_{\beta\mu} + \Gamma^\sigma_{\beta\mu}\Gamma^\alpha_{\alpha\sigma}) - (\partial_\beta\Gamma^\alpha_{\alpha\mu} + \Gamma^\sigma_{\alpha\mu}\Gamma^\alpha_{\beta\sigma}) \tag{5.50}$$

For the sake of simplicity, let consider a Cartesian coordinate system for space associated to the flat Minkowski spacetime. The perturbated metric is given by $g_{\alpha\beta} = \eta_{\alpha\beta} + 2\varepsilon_{\alpha\beta}$ where $\varepsilon_{\alpha\beta} << 1$. We deduce the coefficients of the connection by retaining only the linear terms in $\varepsilon_{\alpha\beta}$. We obtain:

$$\Gamma^\alpha_{\beta\mu} \simeq \eta^{\alpha\nu}\left(\partial_\beta\varepsilon_{\nu\mu} + \partial_\mu\varepsilon_{\beta\nu} - \partial_\nu\varepsilon_{\beta\mu}\right)$$

$$\Gamma^\alpha_{\alpha\mu} \simeq \eta^{\alpha\nu}\left(\partial_\alpha\varepsilon_{\nu\mu} + \partial_\mu\varepsilon_{\alpha\nu} - \partial_\nu\varepsilon_{\alpha\mu}\right)$$

After simplification, we deduce:

$$\mathfrak{R}_{\beta\mu} = \eta^{\alpha\nu}\partial_\alpha\left(\partial_\beta\varepsilon_{\nu\mu} + \partial_\mu\varepsilon_{\beta\nu} - \partial_\nu\varepsilon_{\beta\mu}\right) - \eta^{\alpha\nu}\partial_\beta\left(\partial_\alpha\varepsilon_{\nu\mu} + \partial_\mu\varepsilon_{\alpha\nu} - \partial_\nu\varepsilon_{\alpha\mu}\right)$$

The linearized Ricci curvature tensor holds:

$$\mathfrak{R}_{\beta\mu} = \eta^{\alpha\nu}\partial_\alpha\left(\partial_\mu\varepsilon_{\beta\nu} - \partial_\nu\varepsilon_{\beta\mu}\right) - \eta^{\alpha\nu}\partial_\beta\left(\partial_\mu\varepsilon_{\alpha\nu} - \partial_\nu\varepsilon_{\alpha\mu}\right) \tag{5.51}$$

$$= \partial^\nu\partial_\mu\varepsilon_{\beta\nu} + \partial_\beta\partial^\nu\varepsilon_{\nu\mu} - \partial_\beta\partial_\mu\left(\mathrm{Tr}\varepsilon\right) - \eta^{\alpha\nu}\partial_\alpha\partial_\nu\varepsilon_{\beta\mu} \tag{5.52}$$

The linearized scalar curvature is deduced accordingly:

$$\mathcal{R} := \eta^{\beta\mu}\mathfrak{R}_{\beta\mu}$$

$$= \eta^{\beta\mu}\left[\eta^{\alpha\nu}\partial_\alpha\partial_\mu\varepsilon_{\beta\nu} - \eta^{\alpha\nu}\partial_\beta\left(\partial_\mu\varepsilon_{\alpha\nu} - \partial_\nu\varepsilon_{\alpha\mu}\right)\right] - \eta^{\alpha\nu}\partial_\alpha\partial_\nu\left(\mathrm{Tr}\varepsilon\right)$$

$$= \eta^{\beta\mu}\left[\eta^{\alpha\nu}\partial_\alpha\partial_\mu\varepsilon_{\beta\nu} + \eta^{\alpha\nu}\partial_\beta\partial_\nu\varepsilon_{\alpha\mu}\right] - 2\eta^{\alpha\nu}\partial_\alpha\partial_\nu\left(\mathrm{Tr}\varepsilon\right)$$

$$= 2\left[\partial^\alpha\partial^\nu\varepsilon_{\alpha\nu} - \eta^{\alpha\nu}\partial_\alpha\partial_\nu\left(\mathrm{Tr}\varepsilon\right)\right]$$

with $\mathrm{Tr}\varepsilon := \eta^{\beta\mu}\varepsilon_{\beta\mu}$. $\square$

By considering the group of Lorentz transformations $\left\{J^\alpha_\mu\right\}$, it can be checked that the perturbation of the Minkowskian metric, as for the Minkowskian metric itself is invariant, transform as follows:

$$y^\alpha = J^\alpha_\mu x^\mu \qquad \longrightarrow \qquad \varepsilon'_{\alpha\beta} = J^\mu_\alpha J^\nu_\beta \,\varepsilon_{\mu\nu}$$

5.3.2 Equations of Linearized Gravitation Waves

We deduce the linearized Einstein's equation of gravity $G_{\beta\mu} = 0$ with:

$$G_{\beta\mu} := \mathfrak{R}_{\beta\mu} - (1/2)\mathcal{R}\eta_{\beta\mu}$$

$$= \partial^\nu\partial_\mu\varepsilon_{\beta\nu} + \partial_\beta\partial^\nu\varepsilon_{\nu\mu} - \partial_\beta\partial_\mu\left(\mathrm{Tr}\varepsilon\right) - \eta^{\alpha\nu}\partial_\alpha\partial_\nu\varepsilon_{\beta\mu}$$

$$- \left[\partial^\alpha\partial^\nu\varepsilon_{\alpha\nu} - \eta^{\alpha\nu}\partial_\alpha\partial_\nu\left(\mathrm{Tr}\varepsilon\right)\right]\eta_{\beta\mu}$$

We can re-arrange terms to highlight a D'Alembertain operator and finally to give the equation of linearized gravitation field of vacuum without torsion:

$$G_{\beta\mu} = \underbrace{-\eta^{\alpha\nu}\partial_\alpha\partial_\nu\left[\varepsilon_{\beta\mu} - (\mathrm{Tr}\varepsilon)\eta_{\beta\mu}\right]}_{\text{D'Alembertian}}$$

$$+ \partial^\nu\partial_\mu\varepsilon_{\beta\nu} + \partial_\beta\partial^\nu\varepsilon_{\nu\mu} - \partial_\beta\partial_\mu\left(\mathrm{Tr}\varepsilon\right) - \left(\partial^\alpha\partial^\nu\varepsilon_{\alpha\nu}\right)\eta_{\beta\mu} \tag{5.53}$$

Despite the fact that we have ten equations for ten unknowns, it is not yet possible, at this step, to solve them because we have first to define a coordinate system. Indeed, the decomposition of the metric into two terms as the flat Minkowskian metric $\eta_{\beta\mu}$, and a perturbation $2\varepsilon_{\beta\mu}$ is not unique. Depending on the choice of a coordinate system (x^λ) the shape of the perturbation may be different. Two points of view may be adopted: the first one considers the deformation of the spacetime itself from Minkowskian $\mathcal{M}$ with metric $\eta_{\beta\mu}$ to a non Minkowskian spacetime $\mathcal{B}$ with metric $g_{\beta\mu}(x^\lambda)$. It should be stressed that if the metric $g_{\beta\mu}$ obeys the Einstein's equations on the spacetime $\mathcal{B}$, then the perturbation $\varepsilon_{\beta\mu}$ obeys the linearized equations of gravitation on the Minkowskian spacetime $\mathcal{M}$. At this step, the above equation is too complicated to be solved, then a first method consists in defining the *trace-reversed deformation* of the Minkowskian metric tensor:

$$\bar{\varepsilon}_{\beta\mu} := \varepsilon_{\beta\mu} - (1/2)\mathrm{Tr}(\varepsilon)\eta_{\beta\mu} \qquad \longleftrightarrow \qquad \varepsilon_{\beta\mu} := \bar{\varepsilon}_{\beta\mu} - (1/2)\mathrm{Tr}(\bar{\varepsilon})\eta_{\beta\mu} \tag{5.54}$$

owing that $\mathrm{Tr}(\eta) = \eta^{\alpha\nu}\eta_{\alpha\nu} = 4$. We also obtain $\mathrm{Tr}(\varepsilon) = -\mathrm{Tr}(\bar{\varepsilon})$. Introducing the trace-reversed perturbation within the Einstein tensor equation, we have:

$$-\eta^{\alpha\nu}\partial_\alpha\partial_\nu\bar{\varepsilon}_{\beta\mu} + \partial^\nu\partial_\mu\bar{\varepsilon}_{\beta\nu} + \partial^\nu\partial_\beta\bar{\varepsilon}_{\nu\mu} - \left(\partial^\alpha\partial^\nu\bar{\varepsilon}_{\alpha\nu}\right)\eta_{\beta\mu} = 0 \tag{5.55}$$

Under the assumption of linearized weak gravity, we obtained a system of 10 linear partial differential equations for the trace-reversed variables $\bar{\varepsilon}_{\beta\mu}$.

Physically, the metric perturbation from a small change of originally Minkowskian constant and uniform metric $\eta_{\beta\mu}$ or due to a perturbation of the coordinate system $x^\mu \to x^\mu + \xi^\mu$.[3] To understand the effects of small gauge transformations, we have to use the Lie derivative (retaining only linear terms in ξ):

$$\mathcal{L}_\xi\, g_{\alpha\beta} = \mathcal{L}_\xi\, 2\varepsilon_{\alpha\beta} = 2\xi^\gamma\,\partial_\gamma\varepsilon_{\alpha\beta} + \eta_{\gamma\beta}\,\partial_\alpha\xi^\gamma + \eta_{\alpha\gamma}\,\partial_\beta\xi^\gamma \simeq \partial_\alpha\xi_\beta + \partial_\beta\xi_\alpha := 2\delta\varepsilon_{\alpha\beta} \tag{5.56}$$

The last term expresses the metric perturbation due to an infinitesimal change of coordinates by vector ξ. In terms of the trace-reversed perturbation, we have:

$$2\delta\bar{\varepsilon}_{\alpha\beta} = 2\delta\varepsilon_{\alpha\beta} - (1/2)\mathrm{Tr}\,(2\delta\varepsilon)\,\eta_{\alpha\beta} = \partial_\alpha\xi_\beta + \partial_\beta\xi_\alpha - \left(\partial^\nu\xi_\nu\right)\eta_{\alpha\beta} \tag{5.57}$$

This means that this gauge transformation (local translation vector ξ) leads to a new trace-reversed perturbation $\bar{\varepsilon}'_{\beta\mu} = \bar{\varepsilon}_{\beta\mu} + \partial_\beta\xi_\mu + \partial_\mu\xi_\beta - (\partial^\nu\xi_\nu)\eta_{\beta\mu}$. We have ten equations for the linearized gravity with ten unknowns for the metric perturbation, for either $\varepsilon_{\beta\mu}$ or $\bar{\varepsilon}_{\beta\mu}$. The choice of a coordinate system to solve them introduce four more spacetime variables x^μ. Therefore we need four additional equations

---

[3]In some sense, gauge invariance may be also interpreted as a infinitesimal coordinate transformations.

which are merely the gauge equations. Let introduce the Lorentz gauge (also called Einstein gauge, Hilbert gauge, de Donder gauge or Fock gauge). Starting from the coefficients of connection $\Gamma^{\lambda}_{\mu\nu}(\mathbf{x})$, the Lorentz gauge imposes that the skew-symmetry part of the connection is equal to zero, together with its linearized version:

$$g^{\mu\nu}\,\Gamma^{\lambda}_{\mu\nu} = 0 \qquad \longrightarrow \qquad \eta^{\mu\nu}\,\eta^{\rho\lambda}\left(\partial_\mu\varepsilon_{\rho\nu} + \partial_\nu\varepsilon_{\mu\rho} - \partial_\rho\varepsilon_{\mu\nu}\right) = 0 \qquad (5.58)$$

We deduce the linearized Lorentz gauge $\eta^{\lambda\rho}\partial^\mu\varepsilon_{\rho\mu} - (1/2)\partial^\lambda\mathrm{Tr}(\varepsilon) = 0$. The previous gauge invariance condition is required for both of them. Classically, the so-called *Lorentz gauge* is re-written for linearized gravitation $\partial^\mu\bar{\varepsilon}_{\beta\mu} \equiv 0$. This gauge condition must be also satisfied when the gauge transformations are applied:

$$\partial^\mu\bar{\varepsilon}'_{\beta\mu} = \partial^\mu\left[\bar{\varepsilon}_{\beta\mu} + \partial_\beta\xi_\mu + \partial_\mu\xi_\beta - \left(\partial^\nu\xi_\nu\right)\eta_{\beta\mu}\right] = \partial^\mu\bar{\varepsilon}_{\beta\mu} + \partial_\mu\partial^\mu\xi_\beta \equiv 0 \tag{5.59}$$

The relation has a gauge invariance property if the equation $\partial^\mu\bar{\varepsilon}_{\beta\mu} + \partial_\mu\partial^\mu\xi_\beta \equiv 0$ has solution in ξ. Such is the case, then there are infinity of vector fields that satisfy this condition for any choice $\partial^\mu\bar{\varepsilon}_{\beta\mu}$. Finally, in the case of Lorentz gauge, the linearized equations of gravity take the following form:

$$\eta^{\alpha\nu}\partial_\alpha\partial_\nu\,\bar{\varepsilon}_{\beta\mu} = \Box\,\bar{\varepsilon}_{\beta\mu} = 0 \tag{5.60}$$

This is a wave equation in the Minkowskian spacetime/continuum.

5.3.3 Limit Case of Newton Gravitation

Let consider a gravitational field generated by an isolated distribution of masses in the limit case of Newton–Cartan gravity. Remind that the energy-momentum for dust is given by the $(2, 0)$ type tensor (contravariant components) $T^{\alpha\beta} := \rho_0\,u^\alpha u^\beta$ where ρ_0 is the mass density in a rest frame. For non-relativistic situation where $c >> v^i$, we obtained respectively the components of the stress-energy tensor (from Eq. (4.47)):

$$\begin{cases} T^{00} = \rho_0 u^0 u^0 = \rho \\[2mm] T^{0i} = \rho_0 u^0 u^i = \rho_0\dfrac{1}{c^2}\dfrac{dx^0}{d\tau}\dfrac{d^i}{d\tau} = \rho\dfrac{v^i}{c} \simeq 0 \\[2mm] T^{ij} = \rho_0\dfrac{1}{c^2}\dfrac{dx^i}{d\tau}\dfrac{dx^j}{d\tau} = \rho\dfrac{v^i v^j}{c^2} \simeq 0 \end{cases} \tag{5.61}$$

where ρ is the mass density in a moving frame. The linearized Einstein equation of gravitation (Eq. (5.60)) takes the form of:

$$-\eta^{\alpha\nu}\partial_\alpha\partial_\nu\,\bar{\varepsilon}_{\beta\mu} = \Box\,\bar{\varepsilon}_{\beta\mu} = 8\pi\,T_{\beta\mu} \tag{5.62}$$

The timelike component obeys the Poisson equation $-\Delta \bar{\varepsilon}_{00} = 8\pi\rho$ with all the other components of $\bar{\varepsilon}_{\beta\mu}$ equal to zero, and owing that the derivative with respect to x^0 are neglected due to the slowness of matter motion within this gravitation field. All other components are in fact harmonic and vanishes outside the matter distribution, meaning that at large distance the spacetime is Minkowskian. The may be neglected and put equal to zero. By comparing to the Newtonian potential Φ satisfying the equation $\Delta \Phi = 4\pi\rho$, we deduce that $\bar{\varepsilon}_{00} = -2\Phi$. Coming back to the perturbation term we may write:

$$\varepsilon_{\beta\mu} = \bar{\varepsilon}_{\beta\mu} - \frac{1}{2}\text{Tr}\left(\bar{\varepsilon}\right)\eta_{\beta\mu} = \begin{cases} -\Phi, \ \beta = \mu = 0 \\ \Phi, \ \beta = \mu \neq 0 \\ 0, \text{ otherwise} \end{cases} \tag{5.63}$$

Owing that the conventional Newtonian potential is $\Phi := -m/r$ for a planet with mass m at distance r, we then obtain the linearized metric as:

$$g_{\beta\mu} = \eta_{\beta\mu} + 2\varepsilon_{\beta\mu} = \begin{bmatrix} 1-(2m/r) & 0 & 0 & 0 \\ 0 & -1-(2m/r) & 0 & 0 \\ 0 & 0 & -1-(2m/r) & 0 \\ 0 & 0 & 0 & -1-(2m/r) \end{bmatrix} \tag{5.64}$$

leading to the line element for Cartesian coordinate for the space:

$$ds^2 = [1-(2m/r)]\left(dx^0\right)^2 - [1-(2m/r)]\sum_{i=1}^{i=3}\left(dx^i\right)^2$$

This is an approximate value of line element for a star or planet in the weak-field limit. In other words, this metric corresponds to a gravitation field due to a static distribution of matter in the linear approximation. The Newtonian potential completely determines the metric of the spacetime.

5.3.4 Gravitational Waves

We omit the overline bar for the sake of the simplicity for notation. Let now consider the vacuum solutions of the linearized gravitation equations of Einstein. A recent review on the sources and the technology used for their detection may be found in e.g. Riles (2013). Starting from the D'Alembertian (ten scalar) equations $\Box \bar{\varepsilon}_{\beta\mu} = 0$, we search for solutions taking the form of:

$$\bar{\varepsilon}_{\beta\mu} := \mathfrak{Re}\left(E_{\beta\mu}\ e^{jk_\alpha x^\alpha}\right) = \mathfrak{Re}\left(E_{\beta\mu}\ e^{jk_i x^i}\ e^{jk_0 x^0}\right) \tag{5.65}$$

where k_α is the wave vector. Owing that the D'Alembertian operator $\partial_\alpha \partial^\alpha$ leads to the following equation for exponential function:

$$\partial_\alpha \partial^\alpha \left[\Re e \left(E_{\beta\mu} \, e^{jk_i x^i} \, e^{jk_0 x^0} \right) \right] = -k_\alpha \, k^\alpha \left[\Re e \left(E_{\beta\mu} \, e^{jk_i x^i} \, e^{jk_0 x^0} \right) \right] = 0$$

(5.66)

A non null solution exists if and only if $-k_\alpha \, k^\alpha = 0$ meaning that the wave vector is null. In some sense the gravitational waves propagate at the speed of light. It is usual to define a frequency-like variable $k_0 := -\omega$ and then write the dispersion equation (nullity of the wave vector) as follows (this resembles to classical equation of dispersion in continuum mechanics): $\omega^2 = k_1^2 + k_2^2 + k_3^2$. To this dispersion equation, it should be also reminded the Lorentz gauge $\partial^\mu \overline{\varepsilon}_{\beta\mu} = 0$ leading to four algebraic linear equations:

$$- k^\mu E_{\beta\mu} = 0$$

(5.67)

First, the Lorentz gauge equation eliminates 4 of the degrees of freedom and let 6 degrees in the amplitude $E_{\beta\mu}$. Second, we have seen that considering the infinitesimal transformations (possibly interpreted as a change of coordinates) $x^\mu \to x^\mu + \xi^\mu$ leads to the equation $\partial^\mu \overline{\varepsilon}_{\beta\mu} + \partial_\mu \partial^\mu \xi_\beta \equiv 0$ and has solution in ξ_β whenever $\Box \xi_\beta = -\partial^\mu \overline{\varepsilon}_{\beta\mu}$ for any given $\partial^\mu \overline{\varepsilon}_{\beta\mu}$. Indeed, we can then consider four functions with ξ_β such that $\Box \xi_\beta = 0$. By introducing the particular gauge vector:

$$\xi_\beta := \Re e \left(j B_\beta \, e^{jk_\alpha x^\alpha} \right)$$

(5.68)

which obviously satisfies the D'Alembertian operator $\Box \xi_\beta = -\partial^\mu \overline{\varepsilon}_{\beta\mu} = 0$. The perturbation of the metric due to this change of coordinates holds as previously:

$$\overline{\varepsilon}'_{\beta\mu} = \overline{\varepsilon}_{\beta\mu} + \partial_\beta \xi_\mu + \partial_\mu \xi_\beta - \left(\partial^\nu \xi_\nu \right) \eta_{\beta\mu}$$

(5.69)

Introducing the expression of the ξ_β in this formula gives:

$$E_{\beta\mu} \to E_{\beta\mu} - k_\beta B_\mu - k_\mu B_\beta + k_\nu B^\nu \eta_{\beta\mu}$$

(5.70)

where we can choose the amplitude B_β however we want. Replacing the term $-k^1 = k_0 := -\omega$, we can focus on the following components of $E_{\beta\mu}$ (trace and mixed space-time terms):

$$\begin{pmatrix} (1/2) E_\alpha^\alpha \\ E_{01} \\ E_{02} \\ E_{03} \end{pmatrix}^{\text{new}} = \begin{pmatrix} (1/2) E_\alpha^\alpha \\ E_{01} \\ E_{02} \\ E_{03} \end{pmatrix}^{\text{old}} + \begin{bmatrix} \omega & k_1 & k_2 & k_3 \\ -k_1 & \omega & 0 & 0 \\ -k_2 & 0 & \omega & 0 \\ -k_3 & 0 & 0 & \omega \end{bmatrix} \begin{pmatrix} B_0 \\ B_1 \\ B_2 \\ B_3 \end{pmatrix}$$

(5.71)

where the involved 4×4 matrix is invertible (the determinant is equal to $2\omega^4$). We can thus choose the amplitude B_μ (which is equivalent to choose a new coordinate system) by solving the previous equation as:

$$
\begin{pmatrix} (1/2)E_\alpha^\alpha \\ E_{01} \\ E_{02} \\ E_{03} \end{pmatrix}^{old} + \begin{bmatrix} \omega & k_1 & k_2 & k_3 \\ -k_1 & \omega & 0 & 0 \\ -k_2 & 0 & \omega & 0 \\ -k_3 & 0 & 0 & \omega \end{bmatrix} \begin{pmatrix} B_0 \\ B_1 \\ B_2 \\ B_3 \end{pmatrix} = \begin{pmatrix} 0 \\ 0 \\ 0 \\ 0 \end{pmatrix} \tag{5.72}
$$

because the solution in B_μ is unique. Therefore, we have in sum eight independent conditions for the amplitude $E_{\beta\mu}$ and let two remaining polarizations of the gravitational waves:

$$
\begin{cases} E_{\beta\mu}\, k^\mu = 0 \\ E_{\beta\mu}\, \eta^{\beta\mu} = 0 \\ E_{0\mu} = 0 \end{cases} \tag{5.73}
$$

which can be classified as transverse-traceless gauge conditions. For $\beta = 0$, the first line of Eq. (5.73) combined with the third line of (5.73) induces that $E_{00}\,\omega = E_{0\mu}k_\mu = 0$, and thus $E_{00} = 0$ since $\omega \neq 0$ (presence of wave). The amplitude wave is purely spatial since $E_{00} = 0$ and $E_{0\mu} = 0$. The wave amplitude is traceless $E_{\beta\mu}\,\eta^{\beta\mu} = 0$, and the wave is transverse since $E_{ij}\, k^j = 0$. Let now choose a spatial coordinate such the wave is travelling in the x^3 direction; that is: $k^\mu = (\omega, 0, 0, -\omega)$, which by the way induces $E_{3\mu} = 0$. So, in general we can write the matrix amplitude as:

$$
E_{\beta\mu} = \begin{bmatrix} 0 & 0 & 0 & 0 \\ 0 & E_{11} & E_{12} & 0 \\ 0 & E_{12} & -E_{11} & 0 \\ 0 & 0 & 0 & 0 \end{bmatrix} \tag{5.74}
$$

because $E_{\beta\mu}$ is traceless and symmetric. The metric perturbation is then traceless and perpendicular to the wave vector. The gravitational wave is also called graviton in the language of particle physics. Since the amplitude matrix has only two independent components, it has two polarization states as for massless particles: $E_+ := (1/2)(E_{11} - E_{22})$ and $E_\times := E_{12}$. Discussion about gravitational waves and graviton particles may be found in the literature e.g. Ryder (2009). The perturbed metric associated to the gravitational waves induces the line element:

$$
ds^2 = (dx^0)^2 - \left[1 + \Re e(E_+ e^{jk_\alpha x^\alpha})\right](dx^1)^2 - 2\Re e(E_\times + e^{jk_\alpha x^\alpha})dx^1 dx^2
$$
$$
- \left[1 - \Re e(E_+ e^{jk_\alpha x^\alpha})\right]\left(dx^2\right)^2 + (dx^3)^2 \tag{5.75}
$$

This is the local spacetime metric with perturbation which can strike matter. Consider first the case $E_\times \equiv 0$, the line element holds as:

$$ds^2 = (dx^0)^2 - \left[1 + \Re e(E_+ e^{jk_\alpha x^\alpha})\right](dx^1)^2$$
$$- \left[1 - \Re e(E_+ e^{jk_\alpha x^\alpha})\right]\left(dx^2\right)^2 + (dx^3)^2$$

showing that two particles initially on the diameter of a circle, hit by the gravitational wave will move apart and then move closer together, according to the time evolution $\Re e(E_+ e^{jk_\alpha x^\alpha})$, which is a oscillatory function. For the case $E_+ = 0$ it is usual to rotate the coordinate system to:

$$\tilde{x}^1 = x^1 \cos(\pi/4) + x^2 \sin(\pi/4)$$
$$\tilde{x}^2 = -x^1 \sin(\pi/4) + x^2 \cos(\pi/4)$$

The line element in this new coordinate system holds:

$$ds^2 = (dx^0)^2 - \left[1 + \Re e(E_\times e^{jk_\alpha x^\alpha})\right](d\tilde{x}^1)^2$$
$$- \left[1 - \Re e(E_+ e^{jk_\alpha x^\alpha})\right]\left(d\tilde{x}^2\right)^2 + (d\tilde{x}^3)^2$$

showing that this motion corresponds again to a oscillation of the same type as previously but with respect to $\pi/4$ rotated axes. It conforms to the results stating that the linearized Einstein equations for weak field in relativistic gravitation have wave solutions and then predict the existence of gravitational waves, which is a propagative disturbance of the spacetime itself. The search of gravitational waves has various physical motivations e.g. Riles (2013), and experiments for detecting them began in the sixties. However, up to now, experimental measurements of gravitational waves are still great challenge in cosmology and remain an active research domain e.g. Abbott et al. (2016), a paper with more than 1000 co-authors from LIGO Scientific Collaboration and Virgo Collaboration, where results on the first direct detection of gravitational waves (nearly 100 years after Einstein prediction) based on the measurements of merging of two stellar-mass black holes, is recently reported.

Remark 5.10 The background theory for detecting gravitational waves is based on the concept of geodesic deviation for spinless particle. An alternative method is based on the relativistic top motion. Extension to spinning particle was conducted in e.g. Nieto et al. (2003) to include the effects of the spin and where a covariant formulation of the relativistic top deviation was developed. A Lagrangian formulation of the so-called *Mathisson-Papapetrou equations* in the Riemann–Cartan framework, the basic theory for extended and pole-dipole system, was conducted in e.g. Leclerc (2005). It should nevertheless mentioned that the Lagrangians suggested in Leclerc

(2005) are not covariant although the deduced Euler–Lagrange are the covariant equations of Mathisson-Papapetrou.

5.3.5 Elementary Bases for Measurement of Gravitational Waves

From the amplitude matrix (5.74), we can extract two linear polarizations:

$$
E_{\beta\mu} = \begin{bmatrix} 0\,0 & 0\,0 \\ 0\,1 & 0\,0 \\ 0\,0 & -1\,0 \\ 0\,0 & 0\,0 \end{bmatrix}, \quad
E_{\beta\mu} = \begin{bmatrix} 0\,0\,0\,0 \\ 0\,0\,1\,0 \\ 0\,1\,0\,0 \\ 0\,0\,0\,0 \end{bmatrix},
$$

and two other circular polarizations:

$$
E_{\beta\mu} = \begin{bmatrix} 0\,0 & 0\,0 \\ 0\,1 & i\,0 \\ 0\,i & -1\,0 \\ 0\,0 & 0\,0 \end{bmatrix}, \quad
E_{\beta\mu} = \begin{bmatrix} 0 & 0 & 0\,0 \\ 0 & 1 & -i\,0 \\ 0 & -i & -1\,0 \\ 0 & 0 & 0\,0 \end{bmatrix},
$$

The basic idea behind the experimental measurement of gravitational waves is to consider two test particles, a single test does not feel the waves since it is moving on its geodesic path. Only its coordinates x^μ might be determined when analyzing its motion. For the sake of the simplicity let us consider two particles distant of Δx_1 along the x^1 axis. Say a rectilinear polarized wave orthogonal to the vector separating the two test particles. The proper distance ℓ between them reduces to:

$$
\ell := \int_{x^1=0}^{x^1=\Delta x^1} \sqrt{g_{11}(dx^1)^2} = \int_{x^1=0}^{x^1=\Delta x^1} \sqrt{g_{11}}\,dx^1 = \int_{x^1=0}^{x^1=\Delta x^1} \sqrt{1+h_{11}}\,dx^1
$$

This distance may be approximated by its Taylor expansion:

$$
\ell \simeq \left[1 + (1/2)h_{11}(x^0, x^1, x^2 = 0, x^3 = 0\right] \Delta x^1
$$

This means that the amplitude of the gravitational wave $h_{11}(t)$ may be interpreted as the variation of the proper distance between two test particles. The concept of two test particles is now extended to a set of particles which are on a circle in the presence of gravitational waves. Their motions thus measure the incident gravitational waves according to Fig. 5.6.

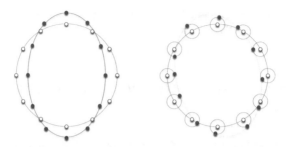

Fig. 5.6 Initial test particles set on a circle are moving: (**a**) on the left when particles are subject to a rectilinear polarized gravitational waves (motion of each particle is linear and harmonic), and (**b**) on the right when the set of particles are subject to a circular polarized gravitational waves (motion of each particle is on a circle centered on the initial position of the particle)

5.3.6 Vacuum Spacetime with Torsion

As an extended version of the Einstein gravitation theory, let us now consider the Einstein–Cartan spacetime without matter. The geometry of the curved spacetime with torsion is assumed to be defined by the usual Einstein–Hilbert Lagrangian:

$$\mathscr{L}_{EH} := \frac{1}{2\chi} \, \mathcal{R}\left[g_{\alpha\beta}, \Gamma^{\gamma}_{\alpha\beta}, \partial_{\lambda}\Gamma^{\gamma}_{\alpha\beta}\right]$$

where we explicitly mention the arguments of the curvature to point out the influence of both the metric components and the connection coefficients. In this subsection, the idea is to decompose the connection as a Levi-Civita connection and the contortion tensor. The next problem for gravitational waves would then be their (in)existence for such a spacetime with torsion. At very particular case will be formulation in the context of teleparallel gravitation theory.

5.3.6.1 Fields Equation for Einstein–Cartan Spacetime

The Eintein–Cartan spacetime is endowed with a metric tensor $g_{\alpha\beta}$, and a connection $\Gamma^{\gamma}_{\alpha\beta}$ with a non zero torsion tensor. The argument we will use is the contortion tensor defined as $\Gamma^{\gamma}_{\alpha\beta} := \overline{\Gamma}^{\gamma}_{\alpha\beta} + \mathcal{T}^{\gamma}_{\alpha\beta}$ where $\overline{\Gamma}^{\gamma}_{\alpha\beta}$ denotes the Levi-Civita connection. Accordingly, the curvature may be decomposed into three contributions:

$$\mathcal{R}^{\gamma}_{\alpha\beta\lambda} = \overline{\mathcal{R}}^{\gamma}_{\alpha\beta\lambda} + \overline{\nabla}_{\alpha}\mathcal{T}^{\gamma}_{\beta\lambda} - \overline{\nabla}_{\beta}\mathcal{T}^{\gamma}_{\alpha\lambda} - (\mathcal{T}^{\gamma}_{\beta\mu}\mathcal{T}^{\mu}_{\alpha\lambda} - \mathcal{T}^{\gamma}_{\alpha\mu}\mathcal{T}^{\mu}_{\beta\lambda}) \tag{5.76}$$

Since we have considered metric compatible connection, the scalar curvature takes the form of:

$$\mathcal{R} = \overline{\mathcal{R}} + \overline{\nabla}_{\alpha}(\mathcal{T}^{\alpha}_{\beta\lambda}g^{\beta\lambda} - \mathcal{T}^{\beta}_{\beta\lambda}g^{\alpha\lambda}) + \delta^{\alpha}_{\gamma}\left(\mathcal{T}^{\gamma}_{\beta\mu}\mathcal{T}^{\mu}_{\alpha\lambda} - \mathcal{T}^{\gamma}_{\alpha\mu}\mathcal{T}^{\mu}_{\beta\lambda}\right)g^{\beta\lambda} \tag{5.77}$$

where we define the following quantities, a vector and a skew-symmetric tensor,

$$
\begin{cases}
W^\alpha := \mathcal{T}^\alpha_{\beta\lambda} g^{\beta\lambda} - \mathcal{T}^\beta_{\beta\lambda} g^{\alpha\lambda} \\
T_{\beta\lambda} := \delta^\alpha_\gamma \left(\mathcal{T}^\gamma_{\beta\mu} \mathcal{T}^\mu_{\alpha\lambda} - \mathcal{T}^\gamma_{\alpha\mu} \mathcal{T}^\mu_{\beta\lambda} \right)
\end{cases}
\tag{5.78}
$$

Let now consider the Hilbert–Einstein action in terms of Riemann curvature and the contortion terms:

$$
\mathscr{S}_{HE} := \frac{1}{2\chi} \int_{\mathscr{B}} \left(\overline{\mathfrak{R}}_{\beta\lambda}\, g^{\beta\lambda} + \overline{\nabla}_\alpha W^\alpha + T_{\beta\lambda}\, g^{\beta\lambda} \right) \omega_n
\tag{5.79}
$$

where the two last terms are calculated by means of (5.78). By the way, it should be mentioned that the vector W^α includes contributions of both metric and contortion. Basically we also should remind that the contortion tensor includes both the metric and torsion tensor. So is the case for $T_{\beta\lambda}$. Let us calculate the variation of the action $\mathscr{S}_{HE}$:

$$
\delta\mathscr{S}_{HE} = \int_{\mathscr{B}} \left[\left(\overline{\mathfrak{R}}_{\beta\lambda} - \frac{\overline{\mathfrak{R}}}{2} g_{\beta\lambda} \right) + \left(T_{\beta\lambda} - \frac{T}{2} g_{\beta\lambda} \right) - \frac{\overline{\nabla}_\alpha W^\alpha}{2} g_{\beta\lambda} \right] \delta g^{\beta\lambda}\, \omega_n
$$
$$
+ \int_{\mathscr{B}} g^{\beta\lambda} \delta^\alpha_\gamma \delta (\mathcal{T}^\gamma_{\beta\mu} \mathcal{T}^\mu_{\alpha\lambda} - \mathcal{T}^\gamma_{\alpha\mu} \mathcal{T}^\mu_{\beta\lambda})\, \omega_n
\tag{5.80}
$$

in which we have defined $T := T_{\beta\lambda} g^{\beta\lambda} = \delta^\alpha_\gamma (\mathcal{T}^\gamma_{\beta\mu} \mathcal{T}^\mu_{\alpha\lambda} - \mathcal{T}^\gamma_{\alpha\mu} \mathcal{T}^\mu_{\beta\lambda}) g^{\beta\lambda}$. If we consider the metric $g_{\beta\lambda}$ and the contortion tensor $\mathcal{T}^\gamma_{\beta\lambda}$ as independent arguments of the Lagrangian, the second line does not include the variation of the metric. Therefore, the first equation of gravitation field holds:

$$
\left(\overline{\mathfrak{R}}_{\beta\lambda} - \frac{\overline{\mathfrak{R}}}{2} g_{\beta\lambda} \right) + \left(T_{\beta\lambda} - \frac{T}{2} g_{\beta\lambda} \right) - \frac{\overline{\nabla}_\alpha W^\alpha}{2} g_{\beta\lambda} = 0
\tag{5.81}
$$

In the absence of torsion tensor for Einstein gravitation, the field equation (5.81) reduces to the classical gravitation equation. For Einstein–Cartan gravitation, the presence of torsion in the spacetime induces a source term as for matter within space, and a kind of non-uniform field of cosmological constant:

$$
\begin{cases}
T^{source}_{\beta\lambda} := T_{\beta\lambda} \\
\Lambda(x^\mu) := -\frac{T}{2} - \frac{\overline{\nabla}_\alpha W^\alpha}{2}
\end{cases}
\tag{5.82}
$$

Remark 5.11 A similar result was obtained in e.g. Bamba et al. (2013) by directly considering a teleparallel gravitation theory where the Lagrangian is assumed to depend on the contortion tensor. These authors suggested to consider a non uniform cosmological constant $\Lambda(x^\mu)$. They have shown that the gravitational waves modes in the teleparallel gravity theory, based on the torsion, are equivalent to that of the Einstein-gravitation, which is based on the curvature.

Recent studies have been done to investigate the influence of the torsion in the Einstein–Cartan gravitation theory e.g. Romero et al. (2016). Similar results are obtained such as for the role of the torsion tensor as a source of gravitation even without matter, and also the possibility to consider the term $\mathcal{W}^\alpha$ as not uniform cosmological constant all over the spacetime. In the present study, we have a explicit and simpler formulation of this flux quantity (expressed as a divergence) in function of the contortion tensor. There is no need to consider the flux as boundary condition even it is directly expressed as a divergence. The second line of the action variation concerns the contortion tensor, and can be written as follows:

$$\left[\left(g^{\beta\lambda}\delta^\alpha_\gamma - g^{\alpha\lambda}\delta^\beta_\gamma \right) \mathcal{T}^\mu_{\alpha\lambda} + \left(g^{\alpha\mu}\delta^\beta_\lambda - g^{\beta\mu}\delta^\alpha_\lambda \right) \mathcal{T}^\lambda_{\alpha\gamma} \right] \delta\mathcal{T}^\gamma_{\beta\mu} = 0$$

Due to arbitrariness of the contortion variation, we deduce the field equation for contortion:

$$\left(g^{\beta\lambda}\delta^\alpha_\gamma - g^{\alpha\lambda}\delta^\beta_\gamma \right) \mathcal{T}^\mu_{\alpha\lambda} + \left(g^{\alpha\mu}\delta^\beta_\lambda - g^{\beta\mu}\delta^\alpha_\lambda \right) \mathcal{T}^\lambda_{\alpha\gamma} = 0 \qquad (5.83)$$

which is an algebraic equation, meaning a constraint equation for the contortion components. The two equations (5.81) and (5.83) constitute the complete set of field equations for both the unknowns metric and contortion. Being an algebraic condition of the contortion tensor, Eq. (5.83) may be better interpreted as a constraint condition namely if a contortion field is assumed a priori.

5.3.6.2 Conservation Laws in Einstein–Cartan Spacetime

First, let us consider a curved spacetime with zero torsion. Then application of Poincaré's gauge invariance leads to the Eulerian variational equation for Hilbert–Einstein action:

$$\delta\mathscr{S}_{HE} = \int_\mathscr{B} \left[(1/2\chi) \left(\overline{\mathfrak{R}}^{\alpha\beta} - (\overline{\mathfrak{R}}/2)g^{\alpha\beta} \right) - T^{\alpha\beta}_{\text{source}} \right] \mathcal{L}_\xi g_{\alpha\beta}\, \omega_n = 0 \qquad (5.84)$$

with: $\mathcal{L}_\xi g_{\alpha\beta} = g_{\alpha\gamma}\overline{\nabla}_\beta\xi^\gamma + g_{\gamma\beta}\overline{\nabla}_\alpha\xi^\gamma$. This classically allows us to deduce the conservation laws by accounting for the Bianchi identity $\overline{\nabla}_\beta G^{\alpha\beta} = 0$,

$$\overline{\nabla}_\beta T^{\alpha\beta}_{\text{source}} = 0 \qquad (5.85)$$

where we remind the definition of the Einstein tensor $G^{\alpha\beta} := \overline{\mathfrak{R}}^{\alpha\beta} - (\overline{\mathfrak{R}}/2)g^{\alpha\beta}$. Second, we now account for the contortion tensor for a Einstein–Cartan spacetime. The field equation (5.81) may be slightly modified to give:

$$\overline{\mathfrak{R}}^{\alpha\beta} - (\overline{\mathfrak{R}}/2)g^{\alpha\beta} + \Lambda\, g^{\alpha\beta} - T^{\alpha\beta}_{\text{source}} = 0 \qquad (5.86)$$

We recognize the usual formulation of relativistic gravitation theory. We nevertheless keep in mind that the cosmological term $\Lambda(x^\mu)$ is in principle neither constant nor uniform but depends on the contortion distribution $\mathfrak{T}^\gamma_{\alpha\beta}(x^\mu)$ within spacetime. The term source of gravitation $T^{\alpha\beta}_{\text{source}}$ depends also on the contortion field. Since we have no boundary terms (the divergence is included in the non uniform cosmological constant), the Poincaré's gauge invariance in the framework of Einstein–Cartan spacetime can be rewritten as follows without further assumption:

$$\delta\mathscr{S}_{HE} = \int_{\mathscr{B}} \left[\overline{\mathfrak{R}}^{\alpha\beta} - (\overline{\mathfrak{R}}/2)g^{\alpha\beta} + \Lambda\, g^{\alpha\beta} - T^{\alpha\beta}_{\text{source}}\right]\mathcal{L}_\xi g_{\alpha\beta}\,\omega_n = 0 \tag{5.87}$$

where the Lie derivative has been derived in the previous chapter, the metric being compatible with the Levi-Civita connection,

$$\mathcal{L}_\xi g_{\alpha\beta} = \overline{\nabla}_\beta \xi_\alpha + \overline{\nabla}_\alpha \xi_\beta + \left(g_{\alpha\gamma}\mathfrak{T}^\gamma_{\rho\beta} + g_{\gamma\beta}\mathfrak{T}^\gamma_{\rho\alpha}\right)\xi^\rho$$

where the contortion tensor is present conversely to the previous Riemannian spacetime. Introduction of this Lie derivative in the variation of Hilbert–Einstein action and applying the integration by parts, together with the field equation (5.86) allows us to obtain the conservation laws in the framework of Einstein–Cartan gravitation theory:

$$\overline{\nabla}_\alpha \left(T^{\alpha\beta}_{\text{source}} - \Lambda\, g^{\alpha\beta}\right) = 0 \tag{5.88}$$

since we have accounted for the first Bianchi identity involving Levi-Civita connection and Einstein tensor $\overline{\nabla}_\alpha G^{\alpha\beta} = 0$. It should be pointed out that the covariant derivative in this conservation law involves the Levi-Civita connection, rather than the initial connection. The tensor $T^{\alpha\beta}_{\text{source}}$ is skew-symmetric according to its definition (5.78).

Remark 5.12 Introducing the flux term $\mathcal{W}^\alpha$ from the definition (5.78) into the conservation law (5.88), we observe that the resulting equation leads to a second-order partial differential equation in which the unknown is the contortion $\mathfrak{T}^\gamma_{\alpha\beta}$ meaning that in such a framework, the torsion tensor can propagate through the spacetime.

For the second conservation law in terms of the contortion, let us define the hypermomentum from the field equation (5.83):

$$\Sigma^{\alpha\beta}_\gamma := \left(g^{\alpha\lambda}\delta^\mu_\gamma - g^{\mu\lambda}\delta^\alpha_\gamma\right)\mathfrak{T}^\beta_{\mu\lambda} + \left(g^{\mu\beta}\delta^\alpha_\lambda - g^{\alpha\beta}\delta^\mu_\lambda\right)\mathfrak{T}^\lambda_{\mu\gamma} \tag{5.89}$$

Then the contribution of the contortion to the Eulerian variation of the Hilbert–Einstein action holds:

$$\delta\mathscr{S}_{HE} = \int_{\mathscr{B}} \dots + \int_{\mathscr{B}} \Sigma^{\alpha\beta}_\gamma\, \mathcal{L}_\xi \mathfrak{T}^\gamma_{\alpha\beta}\,\omega_n$$

where the Lie derivative of the contortion tensor was derived in the previous chapter:

$$\mathcal{L}_\xi \mathcal{T}^\gamma_{\alpha\beta} = \xi^\rho \overline{\nabla}_\rho \mathcal{T}^\gamma_{\alpha\beta} - \mathcal{T}^\rho_{\alpha\beta} \overline{\nabla}_\rho \xi^\gamma + \mathcal{T}^\gamma_{\rho\beta} \overline{\nabla}_\alpha \xi^\rho + \mathcal{T}^\gamma_{\alpha\rho} \overline{\nabla}_\beta \xi^\rho \qquad (5.90)$$

5.3.6.3 Case of Teleparallel Gravitation Theory

Let us now consider a particular case where the Riemannian curvature vanishes $\mathfrak{R}^\gamma_{\alpha\beta\lambda} \equiv 0$. In such a case, the contribution of the contortion $\mathcal{T}^\gamma_{\alpha\beta}$ exactly compensates the vanishing effects of the curvature associated to the Levi-Civita connection $\overline{\mathfrak{R}}^\gamma_{\alpha\beta\lambda} \equiv 0$. The curvature takes the form of:

$$\mathfrak{R}^\gamma_{\alpha\beta\lambda} = \overline{\nabla}_\alpha \mathcal{T}^\gamma_{\beta\lambda} - \overline{\nabla}_\beta \mathcal{T}^\gamma_{\alpha\lambda} - (\mathcal{T}^\gamma_{\beta\mu} \mathcal{T}^\mu_{\alpha\lambda} - \mathcal{T}^\gamma_{\alpha\mu} \mathcal{T}^\mu_{\beta\lambda}) \qquad (5.91)$$

which is a particular case of the previous model of Einstein–Cartan gravitation. Then, up to a divergence term which may be dropped by choosing appropriate boundary conditions, it may be reminded that the Einstein gravitation theory can be described in terms of contortion: It is called the Teleparallel Gravitation Theory e.g. Aldrovandi and Pereira (2013). Up to divergence terms and by using specific boundary conditions, the alternative Lagrangian function reduces to the action, see Eq. (4.107) for more details,

$$\mathscr{S} = (1/2\chi) \int_{\mathscr{M}} \delta^\alpha_\gamma \left(\mathcal{T}^\gamma_{\beta\mu} \mathcal{T}^\mu_{\alpha\lambda} - \mathcal{T}^\gamma_{\alpha\mu} \mathcal{T}^\mu_{\beta\lambda} \right) g^{\beta\lambda} \omega_n \qquad (5.92)$$

where the trace operator contracts the indices α and γ to obtain the Ricci curvature, and the multiplication with the metric $g^{\beta\lambda}$ is applied to calculate the scalar curvature. The Lagrangian function thus depends on the contortion and the metric. It might be possible to use the Minkowski metric to calculate the scalar curvature $\eta^{\beta\lambda}$, although it seems not clear at this point. We now introduce the Lagrangian variation of the contortion field $\Delta \mathcal{T}^\gamma_{\alpha\beta} = \delta \mathcal{T}^\gamma_{\alpha\beta} + \mathcal{L}_\xi \mathcal{T}^\gamma_{\alpha\beta}$ (as a remind in terms of the Eulerian variation and the Lie derivative variation) to obtain the constitutive laws:

$$\Delta \mathscr{S} = \int_{\mathscr{M}} \Delta \left[\mathscr{L} \left(g_{\alpha\beta}, \mathcal{T}^\gamma_{\alpha\beta} \right) \omega_n \right] \qquad (5.93)$$

where the metric $g_{\alpha\beta}$ and the contortion $\mathcal{T}^\gamma_{\alpha\beta}$ are considered as independent arguments of the Lagrangian. By dropping the Riemann curvature term, the field equation is simplified in such a case:

$$\Lambda \, g^{\alpha\beta} - T^{\alpha\beta}_{source} = 0 \qquad (5.94)$$

whereas the conservation law remains unchanged:

$$\overline{\nabla}_\alpha \left(T^{\alpha\beta}_{source} - \Lambda \, g^{\alpha\beta} \right) = 0 \qquad (5.95)$$

owing that no contribution of the curvature occurs even in the general case. It should be stressed that the field equation (5.94) trivially induces the conservation Eq. (5.95). In the framework of teleparallel gravitation theory, the contortion tensor is determined algebraically rather than calculated with partial differential equations. Accordingly, there is no possibility of torsion to propage in Eq. (5.94).

5.3.6.4   Spacetime with Spherical Symmetry

In this illustrating example, we are looking for a spherical symmetric solution of the metric $g_{\alpha\beta}(x^\mu)$ and the contortion $\mathcal{T}^\gamma_{\alpha\beta}(x^\mu)$. Spherical coordinates $(x^\mu) = (x^0 := ct, r, \theta, \varphi)$ are used. Adopting the tetrads approach, we choose the following tetrads, index i for row and index α for column, and their inverse:

$$F^i_\alpha = \begin{bmatrix} R(r) & 0 & 0 & 0 \\ 0 & 1/R(r) & 0 & 0 \\ 0 & 0 & r & 0 \\ 0 & 0 & 0 & r\sin\theta \end{bmatrix}, \quad F^\alpha_i = \begin{bmatrix} 1/R(r) & 0 & 0 & 0 \\ 0 & R(r) & 0 & 0 \\ 0 & 0 & 1/r & 0 \\ 0 & 0 & 0 & 1(r\sin\theta) \end{bmatrix}$$

$$(5.96)$$

The associated induced metric writes $g_{\alpha\beta} := F^i_\alpha \eta_{ij} F^j_\beta$, where $\eta_{ij} :=$ Diag$\{1, -1, -1, -1\}$ denotes the Minkowskian spacetime metric,

$$g_{\mu\nu} = \begin{bmatrix} R^2(r) & 0 & 0 & 0 \\ 0 & -1/R^2(r) & 0 & 0 \\ 0 & 0 & -r^2 & 0 \\ 0 & 0 & 0 & -r^2\sin^2\theta \end{bmatrix}$$

$$(5.97)$$

It is worth to determine the coefficients of the connection $\overline{\Gamma}^\gamma_{\alpha\beta}$ from the metric $g_{\alpha\beta}$. The only non vanishing (symmetric) connection Levi-Civita coefficients are the following:

$$\overline{\Gamma}^0_{01} = \frac{R'(r)}{R(r)} = \overline{\Gamma}^0_{10},$$

$$\overline{\Gamma}^1_{00} = R'(r)R^3(r), \quad \overline{\Gamma}^1_{11} = \frac{R'(r)}{R(r)}, \quad \overline{\Gamma}^1_{22} = -rR^2(r), \quad \overline{\Gamma}^1_{33} = -rR^2(r)\sin^2\theta$$

$$\overline{\Gamma}^2_{21} = \frac{1}{r} = \overline{\Gamma}^2_{12}, \quad \mathcal{T}^2_{33} = -\sin\theta\cos\theta$$

$$\overline{\Gamma}^3_{31} = \frac{1}{r} = \overline{\Gamma}^3_{13}, \quad \overline{\Gamma}^3_{32} = \cot\theta = \overline{\Gamma}^3_{23}$$

We remind the expression of the torsion with the tetrads approach:

$$\aleph_{\alpha\beta}^{\gamma} := F_i^{\gamma}\left(\partial_\alpha F_\beta^i - \partial_\beta F_\alpha^i\right).$$

The only eight non vanishing components of the torsion tensor are:

$$\aleph_{01}^{0} = -\frac{R'(r)}{R(r)} = -\aleph_{10}^{0},$$

$$\aleph_{12}^{2} = \frac{1}{r} = -\aleph_{21}^{2} = \aleph_{13}^{3} = -\aleph_{31}^{3},$$

$$\aleph_{23}^{3} = \cot\theta = -\aleph_{32}^{3}.$$

From the torsion we calculate the contortion tensor. Accordingly, the eight non vanishing components are (we remind that contortion tensor is neither symmetric nor skew-symmetric), all others are equal to zero,

$$\mathcal{T}_{01}^{0} = -\frac{R'(r)}{R(r)},$$

$$\mathcal{T}_{00}^{1} = R'(r)R^3(r), \quad \mathcal{T}_{22}^{1} = rR^2(r), \quad \mathcal{T}_{33}^{1} = rR^2(r)\sin^2\theta$$

$$\mathcal{T}_{21}^{2} = -\frac{1}{r}, \qquad \mathcal{T}_{33}^{2} = \sin\theta\cos\theta$$

$$\mathcal{T}_{31}^{3} = -\frac{1}{r}, \qquad \mathcal{T}_{32}^{3} = -\cot\theta$$

The non vanishing (nonsymmetric) connection coefficients are:

$$\Gamma_{10}^{0} = \frac{R'(r)}{R(r)}, \quad \Gamma_{11}^{1} = -\frac{R'(r)}{R(r)}, \quad \Gamma_{12}^{2} = \frac{1}{r}, \quad \Gamma_{13}^{3} = \frac{1}{r}, \quad \Gamma_{23}^{3} = \cot\theta,$$

$$(5.98)$$

Introducing the constitutive laws, and the expressions of the connection with the contortion tensor allows us to obtain differential equations governing the unknown $R(r)$. It should be stressed that the calculus of these coefficients is also possible by directly using the relation $\Gamma_{\alpha\beta}^{\gamma} := F_i^{\gamma}\partial_\alpha F_\beta^i$. As mentioned in e.g. Ferraro and Fiorini (2011), the teleparallel approach for gravitation, i.e. by assuming a Lagrangian depending on the scalar $\mathcal{T}$, allows to show that the Schwarzschild spacetime may be solution for only very special case where the dependence is linear. A nonlinear dependence $\mathcal{L} = f(\mathcal{T})$ induces other spacetimes.

Remark 5.13 Searching for gravitational waves in the framework of modified teleparallel gravity was done in e.g. Bamba et al. (2013). They explicitly showed that there is an equivalence between teleparallel gravity approach and Einstein approach. In the previous section, Einstein gravitational waves was reminded for self-consistence of the book. The gravitational wave equation within a teleparallel gravitation should be now derived by means of the affine connection with non zero torsion. The use of affine connection different from the classical Christoffel symbols $\overline{\Gamma}^{\gamma}_{\alpha\beta}$ means that the background geometry and by the way the derivative operators are not completely described by only the metric $g_{\alpha\beta}$ but also by another independent tensor filed, the contortion $\mathcal{T}^{\gamma}_{\alpha\beta}$, or equivalently by the torsion tensor $\aleph^{\gamma}_{\alpha\beta}$.

5.4 Geodesic and Autoparallel Deviation for Gravitational Waves

The previous example on gravitational waves illustrates a well-known application of the so-called geodesic deviation equation e.g. Nieto et al. (2007). The measuring of the separation of two neighbored geodesic curves, which are the trajectories of two small points (see Fig. 5.6) in a Riemann spacetime may be evaluated by means of the separation acceleration that we briefly remind in this section. Some previous studies have extended this deviation equation to include the relativistic top moving in a gravitational field e.g. Nieto et al. (2007), or to reformulate the geodesic deviation in terms of teleparallel gravity (Darabi et al. 2015) (in this approach the torsion field is considered to engender the gravity field, instead of the curvature). The contribution of the present subsection is slightly different since we are interested in developing the extension of the deviation equation for geodesics to autoparallel curves where the spacetime is curved with torsion.

5.4.1 Geodesic Equation for Newtonian Mechanics

For Newton mechanics, a geodesic is a curve along which a particle moves as free falling particle. The concept of geodesic deviation is based on the comparison of two geodesic curves in the spacetime (t, x^a) for Newton spacetime and (x^{α}) for Einstein and Einstein–Cartan spacetime. In this section we will consider the formulation of three cases of geodesic (autoparallel for Einstein–Cartan gravitation) curves. But for the present paragraph we remind the basics for geodesic equation and deviation in the framework of Newtonian mechanics within the framework of gravitation field due to a potential $\Phi(x^{\mu})$. For Newton gravitation, the equations of motions are respectively for the two particles:

$$\ddot{x}^a(t) = -\partial^a \Phi(P), \qquad \ddot{z}^a(t) = -\partial^a \Phi(Q) = -\partial^a \Phi(P) + \ddot{\xi}(t) \qquad (5.99)$$

Expanding the gradient of the gravitation potential about the point P gives:

$$-\Phi(Q) = -\Phi(P) - \partial_b \Phi(P)\, \xi^b - \mathcal{O}(\xi)$$

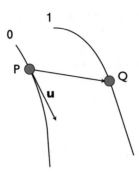

Fig. 5.7 Curves (γ_0) and (γ_1) are geodesics or autoparallel. At each time t of τ, points P and Q are separated by the vector $\xi := \mathbf{PQ}$ depending on the time. The vector $\mathbf{u}$ is a unit vector tangent to the geodesic lines satisfying $\nabla_{\mathbf{u}}\mathbf{u} = 0$. The two vectors satisfy also $\nabla_{\xi}\mathbf{u} = \nabla_{\mathbf{v}\xi}$. t is the parameter along the trajectories of particles, and τ is a parameter along geodesic or autoparallel

This gives the expression of the acceleration of the geodesic deviation:

$$\frac{d^2\xi^a}{dt^2} = -\partial^a\partial_b\Phi\,\xi^a \tag{5.100}$$

which represents the geodesic deviation equation in Newtonian mechanics, and gives the expression of the distance acceleration of two particles falling in a nonuniform gravitation field $\Phi(t, x^a)$. We recognize the components of the curvature tensor introduced in the Newton–Cartan theory of gravitation (4.13) reduced to second-order derivatives of the gravity potential (Fig. 5.7).

To go further, the Lagrangian for classical gravitational field is given by:

$$\mathscr{L} := \rho(x^\mu)\Phi(x^\mu) - \frac{1}{8\pi G}\left(\nabla\Phi(x^\mu)\right)^2$$

Variation of the resulting action with respect to the potential ϕ, and after integrating by parts, the Euler–Lagrange equation holds e.g. Ryder (2009): $\Delta\Phi = -4\pi G\,\rho$ which relates the potential due to the mass density $\rho(x^\mu)$. Consider a spherical body as illustration. Outside a spherically object of mass M, the solution gives the Newtonian gravitational potential $\Phi = GM/r$. As illustration of the geodesic deviation problem (5.100), Greenberg in the seventies found the geodesic deviation equation for a spherical body (earth model) as (Greenberg 1974):

$$\begin{cases} \dfrac{d^2\xi^r}{d\tau^2} - 2\omega_\theta\dfrac{d\xi^\varphi}{d\tau} + 3\omega_\theta^2\,\xi^r = 0 \\[2mm] \dfrac{d^2\xi^\theta}{d\tau^2} + \omega_\theta^2\,\xi^\theta = 0 \\[2mm] \dfrac{d^2\xi^\varphi}{d\tau^2} + 2\omega_\theta^2\,\dfrac{d\xi^r}{d\tau} = 0 \end{cases} \tag{5.101}$$

in a spherical coordinate (r, θ, φ) where he has chosen a circular orbit in the equatorial plane $r_0 = R$, $\theta_0 = \pi/2$, $\varphi_0(\sqrt{m/r^3})\, x^0$. Determination of this particular orbit is obvious. We temporarily use the notation $x^0 := \tau = ct$ to avoid odd notation as $d(x^0)^2$. We denoted $m := GM/c^2$ and $\omega_\theta := m/r^3$. The resolution of the geodesic deviation equation (5.101) is straightforward (Philipp et al. 2015):

$$
\begin{cases}
\xi^r = C_1 + C_2 \sin(\omega_\theta \tau) + C_3 \cos(\omega_\theta \tau) \\[4pt]
\xi^\theta = C_4 \sin(\omega_\theta \tau) + C_5 \cos(\omega_\theta \tau) \\[4pt]
\xi^\varphi = -\dfrac{3}{2} C_1 \omega_\theta\, t + 2\,[C_2 \sin(\omega_\theta \tau) - C_3 \cos(\omega_\theta \tau)] + C_6
\end{cases}
$$

where the constants C_i correspond to various possibilities to perturb the geodesic considered. By the way it was found that the effect of the earth oblateness on the geodesic deviation $\xi(t)$ dominates by far the effect of general relativity (Greenberg 1974). This supports the use of Newtonian mechanics for most past space research and satellite launches. Nevertheless, future satellite mission in faraway space would require the knowledge of relativistic effects of gravitation, particularly on the effects of tidal forces on satellite.

5.4.2   Geodesic Deviation Equation in Riemannian Manifold

First, let consider two geodesic curves in the pseudo-Riemann spacetime $(\mathcal{M}, g_{\alpha\beta})$ denoted by γ_0 and γ_1 respectively. At the same propertime τ, we define the separation four-vector ξ^α of the spacetime $\mathcal{M}$ which connects a point (event) $x^\alpha(\tau)$ of the geodesic γ_0 to a point (event) $x^\alpha(\tau) + \xi^\alpha(\tau)$ of a nearby geodesic γ_1. The separation ξ^α is small in such a way that any expansion of tensor function of ξ^α with respect to ξ^α can be truncated to only the first-order terms. It is reminded that the relativistic acceleration a^α of two material points is defined as the second derivative of the separation vector ξ^α as the two material points move along their respective geodesics. Let define the separation velocity and deduce the separation acceleration as follows:

$$
v^\alpha := u^\beta \overline{\nabla}_\beta \xi^\alpha, \qquad a^\alpha := u^\beta \overline{\nabla}_\beta v^\alpha \tag{5.102}
$$

where $u^\beta := dx^\beta/d\tau$ is the four-vector velocity (timelike vector). From the definition (5.102), we write:

$$
\begin{cases}
v^\alpha = u^\beta \left(\partial_\beta \xi^\alpha + \overline{\Gamma}^\alpha_{\beta\gamma} \xi^\gamma \right) = \dfrac{d\xi^\alpha}{d\tau} + \overline{\Gamma}^\alpha_{\beta\gamma} u^\beta \xi^\gamma \\[8pt]
a^\alpha = u^\beta \left(\partial_\beta v^\alpha + \overline{\Gamma}^\alpha_{\beta\varepsilon} v^\varepsilon \right) = \dfrac{dv^\alpha}{d\tau} + \overline{\Gamma}^\alpha_{\delta\varepsilon} u^\delta v^\varepsilon
\end{cases} \tag{5.103}
$$

where we have introduced the expressions:

$$\frac{d\xi^\alpha}{d\tau} := u^\beta \partial_\beta \xi^\alpha, \qquad \frac{dv^\alpha}{d\tau} := u^\beta \partial_\beta v^\alpha$$

after the general expression $d\varphi/d\tau := \partial f/\partial_\beta \, dx^\beta/d\tau$. First, substituting the separation velocity into the expression of the separation acceleration, gives:

$$a^\alpha = \frac{d^2\xi^\alpha}{d\tau^2} + \frac{d}{d\tau}\left(\overline{\Gamma}^\alpha_{\beta\gamma} u^\beta \xi^\gamma\right) + \overline{\Gamma}^\alpha_{\delta\varepsilon}\left(\frac{d\xi^\varepsilon}{d\tau} + \overline{\Gamma}^\varepsilon_{\beta\gamma} u^\beta \xi^\gamma\right) u^\delta \tag{5.104}$$

Second, let write the geodesic equations for the two curves γ_0 and γ_1:

$$\begin{cases} \dfrac{d^2 x^\alpha}{d\tau^2} + \overline{\Gamma}^\alpha_{\beta\gamma}(x^\mu)\dfrac{dx^\beta}{d\tau}\dfrac{dx^\gamma}{d\tau} = 0 \\[2mm] \dfrac{d^2(x^\alpha + \xi^\alpha)}{d\tau^2} + \overline{\Gamma}^\alpha_{\beta\gamma}(x^\mu + \xi^\mu)\dfrac{d(x^\beta + \xi^\beta)}{d\tau}\dfrac{d(x^\gamma + \xi^\gamma)}{d\tau} = 0 \end{cases} \tag{5.105}$$

Since all components ξ^α are small, we need to retain only the first-order terms after expansion of the second equation. We thus obtain:

$$\frac{d^2\xi^\alpha}{d\tau^2} + \left(\overline{\Gamma}^\alpha_{\beta\gamma} + \overline{\Gamma}^\alpha_{\gamma\beta}\right) u^\beta \frac{d\xi^\gamma}{d\tau} + \frac{\partial \overline{\Gamma}^\alpha_{\beta\gamma}}{\partial x^\delta} u^\beta u^\gamma \xi^\delta = 0 \tag{5.106}$$

Introducing Eq. (5.106) in the separation acceleration leads to the expression:

$$a^\alpha = \left(\overline{\Gamma}^\alpha_{\beta\gamma} - \overline{\Gamma}^\alpha_{\gamma\beta}\right) u^\beta \frac{d\xi^\gamma}{d\tau} + \left(\partial_\delta \overline{\Gamma}^\alpha_{\beta\gamma} - \partial_\gamma \overline{\Gamma}^\alpha_{\beta\delta} + \overline{\Gamma}^\alpha_{\delta\varepsilon}\overline{\Gamma}^\varepsilon_{\beta\gamma} - \overline{\Gamma}^\alpha_{\gamma\varepsilon}\overline{\Gamma}^\varepsilon_{\beta\delta}\right) u^\beta \xi^\gamma u^\delta \tag{5.107}$$

where we have worthily used the orthogonality condition $du^\beta/d\tau = u^\varepsilon \partial_\varepsilon u^\beta = u^\varepsilon \overline{\nabla}_\varepsilon u^\beta - u^\varepsilon \overline{\Gamma}^\beta_{\varepsilon\delta}\xi^\delta$. Conventionally, we then deduce that the separation acceleration is written as follows for Levi-Civita connection in a Riemann manifold e.g. Levi-Civita (1927), Synge (1934):

$$\frac{D^2\xi^\alpha}{D\tau^2} = \overline{\mathfrak{R}}^\alpha_{\delta\gamma\beta} u^\beta u^\delta \xi^\gamma \tag{5.108}$$

The geodesic deviation equation (5.108) due to Levi-Civita for Riemannian space shows that the curvature produces acceleration of the separation between two neighboring geodesics γ_0 and γ_1. This provides a geometrical interpretation of the curvature tensor. The geodesic deviation equation constitutes a fundamental equation for relativistic gravitation since it relates the relativistic acceleration of two nearby particles in presence of gravitation field. In a flat spacetime, the separation will be linear. Equation (5.108) allows us to analyze numerous motions of particles in gravitational field, such as the chaotic behavior of particles orbits but they are not well-suited to study spinning particles, either for microscopic with intrinsic spin or macroscopic bodies with intrinsic spin e.g. Leclerc (2005).

5.4.2.1   Application to Schwarzschild Spacetime

For the sake of the clarity, let us consider the example of Schwarzschild spacetime with the line element:

$$ds^2 = \left(1 - \frac{2m}{r}\right)(dx^0)^2 - \left(1 - \frac{2m}{r}\right)^{-1} dr^2 - r^2 d\theta^2 - r^2 \sin^2\theta d\varphi^2$$

The coefficients of Levi-Civita connection for a Schwarzschild spacetime are directly calculated from the metric (4.33):

$$\overline{\Gamma}^1_{00} = \frac{m}{r^3}(r - 2m), \quad \overline{\Gamma}^0_{01} = \overline{\Gamma}^0_{10} = \frac{m}{r(r - 2m)}, \quad \overline{\Gamma}^1_{11} = -\frac{m}{r(r - 2m)},$$

(5.109)

$$\overline{\Gamma}^2_{12} = \overline{\Gamma}^2_{21} = \frac{1}{r}, \quad \overline{\Gamma}^1_{22} = -(r - 2m),$$

(5.110)

$$\overline{\Gamma}^3_{13} = \overline{\Gamma}^3_{31} = \frac{1}{r}, \quad \overline{\Gamma}^1_{33} = -(r - 2m)\sin^2\theta,$$

(5.111)

$$\overline{\Gamma}^2_{33} = -\sin\theta\cos\theta, \quad \overline{\Gamma}^3_{23} = \overline{\Gamma}^3_{32} = \frac{\cos\theta}{\sin\theta}, \quad \text{others} = 0$$

(5.112)

It extends the symbols of Christoffel of spherical coordinates (2.32) of the $3D$ space. We then obtain the Riemann curvature components by using the definition of the curvature tensor:

$$\overline{\Re}^0_{101} = -\frac{2m}{r^2(r - 2m)}, \quad \overline{\Re}^0_{202} = \frac{m}{r}, \quad \overline{\Re}^0_{303} = \frac{m}{r}\sin^2\theta,$$

$$\overline{\Re}^1_{010} = -\frac{2m}{r^4}(r - 2m), \quad \overline{\Re}^1_{212} = \frac{m}{r}, \quad \overline{\Re}^1_{313} = \frac{m}{r}\sin^2\theta,$$

$$\overline{\Re}^2_{020} = \frac{m}{r^4}(r - 2m), \quad \overline{\Re}^2_{121} = \frac{m}{r^2(r - 2m)}, \quad \overline{\Re}^2_{323} = -\frac{2m}{r}\sin^2\theta$$

$$\overline{\Re}^3_{030} = \frac{m}{r^4}(r - 2m), \quad \overline{\Re}^3_{131} = \frac{m}{r^2(r - 2m)}, \quad \overline{\Re}^3_{232} = -\frac{2m}{r}$$

Staying in the example of satellite launches and their relativistic motions, it is worth to consider only timelike geodesics, which are the trajectories of massive particles at subliminal speed. Anyhow, in order to apply the geodesic deviation equation, we should start search for general forms of geodesic curves in the Schwarzschild spacetime by means of Eq. (4.12) at the initial time. The geodesic equation then takes the following generic form:

$$\frac{Du^\gamma}{D\tau} \equiv \frac{du^\gamma}{d\tau} + \overline{\Gamma}^\gamma_{\mu\nu} u^\mu u^\nu = 0 \qquad (5.113)$$

For our particular case, we obtain:

$$\begin{cases} \dfrac{du^0}{d\tau} + \dfrac{2m}{r(r-2m)} u^0 u^1 = 0 \\[2mm] \dfrac{du^1}{d\tau} + \dfrac{m}{r^3}(r-2m)(u^0)^2 - \dfrac{m}{r(r-2m)}(u^1)^2 \\[2mm] \quad -(r-2m)(u^2)^2 - (r-2m)\sin^2\theta(u^3)^2 = 0 \\[2mm] \dfrac{du^2}{d\tau} + \dfrac{2}{r} u^1 u^2 - \sin\theta\cos\theta(u^3)^2 = 0 \\[2mm] \dfrac{du^3}{d\tau} + \dfrac{2}{r} u^1 u^3 + 2\dfrac{\cos\theta}{\sin\theta} u^2 u^3 = 0 \end{cases} \qquad (5.114)$$

From the Schwarzschild metric (4.33), we also write the line element divided by $d\tau$:

$$\left(\frac{ds}{d\tau}\right)^2 = \left(1 - \frac{2m}{r}\right)(u^0)^2 - \frac{1}{1-2m/r}(u^1)^2 - r^2(u^2)^2 - r^2\sin^2\theta(u^3)^2 \qquad (5.115)$$

where $u^\mu := dx^\mu/d\tau$ denotes the four-velocity ($x^0 := ct$). When we choose the time coordinate as $x^0 := t$, the timelike line element is equal to $ds^2 := c^2 d\tau^2$, whereas if we adopt $x^0 := ct$, then we have $ds^2 = d\tau^2$. The four-velocity is normalized, then the term $ds/d\tau = -1, 0, +1$ takes one of these three constants. Lightlike geodesics $ds/d\tau$ corresponds to massless particles. First, we consider radial geodesic spacetime curves, we can limit to the two first rows and consider $u^2 \equiv 0$, and $u^3 \equiv 0$ which are compatible. Owing that $u^1 := dr/d\tau$, we can integrate the first equation of (5.114) (related to the conservation of energy):

$$\frac{du^0}{u^0} + \frac{2m}{r(r-2m)} dr = 0 \quad \Longrightarrow \quad u^0\left(1 - \frac{2m}{r}\right) = E$$

If there is no initial motion in the spatial directions, the line element (5.115) gives initial four-velocity in the time direction:

$$u^0(0) := \frac{dx^0}{d\tau} = \frac{cdt}{d\tau} = \frac{1}{\sqrt{1 - 2m/r}}, \qquad E = \sqrt{1 - \frac{2m}{r_0}}$$

where r_0 is the initial radial position. For determining the radial component for timelike orbits $ds = d\tau$, we now introduce u^0 into (5.115) to find u^1:

$$1 \equiv \left(\frac{ds}{d\tau}\right)^2 = \left(1 - \frac{2m}{r}\right)(u^0)^2 - \frac{1}{1 - 2m/r}(u^1)^2 \quad \Longrightarrow \quad u^1 = \sqrt{\frac{2m}{r} - \frac{2m}{r_0}}$$

For azimuthal geodesic orbits, we come back to the general Eq. (5.114). The spherical symmetry allows us to consider orbits which remain in a plane by choosing $\theta \equiv \pi/2$, and $u^2(0) = 0$ and so that u^2 remains zero. The previous system (5.114) reduces accordingly. The first equation leads to the same form as previous radial geodesics, and the fourth equation gives u^3 (remind that $u^1 \equiv dr/d\tau$): $u^0 (1 - 2m/r) = E$, and $u^3 = L/r^2$. The solution in u^3 may be rewritten as follows: $r^2 u^3 := r^2 d\varphi/d\tau = L$ which states the conservation on angular momentum, where L is called specific angular momentum. We can compute the component u^1 from the line element by constraining the search to timelike orbits or the null geodesics. To begin with from relation (4.33), we write the line element (5.115) with $\theta \equiv \pi/2$:

$$\left(\frac{ds}{d\tau}\right)^2 = \left(1 - \frac{2m}{r}\right)(u^0)^2 - \left(1 - \frac{2m}{r}\right)^{-1}(u^1)^2 - r^2(u^3)^2$$

For timelike orbits $ds = d\tau$ (timelike geodesics describe the motions of massive particles at subliminal speed) we deduce:

$$u^1 := \frac{dr}{d\tau} = \sqrt{E^2 - \left(1 + \frac{L^2}{r^2}\right)\left(1 - \frac{2m}{r}\right)}$$

For timelike geodesics, this equation can be integrated directly to give elliptic integral. Substituting the expression of u^1 leads to the second derivative of r:

$$u^1 := \frac{dr}{d\tau} \quad \Longrightarrow \quad \frac{d^2r}{d\tau^2} + \frac{m}{r^2} + \frac{L^2}{r^3}\left(\frac{3m}{r} - 1\right) = 0$$

Finally, for circular orbits $r \equiv R$ (meaning that $dr/d\tau = 0$ and $d^2r/d\tau^2 = 0$) we obtain the well-known radius in terms of angular momentum L and Schwarzschild radius from the vanishing of the second derivatives:

$$R = \frac{L^2}{2m}\left(1 + \sqrt{1 - \frac{12m^2}{L^2}}\right), \qquad R = \frac{L^2}{2m}\left(1 - \sqrt{1 - \frac{12m^2}{L^2}}\right) \qquad (5.116)$$

The other solution would be $R = \infty$. Coming back to the equation of geodesic deviation (5.108), we can analyze the deviation acceleration for the circular orbit of radius R (5.116) with the four-velocity:

$$\left\{ u^0 = E \left(1 - \frac{2m}{R} \right)^{-1}, u^1 = \sqrt{E^2 - \left(1 + \frac{L^2}{R^2} \right) \left(1 - \frac{2m}{R} \right)}, u^2 \equiv 0, u^3 = \frac{L}{R^2} \right\}$$

$$(5.117)$$

Solutions of deviation of geodesics in a Schwarzschild spacetime may be found in e.g. Fuchs (1990), Philipp et al. (2015).

Remark 5.14 From Schwarzschild metric (4.33), another way would be considering the Lagrangian: $\mathcal{L} = m^* \left[(1 - 2m/r)(\dot{x}^0)^2 - (1 - 2m/r)^{-1} \dot{r}^2 - r^2 (\dot{\theta}^2 + \sin^2 \dot{\varphi}^2) \right]$. Since no explicit dependence on x^0 and on φ occurs, then the derivative of the Lagrangian with respect to $\dot{x}^0$ and $\dot{\varphi}$ are constant of motions:

$$\left\{ \begin{aligned} m^* \frac{dx^0}{d\tau} \left(1 - \frac{2m}{r} \right) &= E \\ m^* \frac{d\varphi}{d\tau} r^2 \sin^2 \theta &= L \end{aligned} \right. \implies \left\{ \begin{aligned} u^0 &= \frac{E}{m^*} \left(1 - \frac{2m}{r} \right)^{-1} \\ u^\varphi &= \frac{L}{m^*} \frac{1}{r^2 \sin^2 \theta} \end{aligned} \right. \quad (5.118)$$

where E is the energy, and L the analogous of angular momentum in relativistic gravitation. These equations conform to the previous solutions.

Remark 5.15 The radial solution can be re-written as follows:

$$\frac{1}{2} E^2 = \frac{1}{2} \left(\frac{dr}{d\tau} \right)^2 + V(r), \qquad V(r) := \frac{1}{2} - \frac{m}{r} + \frac{L^2}{2r^2} - \frac{mL^2}{r^3} \quad (5.119)$$

where the two first terms are respectively the Newtonian potential and the contribution from angular momentum (same for Newtonian and relativistic gravitation). The third term is the proposed contribution of general relativity. The above equation describes a analogous motion of a particle (unit mass) moving in a one dimensional r with potential $V(r)$. At large distance from the center, Newtonian mechanics matches the relativistic gravitational mechanics.

5.4.3 Autoparallel Deviation in Riemann–Cartan Spacetime

The idea is now to detect the relativistic acceleration of two nearby particles when the spacetime is curved with torsion (Shapiro 2002). First of all, it is worth to remind that in a Riemann–Cartan manifold the deviation from an autoparallel curve is obtained from the definition of the deviation:

$$\frac{D^2\xi^\alpha}{D\tau^2} = u^\gamma \nabla_\gamma \left(u^\beta \nabla_\beta \xi^\alpha \right) \tag{5.120}$$

where the connection have torsion and curvature. In a previous work, Manoff (2001b) proposed a deviation equation by defining a priori deviation operator of the affine connection in the presence of torsion and curvature $\mathcal{L}\Gamma(\xi, \mathbf{u}) := [\mathcal{L}_x i, \nabla_{\mathbf{u}}] - \nabla_{[\xi,\mathbf{u}]}$ to extend the deviation equation to Riemann–Cartan spacetime. In the following, this intrinsic definition can be also used to obtain with a straightforward calculus the result (5.108) on a Riemann manifold. Let now extend to Riemann–Cartan spacetime.

Theorem 5.2 *Let* $(\mathcal{M}, g_{\alpha\beta}, \Gamma^\gamma_{\alpha\beta})$ *a Riemann–Cartan spacetime with* $u^\beta :=$ $dx^\beta/d\tau$, *four-vector velocity (timelike vector), and* ξ^α *the separation between two autoparallel curves* γ_0, *and* γ_1. *We assume the Lie derivative of* u^α *along* ξ *vanishes (as for classical assumption in general relativity)* $\mathcal{L}_\xi u^\alpha \equiv 0$. *Then the acceleration of the separation between* γ_0, *and* γ_1 *takes the form of:*

$$\frac{D^2\xi^\alpha}{D\tau^2} = \tilde{\Re}^\alpha_{\gamma\beta\rho} u^\gamma \xi^\beta u^\rho + \left(\aleph^\gamma_{\beta\gamma} \tilde{\nabla}_\rho u^\alpha - \aleph^\alpha_{\gamma\rho} \tilde{\nabla}_\beta u^\rho \right) u^\gamma \xi^\beta \tag{5.121}$$

where the connection $\tilde{\Gamma}^\gamma_{\alpha\beta} := \Gamma^\gamma_{\alpha\beta} - \aleph^\gamma_{\alpha\beta}$ *is defined from the connection* $\Gamma^\gamma_{\alpha\beta}$ *by substracting its proper torsion.*

Proof First, we have to express the two main assumptions of the theorem. First the vanishing of the Lie derivative in a Riemann–Cartan manifold writes:

$$\mathcal{L}_\xi u^\alpha \equiv 0 \implies \xi^\beta \nabla_\beta = \xi^\beta \nabla_\beta u^\alpha - \xi^\beta \aleph^\alpha_{\beta\mu} u^\mu \tag{5.122}$$

Second, the equation of the autoparallel curves holds (for any index α):

$$u^\gamma \nabla_\gamma u^\alpha = 0 \tag{5.123}$$

Introducing these two equations into the definition of the autoparallel deviation gives:

$$\frac{D^2\xi^\alpha}{D\tau^2} = \xi^\gamma \nabla_\gamma u^\beta \left(\nabla_\beta u^\alpha - \aleph^\alpha_{\beta\mu} u^\mu \right)$$
$$+ \xi^\gamma \aleph^\beta_{\gamma\rho} u^\rho \left(\aleph^\alpha_{\beta\mu} u^\mu - \nabla_\beta u^\alpha \right)$$
$$+ u^\gamma \xi^\beta \nabla_\gamma \nabla_\beta u^\alpha - u^\gamma \xi^\beta \nabla_\gamma \left(\aleph^\alpha_{\beta\mu} u^\mu \right)$$

Factorizing the covariant derivative allows us to rewrite the equation as:

$$\frac{D^2 \xi^\alpha}{D\tau^2} = \xi^\gamma \nabla_\gamma u^\beta \left(\nabla_\beta u^\alpha - \aleph^\alpha_{\beta\mu} u^\mu \right)$$

$$+ \xi^\gamma \aleph^\beta_{\gamma\rho} u^\rho \left(\aleph^\alpha_{\beta\mu} u^\mu - \nabla_\beta u^\alpha \right) + u^\gamma \xi^\beta \nabla_\gamma \left(\nabla_\beta u^\alpha - \aleph^\alpha_{\beta\mu} u^\mu \right)$$

Let now define a new connection defined by $\tilde{\nabla} := \nabla - \aleph$, which is of course a connection on the manifold $\mathcal{M}$. Since we have assumed a metric compatible connection, we can write the coefficients:

$$\tilde{\Gamma}^\gamma_{\alpha\beta} := \overline{\Gamma}^\gamma_{\alpha\beta} + \mathfrak{I}^\gamma_{\alpha\beta} - \aleph^\gamma_{\alpha\beta} = \overline{\Gamma}^\gamma_{\alpha\beta} + D^\gamma_{\alpha\beta} - \Omega^\gamma_{\alpha\beta}$$

The skew-symmetric part of the new connection is merely $\aleph^\gamma_{\alpha\beta} = -\aleph^\gamma_{\alpha\beta}$. We rewrite the separation acceleration as:

$$\frac{D^2 \xi^\alpha}{D\tau^2} = \xi^\gamma \tilde{\nabla}_\gamma u^\beta \tilde{\nabla}_\beta u^\alpha + u^\gamma \xi^\beta \tilde{\nabla}_\gamma \tilde{\nabla}_\beta u^\alpha - \tilde{\aleph}^\alpha_{\gamma\rho} \left(\tilde{\nabla}_\beta u^\rho \right) u^\gamma \xi^\beta \qquad (5.124)$$

Let remind the following relation about second-order derivatives for either ∇ or $\tilde{\nabla}$:

$$\nabla_\gamma \nabla_\beta u^\alpha - \nabla_\beta \nabla_\gamma u^\alpha = \mathfrak{R}^\alpha_{\gamma\beta\rho} u^\rho + \aleph^\mu_{\beta\mu} \nabla_\mu u^\alpha \qquad (5.125)$$

It is therefore straightforward to deduce the relation, by exploiting Leibniz relation for derivative of products and by remarking the autoparallel equation:

$$\frac{D^2 \xi^\alpha}{D\tau^2} = \tilde{\mathfrak{R}}^\alpha_{\gamma\beta\rho} u^\gamma \xi^\beta u^\rho + \left(\tilde{\aleph}^\rho_{\beta\gamma} \tilde{\nabla}_\rho u^\alpha - \tilde{\aleph}^\alpha_{\gamma\rho} \tilde{\nabla}_\beta u^\rho \right) u^\gamma \xi^\beta$$

$$\square$$

For calculating the four-vector u^γ, we remind the definition of auto-parallel curves. Autoparallel curves are piecewise differential curves such that their tangent vectors are parallel along the curves itself. In other words, autoparallel curves are the integral curves of the differential equations e.g. Kleinert (2008):

$$\frac{Du^\gamma}{D\tau} := \frac{du^\gamma}{d\tau} + \Gamma^\gamma_{\mu\nu} u^\mu u^\nu = 0 \qquad (5.126)$$

It should be stressed that the skew-symmetric part of the connection is not involved in this Eq. (5.126). For metric compatible connection, the system becomes after Eq. (4.64):

$$\frac{Du^\gamma}{D\tau} := \frac{du^\gamma}{d\tau} + \overline{\Gamma}^\gamma_{\mu\nu} u^\mu u^\nu + D^\gamma_{\mu\nu} u^\mu u^\nu = 0 \qquad (5.127)$$

where $D^\gamma_{\mu\nu}$ is the symmetric part of the contortion tensor $\mathcal{T}^\gamma_{\mu\nu}$ with respect to the two lower indices. In principle, this relationship involves 40 connection coefficients $\overline{\Gamma}^\gamma_{\mu\nu}$ and other 40 components for $D^\gamma_{\mu\nu}$. Symmetries may reduce the number of components. As illustration, within Einstein–Cartan spacetime with a spherical symmetry where the metric is of Schwarzschild type, the simplest case where the spins of individual particles, or fluid elements composing the continuum are all aligned in the radial direction, only the component $\aleph^0_{23} = -\aleph^0_{32} := \aleph_0$ is not equal to zero e.g. Prasanna (1975b).

Other approaches exist for investigating the influence of Riemann–Cartan geometry on gravitation. As extension of the Schwarzschild metric, non-Riemannian spacetime were investigated for modeling static vacuum spherical symmetric spacetime e.g. Maier (2014). Starting with an Einstein–Hilbert action conforming to classical relativistic gravitation, Maier introduces the covariant version of the contortion tensor as $\mathcal{T}_{\alpha\beta\gamma} := g_{\gamma\beta}\partial_\alpha\Phi - g_{\alpha\beta}\partial_\gamma\Phi$ where $\phi(r)$ is a scalar potential. Investigating and putting apart the source of torsion, he obtains for a spherical symmetry spacetime with torsion the diagonal metric:

$$g_{00} = e^{2\phi}\left(1 - \frac{2m}{r}e^\phi\right), \quad g_{11} = \frac{[1 - r\phi'(r)]}{1 - (2m/r)e^\phi}, \quad g_{22} = -r^2, \quad g_{33} = -r^2\sin^2\theta$$

$$(5.128)$$

where when $\phi(r) \to 0$ then the asymptotic behavior of the metric merges to that of Schwarzschild one, and then to Minkowski flat spacetime. In Maier (2014), the function $\phi(r) = \ln|1 + \alpha/r|$ was chosen.

5.4.3.1 Application to Einstein–Cartan Spacetime Autoparallels

For investigating the influence of the torsion on the deviation equation, let consider the Schwarzschild metric solution of the Einstein field equations surrounding a mass M. The length ds^2 is obtained from (4.33) in the system (t, r, θ, φ):

$$ds^2 = (1 - 2m/r)c^2dt^2 - (1 - 2m/r)^{-1}dr^2 - r^2d\theta^2 - r^2\sin^2\theta d\varphi^2$$

The metric may be written in a worth coordinate system by using a conformal transformation of the radial coordinate: $r := \rho(1 + (m/2\rho))^2$. This allows us to define the expression of the length ds^2 in the coordinate system $(t, \rho, \theta, \varphi)$ e.g. Ryder (2009):

$$ds^2 = \left(\frac{1 - m/2\rho}{1 + m/2\rho}\right)^2 c^2dt^2 - \left(1 + \frac{m}{2\rho}\right)^4\left[d\rho^2 + \rho^2d\theta^2 + \rho^2\sin^2\theta d\varphi^2\right]$$

$$(5.129)$$

On a Riemann–Cartan manifold $\mathcal{M}$ the tangent space of the manifold $T_x\mathcal{M}$ is spanned by the vector base $\{\mathbf{e}_\mu := \partial_\mu\}$, and the dual tangent space by the dual base $\{\mathbf{e}^\nu := dx^\nu\}$ (also called 1-form) when using a coordinate basis e.g. Nakahara (1996). Since $\mathcal{M}$ is endowed with the Schwarzschild metric, there is an alternative to express the metric as: $\mathbf{g} = g_{\mu\nu}\mathbf{e}^\mu \otimes \mathbf{e}^\nu := \eta_{\mu\nu}\theta^\mu \otimes \theta^\nu$, with $\eta_{\mu\nu} := \mathrm{diag}\,\{1, -1, -1, -1\}$ with $(x^0 := ct)$:

$$
\begin{cases}
\theta^0 = (1 - 2m/r)^{1/2}\, dx^0 \\
\theta^1 = (1 - 2m/r)^{-1/2}\, dr \\
\theta^2 = r d\theta \\
\theta^3 = r \sin\theta d\varphi
\end{cases}
\tag{5.130}
$$

which checked to give the Schwarzschild metric. $\{\theta^\mu, \mu = 0, 1, 2, 3\}$ is called non coordinate dual basis. When using the isotropic coordinates $(x^0, \rho, \theta, \varphi)$, the corresponding non coordinate dual basis holds from Eq. (5.129):

$$
\begin{cases}
\theta^0 = T\,(\rho)\, dx^0 \\
\theta^1 = S\,(\rho)\, d\rho \\
\theta^2 = S\,(\rho)\, \rho d\theta \\
\theta^3 = S\,(\rho)\, \rho \sin\theta d\varphi
\end{cases}
\quad \text{with} \quad
\begin{cases}
T(\rho) := \left(\dfrac{1 - m/2\rho}{1 + m/2\rho}\right) \\[2mm]
S(\rho) := (1 + m/2\rho)^2
\end{cases}
\tag{5.131}
$$

Of course both Eqs. (5.130) and (5.131) describe the Schwarzschild metric. We check that Eq. (5.131) has correct behavior asymptotically when $\rho \to \infty$. The use of non coordinate basis holds for Cartan structure equations e.g. Nakahara (1996):

$$
\begin{cases}
d\theta^\mu + \omega^\mu_\nu \wedge \theta^\nu = \aleph^\mu \\
d\omega^\mu_\nu + \omega^\mu_\gamma \wedge \omega^\gamma_\nu = \mathfrak{R}^\mu_\nu
\end{cases}
\tag{5.132}
$$

where $\omega^\mu_\nu := \Gamma^\mu_{\gamma\nu}\,\theta^\gamma$ is called the connection 1-form, $\aleph^\mu := 1/2\,\aleph^\mu_{\gamma\nu}\,\theta^\gamma \wedge \theta^\nu$ the torsion 2-form, and $\mathfrak{R}^\mu_\lambda := 1/2\,\mathfrak{R}^\mu_{\gamma\nu\lambda}\,\theta^\gamma \wedge \theta^\nu$ the curvature 2-form. Cartan structure equations allow one to define the Cartan torsion and curvature by introducing the connection coefficients $\Gamma^\mu_{\nu\gamma} := \overline{\Gamma}^\mu_{\nu\gamma} + \mathcal{T}^\mu_{\nu\gamma}$ where $\mathcal{T}^\mu_{\nu\gamma}$ are the coefficients of the contortion tensor. The symbols of Christoffel $\overline{\Gamma}^\gamma_{\alpha\beta}$ associated to the Schwarzschild metric in terms of isotropic coordinates (5.129) are obtained as, for $\gamma = x^0 = 0$,

$$
\begin{aligned}
&\overline{\Gamma}^0_{00} = 0, \quad &&\overline{\Gamma}^0_{10} = \ln' T, \quad &&\overline{\Gamma}^0_{20} = 0, \quad &&\overline{\Gamma}^0_{30} = 0, \\
&\overline{\Gamma}^0_{01} = \ln' T, \quad &&\overline{\Gamma}^0_{11} = 0, \quad &&\overline{\Gamma}^0_{21} = 0, \quad &&\overline{\Gamma}^0_{31} = 0, \\
&\overline{\Gamma}^0_{02} = 0, \quad &&\overline{\Gamma}^0_{12} = 0, \quad &&\overline{\Gamma}^0_{22} = 0, \quad &&\overline{\Gamma}^0_{32} = 0, \\
&\overline{\Gamma}^0_{03} = 0, \quad &&\overline{\Gamma}^0_{13} = 0, \quad &&\overline{\Gamma}^0_{23} = 0, \quad &&\overline{\Gamma}^0_{33} = 0.
\end{aligned}
$$

for $\gamma = \rho = 1$,

$$\overline{\Gamma}^1_{00} = -\frac{T'T}{S^2}, \; \overline{\Gamma}^1_{10} = 0, \qquad \overline{\Gamma}^1_{20} = 0, \qquad\qquad \overline{\Gamma}^1_{30} = 0,$$
$$\overline{\Gamma}^1_{01} = 0, \qquad \overline{\Gamma}^1_{11} = \ln' S, \; \overline{\Gamma}^1_{21} = 0, \qquad\quad \overline{\Gamma}^1_{31} = 0,$$
$$\overline{\Gamma}^1_{02} = 0, \qquad \overline{\Gamma}^1_{12} = 0, \qquad \overline{\Gamma}^1_{22} = \rho(1 + \rho \ln' S), \; \overline{\Gamma}^1_{32} = 0,$$
$$\overline{\Gamma}^1_{03} = 0, \qquad \overline{\Gamma}^1_{13} = 0, \qquad \overline{\Gamma}^1_{23} = 0, \qquad\qquad \overline{\Gamma}^1_{33} = \rho \sin^2 \theta (1 + \ln' S).$$

for $\gamma = \theta = 2$,

$$\overline{\Gamma}^2_{00} = 0, \; \overline{\Gamma}^2_{10} = 0, \qquad \overline{\Gamma}^2_{20} = 0, \qquad \overline{\Gamma}^2_{30} = 0,$$
$$\overline{\Gamma}^2_{01} = 0, \; \overline{\Gamma}^2_{11} = 0, \qquad \overline{\Gamma}^2_{21} = \ln'(\rho S), \; \overline{\Gamma}^2_{31} = 0,$$
$$\overline{\Gamma}^2_{02} = 0, \; \overline{\Gamma}^2_{12} = \ln'(\rho S), \; \overline{\Gamma}^2_{22} = 0, \qquad \overline{\Gamma}^2_{32} = 0,$$
$$\overline{\Gamma}^2_{03} = 0, \; \overline{\Gamma}^2_{13} = 0, \qquad \overline{\Gamma}^2_{23} = 0, \qquad \overline{\Gamma}^2_{33} = -\sin\theta \cos\theta.$$

and for $\gamma = \varphi = 3$,

$$\overline{\Gamma}^3_{00} = 0, \; \overline{\Gamma}^3_{10} = 0, \qquad \overline{\Gamma}^3_{20} = 0, \quad \overline{\Gamma}^3_{30} = 0,$$
$$\overline{\Gamma}^3_{01} = 0, \; \overline{\Gamma}^3_{11} = 0, \qquad \overline{\Gamma}^3_{21} = 0, \quad \overline{\Gamma}^3_{31} = \ln'(\rho S),$$
$$\overline{\Gamma}^3_{02} = 0, \; \overline{\Gamma}^3_{12} = 0, \qquad \overline{\Gamma}^3_{22} = 0, \quad \overline{\Gamma}^3_{32} = \frac{\cos\theta}{\sin\theta},$$
$$\overline{\Gamma}^3_{03} = 0, \; \overline{\Gamma}^3_{13} = \ln'(\rho S), \; \overline{\Gamma}^3_{23} = \frac{\cos\theta}{\sin\theta}, \; \overline{\Gamma}^3_{33} = 0.$$

In addition to the symbols of Christoffel, the Cartan connection is obtained by adding the contortion tensor (calculated by means of torsion $\aleph^\gamma_{\alpha\beta}$). Mao et al. have derived the most general static and symmetric spherically expressions of torsion, and also the case where the metric and the torsion are generated by a source of rotation (Mao et al. 2007). Focusing only on the case where there is no source of torsion by mass rotation, the additional terms for the previous Levi-Civita connection are obtained by assuming the time translation invariance (nonzero components of torsion have either zero or two temporal indices), the torsion is skew-symmetric in its two covariant indices, and the symmetry under proper rotation, the most general expressions of non vanishing torsion components are e.g. Mao et al. (2007):

$$\aleph^0_{01} = \mathscr{T}_1(\rho)\frac{M}{\rho^2}, \quad \aleph^2_{12} = \mathscr{T}_1(\rho)\frac{M}{\rho^2}, \quad \aleph^3_{13} = \mathscr{T}_2(\rho)\frac{M}{\rho^2}, \tag{5.133}$$

where M is a constant related to a mass, and $\mathscr{T}_1$, and $\mathscr{T}_2$ are arbitrary functions of the radius ρ (Acedo 2015; Mao et al. 2007) (in this illustrative paragraph, they can be set constants for the sake of the simplicity). All other components not related by the skew symmetry are equal to zero. In such a case, the metric is not modified by these torsion components, but the connection coefficients are now changed by the presence of torsion. We discard the influence of any rotating bodies in the remaining of the paragraph. We deduce the additive terms for the contortion tensor. Only few components of the contortion tensor do not vanish:

$$
\begin{cases}
\mathcal{T}^0_{01} = \mathcal{F}_1(\rho)\dfrac{M}{\rho^2} \\[2mm]
\mathcal{T}^1_{00} = \dfrac{S^2(\rho)}{T^2(\rho)}\mathcal{F}_1(\rho)\dfrac{M}{\rho^2}, \quad \mathcal{T}^1_{22} = \dfrac{1}{\rho^2}\mathcal{F}_1(\rho)\dfrac{M}{\rho^2}, \quad \mathcal{T}^1_{33} = \rho^2\sin^2\theta\,\mathcal{F}_2(\rho)\dfrac{M}{\rho^2} \\[2mm]
\mathcal{T}^3_{31} = -\mathcal{F}_2(\rho)\dfrac{M}{\rho^2}
\end{cases}
$$

$$(5.134)$$

All other components are equal to zero.

Remark 5.16 However, it should be stressed that the metric compatibility of the connection is a key assumption since the relation (4.81) has highlighted that the torsion of the spacetime is deduced with an algebraic equation from the covariant derivative of the connection. Therefore, this equation explicitly states that for an a priori compatible connection, the torsion vanishes accordingly.

5.4.3.2   Spinning Particle and Gravitational Waves

In the previous paragraph we deal with the spinless particle immersed within gravitational waves. The theoretical bases in the modelling of spinning on particle motion within gravitation field were obtained by Mathisson e.g. Mathisson (1937), and Papapetrou e.g. Papapetrou (1951), leading to the celebrated Mathisson-Papapetrou equations e.g. Leclerc (2005), Nieto et al. (2007) (the name of equation was set *relativistic top deviation equation* in this later reference). Without going into details, the Mathisson-Papapetrou equations hold:

$$
\begin{cases}
\dfrac{D^2 x^\gamma}{D\tau^2} = \frac{1}{2}\mathfrak{R}^\gamma_{\alpha\beta\lambda}u^\lambda\Omega^{\alpha\beta} \\[3mm]
\dfrac{D\Omega^{\alpha\beta}}{D\tau} = 0
\end{cases}
\tag{5.135}
$$

where $\Omega^{\alpha\beta}$ is the (skew-symmetric) spin tensor of the top which satisfies the Pirani condition of orthogonality $\Omega^{\alpha\beta}u_\beta \equiv 0$. Classical Mathisson-Papapetrou equations may be obtained by means of Lagrangian of extended body:

$$
\mathcal{L} = \frac{m}{2}g_{\alpha\beta}u^\alpha u^\beta - \frac{1}{2}\Gamma^\gamma_{\alpha\beta}g_{\gamma\mu}\Omega^{\alpha\mu}u^\beta
\tag{5.136}
$$

which constitutes the first order expansion of the Riemann gravitation particle $\mathcal{L} := (m/2)g_{\alpha\beta}u^\alpha u^\beta$ around a center of mass of a small extended body (see Leclerc (2005) for details). It should be pointed out that this Lagrangian is not covariant (Christoffel's symbols are not tensors) whereas the associated Euler–Lagrange equations (5.135) are covariant.

Fig. 5.8 When a particle $\mathbf{x} := (x^{\mu})$ has internal structure, in other words is considered as a more or less extended body with its neighborhood $d\mathbf{x}$, it would no longer move along geodesics of the gravitational field. Spinning of the particle modifies its worldline e.g. Nieto et al. (2007)

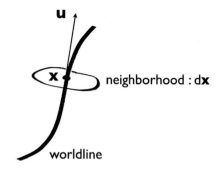

This paragraph is merely a remind of the method that considers a small piece of body characterized by a center of mass and an inertia of rotation (pole-dipole model). Theoretical basis for considering the geodesic of such small extended body is mathematically supported by theorem of Ehlers and Geroch (2004) (see Fig. 5.8). Theoretical basis for analyzing the spinning influence is mainly resumed in the Mathisson-Papapetrou-Dixon equations. Such a theory lies upon multipole method which basically assumes a small rigid neighborhood, as for rigid section in structural mechanics—beams, plates or shells in engineering mechanics e.g. Rakotomanana (2009). In the scope of general relativistic gravitation, the element is called gravitational skeleton. Another point of view would be to consider a Taylor expansion of the metric tensor $g_{\alpha\beta}(x^{\mu} + dx^{\mu})$ in the neighborhood $d\mathbf{x}$ leading to gradient continuum ($\mathscr{B}, g_{\alpha\beta}, \Gamma^{\gamma}_{\alpha\beta}$) or to metric affine spacetime ($\mathscr{M}, \hat{g}_{\alpha\beta}, \hat{\Gamma}^{\gamma}_{\alpha\beta}$), for which "local rigidity" is not assumed. Further extension of the method of multipoles in Riemann–Cartan space may be found in e.g. Leclerc (2005), Mathisson (1937), Papapetrou (1951). However in the present work, we only limit to the influence of spacetime with torsion on the pole particle motion.

Remark 5.17 Further utilization of the theorem with Eq. (5.121) would be the linearization of Eq. (4.159) in terms of metric and contortion tensors to obtain the theoretical equations of extended gravitational waves in presence on torsion of the spacetime. This out of the scope of the present paper.

5.4.3.3 Summary

Geodesic deviation is present for any gravitational theory. We can sketch the analogy between Newton, Einstein, and Einstein–Cartan gravitation in the table below. In the following table we resume the different expressions of the geodesic deviation, t is the usual time—parameter—in classical mechanics whereas τ is the proper time of relativistic theory (Table 5.1).

These three formulae express how the spacetime curvature and torsion influence two nearby geodesic or autoparallel curves, making them converge to or diverge from each other. The right-hand side terms may be considered as tidal forces. The

Table 5.1 Expression of the geodesic and autoparallel deviation equation for Newton (N), Einstein (E), and Einstein–Cartan (EC) theories

Theory	Potential	Geodesic deviation
N	Φ	$\dfrac{d^2\xi^a}{dt^2} = -\partial^a\partial_b\Phi\,\xi^a$
E	$g_{\alpha\beta}$	$\dfrac{D^2\xi^\alpha}{D\tau^2} = -\overline{\mathfrak{R}}^\alpha_{\beta\gamma\rho}u^\gamma\xi^\beta u^\rho$
EC	$g_{\alpha\beta},\ \Gamma^\gamma_{\alpha\beta}$	$\dfrac{D^2\xi^\alpha}{D\tau^2} = -\tilde{\mathfrak{R}}^\alpha_{\beta\gamma\rho}u^\gamma\xi^\beta u^\rho +$ $\left(\tilde{\aleph}^\rho_{\beta\gamma}\tilde{\nabla}_\rho u^\alpha - \tilde{\aleph}^\alpha_{\gamma\rho}\tilde{\nabla}_\beta u^\rho\right)u^\gamma\xi^\beta$

analogies between the tidal forces resulting from the previous three theories appear when we define the following quantities[4]:

$$\mathscr{K}^a_{Nb} := \partial^a\partial_b\Phi,$$

$$\mathscr{K}^\alpha_{E\beta} := \overline{\mathfrak{R}}^\alpha_{\beta\gamma\rho}u^\gamma u^\rho,$$

$$\mathscr{K}^\alpha_{EC\beta} := \tilde{\mathfrak{R}}^\alpha_{\beta\gamma\rho}u^\gamma u^\rho - \left(\tilde{\aleph}^\rho_{\beta\gamma}\tilde{\nabla}_\rho u^\alpha - \tilde{\aleph}^\alpha_{\gamma\rho}\tilde{\nabla}_\beta u^\rho\right)u^\gamma$$

showing that the acceleration of the geodesic deviation takes the form of: $\mathscr{K}^\alpha_\beta\xi^\beta$. For Newtonian gravitation, the tidal forces do not depend on the velocity u^α conversely to Einstein and to Einstein–Cartan gravitation. For each velocity **u** the right hand side of the geodesic deviation equation defines at each point $P \in \mathscr{M}$ a linear map $\xi^\beta \to \mathscr{K}^\alpha_\beta\,\xi^\beta$ of the subspace of $T_P\mathscr{M}$ perpendicular to **u**, and such that the Lie derivative vanishes $[\mathbf{u}, \xi] = 0$.

Remark 5.18 As a final remark on the deviation equations (5.100), (5.108), and (5.121), the vector ξ^α may be physically interpreted as the vector separation of two moving objects (ideally two mass points) near each other, and vector u^γ represents their initial motions. The second term is linear with respect to the separation vector constitutes the influence of the spacetime geometry on this separation acceleration. For Einstein–Cartan spacetime, we again observe and stress that a non curved spacetime with torsion may induce a separation acceleration between the two moving objects.

---

[4]The geodesic deviation equation is also called the Jacobi equation in the framework of differential geometry.

Chapter 6
Topics in Gravitation
and Electromagnetism

6.1 Introduction

First, we remind that derivation of continuum physics equations, namely the formulation of constitutive laws and conservation laws with respect to a given spacetime requires the identification of physical measurable quantities with geometrical variables (metric, torsion, and curvature on the material manifold). It is mandatory that the generation and the evolution of the spacetime and the continuum geometry, both of them are dynamical manifolds with their proper metric, torsion, and curvature, should be specified by means of physical objects, namely material particle, material elements as line, surface, volume, defects, and so on. Again, the tool for deriving constitutive laws and conservation laws from a Lagrangian density lies on the concept of variation, namely the Lagrangian variation and the Eulerian variation (Poincaré invariance).

From the group invariance point of view, the importance of the relation between spacetime geometry and the electromagnetic wave propagation was a starting point to conceive the Minkowski spacetime of special relativity to render compatible the mechanics and the classical electrodynamics. Indeed, classical mechanics is Galilean invariant, special relativity is Lorentz invariant, and classical electromagnetism is also Lorentz invariant. We investigate in this chapter the link between the electromagnetism and the continuum geometry, namely the spacetime geometry. As a slight extension, electromagnetic waves in curved spacetime such as in the framework of relativistic gravitation is important in some extreme situations of gravitation as in astrophysics. For that purpose, it is worth to remind that there are four fundamental forces in physics theory: gravitational, electromagnetic, strong nuclear, and weak nuclear. Gravitation is a weak force whereas electromagnetic force is a strong one. An usual question would whether gravitation affects electromagnetic fields and what would be the level of this interaction if positive. Conversely, what would be the influence of electromagnetism on the gravitation field. Electromagnetic waves, including light wave propagation, are described by Maxwell's equations

© Springer International Publishing AG, part of Springer Nature 2018   239
L. R. Rakotomanana, *Covariance and Gauge Invariance in Continuum Physics*,
Progress in Mathematical Physics 73, https://doi.org/10.1007/978-3-319-91782-5_6

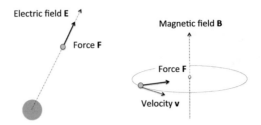

Fig. 6.1 (Left) An electric charge q immerged within an electric field **E** is subject to a force $\mathbf{F} := q\mathbf{E}$; (Right) A magnetic field **B** exerts on an electric charge q a force $\mathbf{F} := q\mathbf{v} \times \mathbf{B}$, where **v** is the velocity of the charge with respect to the magnetic field. The presence of **E** and **B** is in fact felt by their action on an electric charge

within Minkowskian, Riemannian or Riemann–Cartan spacetime. We consider in this chapter some elements of the theory of interaction between gravitation and electromagnetism (Fig. 6.1). The chosen illustrations are motivated by the analysis of interaction between the curvature, and the torsion of the spacetime with the electromagnetic waves.

6.2   Electromagnetism in Minkowskian Vacuum

Electromagnetism theory is built upon two fields: electric field **E**, and magnetic field **B**. Both of them depend on the space coordinate and the time in the general case. Experimental evidence of the two fields mainly lies on their action onto an electric charge. Electric charge q constitutes the basis of electromagnetism theory. Electric charge is a property of bodies at the same level as its mass. The fundamental assumption on the electric charge is that it is conserved. This gives the first equation of the four Maxwell's equations due to Gauss. The flux of the electric vector field, produced by the electric charge q, passing through a closed surface is proportional to the total electric charge contained within that surface. It should be stressed that it is the total charge (algebraic summation of all positive and negative charges), enclosed by the closed surface that is considered in Gauss' law. For the other aspect of electromagnetism, the motion of an electric charge, induces a magnetic flux which is also conserved across a closed surface according to the Gauss law for magnetism. The Gauss' law for magnetic field arises from the assumption that isolated magnetic poles do not exist. Then the magnetic flux passing through a closed surface is equal to zero.

Before going into the derivation of Maxwell's equation, it is worth to remind the notion of proper time in the framework of special relativity. Consider a body/or a reference frame moving with a uniform velocity v with respect to another reference frame $\mathcal{M}$. The proper time is given by (see Eq. (2.76)):

$$d\tau := \sqrt{1 - (v^2/c^2)}dt$$

where the proper time τ along a timelike world line in the spacetime $\mathcal{M}$ is the laps of time measured by a clock following that line. For the sake of the simplicity, the proper time is nothing more than the arc length in $\mathcal{M}$. Since the electromagnetic phenomenae are present in the nature with very high speed, conversely to motion of massive body, the electromagnetic conservation laws are derived using the proper time.

6.2.1 Maxwell's 3D Equations in Vacuum

The general form of the Maxwell's equations is intimately linked to the geometry of the spacetime vacuum defined by the Minkowski metric, they constitute the fundamental basis of classical electrodynamics.

6.2.1.1 General Equations

Vacuum Maxwell's equations are derived in the local coordinates of flat Minkowskian spacetime $\mathcal{M}$ endowed with the metric $\hat{g}_{\mu\nu} := \{+1, -1, -1, -1\}$. Coordinates of the spacetime are denoted $x^\mu := (x^0 = ct, x^1, x^2, x^3)$. Conservation laws are rigorously derived with the derivative with respect to the proper time which is a particular case of the objective derivative defined from the concept of integral invariance of Poincaré (Rakotomanana 2003). The derivative with respect to the proper time τ is then be calculated by applying the transitivity rule:

$$\frac{\partial}{\partial \tau} = \frac{\partial x^0}{\partial \tau} \partial_0 = c \frac{dt}{d\tau} \partial_0 \simeq c \, \partial_0 \tag{6.1}$$

when the velocity is small compared to the light speed $v << c$. Therefore, classical three-dimensional Maxwell's equations take the form of, where the connection of three-dimensional vacuum space is denoted $\hat{\nabla}$, e.g. Gelman (1966), Kleinert (2008), Kovetz (2000) (space is an Euclidean manifold):

$$\begin{cases} \hat{\nabla} \cdot \mathbf{D} = \rho \\ \hat{\nabla} \times \mathbf{H} - c \, \partial_0 \mathbf{D} = \mathbf{J} \\ \hat{\nabla} \cdot \mathbf{B} = 0 \\ \hat{\nabla} \times \mathbf{E} + c \, \partial_0 \mathbf{B} = 0 \end{cases} \tag{6.2}$$

in which **E** and **H** are the electric and magnetic field intensities, whereas **D** (displacement) and **B** (magnetic induction) are the electric and magnetic flux densities. ρ and **J** are the volume charge density and the electric current density respectively (they may be considered as sources of electromagnetic field). The fluxes **D** and **B** are related to the field intensities via the electromagnetic constitutive laws. Notice that the time derivative includes the term $c^2 := (\epsilon_0 \mu_0)^{-1}$, which is the speed of the light and ϵ_0 and μ_0 are respectively two positive, universal constants such that the constitutive laws for electromagnetism hold in the vacuum e.g. Kovetz (2000):

$$\mathbf{D} = \epsilon_0 \mathbf{E}, \qquad \mathbf{B} = \mu_0 \mathbf{H} \tag{6.3}$$

Constants ϵ_0 and μ_0 are called electric permittivity and magnetic permeability of the vacuum space, respectively. The permittivity of a medium determines its response to an applied electric field. The value of permittivity of a vacuum space is $\epsilon_0 \simeq 8.8541878176 \times 10^{-12}$[C/Vm]. The magnetic permeability is deduced from ϵ_0 and c. At first sight, as for the electric flux intensity **D**, we may question why not to use the magnetic flux intensity **B** as the variable to be introduced into the second Maxwell's equation (6.2). In the same way, we may ask why not to use the magnetic flux **D** into the last equations of (6.2). As such, the Maxwell's equations are the conservation laws, and then by introducing theses fluxes in terms of intensity fields **E** and **H** (constitutive laws), we arrive to the partial differential equations where the electric and magnetic intensities are the unknowns. However, we should notice that in reality electric and magnetic fields are experimentally detected by their action of electric charges.

6.2.1.2   Lorentz Force

The Lorentz force due to the electromagnetic field on a material with electric charge e is given by the summation of the two forces induced by electric field and magnetic field:

$$\mathbf{F}_{em} := e\,(\mathbf{E} + \mathbf{v} \times \mathbf{B}) \tag{6.4}$$

where **v** is the velocity of the material point with respect to a reference frame. In the framework of classical mechanics, a charge particle is a material point with which is associated a mass $m > 0$ and a electric charge e positive or negative. In an inertial frame, the (non relativistic) Newton's law governing the motion of such a particle is given by:

$$\frac{d(m\mathbf{v})}{dt} = e\,(\mathbf{E} + \mathbf{v} \times \mathbf{B}) \tag{6.5}$$

The exerted forces on charges involve the electric intensity **E** and the magnetic flux density **B** but not the magnetic intensity **H**. By the way, we should observe that both the magnetic intensity and the electric intensity are involved in the Maxwell's equations only by means of their rotational vector $\hat{\nabla}\times$.

For illustrating the application of the linear momentum equation on the motion of non-relativistic charged particle, let consider a uniform magnetic field $\mathbf{B} = B\,\mathbf{e}_3$ without electric field $\mathbf{E} \equiv 0$ (it is a drastic assumption because an electric field should be necessarily generated with the presence of a moving electric charge). The momentum equation (6.5) simplifies to:

$$\frac{d\mathbf{v}}{dt} = -\Omega\,\mathbf{e}_3 \times \mathbf{v}, \qquad \text{with} \qquad \Omega := \frac{eB}{m}$$

where the term Ω is called the gyration frequency. In component form the velocity of the particle is governed by the system of differential equations:

$$
\begin{cases}
\dot{v}_v^1 = \Omega\,v^2 \\
\dot{v}^2 = -\Omega\,v^1 \\
\dot{v}^3 = 0
\end{cases}
\implies
\begin{cases}
x^1 = x_0^1 + \dfrac{m v_\perp}{eB}\,\sin\left(\Omega t + \alpha\right) \\[2mm]
x^2 = x_0^2 + \dfrac{m v_\perp}{eB}\,\cos\left(\Omega t + \alpha\right) \\[2mm]
x^3 = x_0^3 + v_\parallel t
\end{cases}
\tag{6.6}
$$

where $v_\perp := \|\mathbf{v} - \mathbf{e}_3 \otimes \mathbf{e}_3(\mathbf{v})\|$ is the constant magnitude of the projection of the particle velocity in the plane Ox_1x_2, $v_\parallel = v_3$ (constant), and α is its initial phase. The non-relativistic trajectory of a charged particle under a uniform magnetic reduces to a helix with axis along the magnetic field $\mathbf{B}$ with a cycle $T := 2\pi/\Omega$.

6.2.1.3 Continuity Equation

The first and third equations are the electric and magnetic Gauss' laws. Only the net charge within the enclosed surface matters in the first Gauss' law. To date, the existence of monopole remains questionable. Thus the right hand side of the third Gauss' law is identically zero. The second equation is the Ampère's law (Maxwell introduced the rate of displacement), and the fourth equation the Faraday's law of induction. Constitutive laws (6.3) allows us to obtain the complete set of equations. From the two first equations of (6.2), we deduce the continuity equation:

$$\hat{\nabla} \cdot \mathbf{J} + c\,\partial_0\rho = 0 \tag{6.7}$$

It should be mentioned that for a neutral conductive body in which electrons are in motion but other ionized atoms are fixed, we deduce from Eq. (6.7): $\rho = 0$ and $\hat{\nabla} \cdot \mathbf{J} = 0$, meaning that ρ in fact defines the density of electric free charges. Conversely, if all charges are free of moving, with a velocity $\mathbf{v}$, we can write: $\mathbf{J} = \rho\,\mathbf{v}$ and $\hat{\nabla} \cdot (\rho\mathbf{v}) + c\partial_0\rho = 0$ which corresponds to the continuity equation in the framework of a hydrodynamic flow.

6.2.1.4   Potential Formulation

From Eq. (6.2), we can deduce alternative expression of the electric and magnetic
fields where $\mathbf{A}$ and ϕ are the usual vector and scalar potentials of electromagnetism:

$$\left\{ \begin{array}{l} \hat{\nabla} \cdot \mathbf{B} = 0 \\ \hat{\nabla} \times \mathbf{E} + c\, \partial_0 \mathbf{B} = 0 \end{array} \right. \implies \left\{ \begin{array}{l} \mathbf{B} = \hat{\nabla} \times \mathbf{A} \\ \hat{\nabla} \times (\mathbf{E} + c\, \partial_0 \mathbf{A}) = 0 \end{array} \right. \tag{6.8}$$

The electric and magnetic fields are then defined accordingly:

$$\mathbf{E} := -c\, \partial_0 \mathbf{A} - \hat{\nabla}\phi, \quad \text{and} \quad \mathbf{B} := \hat{\nabla} \times \mathbf{A} \tag{6.9}$$

For illustration, consider a spherical coordinate system (r, θ, φ) associated with the
local base $(\mathbf{f}_r, \mathbf{f}_\theta, \mathbf{f}_\varphi)$. For the sake of the simplicity, let use the normal base $(\mathbf{e}_r :=$
$\mathbf{f}_r, \mathbf{e}_\theta := \mathbf{f}_\theta / r, \mathbf{e}_\varphi := \mathbf{f}_\varphi (r \sin \theta))$. The relations (6.9) take the form of:

$$\mathbf{E} = - \begin{pmatrix} c\partial_0 A_r + \partial_r \Phi \\ c\partial_0 A_\theta + \dfrac{1}{r}\partial_\theta \Phi \\ c\partial_0 A_\varphi + \dfrac{1}{r \sin \theta}\partial_\varphi \Phi \end{pmatrix} \quad \mathbf{B} = \begin{pmatrix} \dfrac{1}{r \sin \theta}\left[\partial_\theta (A_\varphi \sin \theta) - \partial_\varphi A_\theta\right] \\ \dfrac{1}{r}\left[\dfrac{1}{\sin \theta}\partial_\varphi (A_r) - \partial_r (r A_\varphi)\right] \\ \dfrac{1}{r}\left[\partial_r (r A_\theta) - \partial_\theta A_r\right] \end{pmatrix}$$

A spherical symmetric electromagnetic field $\{\Phi(r, t), \mathbf{A}(r, t)\}$ then leads at most to
the non vanishing components (other components are equal to zero):

$$\left\{ \begin{array}{l} E_r = -c\partial_0 A_r - \partial_r \Phi \\ E_\theta = -c\partial_0 A_\theta \\ E_\varphi = -c\partial_0 A_\varphi \end{array} \right. , \qquad \left\{ \begin{array}{l} B_r = \dfrac{\cos \theta}{\sin \theta}\dfrac{A_\varphi}{r} \\ B_\theta = -\dfrac{A_\varphi}{r} - \partial_r A_\varphi \\ B_\varphi = \dfrac{A_\theta}{r} + \partial_r A_\theta \end{array} \right.$$

which show that both electric field and magnetic field might have formally
components along the θ and φ directions too. These expressions nevertheless show
that assumptions of radial fields such as $\mathbf{E}(r, t)$, and $\mathbf{B}(r, t)$ impose the following
conditions:

$$\partial_0 A_\theta = 0, \quad A_\varphi = 0, \quad \frac{A_\theta}{r} + \partial_r A_\theta = 0 \quad (A_\theta = \frac{C}{r})$$

Being the curl of a vector $\mathbf{A}$, the magnetic field $\mathbf{B}$ is thus a solenoidal vector. It is
then usual to define the four-vector potential:

$$A^\nu = (A^0, A^i) := (\phi, A^1, A^2, A^3), \quad A_\mu = \hat{g}_{\mu\nu} A^\nu = (A_0, A_i) \tag{6.10}$$
$$= (\phi, -A^1, -A^2, -A^3)$$

by using the metric of the flat Minkowskian spacetime $\mathcal{M}$. Instead of the electric intensity and the magnetic flux, the unknowns of the Maxwell's equations are now the four-potential vector (A_μ). As illustration, the simplest solution of The Maxwell's equation (6.2) is the electromagnetic field that corresponds to the field generated by a static electric charge Q, assumed to be at the origin of the space,

$$\mathbf{E} = \frac{Q}{4\pi\epsilon_0} \frac{\mathbf{r}}{r^3}, \qquad \mathbf{B} = 0$$

in which $\mathbf{r}$ is the radial vector form the origin of the electric charge Q (static in a flat space) to the point M, and ϵ_0 the electric permittivity of the vacuum flat space. This elementary solution corresponds to the four-potential:

$$A_\mu = \left(\frac{Q}{4\pi\epsilon_0} \frac{1}{r}, 0, 0, 0 \right)$$

Remark 6.1 The vector product does not present any particular difficulty in a three-dimensional Euclidean space $\mathscr{E}$. Further considerations should be done when working with a differentiable manifold. It is worth to remind the notion of ω_n-isomorphism e.g. Rakotomanana (2003) where the product of two vectors ($\hat{\nabla}$ and $\mathbf{A}$ respectively) is defined as a 2-form $\mathbf{B}$ whose the non zero components are such that:

$$\mathbf{B} := \hat{\nabla} \times \mathbf{A} = \epsilon_{ijk} \left(\partial^j A^k \right) \mathbf{e}^i, \qquad B_i = \epsilon_{ijk} \left(\partial^j A^k \right)$$

in which ϵ_{ijk} is the Civita alternating tensor.

$$\epsilon_{ijk} = \begin{cases} +1 \text{ if } ijk = \text{cyclic permutation of } 123 \\ -1 \text{ if } ijk = \text{anticyclic permutation of } 123 \\ 0 \text{ if } ijk = \text{other situations} \end{cases}$$

By using the metric tensor $\hat{g}_{\alpha\beta} := \text{Diag}\{+1, -1, -1, -1\}$, we observe that the sign of both x^j and A^k changes when the indices are lowered (covariant components). Then the place of the indices does matter in the general case and we practically obtain the components of the 2-form:

$$B^1 = \partial_2 A_3 - \partial_3 A_2, \quad B^2 = \partial_3 A_1 - \partial_1 A_3, \quad B^3 = \partial_1 A_2 - \partial_2 A_1 \qquad (6.11)$$

in the basis $\{dx^2 \wedge dx^3, dx^3 \wedge dx^1, dx^1 \wedge dx^2\}$. These relations may be summarized by the correspondence of magnetic variables:

$$\mathbf{B} := B_{jk} \, dx^j \wedge dx^k \qquad \text{and} \qquad B^i := \frac{1}{2}\epsilon^{ijk} B_{jk} \qquad (6.12)$$

and that of electric variables:

$$\mathbf{D} := D_{jk} \, dx^j \wedge dx^k \qquad \text{and} \qquad D^i := \frac{1}{2} \epsilon^{ijk} D_{jk} \qquad (6.13)$$

Relations (6.8) and (6.9) assess the name of $\mathbf{D}$ and $\mathbf{B}$ as "axial vector" (vector product).

Remark 6.2 The presence of vector product conforms to the 2-form nature of the electric displacement $\mathbf{D}$, and the magnetic induction $\mathbf{B}$. From physics insights, we report in Table 6.1 the tensorial nature of each variable. By using the form notation, we can write the electromagnetic strength or also Faraday tensor as a 2-form:

$$\mathcal{F} = E_1 dx^1 \wedge dx^0 + E_2 dx^2 \wedge dx^0 + E_3 dx^3 \wedge dx^0$$
$$+ B^1 dx^2 \wedge dx^3 + B^2 dx^3 \wedge dx^1 + B^3 dx^1 \wedge dx^2 \qquad (6.14)$$

from which we can easily check the correspondence of the components $\mathcal{F}_{\mu\nu}$ with the components of $\mathbf{E}$ and $\mathbf{B}$. In sum the Faraday tensor is built by combining the electric intensity field with the magnetic flux, where the 1-form $E_\mu dx^\mu$, $\mu = 0, 1, 2, 3, 4$ is adapted to fit as a 2-form $E_\mu dx^0 \wedge dx^\mu$, and the magnetic flux is calculated with only space indices.

Remark 6.3 The four-vector potential A_μ defined by (6.10) is not unique since it allows gauge transformations of the form $A_\mu + df$ in which $f(x^\mu)$ is any scalar function.

6.2.1.5 Electromagnetic Waves

For pointing out the existence of wave solutions of the Maxwell's equations, let us account for the relations (6.8) which relate the potential with the electromagnetic

Table 6.1 Electromagnetic variables and their tensor type

Variable	Vector notation	Form notation
Electric field intensity	$\mathbf{E}$	1-form : $E_i \, dx^i$
Magnetic field intensity	$\mathbf{H}$	1-form : $H_i \, dx^i$
Electric flux density	$\mathbf{D}$	2-form : $D_{ij} \, dx^i \wedge dx^j$
Magnetic flux density	$\mathbf{B}$	2-form : $B_{ij} \, dx^i \wedge dx^j$

The electric flux density is also called electric displacement. Due to the skew symmetry of 2-form, the electric and magnetic flux densities are short-handed denoted $\mathbf{D} = D^1 \, dx^2 \wedge dx^3 + D^2 \, dx^3 \wedge dx^1 + D^3 \, dx^1 \wedge dx^2$, and $\mathbf{B} = B^1 \, dx^2 \wedge dx^3 + B^2 \, dx^3 \wedge dx^1 + B^3 \, dx^1 \wedge dx^2$, respectively fields. The first pair of Maxwell's equations becomes:

$$\begin{cases} \epsilon_0 \hat{\nabla} \cdot \left(-c\partial_0 \mathbf{A} - \hat{\nabla}\phi \right) = 0 \\ \hat{\nabla} \times \left(\dfrac{1}{\mu_0} \hat{\nabla} \times \mathbf{A} \right) - c\epsilon_0 \partial_0 \left(-c\partial_0 \mathbf{A} - \hat{\nabla}\phi \right) = 0 \end{cases}$$

The potentials ϕ and $\mathbf{A}$ are still subject to gauge transformations. We worthily choose the Lorenz gauge condition:

$$\hat{\nabla} \cdot \mathbf{A} + \frac{1}{c}\partial_0 \phi \equiv 0 \qquad (6.15)$$

The Lorenz gauge (6.15) is due to the Danish physicist Ludvig V. Lorenz to be not confused with the Dutch physicist Hendrick A. Lorentz. Introducing the gauge condition (6.15) into the previous first pair of Maxwell's equations, and owing that $c^2 := (\epsilon_0\mu_0)^{-1}$, we obtain the following:

$$\begin{cases} -\epsilon_0 \left[c\partial_0 \left(-\dfrac{1}{c}\partial_0 \phi \right) + \hat{\nabla} \cdot \hat{\nabla}\phi \right] = 0 \\ \left[\hat{\nabla} \left(\hat{\nabla} \cdot \mathbf{A} \right) - \hat{\Delta}\mathbf{A} \right] + \partial_0\partial_0 \mathbf{A} + \dfrac{1}{c}\partial_0 \left(\hat{\nabla}\phi \right) = 0 \end{cases}$$

owing the vectorial relations $\hat{\nabla} \times (\hat{\nabla} \times \mathbf{A}) = \hat{\nabla}(\hat{\nabla} \cdot \mathbf{A}) - \hat{\Delta}\mathbf{A}$. We deduce the two equations of electromagnetic wave propagation within the vacuum (symbol $\hat{\Delta}$ denotes here the three-dimensional space Laplacian operator):

$$\begin{cases} \partial_0^2 \phi - \hat{\Delta}\phi = 0 \\ \partial_0^2 \mathbf{A} - \hat{\Delta}\mathbf{A} = 0 \end{cases} \longrightarrow \begin{cases} \partial_t^2 \phi - c^2\, \hat{\Delta}\phi = 0 \\ \partial_t^2 \mathbf{A} - c^2\, \hat{\Delta}\mathbf{A} = 0 \end{cases} \qquad (6.16)$$

which turn into the classical electromagnetic wave equations within vacuum and propagating with the same celerity c, speed of the light. For instance searching for functions of the type $\phi(t, \mathbf{x}) := \Phi(\mathbf{x})T(t)$ and $\mathbf{A}(t, \mathbf{x}) := \mathbf{A}(\mathbf{x})T(t)$ leads to the classical equations:

$$T(t) = A\sin(\omega t) + B\cos(\omega t), \quad \begin{cases} \hat{\Delta}\Phi(\mathbf{x}) - k^2\Phi(\mathbf{x}) = 0 \\ \hat{\Delta}\mathbf{A}(\mathbf{x}) - k^2\mathbf{A}(\mathbf{x}) = 0 \end{cases}, \quad k^2 := \omega^2/c^2$$

$$(6.17)$$

The last equation is the dispersion equation. Analogous solutions are obtained for the potential $\mathbf{A}$. We shortly review hereafter the four-dimensional formulation.

Remark 6.4 Electromagnetic wave propagation equations (6.16) have great interest in the sense that they are locally decoupled. It allows us to use classical methods of wave solutions in an arbitrary coordinate system for each of them, by accounting for boundary conditions.

6.2.1.6   Lorenz Gauge Invariance

The Maxwell's equations in terms of in terms ϕ and $\mathbf{A}$ do not uniquely determine the electric potential and magnetic potential. There is some flexibility in searching for the potential functions. If ϕ_0 and $\mathbf{A}_0$ are solutions, then so are the pair:

$$\phi = \phi_0 + c\partial_0 \Lambda, \qquad \mathbf{A} = \mathbf{A}_0 - \hat{\nabla}\Lambda \qquad (6.18)$$

by checking:

$$\mathbf{E} = -c\partial_0 \left(\mathbf{A}_0 - \hat{\nabla}\Lambda\right) - \hat{\nabla}\left(\phi_0 + c\partial_0 \Lambda\right), \qquad \mathbf{B} = \hat{\nabla} \times \left(\mathbf{A}_0 + \hat{\nabla}\Lambda\right) \qquad (6.19)$$

where $\Lambda(x^\mu)$ is an arbitrary scalar function assumed to be of class $\mathscr{C}^2$, which is called gauge-transformation function. However, it should be reminded that the gauge invariance is satisfied since we assumed:

$$-c\partial_0 \left(\hat{\nabla}\Lambda\right) - \hat{\nabla}\left(c\partial_0 \Lambda\right) = 0, \qquad \text{and} \qquad \hat{\nabla} \times \left(\hat{\nabla}\Lambda\right) = 0$$

Once a gauge (here Lorenz gauge) has been chosen, results obtained for potential have no longer the flexibility as before the choice. These are gauge which fixes the electromagnetic vector potential $\mathbf{A}$. Such is the case for Minkowski and Riemann spacetime, but maybe no longer for Riemann–Cartan spacetime in presence of non vanishing torsion.

6.2.1.7   Energy Conservation of Electromagnetic Waves

The propagation of electromagnetic waves involves evolution of the energy in the course of time. We remind in this paragraph the so-called Poynting's theorem. Let consider a continuum (spacetime or material continuum in the large) $\mathscr{B}$ where electromagnetic fields are defined by the four vectors $\mathbf{E}$, $\mathbf{D}$, $\mathbf{H}$ and $\mathbf{B}$ which verify the Maxwell's equation (6.2). The continuum has a electric resistive behavior in the sense that among the electromagnetic constitutive laws, it verifies Ohm's law stating that the current (motion of electric charge) through a conductor between two points is directly proportional to the electric potential across the two points. At a local point of view, where the material is assumed to be a continuum, Ohm's law of homogeneous and isotropic material holds (Kirchhoffs formulation):

$$\mathbf{J} = \gamma \mathbf{E} \qquad (6.20)$$

where $\mathbf{J}$ is the current density at a given location of the material, $\mathbf{E}$ is the electric field at that point, and γ is the electric conductivity.

Theorem 6.1 (Poynting's Theorem) *Let consider the electric fields* **E**, *and* **D**, *and the magnetic fields* **H**, *and* **B** *which verify the Maxwell's equations, on a resistive continuum* **B**. *Then:*

$$\nabla \cdot (\mathbf{E} \times \mathbf{H}) + \frac{\partial}{\partial t} \left[\frac{1}{2} \left(\mathbf{D} \cdot \mathbf{E} + \mathbf{B} \cdot \mathbf{H} \right) \right] + \mathbf{J} \cdot \mathbf{E} = 0 \tag{6.21}$$

For the terminology, each term is identified as:

1. $\mathbf{S} := \mathbf{E} \times \mathbf{H}$ is called Poynting's vector, quantifying the power leaving the material point (infinitesimal volume surrounding the point)
2. $W^e := \frac{1}{2} \left(\mathbf{D} \cdot \mathbf{E} + \mathbf{B} \cdot \mathbf{H} \right)$ is the electromagnetic energy density inside the infinitesimal volume,
3. $\mathbf{J} \cdot \mathbf{E}$ is the power lost to heat (Joule heating effect). The magnetic field can do no work on the charges.

The Poynting's theorem represents an energy conservation equation for the electromagnetic fields.

Proof Let start with by writing the Maxwell's equation (6.2) and multiplied by fields as follows:

$$\begin{cases} \mathbf{H} \cdot \nabla \times \mathbf{E} = -\mathbf{H} \cdot \partial_t \mathbf{B} \\ \mathbf{E} \cdot \nabla \times \mathbf{H} = \mathbf{E} \cdot \partial_t \mathbf{D} + \mathbf{J} \cdot \mathbf{E} \end{cases}$$

Owing the differential operators relationship:

$$\nabla \cdot (\mathbf{E} \times \mathbf{H}) = \mathbf{H} \cdot \nabla \times \mathbf{E} - \mathbf{E} \cdot \nabla \times \mathbf{H}$$

together with the constitutive laws of the continuum (6.3), we easily obtain:

$$\nabla \cdot (\mathbf{E} \times \mathbf{H}) + \frac{\partial}{\partial t} \left[\frac{1}{2} \left(\epsilon_0 \mathbf{E} \cdot \mathbf{E} + \mu_0 \mathbf{H} \cdot \mathbf{H} \right) \right] + \gamma \mathbf{E} \cdot \mathbf{E} = 0$$

and then the more general form of the Poynting's theorem for a continuum with electric permittivity ϵ, and magnetic susceptibility μ. □

Another formulation of the Poynting's theorem would be:

$$-\frac{\partial W^e}{\partial t} = \nabla \cdot \mathbf{S} + \mathbf{J} \cdot \mathbf{E} \tag{6.22}$$

and after integrating over a finite volume of the continuum $\mathscr{B}$:

$$-\frac{\partial}{\partial t} \int_{\mathscr{B}} W^e \, dv + \int_{\partial \mathscr{B}} \mathbf{S} \cdot \mathbf{n} \, da + \int_{\mathscr{B}} \mathbf{J} \cdot \mathbf{E} \, dv$$

The (opposite of the) rate of the electromagnetic energy W^e in a finite volume $\mathscr{B}$ is equal to the sum of the flux of the Poynting vector $\mathbf{S}$ on the boundary $\partial\mathscr{B}$ and the power lost into heat within $\mathscr{B}$.

6.2.1.8 Four-Dimensional Formulation of Electromagnetic Waves

Let remind the convention for the coordinate system $(x^0 := ct, x^1, x^2, x^3)$. The system of equations (6.16) may be written in a synthetic formulation:

$$\hat{g}^{\alpha\beta}\,\hat{\nabla}_\alpha\hat{\nabla}_\beta A^\mu = 0, \quad \mu = 0, 1, 2, 3, \quad \hat{g}^{\alpha\beta} = \mathrm{diag}\,\{+1, -1, -1, -1\} \quad (6.23)$$

where $\hat{g}^{\alpha\beta}$ is the metric of the Minkowskian spacetime $\mathscr{M}$, $\hat{\nabla}$ a four-dimensional connection, and $A^\mu := (\phi, A^1, A^2, A^3)$ the electromagnetic four-potential. This equation expresses all the Maxwell's equations in one go, in a Lorentz covariant form. It should be again stressed that Lorentz covariance means that it takes the same form in one reference as it does in another (the relative motion being a Lorentz transformation and particularly for boost (2.14)).

As for waves in elastic continuum, electromagnetic waves in vacuum spacetime has the simplest solution of the four-dimensional wave equation is obtained by assuming a plane wave as: $A^\mu = \Re e\left\{\hat{A}^\mu \exp\left(i\kappa_\alpha x^\alpha\right)\right\}$ where $\hat{A}^\mu$ is the wave four-amplitude, and κ_α is a null four-vector such that $\kappa_\alpha\,\kappa^\alpha := 0$. Again, the link with the three dimensional spatial solution and temporal solution is then given by the relationships:

$$\begin{cases} \dfrac{\omega}{c} := \sqrt{\kappa_0\kappa^0} = \sqrt{\kappa_i\kappa^i} \\[2mm] \mathbf{k} := \left(\kappa^1, \kappa^2, \kappa^3\right) \end{cases} \quad (6.24)$$

where ω and $\mathbf{k}$ are the usual frequency and the wave number vector respectively. The light speed c is introduced since the time coordinate was defined as $x^0 := ct$.

Remark 6.5 As for elastic waves, we have the dispersion equation. and the nullity of the four-vector $\kappa_\alpha\kappa^\alpha \equiv 0$ means that κ_α is a lightlike vector.

6.2.2 Covariant Formulation of Maxwell's Equations

Maxwell's equations in vacuum include both conservation laws and the constitutive equations, namely the electric charge conservation, the magnetic flux conservation, the Lorentz force on a charged particle and the linear isotropic and homogeneous constitutive laws (defined by the electric permittivity and the magnetic permeability) of the spacetime.

6.2.2.1 Conservation and Constitutive Laws

Considering again the "axiom of general invariance" of Hilbert e.g. Brading and
Ryckman (2008), the electromagnetic Lagrangian $\mathscr{L}_{EM}$ is assumed to depend
on the electromagnetic potential A_μ and their first derivatives. Accordingly, from
the electromagnetic fields (6.8) and the potential formulation (6.9) the invariant
formulation of electromagnetism theory in the Minkowskian spacetime $\mathscr{M}$ is
obtained by considering a skew symmetric tensor $\mathscr{F}$ to represent both the electric $\mathbf{E}$
and magnetic induction $\mathbf{B}$ fields as the electromagnetic strength also called Faraday
tensor e.g. Ryder (2009). The components of this skew-symmetric tensor is obtained
by means of the definition (6.14):

$$\mathscr{F}_{\mu\nu} := \partial_\mu A_\nu - \partial_\nu A_\mu, \qquad \mathscr{F}_{\mu\nu} = \begin{bmatrix} 0 & -E_1 & -E_2 & -E_3 \\ E_1 & 0 & B^3 & -B^2 \\ E_2 & -B^3 & 0 & B^1 \\ E_3 & B^2 & -B^1 & 0 \end{bmatrix} \qquad (6.25)$$

where the combined electromagnetic field $(\mathbf{E}, \mathbf{B})$ do not transform as three-vectors
but as the six components of the skew-symmetric tensor $\mathscr{F}_{\mu\nu}$. We conform here to
the convention in e.g. Hehl (2008), Obukhov (2008), Hehl and Obukhov (2003).

Remark 6.6 We observe that the calculus of the components of the electromagnetic
tensor is conducted as follows and is coherent with the 2-form definition. For
instance, the component $\mathscr{F}_{01} := \partial_0 A_1 - \partial_1 A_0 = -\partial_0 A^1 - \partial_1 \phi := E^1 = -E_1$ where
we have introduced the contravariant components of the potential $A^\mu = (A^0 =
A_0 = \phi, A^1 = -A_1, A^2 = -A_2, A^3 = -A_3)$. It should be reminded that the three-
dimensional potential vector is defined as $\mathbf{A} := (A^1, A^2, A^2)$. For magnetic field,
on the one hand, we can write accordingly $\mathscr{F}_{12} := \partial_1 A_2 - \partial_2 A_1 := B^3$. On the
other hand, the three dimensional definition is given by $\mathbf{B} := \hat{\nabla} \times \mathbf{A}$ which allows
us to calculate the third component as $B^3 = \partial_1 A_2 - \partial_2 A_1$. We deduce $B^3 = \mathscr{F}_{12}$.
The position of the index may be important for defining the electromagnetic tensor
because the contravariant components are not equal to covariant components in the
framework of Minkowski spacetime.

In this way the skew symmetric tensor $\mathscr{F}_{\mu\nu}$ is chosen as primal variables of the
theory. Let us now define the dual variable $\mathscr{H}^{\mu\nu}$ constructed from the displacement
and the magnetic field as follows (tensor $\mathscr{H}$ will be shortly related to Faraday tensor
by means of tangent tensor $\hat{\Xi}^{\mu\nu\lambda\sigma} := \epsilon_0 \, \hat{g}^{\mu\lambda} \hat{g}^{\nu\sigma}$ defining the constitutive properties
of the vacuum space, and it be will be extended to matter and to general relativity
spacetime hereafter) e.g. Obukhov (2008). Independently on the constitutive laws,
the classical electromagnetism theory considers the electromagnetic excitation as a
two-form $\mathscr{H}_{\mu\nu}$:

$$\mathscr{H} := -H_1 dx^1 \wedge dx^0 - H_2 dx^2 \wedge dx^0 - H_3 dx^3 \wedge dx^0$$
$$+ D^3 dx^1 \wedge dx^2 + D^1 dx^2 \wedge dx^3 + D^1 dx^2 \wedge dx^3 \qquad (6.26)$$

in the same way as the definition of the electromagnetic strength (6.14).

Remark 6.7 The variable we are interested in is in fact the dual variable $\mathcal{H}^{\mu\nu}$ in order to be able to link it with the primal variable $\mathcal{F}_{\alpha\beta}$. To have an intuitive point of view and for better understanding, it is worth to consider the simplest linear constitutive law:

$$\mathcal{H}^{\mu\nu} = \epsilon_0 \sqrt{-\mathrm{Detg}}\; g^{\mu\alpha} g^{\nu\beta} \mathcal{F}_{\alpha\beta}$$

In the Minkowski vacuum spacetime, the calculus is straightforward by using the relation $\epsilon_0\mu_0 \equiv 1$:

$$\mathcal{H}^{\mu\nu} = \epsilon_0 \begin{bmatrix} 0 & E_1 & E_2 & E_3 \\ -E_1 & 0 & B^3 & -B^2 \\ -E_2 & -B^3 & 0 & B^1 \\ -E_3 & B^2 & -B^1 & 0 \end{bmatrix} = \begin{bmatrix} 0 & D^1 & D^2 & D^3 \\ -D^1 & 0 & H_3 & -H_2 \\ -D^2 & -H_3 & 0 & H_1 \\ -D^3 & H_2 & -H_1 & 0 \end{bmatrix}$$

Now going back to arbitrary constitutive laws, we calculate the expression of the dual variable as:

$$\mathcal{H}^{\mu\nu} := \frac{1}{2} \epsilon^{\mu\nu\alpha\beta} \mathcal{H}_{\alpha\beta} \tag{6.27}$$

where $\epsilon^{\mu\nu\alpha\beta}$ is the Levi-Civita tensor. From (6.26), we easily obtain the two contravariant components as exactly as for the linear case:

$$\mathcal{H}^{\mu\nu} = \begin{bmatrix} 0 & D^1 & D^2 & D^3 \\ -D^1 & 0 & H_3 & -H_2 \\ -D^2 & -H_3 & 0 & H_1 \\ -D^3 & H_2 & -H_1 & 0 \end{bmatrix} \tag{6.28}$$

owing again that $\epsilon_0\mu_0 = 1$ (the choice of coordinate system where $x^0 := ct$ allows us to act as if $c = 1$) and the metric of the flat Minkowski spacetime holds $g^{\mu\nu} := \{+1, -1, -1, -1\}$. It conforms to the linear constitutive law but it should again be stressed that in the general case this is in fact considered as a definition of the dual variable independently on the constitutive law.

Previous $3D$ equations (6.2) may be recast in four-dimensional covariant Maxwell equations by using the spacetime connection as e.g. Ryder (2009)[1]:

$$\begin{cases} \hat{\nabla}_\mu \mathcal{H}^{\mu\nu} = J^\nu \\ \hat{\nabla}_\mu \mathcal{F}^{*\mu\nu} = 0 \end{cases} \tag{6.29}$$

where $\mathcal{H}^{\mu\nu}$ denotes the electromagnetic tensor including the electric displacement field and the magnetic field, and $\mathcal{F}^{*\mu\nu} := (1/2)\epsilon^{\mu\nu\kappa\sigma} \mathcal{F}_{\kappa\sigma}$ is the dual of $\mathcal{F}^{\mu\nu}$ ($\epsilon^{\mu\nu\kappa\sigma}$

---

[1] Again, $\hat{\nabla}_\mu$ denotes the four-dimensional connection in the Minkowski spacetime.

being the Levi-Civita tensor $\epsilon^{0123} := +1$). We remind the Civita tensor:

$$\epsilon_{\mu\nu\kappa\sigma} := \begin{cases} +1 & \text{if} & (\mu\nu\kappa\sigma) \text{ is an even permutation of (0123)} \\ -1 & \text{if} & (\mu\nu\kappa\sigma) \text{ is an odd permutation of (0123)} \\ 0 & \text{otherwise} \end{cases} \quad (6.30)$$

Here we give some examples of components where the quadruplet $(\mu, \nu, \alpha, \beta)$ represents the component $\epsilon^{\mu\nu\alpha\beta}$ in order to highlight the indices permutation:

$$\begin{cases} (0, 1, 2, 3) = 1, \ (1, 0, 2, 3) = -1, \ (1, 2, 0, 3) = 1, \ (1, 2, 3, 0) = -1 \\ (0, 1, 2, 3) = 1, \ (0, 1, 3, 2) = -1, \ (0, 3, 1, 2) = 1, \ (3, 0, 1, 2) = -1 \end{cases}$$

To highlight the role of the dual variable, covariant formulation of constitutive laws may be introduced by means of the electromagnetic Lagrangian of the Minkowskian spacetime:

$$\mathscr{L} := -\frac{1}{4} \mathscr{H}^{\mu\nu} \mathscr{F}_{\mu\nu} \quad (6.31)$$

By introducing the definitions (6.25) and (6.28) the three-dimensional formulation of the Lagrangian density function reduces to:

$$\mathscr{L} = \frac{1}{2} (\mathbf{D} \cdot \mathbf{E} - \mathbf{B} \cdot \mathbf{H}) \quad (6.32)$$

For linear constitutive laws, the dual variable $\mathscr{H}^{\mu\nu}$ in the Lagrangian (6.31) defines the constitutive laws of vacuum spacetime, in four-dimensional formulation,

$$\mathscr{H}^{\mu\nu} := \hat{\mathscr{E}}_{v}^{\mu\nu\lambda\sigma} \mathscr{F}_{\lambda\sigma}, \quad \hat{\mathscr{E}}_{v}^{\mu\nu\lambda\sigma} := (\varepsilon_0/\mu_0)^{1/2}\sqrt{-\mathrm{Det}\hat{g}} \left(\hat{g}^{\mu\lambda} \hat{g}^{v\sigma} \right) \quad (6.33)$$

owing that the Minkowskian spacetime of dimension $n = 4$ is endowed with the metric $\hat{g}$. In the framework of $3D$ description, component by component, the electromagnetic constitutive laws of vacuum space hold when considering the theory of e.g. Plebanski (1960) adapted to Minkowskian spacetime, where $\epsilon_0 := 2(\varepsilon_0/\mu_0)^{1/2}$:

$$\begin{cases} \mathbf{D} = \varepsilon_0 \, \mathbf{E} \\ \mathbf{B} = \mu_0 \, \mathbf{H} \end{cases} \implies \begin{cases} D^i = -\dfrac{\epsilon_0\sqrt{-\mathrm{Det}\hat{g}}}{\hat{g}_{00}} \left(\hat{g}^{ij} E_j - \epsilon^{ijk} \hat{g}_{0j} H_k \right) \\ B^i = -\dfrac{\sqrt{-\mathrm{Det}\hat{g}}}{\epsilon_0 \, \hat{g}_{00}} \left(\hat{g}^{ij} H_j + \epsilon^{ijk} \hat{g}_{0j} E_k \right) \end{cases}$$

where $\hat{g}_{ij}$ is the space part of the Minkowski metric with $i, j, k = 1, 2, 3$, and $\epsilon_0\mu_0 := c^{-2} = 1$. In the next subsection, we develop the complete constitutive laws with respect to a moving reference frame with a relative velocity $\mathbf{v}$.

Remark 6.8 To avoid repetition of the procedure, we only apply the variation of the Lagrangian when we will derive the field equations of gravitation-electromagnetic interaction in the cases of Riemann and of Riemann–Cartan framework.

6.2.2.2 Lorentz Covariance and Rules of Variable Transformations

Maxwell's 3D equations (6.2) are formulated in a fixed inertial system where the spatial coordinates (x^i) are defined in a space $\mathbb{R}^3$ and where t denotes the time coordinate that an observer at rest reads on a clock attached to the space. The covariance of the Maxwell's 3D equations with respect to rotations, space reflection, time reversal, and charge conjugation (modification of positive charge to negative charge) may be checked by means of specific (non dynamic) transformations. However, covariance under these groups of non time-dependent transformations is unsatisfactory because the electromagnetic fields depend not only on the space (x^i) but also on the time t. A four-dimensional covariance analysis is worth for instance with respect to Lorentz group of transformations (2.14). Basically, Lorentz transformation relates two observers moving each other with constant relative velocity **v**. Neither electric fields **E** and **D**, nor magnetic fields **H**, and **B** have simple transformation behavior. We remind the Lorentz transformation matrix:

$$\Lambda^i_j = \delta^i_j + \frac{(\gamma - 1)\, v^i v_j}{|\mathbf{v}|^2}, \qquad \Lambda^0_i = \Lambda^i_0 = -\gamma\, v^i, \qquad \Lambda^0_0 = \gamma$$

where v^i are three real (constant) parameters satisfying $|\mathbf{v}|^2 := (v^1)^2 + (v^2)^2 + (v^3)^2 < 1$ (any particle has a velocity lower than the light speed which was set to $c = 1$), and $\gamma := (1 - |\mathbf{v}|^2)^{-1/2}$. The point transformation is:

$$\begin{pmatrix} y^0 \\ y^1 \\ y^2 \\ y^3 \end{pmatrix} = \begin{bmatrix} \Lambda^0_0 & \Lambda^0_1 & \Lambda^0_2 & \Lambda^0_3 \\ \Lambda^1_0 & \Lambda^1_1 & \Lambda^1_2 & \Lambda^1_3 \\ \Lambda^2_0 & \Lambda^2_1 & \Lambda^2_2 & \Lambda^2_3 \\ \Lambda^3_0 & \Lambda^3_1 & \Lambda^3_2 & \Lambda^3_3 \end{bmatrix} \begin{pmatrix} x^0 \\ x^1 \\ x^2 \\ x^3 \end{pmatrix} \tag{6.34}$$

where each component of the transformation Λ^μ_ν does not depend on the coordinate x^α. The transformation rules of variables are given by tensor rule in the general case:

$$\begin{cases} \phi'(y^\mu) = \phi(x^\mu) \\ A'^\mu(y^\alpha) = \Lambda^\mu_\nu(x^\alpha)\, A^\nu(x^\alpha) \\ \mathcal{H}'^{\mu\nu}(y^\alpha) = \Lambda^\mu_\alpha(x^\alpha)\, \Lambda^\nu_\beta(x^\alpha)\, \mathcal{H}^{\alpha\beta}(x^\alpha) \end{cases} \tag{6.35}$$

although Λ^μ_ν does not depend on x^α in our case. It is observed that the system of Maxwell's equations is Lorentz covariant if (and only if) the four-density current $J^\mu(x^\alpha)$ is a four-vector field. To assess the transformation of the electromagnetic fields $(\mathbf{E}, \mathbf{H})$, and $(\mathbf{D}, \mathbf{B})$, it is worth to calculate from the tensorial rules:

$$\tilde{\mathscr{F}}_{\alpha\beta} = \Lambda_\alpha^\mu \Lambda_\beta^\nu \mathscr{F}_{\mu\nu} \tag{6.36}$$

and also:

$$\tilde{\mathscr{H}}^{\alpha\beta} = \Lambda_\mu^\alpha \Lambda_\nu^\beta \mathscr{H}^{\mu\nu} \tag{6.37}$$

For example, we calculate the three components of the magnetic flux as:

$$\begin{cases} \tilde{B}^1 := \mathscr{F}_{23} = \Lambda_2^\mu \Lambda_3^\nu \mathscr{F}_{\mu\nu} = \gamma B^1 \\ -\tilde{B}^2 := \mathscr{F}_{13} = \Lambda_1^\mu \Lambda_3^\nu \mathscr{F}_{\mu\nu} = -\gamma \left(B^2 + v^1 E_3 \right) \\ \tilde{B}^3 := \mathscr{F}_{12} = \Lambda_1^\mu \Lambda_2^\nu \mathscr{F}_{\mu\nu} = \gamma \left(B^3 - v^1 E_2 \right) \end{cases}$$

for the case where only a motion along the direction $\mathbf{e}_1$ is considered. One-dimensional calculus is straightforward. For the general transformation rules, we have to use the formula $\tilde{\mathscr{F}}_{\alpha\beta} = \Lambda_\alpha^\mu \Lambda_\beta^\nu \mathscr{F}_{\mu\nu}$ with the Lorentz boost transformation (2.14) and write:

$$\tilde{E}_i := \mathscr{F}_{0i} = \Lambda_0^\mu \mathscr{F}_{\mu\nu} \Lambda_i^\nu$$

$$= \Lambda_0^0 \mathscr{F}_{0j} \Lambda_i^j + \Lambda_0^j \mathscr{F}_{j0} \Lambda_i^0 + \Lambda_0^j \mathscr{F}_{jk} \Lambda_i^k$$

$$= \gamma \left(\delta_i^j + \frac{\gamma-1}{\|\mathbf{v}\|^2} v^j v_i \right) E_j - \gamma^2 v^j v_i E_j - \gamma v^j \left(\delta_i^k + \frac{\gamma-1}{\|\mathbf{v}\|^2} \right) \epsilon_{jkl} B^l$$

Calculus for the other electric and magnetic fields $(\mathbf{D}, \mathbf{H}, \mathbf{B})$ is analogously treated. For the general motion with the vector $\mathbf{v} = v^i \mathbf{e}_i$ (remind that the time is $x^0 :=ct$), we obtain the Lorentz transformations of the electromagnetic fields, for the intensities:

$$\begin{cases} \tilde{\mathbf{E}} = \gamma \left[\mathbf{E} + \mathbf{v} \times \mathbf{B} - \frac{\gamma-1}{\gamma} \left(\frac{\mathbf{v}}{\|\mathbf{v}\|} \otimes \frac{\mathbf{v}}{\|\mathbf{v}\|} \right) \mathbf{E} \right] \\ \tilde{\mathbf{H}} = \gamma \left[\mathbf{H} - \mathbf{v} \times \mathbf{D} - \frac{\gamma-1}{\gamma} \left(\frac{\mathbf{v}}{\|\mathbf{v}\|} \otimes \frac{\mathbf{v}}{\|\mathbf{v}\|} \right) \mathbf{H} \right] \end{cases} \tag{6.38}$$

and for the fluxes:

$$\begin{cases} \tilde{\mathbf{D}} = \gamma \left[\mathbf{D} + \mathbf{v} \times \mathbf{H} - \frac{\gamma-1}{\gamma} \left(\frac{\mathbf{v}}{\|\mathbf{v}\|} \otimes \frac{\mathbf{v}}{\|\mathbf{v}\|} \right) \mathbf{D} \right] \\ \tilde{\mathbf{B}} = \gamma \left[\mathbf{B} - \mathbf{v} \times \mathbf{E} - \frac{\gamma-1}{\gamma} \left(\frac{\mathbf{v}}{\|\mathbf{v}\|} \otimes \frac{\mathbf{v}}{\|\mathbf{v}\|} \right) \mathbf{B} \right] \end{cases} \tag{6.39}$$

Relations (6.38) and (6.39) are the Lorentz transformations rules of the three-dimensional electromagnetic vectors e.g. Rousseaux (2008) where the medium is moving with relative velocity $\mathbf{v}$ each other. For a linear isotropic and homogeneous medium, the constitutive laws, in the rest frame, are written as:

$$\begin{cases} \mathbf{D} = \epsilon \, \mathbf{E} \\ \mathbf{B} = \mu \, \mathbf{H} \end{cases} \quad \text{and} \quad \begin{cases} \tilde{\mathbf{D}} = \epsilon \, \tilde{\mathbf{E}} \\ \tilde{\mathbf{B}} = \mu \, \tilde{\mathbf{H}} \end{cases} \tag{6.40}$$

where it is essential to observe that the two material constants ε and μ remain the same independently on the motion with velocity $\mathbf{v}$ of the medium. This merely expresses the covariance of the Lagrangian (6.31) which models the electromagnetic behavior of the medium:

$$\mathcal{L} := -\frac{1}{4} \, \mathcal{H}^{\mu\nu} \mathcal{F}_{\mu\nu} = -\frac{1}{4} \, \tilde{\mathcal{H}}^{\mu\nu} \tilde{\mathcal{F}}_{\mu\nu} \tag{6.41}$$

meaning that the quantity $\mathbf{D} \cdot \mathbf{E} - \mathbf{B} \cdot \mathbf{H} \equiv \tilde{\mathbf{D}} \cdot \tilde{\mathbf{E}} - \tilde{\mathbf{B}} \cdot \tilde{\mathbf{H}}$ is Lorentz invariant. A direct calculus in e.g. Rousseaux (2008) allows us to deduce from the relations (6.38), (6.39) and the Minkowski invariance conditions (6.40) the covariant constitutive laws:

$$\begin{cases} \mathbf{D} + \mathbf{v} \times \mathbf{H} = \epsilon \, (\mathbf{E} + \mathbf{v} \times \mathbf{B}) \\ \mathbf{B} - \mathbf{v} \times \mathbf{E} = \mu \, (\mathbf{H} - \mathbf{v} \times \mathbf{D}) \end{cases} \tag{6.42}$$

relating the electric and magnetic fluxes to the electric and magnetic intensities. They are called Minkowski constitutive laws of electromagnetism (6.42) which are independent on the velocity of the medium in the framework of special relativistic theory. They extend the constitutive laws in e.g. Plebanski (1960).

6.2.3   Maxwell's Equations in Terms of Differential Forms

Numerous mathematical descriptions exist for writing the equations governing the electromagnetic field. To extend the Maxwell's equations to continuum media, it is worth to rewrite the equations in terms of differential forms. We consider in this subsection a Riemann continuum. In this way, we define the Hodge operator $*$ after considering again the Civita antisymmetric operator (6.30). The (numerical) contravariant components are then: $\epsilon^{\mu\nu\kappa\sigma} := |\text{Det } \hat{\mathbf{g}}|^{-1} \, \epsilon_{\mu\nu\kappa\sigma}$. Say the spacetime $\mathcal{M}$ of dimension $n = 4$ endowed with the metric $\hat{\mathbf{g}}$, the Hodge linear map $* : \Lambda^{r=2} \rightarrow \Lambda^{n-r=2}$ is the action defined by as follows.[2]

$$\mathcal{F}^{*\mu\nu} := (1/2)\sqrt{|\text{Det}\hat{\mathbf{g}}|}\epsilon^{\mu\nu\kappa\sigma} \, \mathcal{F}_{\kappa\sigma} \tag{6.43}$$

---

[2]*Hodge star operator* The Hodge star operator is the unique linear map on a semi-Riemannian manifold, say $\mathcal{M}$, from r-forms to $n - r$-forms defined by: $\star : \Omega^r(\mathcal{M}) \rightarrow \Omega^{n-r}(\mathcal{M})$, such that for all $(\omega, \omega') \in \Omega^r(\mathcal{M})$, we have $\omega \wedge \omega* := \langle \omega, \omega' \rangle$ where $\langle , \rangle$ is an interior product on $\mathcal{M}$ e.g. Nakahara (1996).

We then obtain the dual of the Faraday tensor:

$$\mathcal{F}^{*\mu\nu} = \begin{bmatrix} 0 & B_1 & B_2 & B_3 \\ -B_1 & 0 & -E_3 & E_2 \\ -B_2 & E_3 & 0 & -E_1 \\ -B_3 & -E_2 & E_1 & 0 \end{bmatrix}$$

The derivation of differential forms formulation of Maxwell's equations by means of these 2-forms merges into the Gauss-Ampère law and the Gauss-Faraday law respectively e.g. Plebanski (1960), Prasanna (1975a), Ryder (2009):

$$d\mathcal{H} = \mathcal{J}, \qquad (d\mathcal{F}^*)^* = 0 \qquad (6.44)$$

Equation (6.44) are equivalent to (6.29), and are equivalent to (6.2) with $\mathcal{J}^\mu = (\rho, J^1, J^2, J^3)$. For a given media, these equations are not complete since the constitutive laws between fluxes and fields should be proposed $\mathbf{D(E, H)}$ and $\mathbf{B(E, H)}$. Indeed, it is also reminded that the covariant and the 2-forms formulation are independent on the constitutive laws. In presence of interfaces between various media and of boundaries, the solving of local equations requires the covariant jump conditions e.g. Itin (2012).

Remark 6.9 The differential forms Maxwell equation (6.44) constitute explicit covariant formulation without requiring neither a metric nor connection, it is thus possible to use them for deriving the Maxwell equations within a manifold with Riemannian metric $g_{\mu\nu}(x^\lambda)$ e.g. Itin (2012), or post-Riemannian spacetime and continua e.g. Puntigam et al. (1997). This is an important starting point for investigating the propagation of electromagnetic waves within matter and / or with gravitation field. Equation (6.44) are typical examples of the Yang-Mills theory e.g. Cho (1976a).

Remark 6.10 The invariant formulation of Maxwell equation (6.44) are mathematically appealing and look synthetic and simple. It should nevertheless observed that they relate electromagnetic variables which are "mixed" in the sense of thermomechanics theory (distinction of primal and dual variables). Indeed, each of the two 2-forms $\mathcal{F}$ (6.25, and $\mathcal{H}$ (6.28) representing the electromagnetic fields are composed of intensity field and flux density: $(\mathbf{E}, \mathbf{B})$ for $\mathcal{F}$, and $(\mathbf{D}, \mathbf{H})$ for $\mathcal{H}$. Physically, this should be handled with cautious when dealing with material constitutive laws.

Remark 6.11 A more complex continuum (spacetime or continuum matter) may be generalized from a geometry point of view. As reminded by e.g. Smalley and Krisch (1992), there is no a priori condition that the Faraday tensor field $\mathscr{F}_{\mu\nu}$ be a two-form. Another way would be the definition of the Faraday tensor as the skew-symmetric part of the gradient $\nabla_\mu A_\nu$ where the connection is affine and may be non symmetric.

6.3   Electromagnetism in Curved Continuum

Various phenomenae in the domain of propagation may have effects on electromagnetic waves: the presence of gravitational field, the presence of media, and the motion of this media. The constitutive properties of the vacuum spacetime (defined by Eq. (6.33)) should be adapted accordingly. Indeed, these constitutive laws are classically identified in a laboratory frame of reference which are assumed to be homogeneous and isotropic, by the effects of electric and magnetic fields on various charges and particles, or systems. They should be worthily changed to account for either the modification of the spacetime environment in presence of gravity for instance, or the propagation of electromagnetic fields within matter.

It is worth to remind the vectorial approach to define electromagnetic constitutive laws, say the electrodynamics in matter. There are two types of material reactions for electric field $\mathbf{E}$ (e.g. Kovetz 2000): (a) conductors have electric charges that are free to move, (b) dielectrics do not have such free electric charges. But when electric field is applied on matter, positively charge nucleus are pushed in the direction of $\mathbf{E}$, and negatively charge in the opposite direction. This polarization is described by a vector field $\mathbf{P} = \epsilon_0 \chi_e \mathbf{E}$ where $\chi_e > 0$ is the electric susceptibility constant. Ferroelectrics are material for which $\mathbf{P} \neq 0$ even in the absence of an electric field. The electric displacement then holds in a general form:

$$\mathbf{D} = \epsilon_0 \mathbf{E} + \mathbf{P} = \epsilon\,\mathbf{E}$$

Magnetic fields $\mathbf{M}$ are created by currents. Current loops with their associated dipole moments exist in materials by two mechanisms: (a) electrons orbiting the nucleus carry angular momentum and act as magnetic dipole moments; (b) electrons carry intrinsic spin, which is a pure quantum effect, and contribute to the magnetic dipole moment: $\mathbf{M} = \frac{1}{\mu_0} \frac{\chi_m}{1+\chi_m} \mathbf{B}$ and:

$$\mathbf{H} := \frac{1}{\mu_0} \mathbf{B} - \mathbf{M} \quad \Longleftrightarrow \quad \mathbf{B} = \mu_0 \mathbf{H} + \mathbf{M} = \mu \mathbf{H}$$

where χ_m is the magnetic susceptibility. Diamagnetic matters have $-1 < \chi_m < 0$, whereas paramagnetic matters $0 < \chi_m$. Ferromagnetic matters (or magnets) may have $\mathbf{M} \neq 0$ even if $\mathbf{B} = 0$. For diamagnets and paramagnets we can write $\mathbf{B} = \mathbb{M}\,\mathbf{H}$ accordingly where $\mathbb{M}$ is the magnetic permeability tensor.

Remark 6.12 Relations $\mathbf{D} = \epsilon \mathbf{E}$ and $\mathbf{B} = \mu \mathbf{H}$ are the constitutive equations of the electromagnetic matter in a particular reference frame, where ϵ and μ are the electric and magnetic parameters of the matter. Covariant constitutive equations with respect to an arbitrary reference frame are given by (6.42). For this particular case, it suffices to replace ϵ_0 and μ_0 by ϵ and μ.

6.3.1 Maxwell's Equations and Constitutive Laws

Maxwell's equations which are the conservation laws, independent of the continuum where electromagnetic waves propagate, are reformulated and constitutive laws are explicitly derived.

6.3.1.1 Maxwell's Equations

Let a continuum $\mathscr{B}$ which may be not a Minkowski spacetime $\mathscr{M}$, each of continuum has its proper metric and connection. We consider the propagation of electromagnetic field within the continuum $\mathscr{B}$. The continuum $\mathscr{B}$ may be also considered as an Einstein or Einstein–Cartan spacetime with gravitation. However in this subsection the symbol ∇ denotes the three-dimensional connection in the space and $\overline{\nabla}$ the Levi-Civita four-dimensional connection. The same form as for the basic equations (6.2) and (6.29) governing the electromagnetic waves hold within matter or within a spacetime with gravitation written in a source-free from e.g. Fernandez-Nunez and Bulashenko (2016):

$$
\begin{cases}
\nabla \cdot \mathbf{D} = 0 \\
\nabla \times \mathbf{H} - c\, \partial_0 \mathbf{D} = 0 \\
\nabla \cdot \mathbf{B} = 0 \\
\nabla \times \mathbf{E} + c\, \partial_0 \mathbf{B} = 0
\end{cases}
\quad \text{or} \quad
\begin{cases}
\overline{\nabla}_\mu \mathcal{H}^{\mu\nu} = 0 \\
\overline{\nabla}_\mu \mathcal{F}^{*\mu\nu} = 0
\end{cases}
\tag{6.45}
$$

where the connection ∇ is referred to continuum matter $\mathscr{B}$ than to the spacetime $\mathscr{M}$. In the framework of Einstein gravitation (curved but torsionless), the extension of Minkowski Faraday tensor (6.25) is determined from a four-vector potential by means of the Riemannian connection (torsionless and metric compatible):

$$
\mathcal{F}_{\mu\nu} := \overline{\nabla}_\mu A_\nu - \overline{\nabla}_\nu A_\mu = \partial_\mu A_\nu - \partial_\nu A_\mu
\tag{6.46}
$$

As a first example, we consider a field of gravitation characterized by a non Minkowskian metric $g_{\alpha\beta}(x^\mu)$ e.g. Leonhardt and Philbin (2006), Fernandez-Nunez and Bulashenko (2016). By analogy to the electromagnetic waves in vacuum, we start with the field of 2-form (same shape of contravariant components as in the vacuum):

$$
\mathcal{H}^{\mu\nu} =
\begin{bmatrix}
0 & D^1 & D^2 & D^3 \\
-D^1 & 0 & H_3 & -H_2 \\
-D^2 & -H_3 & 0 & H_1 \\
-D^3 & H_2 & -H_1 & 0
\end{bmatrix}
$$

where the "physical" variables are the electric flux (displacement) and the magnetic field.

From this quantity, we use hereafter the Riemannian metric $g_{\mu\nu}$, as for relativistic gravitation theory, instead of Minkowskian $\hat{g}_{\mu\nu}$ one, the associated Faraday tensor takes the form of (inversion): $\mathcal{F}_{\mu\nu} = (\epsilon_0\sqrt{-\mathrm{Detg}})^{-1} g_{\mu\lambda} g_{\nu\sigma} \, \mathcal{H}^{\lambda\sigma}$.

6.3.1.2 Constitutive Equations in Presence of Rotating Body

In presence of spacetime or continuum curvature, generated by non uniform metric tensor, electrostatics and magnetostatics are coupled each other. By analogy to the vacuum Minkowski spacetime, we may write the component by component electromagnetic constitutive laws in three-dimensional formulation e.g. Leonhardt and Philbin (2006), Plebanski (1960):

$$
\begin{cases}
D^i = -\dfrac{\epsilon_0\sqrt{-\mathrm{Detg}}}{g_{00}} \left(g^{ij}\, E_j - \epsilon^{ijk}\, g_{0j}\, H_k \right) \\[4mm]
B^i = -\dfrac{\sqrt{-\mathrm{Detg}}}{\epsilon_0 g_{00}} \left(g^{ij}\, H_j + \epsilon^{ijk}\, g_{0j}\, E_k \right)
\end{cases}
$$

where the components of the metric tensor $g_{\alpha\beta}$ and its inverse $g^{\alpha\beta}$ are a priori given, or more generally calculated from the presence of massive bodies in the case of relativistic gravitation. They depend on the point event x^μ. We will consider in the next section the mutual interaction of electromagnetism and gravitation by using a variational procedure. The same method can be used to define the inverse of constitutive laws.

For linear matter, the above electromagnetic constitutive laws with respect to a rest frame take the general form of (see Plebanski 1960):

$$
\mathbf{D} = \varXi\, \mathbf{E} - \varGamma \times \mathbf{H}, \qquad \mathbf{B} = \mathbb{M}\, \mathbf{H} + \varGamma \times \mathbf{E} \tag{6.47}
$$

where $\varXi$ and $\mathbb{M}$ are respectively the electric permittivity and the magnetic permeability tensors. The time-dependent vector $\varGamma$ defines the coupling between the electric and magnetic fields (see (6.42)) for the case of moving frame.

As illustration, by considering a set of material points of total mass $M := \sum m_\alpha$ with a constant angular momentum $\mathbf{J} := \sum \mathbf{GM}_\alpha \times \mathbf{v}_\alpha$ (G is the mass center of material point m_α at position M_α) and at large distance r from the set of material points, the approximate gravitational metric differs from the Minkowskian one to give the so-called Landau's formulae e.g. Hartle and Sharp (1967), Landau and Lifchitz (1971) (here with the convention $\hat{g}_{\alpha\beta} := \mathrm{diag} = \{+1, -1, -1, -1\}$):

$$
\begin{cases}
g_{00} = 1 - \dfrac{2m}{r} & \simeq 1 - \dfrac{2m}{r} + \mathcal{O}\left(\dfrac{1}{r^2}\right) \\[4mm]
g_{ij} = -\delta_{ij}\left(1 - \dfrac{2m}{r}\right) & \simeq -\delta_{ij}\left(1 + \dfrac{2m}{r}\right) + \mathcal{O}\left(\dfrac{1}{r^2}\right) \\[4mm]
g_{0i} = -\dfrac{2m}{r^3}\,\epsilon_{ijk}x^j\,j^k = g_{i0} & \simeq -\dfrac{2m}{r^3}\,\epsilon_{ijk}x^j\,j^k + \mathcal{O}\left(\dfrac{1}{r^3}\right)
\end{cases}
\tag{6.48}
$$

in which we have defined the Schwarzschild radius m and the adimensional angular momentum:

$$m := \frac{GM}{c^2}, \qquad j^k := \frac{J^k}{Mc} \tau \tag{6.49}$$

The gravitational field engendered by a rotating massive body at the origin is thus characterized by the approximated metric (6.48) which merged into a Minkowski spacetime at very large distance. The propagation of electromagnetic waves is influenced by the gravitation only at the vicinity of the massive body. The quantitative role of the angular momentum is not of the same order as that of the mass m. It should be however reminded that (6.48) are not the exact external solutions of Einstein field equation due to rotating body. They are only approximated conversely to Kerr solutions e.g. Ryder (2009).

Remark 6.13 It is worth to mention that the off-diagonal terms g_{0i} expresses the time averages of the original components g_{0i} over the period of motion τ of the matter (Plebanski 1960). Otherway, the physical dimension of the original off-diagonal terms in Eq. (6.49) is $[1/s]$.

Remark 6.14 Physically, constitutive laws express the electromagnetic flux densities **D** (electric displacement) and **B** (magnetic induction) in terms of electromagnetic field intensities **E** (electric field) and **H** (magnetic field), see Table 6.1 for classification.

The $3D$ constitutive equations (6.47) may be written as four-dimensional constitutive laws:

$$\begin{cases} D^\alpha = \Xi^{\alpha\beta} E_\beta + \epsilon^{\alpha\beta\gamma} \Gamma_\beta H_\gamma \\ B^\alpha = \mathbb{M}^{\alpha\beta} H_\beta - \epsilon^{\alpha\beta\gamma} \Gamma_\beta E_\gamma \end{cases} \tag{6.50}$$

where $\epsilon^{\alpha\beta\gamma}$ is the Levi-Civita alternating tensor. Plebanski calculated the equivalent electric permittivity and magnetic permeability induced by a metric-gravitational field $g_{\alpha\beta}$ and determined in Plebanski (1960) as:

$$\Xi^{\alpha\beta} = \mathbb{M}^{\alpha\beta} = -\sqrt{\text{Det}(-\mathbf{g})} \, \frac{g^{\alpha\beta}}{g_{00}}, \qquad \Gamma_\alpha = \frac{g_{0\alpha}}{g_{00}} \tag{6.51}$$

Remark 6.15 We observed that the Plebanski form for electromagnetic constitutive equations (6.50) conform more the way of formulating constitutive laws in continuum mechanics models of matter where fluxes **D** and **B** (i.e. stress for mechanics) are expressed in terms of intensities **E** and **H** (i.e. strain for mechanics).

6.3.1.3 Electric Three-Dimensional Wave Equations

For the static spacetime metric, as Schwarzschild metric case, vector Γ vanishes. If we consider a monochromatic wave (sufficiently narrow bandwidth frequency), the combination of the Maxwell's equations and the constitutive laws (where Γ is assumed to be null) leads to the time-harmonic electromagnetic wave propagation e.g. Fernandez-Nunez and Bulashenko (2016):

$$\overline{\nabla} \times \left[\mathbb{M}^{-1} \left(\overline{\nabla} \times \mathbf{E} \right) \right] - \omega^2 \, \boldsymbol{\Xi} \, \mathbf{E} = 0 \tag{6.52}$$

where the connection ∇ represents the three-dimensional connection of the space.

Propagation of electromagnetic waves in continuum is also modified as the motion of the continuum itself engenders effective gravitational field e.g. Leonhardt and Piwnicki (2000). For instance, optical effects of moving media have been known as earlier as the work of Fresnel in 1818. In addition to the drag effects and as for non-Euclidean metrics such as Schwarzschild metric, the continuum motion is a source of bending when the four-dimensional aspect of the spacetime is considered. Gordon in 1920 noticed that moving media induces effective spacetime non-Minkowskian metric influencing electromagnetic fields e.g. Leonhardt and Philbin (2006). The constitutive relations of vacuum spacetime are no longer valid. In Riemannian manifold (case of electromagnetic waves within Einstein gravitational fields), the Maxwell's equations within matter remain the same as (6.44) e.g. Frankel (1997), Puntigam et al. (1997): $d\mathcal{H} = \mathcal{J}$, and $(d\mathcal{F}^*)^* = 0$ after substituting the appropriate conservation laws, for instance (6.50). It can thus be observed that electromagnetism within matter modelled by Riemannian manifold is also described by equations of Yang-Mills theory.

6.3.2 Variational Method and Covariant Maxwell's Equations

We have seen that the vacuum constitutive laws introduce both the metric (Minkowski) and physical properties as electric permittivity ϵ_0 and magnetic permeability μ_0 with the homogeneity and isotropy assumption of the space and time.

6.3.2.1 Lagrangian and Constitutive Laws

For real-world materials and media, the constitutive relations are not linear at all. Physical parameter may depend on the location. In a general manner, constitutive equations relate the fluxes $\mathbf{D}$ and $\mathbf{B}$ en terms of fields $\mathbf{E}$ and $\mathbf{H}$. Usual texts of electromagnetic theory may mixed variables. Linear electromagnetic constitutive laws for spacetime and for matter may be integrated within a quadratic Lagrangian function and can be written as:

$$\mathscr{L} = -\frac{1}{4} \Xi^{\mu\nu\alpha\beta} \mathscr{F}_{\mu\nu} \mathscr{F}_{\alpha\beta} \tag{6.53}$$

where the tensor $\Xi^{\mu\nu\alpha\beta}(g_{\lambda\gamma}, u_\lambda, \cdots)$ characterizes the electromagnetic properties of the media.

Some particular materials may be considered. Physically, a polarizable and magnetizable continuum matter may be characterized by polarization **P** and magnetization **M**. The three dimensional "physics" constitutive laws hold:

$$\begin{cases} \mathbf{D} = \varepsilon\, \mathbf{E} + \mathbf{P} \\ \mathbf{B} = \mu\, (\mathbf{H} + \mathbf{M}) \end{cases} \tag{6.54}$$

where the polarization **P** and magnetization **M** are not vectors but are rather the components of a two-forms in the four-dimensional description:

$$\mathscr{M}^{\mu\nu} := \begin{bmatrix} 0 & P^1 & P^2 & P^3 \\ -P^1 & 0 & -M_3 & M_2 \\ -P^2 & M_3 & 0 & -M_1 \\ -P^3 & -M_2 & M_1 & 0 \end{bmatrix} \tag{6.55}$$

From (6.54), the four-dimensional constitutive laws thus take the form of e.g. Obukhov (2008):

$$\mathscr{H}^{\mu\nu} = \mathscr{H}_m^{\mu\nu} + \mathscr{M}^{\mu\nu} \tag{6.56}$$

with:

$$\mathscr{H}_m^{\mu\nu} = \varepsilon \left(g^{\mu\alpha} g^{\nu\beta}\right) \mathscr{F}_{\alpha\beta}, \qquad \mathscr{M}^{\mu\nu} = \Xi_{em}^{\mu\nu\alpha\beta} \mathscr{F}_{\alpha\beta} \tag{6.57}$$

Tensor $\mathscr{H}_m^{\mu\nu}$ is the excitation field in the matter by analogy to the vacuum equation (6.33). The electromagnetic susceptibility tensor $\Xi_{em}^{\mu\nu\alpha\beta}$ defines the properties of the matter, and it also describes both the electric polarization and the magnetization of the matter when it is submitted to an electromagnetic field.

6.3.2.2 Gravito-Electromagnetism in Einstein Spacetime

For the sake of the simplicity, let us consider the simplest example of action for free electromagnetic field without sources and occurring within a Riemann spacetime (curved):

$$\mathscr{S} := \int_{\mathscr{M}} \mathscr{L}\omega_n \quad \text{with} \quad \mathscr{L} := -\frac{1}{4}\mathscr{F}^{\mu\nu}\,\mathscr{F}_{\mu\nu} + \frac{1}{2\chi}\overline{\mathscr{R}} \tag{6.58}$$

The Faraday tensor (here the electromagnetic field is minimally coupled to the gravitation in a Einstein spacetime via Levi-Civita connection) and the scalar curvature are defined by the relationships[3]:

$$\mathcal{F}_{\mu\nu} := \overline{\nabla}_\mu A_\nu - \overline{\nabla}_\nu A_\mu, \qquad \mathcal{R} := g^{\mu\nu}\mathfrak{R}_{\mu\nu} \tag{6.59}$$

where the Faraday tensor $\mathcal{F}^{\mu\nu}$ is calculated with the connection with zero torsion. First, the Lagrangian variation of the action (6.58) allows us to obtain the expression:

$$\begin{aligned}
\Delta\mathscr{S} = \int_{\mathcal{M}} \Bigg\{ &-\frac{1}{2}\mathcal{F}^{\mu\nu}\Delta\mathcal{F}_{\mu\nu} + \frac{1}{4}\mathcal{F}_{\mu\nu}\left(g^{\mu\lambda}\mathcal{F}^{\rho\nu} + \mathcal{F}^{\mu\rho}g^{\lambda\nu}\right)\Delta g_{\lambda\rho} \\
&+ \frac{1}{2\chi}\left(\mathfrak{R}^{\lambda\rho} - \frac{\mathcal{R}}{2}g^{\lambda\rho}\right)\Delta g_{\lambda\rho} + \frac{1}{8}\mathcal{F}^{\mu\nu}\,\mathcal{F}_{\mu\nu}\,g^{\lambda\rho}\Delta g_{\lambda\rho} \\
&+ \frac{1}{2\chi}g^{\mu\nu}\left[\overline{\nabla}_\lambda\left(\Delta\overline{\Gamma}^\lambda_{\mu\nu}\right) - \overline{\nabla}_\mu\left(\Delta\overline{\Gamma}^\lambda_{\lambda\nu}\right)\right] \Bigg\}\,\omega_n
\end{aligned}$$

In the variational of the previous Lagrangian (6.58), it is worth to remind that the independent variations of the metric and the four-potential vector should be performed. The Lagrangian variation of the Faraday tensor takes the form of:

$$\Delta\mathcal{F}_{\mu\nu} = \overline{\nabla}_\mu(\Delta A_\nu) - \overline{\nabla}_\nu(\Delta A_\mu) \tag{6.60}$$

This relation is obtained by directly writing:

$$\begin{aligned}
\Delta\mathcal{F}_{\mu\nu} &= \Delta\left(\partial_\mu A_\nu - \overline{\Gamma}^\rho_{\mu\nu}A_\rho\right) - \Delta\left(\partial_\nu A_\mu - \overline{\Gamma}^\rho_{\nu\mu}A_\rho\right) \\
&= \left(\partial_\mu\Delta A_\nu - \overline{\Gamma}^\rho_{\mu\nu}\Delta A_\rho - \Delta\overline{\Gamma}^\rho_{\mu\nu}A_\rho\right) - \left(\partial_\nu\Delta A_\mu - \overline{\Gamma}^\rho_{\nu\mu}\Delta A_\rho - \Delta\overline{\Gamma}^\rho_{\nu\mu}A_\rho\right)
\end{aligned}$$

accounting for that the variation of the geometric structure, say $\Delta\overline{\Gamma}^\rho_{\mu\nu}$, induces a variation of the field $\Delta\mathcal{F}_{\mu\nu}$. Second, the two systems of conservation laws associated to the unknown primal variables (say the four-vector potential A_μ, and the Riemannian metric $g_{\mu\nu}$) are derived by varying the Lagrangian along the Lie-derivative variations $\mathcal{L}_\xi A_\mu$, and $\mathcal{L}_\xi g_{\mu\nu}$. Shifting the divergence terms at the boundary of the continuum and assuming a zero divergence at this boundary allow us to obtain the conservation laws governing the gravitation interacting with

---

[3]In the remaining part of this subsection the symbol $\overline{\nabla}$ represents the Levi-Civita connection associated to the nonuniform metric tensor of the continuum manifold.

electromagnetism. We can rearrange the Lagrangian variation of the action to give:

$$\Delta \mathscr{S} = \int_{\mathcal{M}} \overline{\nabla}_\nu \mathcal{F}^{\mu\nu} \, \Delta A_\mu \, \omega_n + \int_{\mathcal{M}} \left[\frac{1}{2\chi} \left(\overline{\mathfrak{R}}^{\lambda\rho} - \frac{\overline{\mathcal{R}}}{2} g^{\lambda\rho} \right) \right.$$

$$\left. + \frac{1}{8} \mathcal{F}^{\mu\nu} \mathcal{F}_{\mu\nu} \, g^{\lambda\rho} + \frac{\mathcal{F}_{\mu\nu}}{4} \left(g^{\mu\lambda} \mathcal{F}^{\rho\nu} + \mathcal{F}^{\mu\rho} g^{\lambda\nu} \right) \right] \Delta g_{\lambda\rho} \, \omega_n \qquad (6.61)$$

owing that the Faraday tensor is in fine expressed in terms of the potential A_μ by means of Eq. (6.59). To ensure the existence of electromagnetic field. Due to the arbitrariness of the metric and potential variations, we obtain the classical (and covariant) Einstein–Maxwell's equations:

$$\begin{cases} \overline{\nabla}_\nu \mathcal{F}^{\mu\nu} = 0 \\ \dfrac{1}{2\chi} \left(\overline{\mathfrak{R}}^{\lambda\rho} - \dfrac{\overline{\mathcal{R}}}{2} g^{\lambda\rho} \right) + \dfrac{1}{8} \mathcal{F}^{\mu\nu} \mathcal{F}_{\mu\nu} \, g^{\lambda\rho} + \dfrac{\mathcal{F}_{\mu\nu}}{4} \left(g^{\mu\lambda} \mathcal{F}^{\rho\nu} + \mathcal{F}^{\mu\rho} g^{\lambda\nu} \right) = 0 \end{cases}$$

$$(6.62)$$

where the first equation is the covariant Maxwell's equations in a Riemann continuum, such as curved spacetime or curved continuum matter. The second equation governs the interaction of the electromagnetism to the gravitation. The unknowns in first term of the second equation are the spacetime metric. The electromagnetic source (including both the second and the third terms) in the second equation constitutes the energy-momentum tensor. They influence the gravitation field and vice versa the metric field has also some influence on the electromagnetic field via the Levi-Civita covariant derivative $\overline{\nabla}$.

6.3.2.3 Electromagnetic Four-Dimensional Wave Equations

The Maxwell's equation (6.62) (first row) is used to analyze the electromagnetic wave propagation within a Riemann continuum. Let consider a spacetime $\mathcal{M}$ endowed with a metric $g_{\alpha\beta}$ and a Levi-Civita connection $\overline{\Gamma}^\gamma_{\alpha\beta}$. The first equation may be re-written as follows:

$$\overline{\nabla}_\nu \mathcal{F}^{\mu\nu} = \overline{\nabla}_\nu \left(g^{\mu\alpha} g^{\alpha\nu} \mathcal{F}_{\alpha\beta} \right)$$

$$= \overline{\nabla}_\nu \left[g^{\mu\alpha} \overline{\nabla}_\alpha A^\nu - g^{\nu\beta} \overline{\nabla}_\beta A^\mu \right]$$

$$= g^{\mu\alpha} \left[\overline{\nabla}_\alpha \overline{\nabla}_\nu A^\nu + \mathfrak{R}^\nu_{\nu\alpha\gamma} A^\gamma \right] - g^{\nu\beta} \overline{\nabla}_\nu \overline{\nabla}_\beta A^\mu = 0$$

where we have used the Schouten's relations (5.21) with a zero torsion. The Maxwell's equations include a classical wave part, a divergence term, and the contribution of the Ricci curvature of the continuum:

$$- g^{\nu\beta}\overline{\nabla}_\nu\overline{\nabla}_\beta A^\mu + g^{\mu\alpha}\overline{\nabla}_\alpha\overline{\nabla}_\nu A^\nu + g^{\mu\alpha}\overline{\mathfrak{R}}_{\alpha\gamma}A^\gamma = 0 \tag{6.63}$$

The first term expresses a D'Alembertian operator. The second term may be dropped if we assume a null divergence $\overline{\nabla}_\nu A^\nu = 0$ as a gauge condition (Lorenz gauge). We then obtain the electromagnetic wave propagation equation within curved continuum:

$$- g^{\nu\beta}\overline{\nabla}_\nu\overline{\nabla}_\beta A^\mu + g^{\mu\alpha}\overline{\mathfrak{R}}_{\alpha\gamma}A^\gamma = 0 \tag{6.64}$$

in which we notice the direct influence of the gravitation (represented by the Ricci curvature) on the electromagnetic wave propagation. In the following we will consider an extension of Eq. (6.64) in the framework of Riemann–Cartan continuum (also called Einstein–Cartan gravitation).

Let now consider the dispersion equation. Remind that in a covariant formulation, space and time are merged into a spacetime where each event is defined by four-coordinate $x^\mu = (x^0 := ct, x^1, x^2, x^3)$. The normalization with c allows us to work with only space coordinate. Similarly, the spectral domain parameters are combined into a covariant vector $\kappa_\mu = (\omega/c, k_1, k_2, k_3) = (\omega/c, -k^1, -k^2, -k^3)$. The phase is thus written compactly as a inner product: $\kappa_\mu x^\mu = \omega t - k_1 x^1 - k_2 x^2 - k_3 x^3 = \omega t - \mathbf{k} \cdot \mathbf{x}$. Searching solutions as plane wave $A^\mu = \hat{A}^\mu e^{\kappa_\sigma x^\sigma}$ leads to the vector equation of electromagnetic waves within a curved spacetime as[4]:

$$\left[-\delta^\mu_\nu g^{\sigma\beta}\kappa_\sigma\kappa_\beta + g^{\mu\sigma}\kappa_\sigma\kappa_\nu + g^{\mu\alpha}\overline{\mathfrak{R}}_{\alpha\nu}\right] A^\nu = 0$$

owing the derivative formula $\overline{\nabla}_\beta \left(\hat{A}^\mu e^{\kappa_\sigma x^\sigma}\right) = \kappa_\beta \hat{A}^\mu e^{\kappa_\sigma x^\sigma}$ and the other derivatives take analogous formulation. Non zero four-vector A^ν exists if and only if the determinant of the matrix vanishes. This permits to deduce the dispersion equation within a curved spacetime:

$$\mathrm{Det}\left[-\delta^\mu_\nu g^{\sigma\beta}\kappa_\sigma\kappa_\beta + g^{\mu\sigma}\kappa_\sigma\kappa_\nu + g^{\mu\alpha}\overline{\mathfrak{R}}_{\alpha\nu}\right] = 0 \tag{6.65}$$

For physical interpretation in the framework of three-dimensional formulation, dispersion relations reduce to algebraic equations that merely relate the wave frequency ω and the wave number vector $\mathbf{k}$.

---

[4]Care should be taken since suggesting the coordinate dependence $e^{\kappa_\sigma x^\sigma}$ implicitly assumes that time and space are separated (method of variable separation). The covariance of the formulation is expected to allow us to derive the dispersion equation in any inertial frame in the framework of special relativity.

6.3.2.4 Field Equations and Conservation Laws

The second row of system (6.62) is the field equation which extends the Einstein equation for vacuum spacetime, where the term represents the energy momentum analogous of the Maxwell energy-momentum for the space part, and with nonsymmetric property when considering the timelike part:

$$T^{\lambda\rho} := -\frac{1}{4}\mathcal{F}^{\mu\nu}\mathcal{F}_{\mu\nu}\,g^{\lambda\rho} - \frac{\mathcal{F}_{\mu\nu}}{2}\left(g^{\mu\lambda}\mathcal{F}^{\rho\nu} + \mathcal{F}^{\mu\rho}g^{\lambda\nu}\right) \tag{6.66}$$

It is the Minkowski energy-momentum tensor due to electromagnetic field. It modifies the gravitational field as source whereas the spacetime modifies the electromagnetic field according to (6.64). The temporal component of the energy-momentum (6.66) holds:

$$T^{00} = \frac{1}{2}\left(D^i E_i + B^i H_i\right) \tag{6.67}$$

which is exactly the electromagnetic energy density $T^{00} = (1/2)\,(\mathbf{D}\cdot\mathbf{E} + \mathbf{B}\cdot\mathbf{H}) :=$ $\mathscr{E}$ in a three-dimensional formulation. By introducing the electromagnetic tensors (6.25) and (6.28) into the expression of the energy-momentum tensor, we have the following particular cases:

$$\begin{cases} T^{01} = E_2 H_3 - E_3 H_2 \\ T^{02} = E_3 H_1 - E_1 H_3 \\ T^{03} = E_1 H_2 - E_2 H_1 \end{cases} \qquad \begin{cases} T^{10} = D^2 B^3 - D^3 B^2 \\ T^{20} = D^3 B^1 - D^1 B^3 \\ T^{30} = D^1 B^2 - D^2 B^1 \end{cases}$$

showing again that the $\{T^{01}, T^{02}, T^{03}\}$ are the components of the vector $\mathbf{E} \times \mathbf{H}$, whereas $\{T^{10}, T^{20}, T^{30}\}$ are the components of the vector $\mathbf{D} \times \mathbf{B}$, also called Minkowski momentum density e.g. Milonni and Boyd (2010), in the three-dimensional formulation. This highlights that the Minkowski tensor $T^{\mu\nu}$ is not symmetric when considering the time index 0. It is worth to express the energy momentum as:

$$T_M^{\mu\nu} = \begin{bmatrix} \mathscr{E} & \mathbf{E} \times \mathbf{H} \\ \mathbf{D} \times \mathbf{B} & \mathbf{T}^M \end{bmatrix} \tag{6.68}$$

where $\mathscr{E}$ is the energy, and $\mathbf{T}^M$ is the Maxwell tensor with contravariant components T^{ij}. The (nonsymmetric) energy-momentum such defined is called Minkowski energy momentum.

Remark 6.16 The Poynting vector $\mathbf{S} := \mathbf{E} \times \mathbf{H}$ (originally discovered by JH Poynting in 1884 and denoted this form) represents the rate of energy in the i-direction as mentioned in the formulation T^{0i}. The definition of the Poynting vector as the vector product of the electric field $\mathbf{E}$ with the magnetic field $\mathbf{H}$ is valid in a continuum such as the spacetime or a continuum matter.

Remark 6.17 Due to the nonsymmetry of the Minkowski energy-momentum, other tensors have been considered in the past, the most known is the symmetric Abraham energy momentum tensor which can be written as:

$$T_A^{\mu\nu} = \begin{bmatrix} \mathscr{E} & \mathbf{E} \times \mathbf{H} \\ \mathbf{E} \times \mathbf{H} & \mathbf{T}^M \end{bmatrix} \tag{6.69}$$

$T_M^{\mu\nu}$ and $T_A^{\mu\nu}$ are by far the most cited and considered. They have been also compared and constitute a source of controversial debate since 100 of years. It is not yet closed. Among the numerous results for supporting the choice of the one or the other, experimental measurements of radiation pressure of light on matter have shown the complementarity of these two energy-momenta, see for instance (Griffiths 2011) for mainly macroscopic electromagnetism, and Milonni and Boyd (2010) for physical interpretations of the two energy-momentum on the radiation pressure of light on a dielectric. Radiation pressure is the mechanical pressure exerted on a surface in the direction of propagation by an incident electromagnetic radiation.

For completeness, the spatial components of the Maxwell energy-momentum take the form of for diagonal contributions:

$$\begin{cases} T^{11} = -\dfrac{1}{2}\left(+D^1 E_1 - D^2 E_2 - D^3 E_3\right) - \dfrac{1}{2}\left(+B^1 H_1 - B^2 H_2 - B^3 H_3\right) \\[2mm] T^{22} = -\dfrac{1}{2}\left(-D^1 E_1 + D^2 E_2 - D^3 E_3\right) - \dfrac{1}{2}\left(-B^1 H_1 + B^2 H_2 - B^3 H_3\right) \\[2mm] T^{33} = -\dfrac{1}{2}\left(-D^1 E_1 - D^2 E_2 + D^3 E_3\right) - \dfrac{1}{2}\left(-B^1 H_1 - B^2 H_2 + B^3 H_3\right) \end{cases}$$

For off-diagonal spatial terms we obtain (remind that $T^{\mu\nu}$ is symmetric):

$$\begin{cases} T^{12} = -\left(D^2 E_1 + B^2 H_1\right) \\ T^{23} = -\left(D^3 E_2 + B^3 H_2\right) \\ T^{31} = -\left(D^1 E_3 + B^1 H_3\right) \end{cases}$$

By considering the Eulerian variation (Lie derivative variation $\mathcal{L}_\xi g_{\lambda\rho}$) on the Riemann continuum we obtain the conservation equation as follows:

$$\overline{\nabla}_\rho \left(\overline{\mathfrak{R}}^{\lambda\rho} - \frac{\overline{\mathcal{R}}}{2} g^{\lambda\rho}\right) = 0 \qquad \Longrightarrow \qquad \overline{\nabla}_\rho T^{\lambda\rho} = 0$$

since we have seen that the covariant derivative of the Einstein tensor vanishes by using the Bianchi relationships (see Eq. (4.73)).

It is now essential to introduce the energy-momentum tensor in the conservation laws. Notice that the spatial part of the Minkowski energy-momentum tensor is also called Maxwell stress tensor. The conservation laws take the form of:

$$\begin{cases} \overline{\nabla}_\rho T^{0\rho} = \overline{\nabla}_0 T^{00} + \overline{\nabla}_i T^{0i} \\ \overline{\nabla}_\rho T^{i\rho} = \overline{\nabla}_0 T^{i0} + \overline{\nabla}_j T^{ij} \end{cases} \tag{6.70}$$

We recognize that the first row reduces to the Poynting's theorem (6.21) when considering an isotropic and homogeneous continuum, where the Joule effect is not present because we did not introduce it in the Lagrangian function $\mathscr{L}$. The second row represents the "force equilibrium" where the time derivative of the Poynting's vector compensate the divergence of the Maxwell's stress tensor.

Going back to the system of equation (6.62), it is interesting to re-formulate the second row to give:

$$\frac{1}{\chi} \left(\overline{\mathfrak{R}}^{\lambda\rho} - \frac{\overline{\mathcal{R}}}{2} g^{\lambda\rho} \right) = T_M^{\lambda\rho} = \frac{1}{2} \left(T_M^{\lambda\rho} + T_M^{\rho\lambda} \right) + \frac{1}{2} \left(T_M^{\lambda\rho} - T_M^{\rho\lambda} \right) \tag{6.71}$$

where the left-hand side of the equation is symmetric whereas the right-hand side is not. This induces that the skew-symmetric part of the Minkowski energy-momentum $T_M^{\lambda\rho}$ does not contribute to bend the spacetime and then has no influence on the gravitation field. In a vacuum Minkowskian spacetime remind however that $\epsilon_0 \mu_0 = c^2 = 1$, for we adopt a coordinate system $(x^0 := ct, x^1, x^2, x^3)$. In such a case the two energy-momenta merge and the Abraham energy-momentum coincides with the Minkowski energy-momentum, problems only arise when electromagnetism interact with continuum matter.

Starting from Eq. (6.71) which relates the electromagnetic fields $T^{\lambda\rho}$ as source of the bending of the spacetime, we can multiply this equation by the covariant components of the metric $g_{\lambda\rho}$ to obtain without difficulty the Ricci curvature and then the curvature of the spacetime:

$$-\overline{\mathcal{R}} = \mathscr{T} := g_{\lambda\rho} T_M^{\lambda\rho} = -2\chi \mathscr{F}_{\lambda\rho} \mathscr{F}^{\lambda\rho} = \chi \, (\mathbf{D} \cdot \mathbf{E} - \mathbf{B} \cdot \mathbf{H}) \tag{6.72}$$

which is exactly χ times twice of the electromagnetic part of the Lagrangian function.

Remark 6.18 The debate of physicists and mathematicians about the choice of energy-momentum draw back to the beginning of relativistic theory more than 100 years ago. At least two factors may constitute the reasons of its revival nowadays: the increasing role of optics in modern communication technology, and the legitimate seek of consistent physics theory. Previous studies suggest that the two energy-momenta (6.68) and (6.69) are in fact correct but in different circumstances e.g. Milonni and Boyd (2010). They have their own physical interpretations. Anyway, the non equilibrium of the skew-symmetric part of the Minkowski energy momentum (6.68) may be source of indeterminacy of solutions of the field equations, namely for the angular momentum.

It is reminded that the Lorentz force and the four-dimensional form of conservation laws can be directly deduced from the Maxwell's equations. For the sake of the completeness, we report here below the sketch of proof. Let us start with the Maxwell's equation (6.2) by multiplying the lines (1) and (3) respectively by **E** and by **H** (both 1-forms), and by cross-multiplying the lines (2) and (4) by **B** and **D** respectively:

$$\begin{cases} (\hat{\nabla} \cdot \mathbf{D})\mathbf{E} = \rho\mathbf{E} \\ \mathbf{B} \times (\hat{\nabla} \times \mathbf{H} - c\,\partial_0\mathbf{D}) = \mathbf{B} \times \mathbf{J} \\ (\hat{\nabla} \cdot \mathbf{B})\mathbf{H} = 0 \\ \mathbf{D} \times (\hat{\nabla} \times \mathbf{E} + c\,\partial_0\mathbf{B}) = 0 \end{cases} \tag{6.73}$$

The sum of the first and third lines minus the second and the last rows gives a necessary condition in which we remark the Minkowski energy momentum:

$$\mathbf{F}_E + \mathbf{F}_M - c\partial_0\,(\mathbf{D} \times \mathbf{B}) = \rho\mathbf{E} + \mathbf{J} \times \mathbf{B}$$

where the right-hand side of the equation is merely the so-called Lorentz force, and where we have defined the electric and magnetic internal forces:

$$\begin{cases} \mathbf{F}_E := (\hat{\nabla} \cdot \mathbf{D})\mathbf{E} - \mathbf{D} \times (\hat{\nabla} \times \mathbf{E}) \\ \mathbf{F}_M := (\hat{\nabla} \cdot \mathbf{B})\mathbf{H} - \mathbf{B} \times (\hat{\nabla} \times \mathbf{H}) \end{cases} \tag{6.74}$$

We remind some obvious vectorial relationships (proof may be easily derived with component form) (same relations may be established with **B** and **H**):

$$\begin{cases} \hat{\nabla}(\mathbf{D} \cdot \mathbf{E}) = (\hat{\nabla}\mathbf{D})\mathbf{E} + (\hat{\nabla}\mathbf{E})\mathbf{D} \\ \hat{\nabla} \cdot (\mathbf{E} \otimes \mathbf{D}) = (\hat{\nabla}\mathbf{D})\mathbf{E} + \hat{\nabla}_{\mathbf{D}}\mathbf{E} \\ \mathbf{D} \times (\hat{\nabla} \times \mathbf{E}) = (\hat{\nabla}\mathbf{E})\mathbf{D} - \hat{\nabla}_{\mathbf{D}}\mathbf{E} \end{cases}$$

We introduce these relations into the previous equation. We recover the following formulas for the internal electromagnetic forces in the vectorial form:

$$\begin{cases} \mathbf{F}_E := \hat{\nabla} \cdot \left(\mathbf{E} \otimes \mathbf{D} - \dfrac{1}{2}\,(\mathbf{E} \cdot \mathbf{D})\,\mathbb{I}\right) - \dfrac{1}{2}\left[(\hat{\nabla}\mathbf{D})\mathbf{E} - (\hat{\nabla}\mathbf{E})\mathbf{D}\right] \\ \mathbf{F}_M := \hat{\nabla} \cdot \left(\mathbf{H} \otimes \mathbf{B} - \dfrac{1}{2}\,(\mathbf{H} \cdot \mathbf{B})\,\mathbb{I}\right) - \dfrac{1}{2}\left[(\hat{\nabla}\mathbf{B})\mathbf{H} - (\hat{\nabla}\mathbf{H})\mathbf{B}\right] \end{cases} \tag{6.75}$$

Now, we define the three-dimensional spatial part of the (non-symmetric) Maxwell stress and the Maxwell force density respectively as:

$$\begin{cases} \mathbf{T}_{\text{Maxwell}} := \mathbf{E} \otimes \mathbf{D} + \mathbf{H} \otimes \mathbf{B} - \dfrac{1}{2}\,(\mathbf{E} \cdot \mathbf{D} + \mathbf{H} \cdot \mathbf{B}) \\ \mathbf{f}_{\text{Maxwell}} := \underbrace{\rho\mathbf{E} + \mathbf{J} \times \mathbf{B}}_{\text{Lorentz force}} + \dfrac{1}{2}\left[(\hat{\nabla}\mathbf{D})\mathbf{E} - (\hat{\nabla}\mathbf{E})\mathbf{D}\right] + \dfrac{1}{2}\left[(\hat{\nabla}\mathbf{B})\mathbf{H} - (\hat{\nabla}\mathbf{H})\mathbf{B}\right] \end{cases}$$

$$\tag{6.76}$$

These are the three-dimensional version of the four-dimensional electromagnetic Minkowski energy momentum (6.68). The total Maxwell force density $\mathbf{f}_{\text{Maxwell}}$ includes the so-called Lorentz force and the Helmholtz's force. The Lorentz force $\rho \mathbf{E} + \mathbf{J} \times \mathbf{B}$ is the total force exerted by the electromagnetic field on the material continuum. In sum, we can deduce from the Maxwell's equations the momentum balance equation and the continuity equation. Indeed, by considering the quantities (6.75) and (6.76), we obtain:

$$\hat{\nabla} \cdot \mathbf{T}_{\text{Maxwell}} = \mathbf{f}_{\text{Maxwell}} + c\partial_0 \left(\mathbf{D} \times \mathbf{B}\right) \tag{6.77}$$

When the medium is homogeneous and the electric permittivity and the magnetic susceptibility are uniform, the Helmholtz's force is equal to zero.

Remark 6.19 It should be stressed again that the Lorentz force is not an a priori constitutive law since it is expressed only in terms of electromagnetic external field where the charged body is in immersion. It is rather deduced from non homogeneous Maxwell's equations. However, it should be remarked that the Lorentz force remains an approximation for small charges and generated currents which allows us to neglect the self-interaction.

6.4 Electromagnetism in Curved Continuum with Torsion

Analysis of electromagnetic fields in presence of extremely massive gravitation remains a relevant topic in relativistic astrophysics. Propagation of electromagnetic waves governed by Maxwell's equation (6.44) within a curved spacetime constitutes a fundamental basis for studying signals received from neutron stars and black holes to name but a few in astrophysics. Other methods consist in measuring the signal due to gravitational waves. It is now admitted that the influence of the non-Minkowskian metric of the curved spacetime is much stronger on the electromagnetic field $\mathcal{F}_{\mu\nu}(x^\alpha)$ than the influence of this field on the bending of the spacetime $\mathcal{M}$. In this section we consider the gravitation electromagnetism interaction within a Riemann–Cartan continuum endowed with metric $g_{\alpha\beta}(x^\mu)$ and connection $\Gamma^\gamma_{\alpha\beta}(x^\mu)$.

6.4.1 Electromagnetic Strength (Faraday Tensor)

First, we remind that interaction of Einstein gravitation and electromagnetism was considered in a curved but torsionless Riemannian spacetime e.g. Fernandez-Nunez and Bulashenko (2016). It is usually assumed the case where the electromagnetic field is of the order of small perturbation of the spacetime metric. Only the influence of the metric on electromagnetic field is mostly accounted for, not the converse. Second, the influence of the Riemann–Cartan geometry on the electromagnetic field

is not so easy. A free electromagnetic field is suggested to not produce torsion e.g.
Hehl et al. (1976), and there is in principle no contribution from torsion in Maxwell
equations. When a strong magnetic field coexists with matter distribution, there is
however a possibility to induce spin polarization of individual particles composing
the continuum matter e.g. Prasanna (1975a). Some authors have even suggested
that torsion play a keyrole in electromagnetism when considering electromagnetic
field within continuum with torsion e.g. Hammond (1989), Poplawski (2009). They
propose that the electromagnetic potential is represented by the torsion vector $A_\alpha :=$
$\aleph_\alpha = \aleph^\beta_{\alpha\beta}$. The influence of torsion tensor as cosmic dislocation (singularity of the
curvature tensor) was investigated in e.g. Dias and Moraes (2005), or some material
defects as screw dislocations (Fumeron et al. 2015), or fluids with spin density
e.g. Schutzhold et al. (2002). They have included the 2-form torsion with one non-
vanishing component as $2\pi\beta\delta^2(r)\, dr \wedge d\theta$ in the metric of the spacetime, then derive
Maxwell's equations in a cylindrical coordinates to solve two interesting cases: the
electric field of a line charge, and the magnetic field of the line current. However,
they seemed to assume a connection based only a metric but do not consider the
contortion tensor for covariant derivation of Maxwell's equations in the framework
of Riemann–Cartan geometry. Formulation of Maxwell's equations by means of
differential forms may be not equivalent to formulation by means connection in
Riemann–Cartan spacetime or continuum e.g. Vandyck (1996). In a Riemann–
Cartan spacetime, the Faraday tensor is calculated as follows e.g. Prasanna (1975a),
Smalley (1986): $\mathcal{F}_{\mu\nu} := \nabla_\mu A_\nu - \nabla_\nu A_\mu = \partial_\mu A_\nu - \partial_\nu A_\mu + \aleph^\rho_{\mu\nu} A_\rho$. It is rather
different if calculated by means of an exterior derivative of the 1-form $\mathbf{A} = (A_\mu)$
e.g. Nakahara (1996), Prasanna (1975a):

$$\mathcal{F} := d\mathbf{A} \qquad \Longrightarrow \qquad \mathcal{F}_{\mu\nu} = \partial_\mu A_\nu - \partial_\nu A_\mu \qquad (6.78)$$

(the vector base is assumed to satisfy the Frobenius theorem) where, in such a
case, we have exactly the same form as the Minkowski (6.25), and Riemann (6.46)
Faraday tensor. In this framework, two of the Maxwell's equations $d\mathcal{F} = 0$ would
be expected since the Faraday tensor 2-form $\mathcal{F}$ is exact, say $\mathcal{F} := d\mathbf{A}$, and hence
closed, $d\mathcal{F} = d(d\mathbf{A}) = 0$. To investigate electromagnetic waves within curved
continuum matter with torsion (which may be considered as a Riemann–Cartan
manifold), it is then assumed that the electromagnetic field is described by an
electromagnetic 2-form $\mathcal{F}_{\mu\nu}$. It constitutes an extended model of electromagnetism
within curved spacetime as earlier as in e.g. Plebanski (1960), and in the framework
of differential forms e.g. Frankel (1997), Prasanna (1975a). Prasanna (1975a) has
derived the Maxwell equations in a Riemann–Cartan spacetime. In the following,
we would like to derive the Maxwell's equations in a curved manifold with torsion
$\mathcal{M}$. By using a formalism based on exterior calculus, Maxwell's equation (6.44)
were established for various spacetimes (Minkowski, Riemann, and almost post-
Riemann) (Puntigam et al. 1997) where they considered as basic axioms the
conservation of electric charge and the conservation of magnetic flux. This allows

them to put aside the connection structure of the spacetime. Third, either for metric-based energy, or metric-torsion based energy, it is worth to define Lagrangian $\mathscr{L}(\mathcal{F}_{\mu\nu}, u^{\mu}, \mathcal{M}^{\mu\nu}, \mathcal{J}^{\mu}, A_{\mu}, \cdots)$ associated to the electromagnetic fields when we face the question of variational formulation.

Remark 6.20 To relate electromagnetism with relativistic gravitation, it is interesting to remind that application of the gauge invariance principle for the group of translation (corresponding to torsion) of the spacetime $\mathcal{M}$ with Yang-Mills type Lagrangian, quadratic in the field strengths $\mathcal{F}_{\mu\nu}$ (as for electromagnetism), allows us to deduce the usual Einstein's theory of gravitation, based on the Einstein–Hilbert action e.g. Cho (1976a).

6.4.2 Electromagnetism Interacting with Gravitation

We derive in this subsection the equations governing the interaction of electromagnetism with a Riemann–Cartan continuum.

6.4.2.1 Different Lagrangians in Electromagnetism

A priori definition of a Lagrangian is preferred here rather than directly considering local equations of Maxwell and introducing at a second step the constitutive laws. This is extension of linear constitutive laws previously investigated (6.50). Some Lagrangian functions have been proposed for describing the evolution of electromagnetic fields:

1. a 4-form of the Lagrangian which is Poincaré invariant for moving medium with 4-velocity u^{μ} e.g. Obukhov (2008), Schutzhold et al. (2002):

$$\mathscr{L} = -\frac{1}{4}\mathcal{F}^{\mu\nu}\,\mathcal{F}_{\mu\nu} - \frac{\varepsilon - 1}{2}\mathcal{F}_{\mu\lambda}u^{\lambda}\,\mathcal{F}^{\mu\nu}u_{\nu} \tag{6.79}$$

This Lagrangian function is based on the finding of Gordon that the propagation of electromagnetic waves in a moving dielectric continua has analogy with the propagation of electromagnetic waves in a pseudo-Riemannian curved spacetime. Gordon introduced an effective metric $g^{\mu\nu} := \hat{g}^{\mu\nu} + (\varepsilon - 1)u^{\mu}u^{\nu}$ with $\hat{g}^{\mu\nu}$ the metric of Minkowski spacetime where the continuum is flowing. The Lagrangian is merely a alternative formulation of

$$\mathscr{L} = -\frac{1}{4}g^{\mu\alpha}g^{\nu\beta}\mathcal{F}_{\alpha\beta}\,\mathcal{F}_{\mu\nu}$$

2. a 4-form of Lagrangian in the presence of magnetization tensor $\mathcal{M}$, a current four-vector $\mathcal{J}$, a four-vector potential $\mathcal{A}$ e.g. Smalley and Krisch (1992) and references herein:

$$\mathscr{L} = -\frac{1}{4}\mathcal{F}^{\mu\nu}\,\mathcal{F}_{\mu\nu} + \frac{1}{2}\mathcal{F}_{\mu\nu}\,\mathcal{M}^{\mu\nu} - \mathcal{J}^{\mu}\mathcal{A}_{\mu} \tag{6.80}$$

In this reference, the electromagnetic field is defined as $F_{\mu\nu} := \nabla_{\mu}A_{\nu} - \nabla_{\nu}A_{\mu}$ where the connection is not torsionless. In such a way, there is minimal coupling between the electromagnetic field and the Riemann–Cartan spacetime geometry. The Faraday tensor is not necessarily a 2-form obtained by an exterior derivative when considering a Einstein–Cartan spacetime.

3. A model including both the electromagnetic field and gravitation field is obtained with the Lagrangian:

$$\mathscr{L} = -\frac{1}{4}\mathcal{F}^{\mu\nu}\,\mathcal{F}_{\mu\nu} - \mathcal{J}^{\mu}\mathcal{A}_{\mu} + \frac{1}{2\chi}\mathcal{R} \tag{6.81}$$

where χ involved in the Einstein–Hilbert Lagrangian is a constant for normalizing the Lagrangian. For the Maxwellian part of the Lagrangian, the raising and lowering indices are done by mean of the gravitational metric $g_{\mu\nu}(x^{\alpha})$.[5]

4. In the general case of general relativity, the metric is no longer a Minkowskian since $g_{\mu\nu} = g_{\mu\nu}(x^{\gamma})$ depend on the coordinates of the spacetime. In an attempt to formulate an unified theory of gravitation and electromagnetism, following the idea of Ferraris and Kijowski, Chrusciel proposed a Lagrangian function having the form of e.g. Chrusciel (1984):

$$\mathscr{L} = \mathscr{L}\left(\mathcal{F}_{\mu\nu}, \mathcal{K}_{\mu\nu}\right), \qquad \mathcal{F}_{\mu\nu} := \mathfrak{R}^{\lambda}_{\mu\nu\lambda} \quad \text{and} \quad \mathcal{K}_{\mu\nu} := \frac{1}{2}\left(\mathfrak{R}^{\lambda}_{\lambda\mu\nu} + \mathfrak{R}^{\lambda}_{\lambda\nu\mu}\right)$$

---

[5] *Physics background*: This continuum version is the extension of Lagrangian for particles within Minkowskian spacetime. For physical particles with relativistic speeds, the action of a charged particle e moving within such a spacetime with electromagnetic field A_{μ} takes the form of e.g. Kovetz (2000):

$$\mathscr{S} := \int \left[-mc^2 - e\,u^{\mu}A_{\mu}\right]d\tau, \quad d\tau := dt\sqrt{1 - v^2/c^2}, \quad u^{\mu} : (1, v^i)\left(1 - v^2/c^2\right)^{-1},$$

$$A_{\mu} : (-\phi, A_i)$$

where the Lagrangian function in terms of three dimensional variables (integration with respect to dt), with its Euler–Lagrange equation (Heaviside-Lorentz equation) hold:

$$\mathscr{L} = -mc^2\sqrt{1 - v^2/c^2} + e\,(\mathbf{v} \cdot \mathbf{A} - \phi) \implies \frac{d}{dt}\left(\frac{m\mathbf{v}}{\sqrt{1 - |\mathbf{v}|^2/c^2}}\right) = e\,(\mathbf{E} + \mathbf{v} \times \mathbf{B}) \tag{6.82}$$

modelling the charged particle e motion under given electromagnetic field $(\mathbf{E}, \mathbf{B})$.

where the torsion tensor does not vanish. This is based on the idea that the presence of electromagnetic fields bends the spacetime and therefore induces a non vanishing curvature. He assumed that the Lagrangian which depends only upon the tensors $\mathcal{F}_{\mu\nu}$, and $\mathcal{K}_{\mu\nu}$ (symmetric part of the Ricci curvature tensor) provides the basis of an unified theory of electromagnetism and gravitation. The use of the curvature of the spacetime as the only one variable to sketch electromagnetism and gravitation might be questionable. Nevertheless, it could be checked that the two-covariant skew-symmetric tensor $\mathcal{F}_{\mu\nu}$ satisfies the first set of Maxwell equations (Bianchi equations), and thus could be suggested as the combined electromagnetic field e.g. Hammond (1989), whereas the symmetric tensor obtained from Ricci curvature capture the gravitational fields.

Implicitly the model proposed by Chrusciel suggests that the origin of the electromagnetism comes from the skew-symmetric part of the spacetime connection, and then of the torsion tensor.

6.4.2.2 Variation Procedure

For the sake of the simplicity, let us consider the simplest example of action for free electromagnetic field without sources and occurring within a Einstein–Cartan spacetime (curved with torsion):

$$\mathcal{S} := \int_{\mathcal{M}} \mathcal{L}\,\omega_n \quad \text{with} \quad \mathcal{L} := -\frac{1}{4}\mathcal{F}^{\mu\nu}\,\mathcal{F}_{\mu\nu} + \frac{1}{2\chi}\mathcal{R} \tag{6.83}$$

For the variation of the Lagrangian (6.83), it is worth to remind that the metric tensor and the torsion tensor are independents primal variables in addition to the electromagnetic four-potential. The Faraday tensor (here the electromagnetic field is minimally coupled to the gravitation in a Einstein–Cartan spacetime via the torsion of the connection) and the scalar curvature are defined by the relationships:

$$\mathcal{F}_{\mu\nu} := \nabla_\mu A_\nu - \nabla_\nu A_\mu = \overline{\nabla}_\mu A_\nu - \overline{\nabla}_\nu A_\mu + \aleph^\rho_{\mu\nu} A_\rho, \qquad \mathcal{R} := g^{\mu\nu}\mathfrak{R}_{\mu\nu} \tag{6.84}$$

where the contravariant components of the Faraday tensor $\mathcal{F}^{\mu\nu}$ are calculated by means the connection with non zero torsion e.g. Smalley and Krisch (1992). It should be observed that the definition (6.78) in Riemann spacetime holds for both Euclidean and (pseudo)-Riemannian and also proposed in some post-Riemannian spacetimes e.g. Puntigam et al. (1997). As extension the definition (6.84) is valid for both Euclidean, Riemannian and Riemann–Cartan spacetime. This again illustrates the fact that the extension of physical variables as $\mathcal{F}_{\mu\nu}$ can be done in many ways (as a 2-form in e.g. Puntigam et al. (1997) or as a twice the skew-symmetric part of the gradient in e.g. Smalley and Krisch 1992).

The first and second equations show that the Lagrangian variation of this 2-form and curvature include both the variation of the potential A_μ, the variation of the Riemann metric $g_{\alpha\beta}$, and also the variation of the connection $\Gamma^\gamma_{\alpha\beta}$. Here, the

Lagrangian we consider includes Yang-Mills electromagnetic part and Einstein–Hilbert gravitation part e.g. Charap and Duff (1977). As a first step, the Lagrangian variation of the action (6.83) allows us to obtain the expression:

$$\Delta \mathscr{S} = \int_{\mathscr{M}} \left\{ -\frac{1}{2} \mathscr{F}^{\mu\nu} \Delta \mathscr{F}_{\mu\nu} + \frac{1}{4} \mathscr{F}_{\mu\nu} \left(g^{\mu\lambda} \mathscr{F}^{\rho\nu} + \mathscr{F}^{\mu\rho} g^{\lambda\nu} \right) \Delta g_{\lambda\rho} \right.$$
$$+ \frac{1}{2\chi} \left(\mathscr{R}^{\lambda\rho} - \frac{\mathscr{R}}{2} g^{\lambda\rho} \right) \Delta g_{\lambda\rho} + \frac{1}{8} \mathscr{F}^{\mu\nu} \mathscr{F}_{\mu\nu} g^{\lambda\rho} \Delta g_{\lambda\rho}$$
$$+ \left. \frac{1}{2\chi} g^{\mu\nu} \left[\nabla_\lambda \left(\Delta \Gamma^\lambda_{\mu\nu} \right) - \nabla_\mu \left(\Delta \Gamma^\lambda_{\lambda\nu} \right) - \aleph^\rho_{\lambda\mu} \Delta \Gamma^\lambda_{\rho\nu} \right] \right\} \omega_n$$

where the last equation may be derived from Palatini relation e.g. Lichnerowicz (1955), Rakotomanana (2003). The Lagrangian variation of the Faraday tensor takes the form of:

$$\Delta \mathscr{F}_{\mu\nu} = \nabla_\mu (\Delta A_\nu) - \nabla_\nu (\Delta A_\mu) + \Delta \aleph^\rho_{\mu\nu} A_\rho \qquad (6.85)$$

This relation is obtained by directly writing:

$$\Delta \mathscr{F}_{\mu\nu} = \Delta \left(\partial_\mu A_\nu - \Gamma^\rho_{\mu\nu} A_\rho \right) - \Delta \left(\partial_\nu A_\mu - \Gamma^\rho_{\nu\mu} A_\rho \right)$$
$$= \left(\partial_\mu \Delta A_\nu - \Gamma^\rho_{\mu\nu} \Delta A_\rho - \Delta \Gamma^\rho_{\mu\nu} A_\rho \right) - \left(\partial_\nu \Delta A_\mu - \Gamma^\rho_{\nu\mu} \Delta A_\rho - \Delta \Gamma^\rho_{\nu\mu} A_\rho \right)$$

accounting for that the variation of the geometric structure, say $\Delta \aleph^\rho_{\mu\nu}$, induces a variation of the field $\Delta \mathscr{F}_{\mu\nu}$. At a second step, the three systems of conservation laws associated to the unknown primal variables (say the four-vector potential A_μ, the Riemannian metric $g_{\mu\nu}$, and the torsion $\aleph^\rho_{\mu\nu}$) are derived by varying the Lagrangian along the Lie-derivative variations $\mathcal{L}_\xi A_\mu$, $\mathcal{L}_\xi g_{\mu\nu}$, and $\mathcal{L}_\xi \aleph^\rho_{\mu\nu}$.

Remark 6.21 It should be noticed that the arbitrariness of the potential variation ΔA_μ should be used instead of the Faraday tensor $\Delta \mathscr{F}_{\mu\nu}$ in order to ensure the existence of electromagnetic field.

6.4.2.3 Field Equations

The variation of the electromagnetic strength (6.85) plays a keyrole in the present work since it allows us to couple the torsion of the continuum material and the electromagnetic field.

Now we factorize the variation with respect to the Lagrangian variations of the electromagnetic potential ΔA_μ, the metric $\Delta g_{\lambda\rho}$, and the connection $\Delta \Gamma^\lambda_{\mu\nu}$ respectively. The presence of the term $\Delta \Gamma^\lambda_{\mu\nu}$ means that the torsion and curvature may evolve since they are independent primal variables of the theory. By shifting divergence terms at the boundary of the continuum $\mathscr{M}$ we can rearrange the

Lagrangian variation of the action to give:

$$\Delta \mathscr{S} = \int_{\mathscr{M}} \nabla_\nu \mathscr{F}^{\mu\nu} \, \Delta A_\mu \, \omega_n$$
$$+ \int_{\mathscr{M}} \left[\frac{1}{2\chi} \left(\mathfrak{R}^{\lambda\rho} - \frac{\mathscr{R}}{2} g^{\lambda\rho} \right) \right.$$
$$\left. + \frac{1}{8} \mathscr{F}^{\mu\nu} \mathscr{F}_{\mu\nu} \, g^{\lambda\rho} + \frac{\mathscr{F}_{\mu\nu}}{4} \left(g^{\mu\lambda} \mathscr{F}^{\rho\nu} + \mathscr{F}^{\mu\rho} g^{\lambda\nu} \right) \right] \Delta g_{\lambda\rho} \, \omega_n$$
$$- \int_{\mathscr{M}} \left((\mathscr{F}^{\mu\nu} - \mathscr{F}^{\nu\mu}) A_\lambda + \frac{1}{\chi} g^{\rho\nu} \aleph^\mu_{\lambda\rho} \right) \Delta \Gamma^\lambda_{\mu\nu} \, \omega_n \qquad (6.86)$$

owing that the Faraday tensor is in fine expressed in terms of the potential A_μ by means of Eq. (6.84). Due to the arbitrariness of the variation of primal variables, this variational principle allows us to deduce the system of partial differential equations:

$$\begin{cases} \nabla_\nu \mathscr{F}^{\mu\nu} = 0 \\ \dfrac{1}{2\chi} \left(\mathfrak{R}^{\lambda\rho} - \dfrac{\mathscr{R}}{2} g^{\lambda\rho} \right) + \dfrac{1}{8} \mathscr{F}^{\mu\nu} \mathscr{F}_{\mu\nu} \, g^{\lambda\rho} + \dfrac{\mathscr{F}_{\mu\nu}}{4} \left(g^{\mu\lambda} \mathscr{F}^{\rho\nu} + \mathscr{F}^{\mu\rho} g^{\lambda\nu} \right) = 0 \\ (\mathscr{F}^{\mu\nu} - \mathscr{F}^{\nu\mu}) A_\lambda + \dfrac{1}{\chi} g^{\rho\nu} \aleph^\mu_{\lambda\rho} = 0 \end{cases}$$

$$(6.87)$$

where we notice a slightly extension of the fields equations in Charap and Duff (1977) for Riemann–Cartan spacetime. The first row of the system (6.87) expresses the Maxwell's equations in Riemann–Cartan vacuous spacetime, and it should be stressed that in this Lagrangian (model), the potential A_μ may be apparently calculated independently on the gravitation (except eventual coupling at the boundary $\partial \mathscr{M}$). Once again, we remind that the connection approach, for example in matter (6.45), is equivalent to the differential form approach (6.44) when the spacetime is Riemannian without torsion (Vandyck 1996), or when the non metricity of the connection is traceless. The Maxwell's equations of the system (6.87) show that the connection approach with torsion is "naturally" deduced from a variation principle, and the same form as the connection approach is obtained. It is worth to give some details about Eq. (6.87) by writing:

$$\nabla_\nu \mathscr{F}^{\mu\nu} = \nabla_\nu \left(g^{\mu\alpha} \mathscr{F}_{\alpha\beta} g^{\alpha\nu} \right) = \left(g^{\mu\alpha} g^{\nu\beta} - g^{\mu\beta} g^{\nu\alpha} \right) \nabla_\nu \nabla_\alpha A_\beta = 0$$

owing that the connection is metric compatible, then $\nabla_\gamma g_{\alpha\beta} \equiv 0$ and that the skew-symmetric Faraday tensor is defined as $\mathscr{F}_{\alpha\beta} := \nabla_\alpha A_\beta - \nabla_\beta A_\alpha$ (see Eq. (6.84)). This later formulation points out the wave characteristic of the electromagnetic field in which the unknowns are the potential components $A_\beta = (\phi, A_1, A_2, A_3)$.

6.4.2.4 Electromagnetic Wave Equation

The Maxwell's equation (6.87) (first row) may be used to analyze the electromagnetic wave propagation within a curved spacetime with torsion. Let consider a spacetime $\mathcal{M}$ endowed with a metric $g_{\alpha\beta}$ and a connected with $\Gamma^{\gamma}_{\alpha\beta}$, this later is compatible with the metric. Maxwell's equations may be re-written as follows:

$$
\begin{aligned}
\nabla_{\nu}\mathcal{F}^{\mu\nu} &= \nabla_{\nu}\left(g^{\mu\alpha}g^{\alpha\nu}\mathcal{F}_{\alpha\beta}\right) \\
&= \nabla_{\nu}\left[g^{\mu\alpha}\nabla_{\alpha}A^{\nu} - g^{\nu\beta}\nabla_{\beta}A^{\mu}\right] \\
&= g^{\mu\alpha}\left[\nabla_{\alpha}\nabla_{\nu}A^{\nu} - \aleph^{\gamma}_{\nu\alpha}\nabla_{\gamma}A^{\nu} + \mathfrak{R}^{\nu}_{\nu\alpha\gamma}A^{\gamma}\right] - g^{\nu\beta}\nabla_{\nu}\nabla_{\beta}A^{\mu} = 0
\end{aligned}
$$

where we have used the Schouten's relations (5.21). By arranging the previous relationships, we notice that the Maxwell's equations include, as for elastic wave propagation, a classical wave part, a divergence term, and the contribution of the torsion and the Ricci curvature of the spacetime:

$$
- g^{\nu\beta}\nabla_{\nu}\nabla_{\beta}A^{\mu} + g^{\mu\alpha}\nabla_{\alpha}\nabla_{\nu}A^{\nu} - g^{\mu\alpha}\aleph^{\gamma}_{\nu\alpha}\nabla_{\gamma}A^{\nu} + g^{\mu\alpha}\mathfrak{R}_{\alpha\gamma}A^{\gamma} = 0 \tag{6.88}
$$

The first term expresses merely a D'Alembertian operator. The second term may be dropped if we assume a null divergence as a gauge condition. As for Riemann spacetime, the search of solutions as $A^{\mu} = \hat{A}^{\mu}e^{\kappa_{\sigma}x^{\sigma}}$ provides the dispersion equation:

$$
\mathrm{Det}\left[-\delta^{\mu}_{\nu}g^{\sigma\beta}\kappa_{\sigma}\kappa_{\beta} + g^{\mu\sigma}\kappa_{\sigma}\kappa_{\nu} - g^{\mu\alpha}\aleph^{\gamma}_{\nu\alpha}\kappa_{\gamma} + g^{\mu\alpha}\mathfrak{R}_{\alpha\nu}\right] = 0 \tag{6.89}
$$

Remark 6.22 In classical electromagnetism theory, null condition of the divergence is called Lorenz condition which fixes the electromagnetic potential. The condition $\nabla_{\nu}A^{\nu} \equiv 0$ extends the usual Lorenz condition in the framework of Riemann–Cartan geometry, more specifically in the way of Gauss units system.

For a non curved spacetime without torsion, the electromagnetic wave propagation equation reduces to $\square\, A^{\mu} = 0$. The third term introduces a first covariant derivative which leads to a diffusion of the wave (spacetime attenuation), and the last term points out a breathing mode whenever the boundary conditions allow it. What should be observed too is that the torsion and Ricci curvature influence the wave propagation linearly. It should be stressed that the spacetime geometry and *in fine* the gravitation is in fact tightly linked to the electromagnetism phenomenon. This may not be perceived at a first sight.

In sum, the second row of the system (6.87) gives the coupling equation of the electromagnetic field and the gravitational field one. The electromagnetic terms act as a source-term for the gravitation. They act as a kind of electromagnetic energy generating evolution of the spacetime metric. We recognize the Einstein field equation in the absence of the electromagnetic field. Despite its apparent relative simplicity, the system of partial differential equations (6.87) remains complex since the connection, and by the way the Ricci and total curvatures, includes both the (gravitational) metric $g_{\mu\nu}(x^\lambda)$ and the contortion tensor $\mathcal{T}^\gamma_{\mu\nu}(x^\lambda)$. The use of dispersion equation (6.89) allows us to "drop" the covariant derivative although it is always present implicitly.

6.4.2.5 Electromagnetism and Spacetime Defects

The third row of (6.87) gives the equation to calculate the torsion field. It is striking its analogy with the result obtained by Fernando et al. (2012) by considering a particular Riemann–Cartan spacetime and working with contortion tensor. It is a link between electromagnetic fields and the torsion of the spacetime. What is interesting is that the electromagnetic field allows us to calculate with an algebraic explicit formula the torsion field by means of the third row. Once the torsion is obtained, we can apply covariant derivative within Riemann–Cartan geometry. By multiplying with $g_{\nu\sigma}$, the explicit formula for calculating the torsion is obtained accordingly by means of an algebraic relation:

$$\aleph^\mu_{\lambda\sigma} = -\chi\, g_{\sigma\nu} \left(\mathcal{F}^{\mu\nu} - \mathcal{F}^{\nu\mu}\right) A_\lambda = -2\chi\, g_{\sigma\nu} \left(g^{\mu\alpha} g^{\nu\beta} - g^{\mu\beta} g^{\nu\alpha}\right) A_\lambda \nabla_\alpha A_\beta$$

$$(6.90)$$

owing the expression of the electromagnetic strength in terms of potential. It may be noticed that the contribution of the electromagnetic potential to the torsion field is of second order "$A_\lambda \nabla_\alpha A_\beta$".

Remark 6.23 The investigation of the interaction of electromagnetic masses with Einstein–Cartan spacetime was done by numerous authors for charged and spinning "static" dust (static means here no displacement of the center of mass), for perfect fluids with spin density e.g. Smalley and Krisch (1992). It was shown that by analyzing the solutions of Maxwell's equations, the torsion field together with the spin of Einstein–Cartan gravitation theory may be suggested as produced by the electromagnetic field e.g. Tiwari and Ray (1997). Paraphrasing these authors, it was concluded that in the absence of electromagnetic fields, the body has a vanishing spin density which itself is associated to the spacetime torsion e.g. Hehl and von der Heyde (1973). The third algebraic equation of (6.87) conforms this conclusion concerning the torsion field.

Remark 6.24 What could be relevant is that the torsion field of the Riemann–Cartan vacuum spacetime may be explicitly calculated. In other words and from the physical point of view, torsion is thought to be generated by the electromagnetic field demonstrated here by assuming the simplest Lagrangian of the type (6.83).

Remark 6.25 From the system of equations governing the electromagnetism interacting with gravitation (6.87), we notice that the electromagnetic energy-momentum in a vacuum has the same shape as for as for electromagnetic within a continuum matter (Obukhov and Hehl 2003):

$$T^{\lambda\rho} = \frac{1}{4}\mathfrak{F}^{\mu\nu}\mathfrak{F}_{\mu\nu}\, g^{\lambda\rho} - \frac{\mathfrak{F}_{\mu\nu}}{2}\left(g^{\mu\lambda}\mathfrak{F}^{\rho\nu} + \mathfrak{F}^{\mu\lambda}g^{\rho\nu}\right) \tag{6.91}$$

This is a nonsymmetric Minkowski (canonical) energy-momentum tensor e.g. Obukhov (2008) for the free electromagnetic field occurring within spacetime. For electromagnetic field without sources and occurring within a Einstein–Cartan continuum (curved and non zero torsion) matter, the Lagrangian (6.83) is slightly extended to:

$$\mathscr{S} := \int_{\mathscr{M}} \mathscr{L}\omega_n \quad \text{with} \quad \mathscr{L} := -\frac{1}{4}\mathcal{H}^{\mu\nu}\,\mathfrak{F}_{\mu\nu} + \frac{1}{2\chi}\mathcal{R} \tag{6.92}$$

where the constitutive laws are given by Eq. (6.28). The same developments as previously may be conducted to analyze the behavior of the matter electromagnetic interaction. There is a controversy between the version of Minkowski and that of Abraham, not deduced from a Lagrangian. We do not enter into this long last debate, which was done in the past. We have just to remind that the Minkowski version is defined in the framework of Lagrange-Noether conforming to the invariance approach we adopt in this work. Obukhov and Hehl suggested the adoption of the Minkowski like version (6.91) which is motivated by the Lagrangian axiomatic approach, and by the experimental evidence conducted in the past by Walker and Walker (which is based on experimental measurements of dielectric disk placed in a crossed oscillating radial electric and longitudinal magnetic fields), and James (which is based on a similar experimental jig but with radial electric field and azimuthal magnetic field) (Obukhov and Hehl 2003). For illustrating, consider the continuum matter with isotropic constitutive laws $\mathcal{H}^{\mu\nu}(\mathfrak{F}_{\alpha\beta})$ by analogy to (6.33):

$$\mathcal{H}^{\mu\nu} := \Xi^{\mu\nu\alpha\beta}\,\mathfrak{F}_{\alpha\beta}, \qquad \Xi^{\mu\nu\alpha\beta} := \varepsilon\left(g^{\mu\alpha}g^{\nu\beta}\right) \tag{6.93}$$

where the electromagnetic constant ε is assumed uniform within the matter. Metric components $g^{\mu\nu}(x^{\lambda})$ are point dependent within the matter.

Remark 6.26 In the equations we have previously developed, the torsion does not propagate. In order to account for the torsion propagation, i.e., a well-known method would be to add a scalar bilinear term of the covariant derivatives of the torsion e.g. Hammond (1987). Hammond in this reference has interestingly shown a physical interpretation that the trace of the torsion $\aleph_{\nu} := \aleph^{\mu}_{\nu\mu}$ can be considered as the electromagnetic four-potential, and the skew-symmetric part of the Ricci curvature tensor as proportional to the electromagnetic Faraday tensor. For that purpose, he has considered the Lagrangian:

$$\mathscr{S} := \int_{\mathscr{M}}\left(\frac{1}{\chi}\mathcal{R} + a\,\mathcal{G}_{\mu\nu}\mathcal{G}^{\mu\nu}\right) \quad \text{with} \quad \mathcal{G}_{\mu\nu} := \partial_{\mu}\aleph_{\nu} - \partial_{\nu}\aleph_{\mu}$$

where the independent variations of the metric and the torsion are the rules. In his approach the electromagnetic variables are deduced from spacetime geometry. Our approach in the present book is slightly different.

6.5 Einstein–Cartan Gravitation and Electromagnetism

In this section we deal with some particular solutions of the system (6.87) in presence of electric charges and non vanishing torsion field. We consider some special examples of spacetimes with spherical symmetry for illustration. The first example relates the influence of an electric charge on the spacetime geometry, without torsion. The second example illustrates the influence of the torsion on the geometry of the spacetime. The two examples are merged into one.

6.5.1 Reissner-Nordström Spacetime

The Reissner-Nordström metric corresponds to the static solution of (6.87) allowing us to determine the gravitational field of a punctual spherical non-rotating charge Q, contained in body of mass m. The torsion is assumed to be zero. We remind that the punctual electric charge Q is assumed to be at rest at the origin. The goal of this subsection is to calculate the gravitational effect of a body with electric charge.[6]

6.5.1.1 Classical Reissner-Nordström Metric (1916–1918)

Let consider a static gravitation and electromagnetic field with a spherical symmetry governed by the system of equation (6.87). The static condition requires that the metric is independent of the time x^0, and that the time reversal does not modify the metric. However, it is not necessary to assume a static metric. It can be deduced. We can search for spacetime metric defined by the line element:

$$ds^2 := e^{2\nu} (dx^0)^2 - e^{2\mu} dr^2 - r^2 d\theta^2 - r^2 \sin^2 \theta d\varphi^2 \tag{6.94}$$

where the two scalar functions $\nu(r)$ and $\mu(r)$ depend only on the radius r. This line element constitutes the most general form of a spherical symmetric line element. For the sake of the simplicity, we directly consider here a non transient case.

---

[6]In astrophysics, most planets as stars being electrically neutral, the Reissner-Nordström may be considered as only an academic exercise, although interesting, rather than a realistic and relevant field of gravitation.

To begin with, we only consider the case where torsion vanishes which reduces the model to the classical Reissner-Nordström metric e.g. Ryder (2009). The only non vanishing coefficients of the Levi-Civita connection $\overline{\nabla}$ are the symbols of Christoffel calculated by means of the partial derivative of the metric components:

$$\overline{\Gamma}^0_{01} = \overline{\Gamma}^0_{10} = v'$$

$$\overline{\Gamma}^1_{00} = v' e^{2v-2\mu}, \quad \overline{\Gamma}^1_{11} = \mu', \quad \overline{\Gamma}^1_{22} = -r e^{-2\mu}, \quad \overline{\Gamma}^1_{33} = -r e^{-2\mu} \sin^2 \theta$$

$$\overline{\Gamma}^2_{12} = \overline{\Gamma}^2_{21} = \frac{1}{r}, \quad \overline{\Gamma}^2_{33} = -\sin \theta \cos \theta$$

$$\overline{\Gamma}^3_{13} = \overline{\Gamma}^3_{31} = \frac{1}{r}, \quad \overline{\Gamma}^3_{23} = \overline{\Gamma}^3_{32} = \cot \theta$$

The non vanishing components of the Ricci curvature $\overline{\Re}_{\alpha\beta} := \overline{\Re}^\gamma_{\gamma\alpha\beta}$ hold:

$$\begin{cases} \overline{\Re}_{00} = e^{2v-2\mu} \left[v'' + (v')^2 - v'\mu' + (2v'/r) \right] \\ \overline{\Re}_{11} = -\left[v'' + (v')^2 - v'\mu' + (2\mu'/r) \right] \\ \overline{\Re}_{22} = e^{-2\mu} \left(-1 + r \mu' - r v' \right) + 1 \\ \overline{\Re}_{33} = e^{-2\mu} \left[\left(-1 + r \mu' - r v' \right) + 1 \right] \sin^2 \theta \end{cases}$$

the off-diagonal components all vanish. The Einstein tensor $\overline{G}_{\alpha\beta} := \overline{\Re}_{\alpha\beta} - (\overline{\Re}/2)g_{\alpha\beta}$ has also four non vanishing diagonal components:

$$\begin{cases} \overline{G}_{00} = e^{2v} \left[e^{-2\mu} \left(\frac{2\mu'}{r} - \frac{1}{r^2} \right) + \frac{1}{r^2} \right] \\ \overline{G}_{11} = e^{2\mu} \left[e^{-2\mu} \left(\frac{2v'}{r} + \frac{1}{r^2} \right) - \frac{1}{r^2} \right] \\ \overline{G}_{22} = r^2 e^{-2\mu} \left[v'' + (v')^2 - v'\mu' + \frac{v' - \mu'}{r} \right] \\ \overline{G}_{33} = r^2 e^{-2\mu} \left[v'' + (v')^2 - v'\mu' + \frac{v' - \mu'}{r} \right] \sin^2 \theta \end{cases} \qquad (6.95)$$

Remark 6.27 It is important to mention that the components (6.95) are the projection of the Einstein tensor onto the dyadic: $\{dx^0 \otimes dx^0, dr \otimes dr, d\theta \otimes d\theta, d\varphi \otimes d\varphi\}$. It is sometimes worth to define the coframe for the tensor decomposition:

$$\left\{ \theta^0 := e^v dx^0, \theta^1 := e^\mu dr, \theta^2 := r d\theta, \theta^3 := r \sin \theta d\varphi \right\} \qquad (6.96)$$

It could be checked that for a vacuum spacetime without torsion, the Einstein tensor vanishes and by means of the two first field equations (6.95), we deduce easily that $v' + \mu' = 0$. In the particular case when the electromagnetic field $\mathcal{F}^{\mu\nu} \equiv 0$ is equal to zero in the relation (6.87), and when requiring additionally a flat Minkowskian spacetime at $r \to \infty$, we obtain the classic Schwarzschild metric characterized by the line element ($2m$ being a constant of integration):

$$ds^2 = \left(1 - \frac{2m}{r}\right)(dx^0)^2 - \left(1 - \frac{2m}{r}\right)^2 dr^2 - r^2\, d\theta^2 - r^2 \sin^2\theta\, d\varphi^2 \qquad (6.97)$$

Without electromagnetic field, the torsion tensor is also equal to zero according to (6.87).[7] In addition to gravitation field induced by the Schwarzschild metric, we add now the influence of a electrostatic charge. We thus consider a massive body of mass M with spherical symmetry corresponding to the Schwarzschild radius m and with the electric charge Q. We suggest first the Lagrangian within spacetime (6.83). To go further, let now assume the existence of a radial electric field E so that the Faraday electromagnetic field holds:

$$\mathcal{F}_{\mu\nu} := E \begin{bmatrix} 0 & -1 & 0 & 0 \\ 1 & 0 & 0 & 0 \\ 0 & 0 & 0 & 0 \\ 0 & 0 & 0 & 0 \end{bmatrix} \implies \mathcal{F}^{\mu\nu} := E \begin{bmatrix} 0 & 1 & 0 & 0 \\ -1 & 0 & 0 & 0 \\ 0 & 0 & 0 & 0 \\ 0 & 0 & 0 & 0 \end{bmatrix} \qquad (6.98)$$

by using the Riemannian metric $g_{\mu\nu}$ to uprise the indices. In addition, consider a torsionless Riemannian manifold. Introduction of this particular electromagnetic field into the relations (6.87), and owing that the Maxwell's equations reduce to two components:

$$\begin{cases} \overline{\nabla}_\nu \mathcal{F}^{0\nu} = \partial_0 E = 0 \\ \overline{\nabla}_\nu \mathcal{F}^{1\nu} = \partial_r E + \left(\nu' + \mu' + \dfrac{2}{r}\right) E = 0 \end{cases}$$

we obtain the system of governing equations for the gravito-electromagnetic field. The first row expresses that the field is stationary whereas the second allows us to explicitly obtain the electric field. The second row of the system (6.87) leads to the coupling of gravitation and electromagnetic fields in this particular case:

$$\begin{cases} \left[e^{-2\mu}\left(\dfrac{2\mu'}{r} - \dfrac{1}{r^2}\right) + \dfrac{1}{r^2}\right] = \dfrac{3\chi}{4} E^2 \\[2mm] \left[e^{-2\mu}\left(\dfrac{2\nu'}{r} + \dfrac{1}{r^2}\right) - \dfrac{1}{r^2}\right] = -\dfrac{3\chi}{4} E^2 \\[2mm] \left[\nu'' + (\nu')^2 - \nu'\mu' + \dfrac{\nu' - \mu'}{r}\right] = 0 \\[2mm] E' + \left(\nu' + \mu' + \dfrac{2}{r}\right) E = 0 \end{cases} \qquad (6.99)$$

---

[7]For instance in the equation $G_{00} = 0$, we define $U(r) := e^{-2\mu(r)}$ to arrive to the differential equation $U'(r) + U(r)/r = 1/r$. We remark that $U(r) := 1$ is a particular solution of the non homogeneous equation. The second solution may be found by a change of variable $V(r) := U(r) - 1$.

where the unknowns are the three functions $v(r)$, $\mu(r)$ and $E(r)$. Summation of two first equations (6.99) gives $v' + \mu' = 0$. This allows us to have the solution for electric field which is the case where the electric field surrounding a non rotating charge Q (a constant of integration) depends on the distance r as: $E(r) := Q/r^2$. For the general solution of the first homogeneous equation of (6.99), as for the case where the electric field is set equal to zero, we may define a variable change to have $e^{-2\mu} := 1 - V(r)$ and deduce the differential equation replacing the first row of system (6.99):

$$\frac{V'(r)}{r} + \frac{V(r)}{r^2} = 0 \quad \Longrightarrow \quad V(r) = -\frac{2m}{r}$$

We thus deduce:

$$e^{2v} = e^{-2\mu} = 1 - \frac{2m}{r} + \frac{3\chi}{4} \frac{Q^2}{r^2} \tag{6.100}$$

where $2m$ is also constant of integration ensuring to obtain Minkowskian spacetime for large distance r from the origin. We recognize components of the Reissner-Nordström metric. However, care should be taken since in view of the entire system of equations (6.99), it is observed that we have four equations with three unknowns, then the compatibility of all of them is mandatory. The resulting metric takes the form of:

$$ds^2 = \left(1 - \frac{2m}{r} + \frac{3\chi}{4} \frac{Q^2}{r^2}\right) (dx^0)^2$$

$$- \left(1 - \frac{2m}{r} + \frac{3\chi}{4} \frac{Q^2}{r^2}\right)^{-1} dr^2 - r^2 d\theta^2 - r^2 \sin^2\theta d\varphi^2 \tag{6.101}$$

This gives the metric describing the spherically symmetric spacetime satisfying the Einstein field equation in a region with no matter but with a radial electric field generated by a charge Q.

6.5.1.2 Electric Charge and Spacetime Torsion

A next question would be the eventually role of the electric charge on the creation of spacetime torsion that seems to be suggested by the third equation of the system (6.87). This is legitimate since we have assumed but not deduced that the torsion field is equal to zero for the classical Reissner-Nordström metric. Now we have to check if this is only the possible situation.

Remark 6.28 Adding the contortion tensor to the symbols of Christoffel leads to the connection coefficient, the Riemann–Cartan curvature together with Ricci curvature are obtained accordingly. We nevertheless observe that the system of equations

(6.87) governing both gravitation and electromagnetism is highly coupled in the framework of Riemann–Cartan physics. The third row of (6.87) is expected to allow us to determine the torsion field by means of the algebraic equation from the electromagnetism variable.

The electromagnetic field is expected to be a potential generator of torsion in the spacetime. We have deduced by means of Eq. (6.87) not only the possibility of their coupling but also the determination of torsion in terms of the electromagnetic strength of Faraday. The torsion is produced according to quite general relationship:

$$g^{\rho\nu}\,\aleph^{\mu}_{\lambda\rho} = \chi\left(\mathcal{F}^{\nu\mu} - \mathcal{F}^{\mu\nu}\right)A_{\lambda} \qquad (6.102)$$

For the particular example of spherical symmetric electrostatic field we previously treated, care should be however taken since the two non vanishing components of the torsion in the presence of the particular electromagnetic field (6.98) are:

$$\aleph^{0}_{01} = -2\chi e^{-2\mu(r)}E(r)A_{0}(r) = -\aleph^{0}_{01}, \qquad \aleph^{1}_{00} = 2\chi e^{-2\nu(r)}E(r)A_{0}(r)$$

For instance, assuming a priori that the electric field is given as $E(r) := Q/r^2$ seems to enforce a solution which rigorously should be a priori solved by means of Maxwell's equations e.g. Puntigam et al. (1997). The skew-symmetry of the second equation imposes that it is true if and only if $A_0(r) \equiv 0$ or $E(r) \equiv 0$. In other words, a electric field alone $E(r)$ cannot produce torsion. Of course a systematic investigation of the four-potential $A_\mu := (\phi, A_1, A_2, A_3)$ should be conducted. Nevertheless, description with teleparallel gravity theory in the Weitzenböck spacetime with non vanishing torsion but zero curvature is not the only path to couple electromagnetism and gravitation as suggested in e.g. de Andrade and Pereira (1999). The presence of the torsion in the governing equations is essentially due to the Palatini relation (variation of the curvature), and its consequence is the presence of additional skew-symmetric term in the variation of the Faraday tensor $\mathcal{F}^{\nu\mu}$ e.g. Frankel (1997), Nakahara (1996). This seems to conform to the finding of de Andrade and Pereira (1999) which has shown, by using the teleparallel description of gravitation, that electromagnetic field is also able to produce torsion by reading Eq. (6.102) from right to left.

6.5.2 Schwarzschild Anti-de Sitter ($\mathscr{A}d\mathscr{S}$) Spacetimes

One of the major actual problems in the theory of physics and cosmology concerns the presence and the smallness of the amplitude of the cosmological constant Λ. The accounting of the dark energy allows us to include the accelerated expansion of the universe by adding a cosmological constant $\Lambda \simeq\leq 10^{-52}[\mathrm{m}^{-2}]$. It is proven that the static spherical symmetric solutions of the Einstein's equation with negative cosmological constant $\Lambda < 0$ lead to the so-called Anti-de Sitter spacetime in the

Fig. 6.2 A spherical massive body source of the curvature and the torsion of surrounding spacetime. Only $\aleph^0_{23} = 2\aleph_0(r) = -\aleph^0_{32}$ is non zero whereas all other components vanish

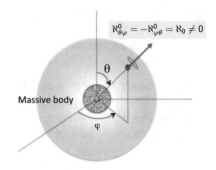

framework of Einstein gravitation theory. For attempting to overcome this problem, we adopt a different path in this work by considering rather a Einstein–Cartan spacetime with uniform torsion to arrive at the $\mathscr{A}d\mathscr{S}$ spacetime. Comparing the curvature formalism and the torsion formalism to describe the gravitation field, it was shown in e.g. de Andrade and Pereira (1999) that the description in terms of curvature does not allow us to introduce the torsion without violating the gauge invariance, but a teleparallel gravitation based on the torsion variable enables to show that the gravitational field interacts with the electromagnetic field by choosing appropriate covariant derivative (in turn based on the Levi-Civita connection) to define the Faraday tensor.

6.5.2.1 Torsion of Spacetime with Spherical Symmetry

The objective of this paragraph is to define the simplest example of Einstein–Cartan gravitation model with spherical symmetry (Fig. 6.2). For this purpose, let now consider on a Riemann–Cartan spacetime a particular field of torsion $\aleph^\gamma_{\alpha\beta}(x^\mu)$ for which the non vanishing components are only $\aleph^0_{23} = 2\aleph_0(r) = -\aleph^0_{32}$ (the other components are zero) as in e.g. Prasanna (1975b). However, we do not consider in this example the problem of stationary spinning fluid as in Prasanna (1975b), we directly consider an empty curved and non zero torsion spacetime modeled by a Riemann–Cartan manifold. Physically, this torsion field may be interpreted as the consequence of spins of particle fluids that are aligned in the radial direction and where particles are assumed to be static. It should be mentioned the decomposition means $\aleph = \aleph^0_{23}\theta^2 \otimes \theta^3 + \aleph^0_{32}\theta^3 \otimes \theta^2$. Nevertheless, we have not to relate this torsion field with the fluid spin in this subsection. We consider the torsion at its own. The Einstein tensor $G_{\mu\nu}$ is calculated by means of the contortion tensor $\mathcal{T}^\gamma_{\alpha\beta}$ and the curvature tensor $\mathfrak{R}^\gamma_{\alpha\beta\lambda}$ calculated from the relations (2.44) and (4.99) respectively. The non zero components of the contortion tensor are (in the basis $\{\theta^0, \theta^1, \theta^2, \theta^3\}$):

$$
\begin{cases}
\mathcal{T}^0_{23} = \Omega^0_{23} = \aleph_0, & \mathcal{T}^0_{32} = -\mathcal{T}^0_{23} \\
\mathcal{T}^2_{03} = g^{22}g_{00}\,\Omega^0_{23} = g^{22}g_{00}\,\aleph_0, & \mathcal{T}^2_{30} = \mathcal{T}^2_{03} \\
\mathcal{T}^3_{02} = g^{33}g_{00}\,\Omega^0_{32} = -g^{22}g_{00}\,\aleph_0, & \mathcal{T}^3_{20} = \mathcal{T}^3_{02}
\end{cases}
\tag{6.103}
$$

Introducing these equations in the relation (4.99) gives the Ricci curvature:

$$\mathfrak{R}_{\beta\lambda} := \mathfrak{R}^{\alpha}_{\alpha\beta\lambda} = \overline{\mathfrak{R}}_{\beta\lambda} + \overline{\nabla}_{\alpha}\mathfrak{T}^{\alpha}_{\beta\lambda} - \overline{\nabla}_{\beta}\mathfrak{T}^{\alpha}_{\alpha\lambda} + \mathfrak{T}^{\alpha}_{\alpha\mu}\mathfrak{T}^{\mu}_{\beta\lambda} - \mathfrak{T}^{\alpha}_{\beta\mu}\mathfrak{T}^{\mu}_{\alpha\lambda}$$

A straightforward calculus gives the non zero components:

$$\mathfrak{R}_{00} = \overline{\mathfrak{R}}_{00} - 2g^{22}g^{33}g^{2}_{00}\,\Omega^{0}_{23}\Omega^{0}_{32},$$
$$\mathfrak{R}_{11} = \overline{\mathfrak{R}}_{11}, \quad \mathfrak{R}_{22} = \overline{\mathfrak{R}}_{22}, \quad \mathfrak{R}_{33} = \overline{\mathfrak{R}}_{33}$$

Projection of the additional term onto the basis $\{dx^0, dr, d\theta, d\varphi\}$ gives the component (Prasanna 1975b): $\mathfrak{R}_{00} = \overline{\mathfrak{R}}_{00} + 2e^{2\nu}\aleph^2_0$ since $\theta^0 := e^{\nu}dx^0$, $\theta^2 := rd\theta$ and $\theta^3 := r\sin\theta d\varphi$. Only the time component is influenced by the torsion tensor in such a case. The Einstein tensor is obtained accordingly by considering this additional term in the relations (6.95). Projection of the Einstein tensor onto the co-basis gives:

$$
\begin{cases}
G_{00} = e^{2\nu}\left[e^{-2\mu}\left(\dfrac{2\mu'}{r} - \dfrac{1}{r^2}\right) + \dfrac{1}{r^2} + \aleph^2_0\right] \\[3mm]
G_{11} = e^{2\mu}\left[e^{-2\mu}\left(\dfrac{2\nu'}{r} + \dfrac{1}{r^2}\right) - \dfrac{1}{r^2} - \aleph^2_0\right] \\[3mm]
G_{22} = r^2\,e^{-2\mu}\left[\nu'' + (\nu')^2 - \nu'\mu' + \dfrac{\nu' - \mu'}{r} - \aleph^2_0\right] \\[3mm]
G_{33} = r^2\,e^{-2\mu}\left[\nu'' + (\nu')^2 - \nu'\mu' + \dfrac{\nu' - \mu'}{r} - \aleph^2_0\right]\sin^2\theta
\end{cases}
\tag{6.104}
$$

since $G_{\mu\nu} := \mathfrak{R}_{\mu\nu} - (\mathcal{R}/2)g_{\mu\nu} = \overline{G}_{\mu\nu} - (\mathcal{R} - \overline{\mathcal{R}})/2\,g_{\mu\nu}$. In a Einstein–Cartan vacuum spacetime, the field equation $G_{\mu\nu} = 0$ permits to obtain by adding the two first equation of (6.104): $\nu' + \mu' = 0$ to give $e^{\mu} = e^{-\nu}$ by choosing again worth asymptotic behavior at $r \to \infty$. Taking the first equation and owing that $e^{2\nu} \neq 0$, we have for a vacuum:

$$e^{-2\mu}\left(\frac{2\mu'}{r} - \frac{1}{r^2}\right) + \frac{1}{r^2} + \aleph^2_0 = 0$$

by making a change of variable $U(r) := e^{-2\mu}$ the equation reduces to: $[rU(r)]' = 1 + \aleph^2_0(r)\,r^2$. In a general case, the field $\aleph^2_0(r)$ is unknown and its resolution is coupled with other unknowns in the system (6.104). If we assume a uniform torsion field $\aleph_0(r) = \aleph_0$, for the sake of the simplicity, we directly obtain the solution:

$$U(r) = 1 - \frac{2m}{r} + \frac{\aleph^2_0}{3}r^2 \tag{6.105}$$

by choosing the constant of integration $2m$ (black hole radius). The constant $2m := 2GM/c^2$ reduces to the Schwarzschild radius of a massive body M. We recognize here an Schwarzschild-anti-de Sitter ($\mathscr{A}d\mathscr{S}$) spacetime metric where $\aleph_0^2(r) := \Lambda > 0$ (see Eq. (4.43)) is a kind of *cosmological constant*, which is a particular case of Kottler spacetime metric with line element:

$$ds^2 = \left(1 - \frac{2m}{r} + \frac{\Lambda}{3}r^2\right)(dx^0)^2$$

$$- \left(1 - \frac{2m}{r} + \frac{\Lambda}{3}r^2\right)^{-1} dr^2 - r^2 d\theta^2 - r^2 \sin^2\theta d\varphi^2 \qquad (6.106)$$

Remark 6.29 As a finding, it should be mentioned that the anti-de Sitter spacetime metric constitutes a solution of the Einstein–Cartan field equations with an analogy of an attractive cosmology constant. In fact, this seems unphysical but could be considered as a regularization of the gravitation field at large distance which is not flat for $r \to \infty$. The horizon of the Anti-de Sitter spacetime is merely related to the torsion field $\ell^2 := 3/\aleph_0^2$ suggesting that the amplitude of the torsion tensor is very great. In view of the physical interpretation of the torsion tensor as the density of non smoothness of fields (density of dislocations in the context of theory of dislocations), it suggests that the spacetime has very high defects density as the spacetime is merely considered as a set of microcosms (see Fig. 6.3 for a schematic picture of the spacetime).

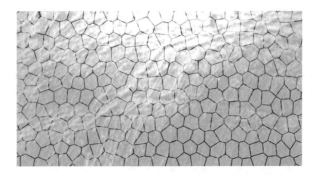

Fig. 6.3 Scheme of an affinely connected manifold with torsion. The Riemann–Cartan spacetime $\mathscr{RC}$ is assumed to be a set of infinitesimal "microcosms" linked each other by metric compatible connection $\Gamma^\gamma_{\alpha\beta} = D^\gamma_{\alpha\beta} + \mathfrak{T}^\gamma_{\alpha\beta}$. Riemann–Cartan continuum assumes metric compatibility meaning that the metricity $Q^\gamma_{\alpha\beta} := \nabla^\gamma g_{\alpha\beta} \equiv 0$ on $\mathscr{RC}$. Torsion field characterizes the density of defects locally (defined here by the dislocations between "microcosms" on the figure e.g. Rakotomanana 2003). For short, dislocation is a line defect, the density is defined as the total length of dislocation per unit volume [m/m^3] = [m^{-2}]. It is the number of dislocation lines intersecting a unit area. For a lose comparison, dislocation density is usually of the order of 10^{10}[m^{-2}] in a metal, increasing up to $\simeq 10^{16}$[m^{-2}] after work hardening and fatigue. In the present work, we expect for the spacetime a very small density of defects as $\aleph_0 \simeq 10^{-26}$[m^{-2}] for the spacetime, meaning *in fine* to a kind of smooth dusts cloud

Capozziello and his co-workers already investigated the relation between the cosmological constant Λ with the torsion field by considering Einstein–Cartan spacetime. They take into account a torsion field $\aleph \neq 0$ which leads to a negative pressure contribution in the cosmological dynamics and therefore to an accelerated behavior of universe. They stated that the presence of torsion has the same effect as a cosmological constant e.g. Capoziello et al. (2003). In some sense, a spin density acts as a source of torsion and then replaces the dark matter concept. In the same context of Einstein–Cartan gravitation, a more recent work has nevertheless contradicted the previous results in Capoziello et al. (2003), by confronting them with the supernovae data from Hubble diagram (Schücker and Tilquin 2012). Independently on the experimental measurements, the present work attempts to show that the torsion tensor by considering an Einstein–Cartan gravitation is related to the cosmological constant for the Anti-de Sitter spacetime in Eq. (6.106). Further studies should be done for other spacetimes involving the cosmological constant.

Some authors consider the cosmological constant as one of the major problems for the modern theory of cosmology and physics. Quasi-exhaustive review is available in a recent of multiple-authors paper (Bull et al. 2016) about the relevance of the non zero cosmological constant, namely about the very unnatural small value of Λ and its introduction as a quasi-empirical constant to capture the dark energy in the framework of ΛCDM theory (a cosmological theory based on the existence of Cold Dark Matter and cosmological constant Λ). It is well known that the ΛCDM model lies upon the assumption of the existence of dark matter, dark energy and the inflation field, which induces some controversies. On the one hand, this model turns out to be predictive and relatively robust with respect to old and recent cosmological experiments in a large extend (although very large scale lengths greater than Solar System need further experimental observations, and very small scale lengths at the quantum level are out of reach). On the other hand, the incorporation of a unique extremely low and unnatural value of the cosmological constant Λ, particularly in the Planck units, and why not a zero value is not satisfactory from theoretical point of view even if the ΛCDM model remains an effective (phenomenological) model. Despite its successful to explain most of cosmological problems, the definition of a cosmological constant as an ad hoc parameter and without a link with the small scale length physics phenomenae might be also unsatisfactory for physics point of view.

Several ways may be proposed to build gravitation theory to overcome the problem of Λ in the ΛCDM model among them the increasing the spacetime dimension (greater than 4), the introduction non-local gravitation, and the development of modified gravitation. In the present work, we would like to remain as simplest as possible. The accounting of the local topology defects of the spacetime, or more generally the accounting of theory of microcosms e.g. Gonseth (1926) which are not metrically connected but rather affinely connected, and then based on the presence of a non vanishing torsion field $\aleph^{\gamma}_{\alpha\beta}$ in the spacetime allows us to derive the Schwarzschild-Anti-de Sitter spacetime by adding in fine a contortion tensor as primal variables other than the usual metric tensor. Some remarks should be done. The negativity of $\Lambda < 0$ leads to a universe re-collapsing rather than expansion, nevertheless the $\mathscr{A}d\mathscr{S}$/CFT conjecture is hopefully a guiding result to link gravitation with quantum physics (Maldacena 1998). The equality $\aleph_0^2 := \Lambda$ seems very interesting owing that a torsion field $\aleph_{\alpha\beta}^{\gamma}$ on a manifold captures the local density of defects on a manifold e.g. Rakotomanana (1997). Extremely low values of Λ means no more than a very high density of defects. In some sense, the basic equation of the Einstein relativistic gravitation is exactly obtained to account for the spherical symmetric tensor metric with Λ but the physics interpretation is completely different. The torsion field is a local variable and has very clear geometrical local interpretation for the spacetime and not for any other dark matter and dark energy assumption. The covariance theorem we stated in the first part of this work allows us to go beyond the Lovelock's theorem on the uniqueness of Einstein tensor e.g. Lovelock (1969) by building Lagrangian function depending on the metric, the torsion and the curvature rather than on the metric and its Levi-Civita curvature only.

We remind that for non zero torsion field, it has been shown (see also previous chapter) that the torsion field may be considered as a source and engenders a non uniform cosmological constant $\Lambda(x^{\mu})$.

Remark 6.30 It is interesting to remark that accounting of a uniform radial torsion field allows us to obtain the anti-de Sitter spacetime metric which is also solution of the Einstein equation with a attractive cosmology constant $\Lambda := \aleph_0^2$. It highlights the role of a torsion field in the theory of gravitation at least at the same level as the cosmological constant, which defines is the value of the energy mean density of the vacuum of spacetime. The attractiveness of anti-de Sitter spacetime with metric (6.106) can also be interpreted as an empty spacetime with negative energy, which causes this spacetime to collapse in on itself. The interest of $\mathscr{A}d\mathscr{S}$ spacetime is mostly in the domain of quantum physics, thermodynamics of black holes, and the correspondence between $\mathscr{A}d\mathscr{S}$ and Conformal Field Theory (CFT) and high energy processes. Nevertheless, we think at a modest level that the accounting of Einstein–Cartan spacetime torsion may give new insights for future relation of the $\mathscr{A}d\mathscr{S}$, the torsion field and the quantum mechanics.

6.5.3 Extension to Electromagnetism-Matter Interaction

Interaction of electromagnetic waves with matter is a quite complicated from physics point of view. Depending on the scale length observation, matter should be considered as comprised of atoms. At another level, say subatomic, matter is mostly composed of empty space. At a continuum level, spacetime is assumed to be filled bay continuum matter. In this subsection, we are interested in modelling the interaction of electrodynamics with matter with non uniform properties as slight extension of the model (6.92). Let thus consider the interaction of electromagnetism with a continuum matter with constitutive laws:

$$\mathcal{H}^{\mu\nu} = \varXi^{\mu\nu\alpha\beta}\mathcal{F}_{\alpha\beta} \qquad (6.107)$$

where the tensor $\varXi^{\mu\nu\alpha\beta}(g_{\lambda\rho}, \aleph^{\gamma}_{\sigma\rho})$ depend on the metric and the torsion tensor. Metric and torsion argument are considered to ensure covariance.

6.5.3.1 Basic Equations of Interaction

We are working within an Einstein–Cartan spacetime which is a curved continuum with torsion in this subsection. There is additionally an electromagnetic field. The spacetime curvature explicitly appears in the Lagrangian as that of Einstein–Hilbert. Let consider now the Lagrangian (6.92) where the electromagnetic field interacts with a continuum matter:

$$\mathscr{L} := -\frac{1}{4}\mathcal{H}^{\mu\nu}\,\mathcal{F}_{\mu\nu} + \frac{1}{2\chi}\mathcal{R} \tag{6.108}$$

in which the constitutive laws hold $\mathcal{H}^{\mu\nu} = \varXi^{\mu\nu\alpha\beta}\mathcal{F}_{\alpha\beta}$. The only difference with the previous developments is the contribution of the term: $\varXi^{\mu\nu\alpha\beta} := \varepsilon\, g^{\mu\alpha} g^{\nu\beta}$ which includes the material properties and constitutes a tangent tensor. We remind that the influence of the physical properties of the matter is pointed by the presence of $\varepsilon := \varepsilon_r\varepsilon_0$ which is the electric permittivity of the material. The other terms in function of the metric components are related to the geometry of the medium. First after application of a Lagrangian variation $\varDelta$,

$$\varDelta\mathscr{S} = \int_{\mathscr{M}} \left\{ -\frac{1}{2}\mathcal{H}^{\mu\nu}\varDelta\mathcal{F}_{\mu\nu} - \frac{1}{4}\varDelta\varXi^{\mu\nu\alpha\beta}\,\mathcal{F}_{\mu\nu}\mathcal{F}_{\alpha\beta} \right.$$
$$+ \frac{1}{2\chi}\left(\mathfrak{R}^{\lambda\rho} - \frac{\mathcal{R}}{2}g^{\lambda\rho}\right)\varDelta g_{\lambda\rho} + \frac{1}{8}\mathcal{H}^{\mu\nu}\,\mathcal{F}_{\mu\nu}\,g^{\lambda\rho}\varDelta g_{\lambda\rho}$$
$$+ \left. \frac{1}{2\chi}g^{\mu\nu}\left[\nabla_{\lambda}\left(\varDelta\varGamma^{\lambda}_{\mu\nu}\right) - \nabla_{\mu}\left(\varDelta\varGamma^{\lambda}_{\lambda\nu}\right) - \aleph^{\rho}_{\lambda\mu}\varDelta\varGamma^{\lambda}_{\rho\nu}\right]\right\}\omega_n$$

for which all divergence terms may be dropped by assuming a zero flux at the boundary of the medium $\partial\mathscr{M}$. The Lagrangian variation of the tangent tensor of constitutive laws holds:

$$\varDelta\varXi^{\mu\nu\alpha\beta}\,\mathcal{F}_{\mu\nu}\mathcal{F}_{\alpha\beta} = \frac{\partial\varXi^{\mu\nu\alpha\beta}}{\partial g_{\lambda\rho}}\varDelta g_{\lambda\rho}\,\mathcal{F}_{\mu\nu}\mathcal{F}_{\alpha\beta} + \frac{\partial\varXi^{\mu\sigma\alpha\beta}}{\partial\aleph^{\lambda}_{\rho\nu}}\varDelta\aleph^{\lambda}_{\rho\nu}\,\mathcal{F}_{\mu\sigma}\mathcal{F}_{\alpha\beta} \tag{6.109}$$

The previous variation form becomes:

$$\varDelta\mathscr{S} = \int_{\mathscr{M}} \left\{ -\frac{1}{2}\mathcal{H}^{\mu\nu}\varDelta\mathcal{F}_{\mu\nu} - \frac{1}{4}\left(\frac{\partial\varXi^{\mu\nu\alpha\beta}}{\partial g_{\lambda\rho}}\,\mathcal{F}_{\mu\nu}\mathcal{F}_{\alpha\beta}\right)\varDelta g_{\lambda\rho} \right.$$
$$+ \frac{1}{2\chi}\left(\mathfrak{R}^{\lambda\rho} - \frac{\mathcal{R}}{2}g^{\lambda\rho}\right)\varDelta g_{\lambda\rho} + \frac{1}{8}\mathcal{H}^{\mu\nu}\,\mathcal{F}_{\mu\nu}\,g^{\lambda\rho}\varDelta g_{\lambda\rho}$$

$$-\frac{1}{4}\left[\left(\frac{\partial\Xi^{\mu\sigma\alpha\beta}}{\partial\aleph^\lambda_{\rho\nu}}-\frac{\partial\Xi^{\mu\sigma\alpha\beta}}{\partial\aleph^\lambda_{\nu\rho}}\right)\mathscr{F}_{\mu\sigma}\mathscr{F}_{\alpha\beta}\right]\Delta\Gamma^\lambda_{\rho\nu}$$

$$\left.+\frac{1}{2\chi}g^{\mu\nu}\left[\nabla_\lambda\left(\Delta\Gamma^\lambda_{\mu\nu}\right)-\nabla_\mu\left(\Delta\Gamma^\lambda_{\lambda\nu}\right)-\aleph^\rho_{\lambda\mu}\Delta\Gamma^\lambda_{\rho\nu}\right]\right\}\omega_n$$

Second after application of Lie derivative variation, the field equations governing the interaction of the electromagnetic and the gravitation are obtained accordingly by varying arbitrarily the primal variables:

$$\begin{cases} & \nabla_\nu\mathcal{H}^{\mu\nu}=0 \\[4pt] \frac{1}{2\chi}\left(\Re^{\lambda\rho}-\frac{\mathcal{R}}{2}g^{\lambda\rho}\right)+\frac{1}{8}\mathcal{H}^{\mu\nu}\mathscr{F}_{\mu\nu}\,g^{\lambda\rho}-\frac{1}{4}\left(\frac{\partial\Xi^{\mu\nu\alpha\beta}}{\partial g_{\lambda\rho}}\,\mathscr{F}_{\mu\nu}\mathscr{F}_{\alpha\beta}\right)=0 \\[4pt] \frac{1}{4}\left(\frac{\partial\Xi^{\mu\sigma\alpha\beta}}{\partial\aleph^\lambda_{\rho\nu}}-\frac{\partial\Xi^{\mu\sigma\alpha\beta}}{\partial\aleph^\lambda_{\nu\rho}}\right)\mathscr{F}_{\mu\sigma}\mathscr{F}_{\alpha\beta}+\left(\mathcal{H}^{\mu\nu}-\mathcal{H}^{\nu\mu}\right)A_\lambda+\frac{1}{\chi}g^{\rho\nu}\aleph^\mu_{\lambda\rho}=0 \end{cases}$$

$$(6.110)$$

which slightly extends the vacuum equation (6.87). Instead of the contravariant components of the electromagnetic field, we introduce here the dual variables $\mathcal{H}$. The first row of the system (6.110) expresses the Maxwell equation:

$$\nabla_\nu\mathcal{H}^{\mu\nu}=\epsilon g^{\mu\alpha}g^{\nu\beta}\nabla_\nu\mathscr{F}_{\alpha\beta}=0$$

since the connection is compatible with the metric tensor and the material properties assumed to be uniform within the volume of reference.

To go further let us consider the first and third equations of the system (6.110). As for the Maxwell's equations within vacuum spacetime, the above equation may be formulated by means of four-potential vector A^μ by introducing properties of curvature tensor, the metricity of the connection and the Lorenz gauge ($\nabla_\nu A^\nu\equiv0$). The second equation of (6.110) should be considered as an equation allowing us to calculate the interaction of the gravitation and the electromagnetic field. Once the electromagnetic field determined, the gravity is calculated. The third equation may be re-arranged to isolate the torsion. The two equations thus give, by assuming a null divergence for the potential $\nabla_\nu A^\nu=0$,

$$\begin{cases} -g^{\nu\beta}\nabla_\nu\nabla_\beta A^\mu-g^{\mu\alpha}\aleph^\gamma_{\nu\alpha}\nabla_\gamma A^\nu+g^{\mu\alpha}\Re_{\alpha\gamma}A^\gamma=0 \\[4pt] 2\epsilon\chi\left(\nabla^\gamma A_\alpha-\nabla_\alpha A^\gamma\right)A_\beta=\aleph^\gamma_{\alpha\beta} \end{cases}$$

$$(6.111)$$

where, in the Maxwell's equations, the first term represents a wave equation, the second term a diffusion contribution due to the torsion field, and the last term with the Ricci curvature introduces a "breathing" mode due to the non vanishing of curvature tensor. It should be reminded that the torsion and the curvature might be attributed to the abrupt gradients of scalar field and vector field within the matter respectively. It should be pointed out that the torsion field is of second-order with respect to the potential A^α.

Remark 6.31 By proposing a particular Riemann–Cartan spacetime structure, and working with contortion tensor rather than with the torsion tensor, a similar and interesting relation as the second row of the system (6.111) is obtained in Fernando et al. (2012) without dealing with a variational formulation.

Remark 6.32 Throughout this section, we have define the Faraday tensor as $\mathscr{F}_{\alpha\beta} := \nabla_\alpha A_\beta - \nabla_\beta A_\alpha$ where the connection $\Gamma^\gamma_{\alpha\beta}$ has non torsion. The $U(1)$ gauge invariance of Maxwell's equations is violated without cautions in such a way. Indeed by modifying the potential as $A_\beta \to A_\beta + g_{\beta\gamma}\nabla^\gamma \Lambda$ where $\Lambda(x^\mu)$ is an arbitrary function, we get:

$$\mathscr{F}_{\alpha\beta} := \nabla_\alpha A_\beta - \nabla_\beta A_\alpha \to \nabla_\alpha A_\beta - \nabla_\beta A_\alpha + \underbrace{\left(\nabla_\alpha \nabla_\beta \Lambda - \nabla_\beta \nabla_\alpha \Lambda\right)}_{= -\aleph^\gamma_{\alpha\beta}\nabla_\gamma \Lambda}$$

where the term in brackets vanishes if and only if the torsion is zero or the function Λ is covariantly uniform. Some previous authors propose to define $\mathscr{F}_{\alpha\beta} := \overline{\nabla}_\alpha A_\beta - \overline{\nabla}_\beta A_\alpha$ as electromagnetic tensor even in Riemann–Cartan geometry e.g. Smalley (1986), de Andrade and Pereira (1999).

Remark 6.33 At the end of this paragraph, we remind that the present Lagrange-Noether method enables to obtain a symmetric energy-momentum which differs from the original Minkowski energy-momentum one e.g. Obukhov and Hehl (2003). We also point out the possibility that there is no need of additional energy density component to the universe (explicitly by Λ), but rather that the accounting for (uniform) torsion field $\aleph_0$ modifies the equations of general relativity and by the way the change of Minkowski spacetime ($\mathscr{M}$) to Anti-de Sitter spacetime ($\mathscr{A}d\mathscr{S}$), owing that the $\mathscr{A}d\mathscr{S}$/CFT correspondence may be considered as a link between gravitation and quantum physics (Maldacena 1998). Indeed, this is conform to the interest on studying quantum fields in an Anti-de Sitter ($\mathscr{A}d\mathscr{S}$) spacetime.

6.5.3.2 Bridge to Riemann–Cartan Medium due to Electromagnetism

In a paper Oprisan and Ziet obtained the Reissner-Nordström metric by solving the Einstein–Cartan equation fields in a Tele Parallel Gravitation theory (Oprisan and Zet 2006). They have considered TPG with the basic arguments: the tetrads F^i_μ and the spin connection $\omega^{\mu\nu}_i$ from which they introduced classical formulation of the metric and the connection coefficients and then the torsion and the curvature tensor: $g_{\mu\nu} := g_{ij}F^i_\mu F^j_\nu$, $\Gamma^\rho_{\mu\nu} := F^\rho_i \partial_\mu F^i_\nu$. Then looking for spherically symmetric (and static) solution form metric they arrive to the Reissner-Nordström metric solution.

Now from the system of equations governing the interaction of gravitation and the electromagnetism (6.110), it is worth to focus on the third equation and remind it as (for the sake of the simplicity we assume that the tangent tensor does not depend on the torsion):

$$\aleph^{\gamma}_{\alpha\beta} = 2\epsilon\chi \left(\nabla^{\gamma} A_{\alpha} - \nabla_{\alpha} A^{\gamma} \right) A_{\beta} \tag{6.112}$$

which suggests that the electromagnetic field is potentially a generator of a skew-symmetric contribution, that is torsion, of the affine connection. In other words, electromagnetism interacting with gravity is source of change of Riemann continuum to Riemann–Cartan continuum by means of Eq. (6.112). The present study concludes that the use of TPG is not mandatory to study the interaction of electromagnetism with gravitation. We have considered in the present study the Hilbert–Einstein action (which is the simplest case among numerous gravitation theories) to relate electromagnetism and gravitation with the extended spacetime with torsion.

6.5.4 Geodesics in a Anti-de Sitter Spacetime

We have shown that a uniform distribution of torsion field $\aleph_0$ leads to an anti-de Sitter spacetime with the metric (6.106) with nevertheless a difference that the obtained spacetime has non vanishing torsion field. In this model, the torsion field acts anyhow like a vacuum energy (cosmological constant) which may be considered in some sense as a variable unifying some aspects of physics as superstring, cosmology and astrophysics. If we are interested to search for geodesics of the spacetime rather than on his autoparallels, we can again consider an analogous Lagrangian as for Schwarzschild:

$$\mathcal{L} = m^* \left[\left(1 - \frac{2m}{r} + \frac{\aleph^2_0}{3} r^2 \right) (\dot{x}^0)^2 \right. \tag{6.113}$$
$$\left. - \left(1 - \frac{2m}{r} + \frac{\aleph^2_0}{3} r^2 \right)^{-1} \dot{r}^2 - r^2 \dot{\theta}^2 - r^2 \sin^2\theta \dot{\varphi}^2 \right]$$

of a point of mass m^* in motion within this $(\mathscr{A}d\mathscr{S})$ spacetime. We remind that $\aleph^2_0$ characterizes the intensity of the torsion field. Solving the Euler–Lagrange associated to the $(\mathscr{A}d\mathscr{S})$ metric leads to the geodesics. The dot means derivative with respect to proper time τ. Since the Lagrangian $\mathscr{L}$ does not explicitly depend on the x^0 and φ, there are two conserved quantities: the energy associated to time variable x^0, and the angular momentum associated to φ, as for Eq. (5.118). The derivative of the Lagrangian with respect to $\dot{x}^0$ and $\dot{\varphi}$ are constant of motions:

$$\begin{cases} m^* \dfrac{dx^0}{d\tau}\left(1 - \dfrac{2m}{r} + \dfrac{\aleph_0^2}{3}r^2\right) = E \\ m^* \dfrac{d\varphi}{d\tau} r^2 \sin^2\theta = L \end{cases} \implies \begin{cases} u^0 = \dfrac{E}{m^*}\left(1 - \dfrac{2m}{r} + \dfrac{\aleph_0^2}{3}r^2\right)^{-1} \\ u^\varphi = \dfrac{L}{m^*}\dfrac{1}{r^2 \sin^2\theta} \end{cases}$$

(6.114)

where, again, E is the energy, and L the analogous of angular momentum in relativistic gravitation. For the sake of the simplicity, if we consider initial conditions as $\theta = \pi/2$ and $u^\theta := \dot{\theta} = 0$, the angular momentum induces $u^\varphi = L/mr^2$. This means that the motion remains confined in the plane $\theta = \pi/2$. We then deduce:

$$\mathscr{L} = \dfrac{E^2}{m^*}\left(1 - \dfrac{2m}{r} + \dfrac{\aleph_0^2}{3}r^2\right)^{-1} - m^*\left(1 - \dfrac{2m}{r} + \dfrac{\aleph_0^2}{3}r^2\right)^{-1}\dot{r}^2 - \dfrac{L^2}{m^* r^2}$$

(6.115)

where we can consider respectively the massive particle $\mathscr{L} = \epsilon = 1$, and non-massive particles (photons) $\mathscr{L} = \epsilon = 0$. This allows us to calculate the four-velocity $u^r := \dot{r}$ for each case:

$$u^r = \pm\sqrt{\dfrac{E^2}{m^{*2}} - \left(1 - \dfrac{2m}{r} + \dfrac{\aleph_0^2}{3}r^2\right)\left(\epsilon + \dfrac{L^2}{m^*}\dfrac{1}{r^2}\right)}$$

(6.116)

provided the term under square root is positive. As for the case of Schwarzschild metric without torsion, it is worth to define an effective potential:

$$V_{eff}(r) := \left(1 - \dfrac{2m}{r} + \dfrac{\aleph_0^2}{3}r^2\right)\left(\epsilon + \dfrac{L^2}{m^*}\dfrac{1}{r^2}\right)$$

(6.117)

Again, as for the Schwarzschild metric, circular orbits $u^r := \dot{r} \equiv 0$, for also each case were determined in a systematic way in e.g. Hackmann and Lämmerzal (2008) ($m^* = 1$ in this reference). Orbits are classified with respect to the energy E and the angular momentum L, and where the cosmological constant Λ is considered as parameter. The effective potential (6.117) permits to solve two particular cases of radial motions $L = 0$ and bound orbits $L \neq 0$.

Remark 6.34 Extension of the geodesic deviation equation to Riemann–Cartan spacetime was already done in the past e.g. Manoff (2001b). In the present paper, we slightly modify the formulation by highlighting a second term which is expressed explicitly in terms of torsion. This constitutes an extension of the geodesic deviation of Einstein relativistic gravitation to autoparallel deviation of Einstein–Cartan gravitation. Basically Eq. (5.121) includes two terms: the first one corresponds to the bending of spinless particles moving within a gravitational field, whereas the second term one acts on the particle as a consequence of the torsion of the spacetime. It is observed that for a non curved, but gravitation with torsion, there is a separation of autoparallel curves. At least from the theoretical point of view, the possibility to detect the torsion field appears in Eq. (5.121). This possibility was sketched as perspective in Nieto et al. (2007) where the covariant form of the relativistic top deviation was stated, in presence of spin. This extension of deviation equation explicitly demonstrates that even in a non curved spacetime, the torsion influences the path of the spinning particles. This gives an answer to the third argument advocated by Manoff's paper (Manoff 2001b).

Remark 6.35 It should be mentioned that the geodesics we found are not the geodesics of the Einstein–Cartan spacetime since the symmetric part of the contortion tensor does not vanish. We remind the integral curves defined by both the Levi-connection coefficients from metric (6.113) and the symmetric part of the contortion tensor (6.103):

$$
\begin{cases}
g_{00} := \left(1 - \dfrac{2m}{r} + \dfrac{\aleph_0^2}{3}r^2\right) \\[2ex]
g_{11} := -\left(1 - \dfrac{2m}{r} + \dfrac{\aleph_0^2}{3}r^2\right)^{-1}, \\[2ex]
g_{22} := -r^2 \\[1ex]
g_{33} := -r^2 \sin^2\theta
\end{cases}
\tag{6.118}
$$

$$
\begin{cases}
\mathcal{T}_{23}^0 = \Omega_{23}^0 = \aleph_0, & \mathcal{T}_{32}^0 = -\mathcal{T}_{23}^0 \\[1ex]
\mathcal{T}_{03}^2 = g^{22}g_{00}\,\Omega_{23}^0 = g^{22}g_{00}\,\aleph_0, & \mathcal{T}_{30}^2 = \mathcal{T}_{03}^2 \\[1ex]
\mathcal{T}_{02}^3 = g^{33}g_{00}\,\Omega_{32}^0 = -g^{22}g_{00}\,\aleph_0, & \mathcal{T}_{20}^3 = \mathcal{T}_{02}^3
\end{cases}
\tag{6.119}
$$

projected onto different bases (take care). The are introduced into Eq. (5.127):

$$
\frac{Du^\gamma}{D\tau} := \frac{du^\gamma}{d\tau} + \overline{\Gamma}^\gamma_{\mu\nu}u^\mu u^\nu + D^\gamma_{\mu\nu}u^\mu u^\nu = 0
\tag{6.120}
$$

Considering the influence of the torsion in the spacetime model remains an open question even in the framework of classical physics (in the sense of no quantum approach). Focusing on the gravitation and electromagnetic interaction, on the one hand, it was already observed by e.g. Kleinert (2008) that the third equation of the system (6.110) would induce that background electromagnetic radiation (microwave) of the universe would create a non propagating torsion field. This suggests that the spacetime may be better modeled by a Riemann–Cartan manifold rather than by a Riemann manifold one. On the other hand, the amplitude of the torsion field (density of spacetime defects) would be extremely "unnatural" small since it is equal to the cosmological constant, and even be out of reach of experimental measurements as reminded by e.g. Bull et al. (2016), Hehl et al. (1976), Kleinert (2008). However, the previous finding (6.105) that the torsion $\aleph_0$ may be directly related to the cosmological constant Λ may give new interesting insights for further indirect measurement of this background torsion field, becoming a link between local spacetime properties and the large scale universe. The dark energy can be described by means of unknown contribution as cosmological constant Λ, and it was suggested in the past if geometric arguments might be used to tackle the dark energy problem e.g. Bamba et al. (2013). In this reference, the existence of other gravitational waves than those based on Einstein curvature based fields is investigated by means of teleparallel gravitation theory. Again in this previous reference, the authors have shown that gravitational waves in the framework of teleparallel gravity were the same as those in the framework of classical relativistic gravitation of Einstein. In the present book, we have a slightly different approach since we have shown that the cosmological constant Λ, at least for negative values, may be approached in a Riemann–Cartan spacetime. In fine, the cosmological dark energy may have both unknown contributions origin and geometric origin too.

Combining the results of the two Chaps. 5 and 6, we may suggest on the other hand that electromagnetic field is potentially source of torsion, and that torsion is itself potentially source of the cosmological constant, the dark energy. These two suggestions are supported by simple and explicit examples.

6.6 Summary on Gravitation-Electromagnetism Interaction

In this section, we mainly consider the intimate link between the electrodynamics and the geometry of the continuum where the electromagnetic waves are propagating. Considering a very simple shape of the Lagrangian (the same form for all the models), we extend the geometry structure from the flat spacetime to curved and then non zero torsion and curved spacetimes.

Relativistic gravitation theory is commonly based on the design of the spacetime geometry, namely the definition (or the determination) of the metric tensor $g_{\alpha\beta}(x^\mu)$ in the presence of massive body which causes the gravitational forces. The spacetime metric underlying the electromagnetism theory leading to the Maxwell's equations was initially based on the uniform Minkowski metric. In order to analyze the interaction of electromagnetism and the gravitation, the development of the Maxwell's equations within curved continuum shows the electromagnetism-gravitation mutual influence by means of the geometry characterized by metric, Levi-Civita connection, and associated Ricci curvature. When dealing with the so called second gradient continuum, where abrupt gradients of physical properties may occur, the extension of the Maxwell's equations, namely the resulting wave propagation, is necessary to account for the non zero torsion $\aleph_{\alpha\beta}^\gamma \neq 0$ and non zero curvature $\Re_{\alpha\beta\lambda}^\gamma \neq 0$. Among numerous approaches, the use of Riemann–Cartan manifold as underlying geometrical structure is worth e.g. Fernando et al. (2012).

Table 6.2 Theories of electromagnetism interacting with gravitation in curved spacetimes: Minkowski (flat), and Riemann (curved)

Minkowski		Special relativity
Spacetime metric	$g_{\alpha\beta} := \{+, -, -, -\}$	$g := \sqrt{|\text{Det}g_{\alpha\beta}|}$
Electromagnetic tensor	$\Gamma^{\gamma}_{\alpha\beta} \equiv 0$	$\mathcal{F}_{\alpha\beta} = \partial_{\alpha}A_{\beta} - \partial_{\beta}A_{\alpha}$
Constitutive laws	$\mathcal{L} := -\dfrac{1}{4}\mathcal{F}^{\alpha\beta}\mathcal{F}_{\alpha\beta}\, g$	$\mathcal{F}^{\alpha\beta} = \epsilon_0\, g^{\alpha\mu}g^{\beta\nu}\mathcal{F}_{\mu\nu}$
Conservation laws	$\partial_{\beta}\mathcal{F}^{\alpha\beta} = 0$	$g^{\nu\beta}\partial_{\nu}\partial_{\beta}A^{\mu} = 0$
Riemann		Einstein gravitation
Spacetime metric	$g_{\alpha\beta} := g_{\alpha\beta}(x^{\mu})$	$g := \sqrt{|\text{Det}g_{\alpha\beta}|}$
Electromagnetic tensor	$\overline{\Gamma}^{\gamma}_{\alpha\beta}$ (Levi-Civita)	$\mathcal{F}_{\alpha\beta} = \overline{\nabla}_{\alpha}A_{\beta} - \overline{\nabla}_{\beta}A_{\alpha}$
Constitutive laws	$\mathcal{L} := -\dfrac{1}{4}\mathcal{F}^{\alpha\beta}\mathcal{F}_{\alpha\beta}\, g$	$\mathcal{F}^{\alpha\beta} = \epsilon_0\, g^{\alpha\mu}g^{\beta\nu}\mathcal{F}_{\mu\nu}$
Conservation laws	$\overline{\nabla}_{\beta}\mathcal{F}^{\alpha\beta} = 0$	$g^{\nu\beta}\overline{\nabla}_{\nu}\overline{\nabla}_{\beta}A^{\mu} - g^{\mu\alpha}\mathfrak{R}_{\alpha\gamma}A^{\gamma} = 0$

We report on the table below the overview of interaction of electromagnetic waves with various spacetimes from the simplest to the complicated ones. Table 6.2 displays most classical approaches for analyzing the interaction of the electromagnetic waves with gravitation and more generally the geometric structure of spacetimes underlying the gravitation e.g. Ryder (2009).

For the first two spacetimes, the basic geometric variable is the spacetime metric tensor: uniform for Minkowski (Special Relativity) and depending on the coordinates x^{μ} for Riemann spacetime (Relativistic Gravitation). We observe that the Minimal Coupling Procedure induces the interaction of the electromagnetic waves with the spacetime curvature, by means of the Ricci curvature. This is pointed out by the shape of the field equation e.g. de Andrade and Pereira (1999), Smalley (1986).

Relativistic gravitation may be also approached by using the Tele Parallel Gravitation, the tetrads are introduced prior to the metric, then the electromagnetic tensor is defined by means of the Levi-Civita connection associated to the induced metric tensor. The electromagnetic waves interact with the spacetime via the Ricci curvature (calculated from the change of curvature defined with the contortion tensor). The last model in the framework of Riemann–Cartan Gravitation (e.g. Sotiriou and Liberati 2007) may highlight some problems of gauge invariance since the electromagnetic tensor does not satisfy the Lorenz gauge invariance (say $U(1)$ gauge invariance) e.g. Puntigam et al. (1997). This may hurt at first sight, however, more investigations should be conducted since the concept of magnetic monopole enters into the discussion because the Gauss law on magnetic flux should be re-analyzed in such a case e.g. Fernando et al. (2012). The two non zero torsion and /or curved spacetimes are displayed in Table 6.3. It is striking that Eq. (6.110) we obtained, is analogous to the particular contortion tensor $\mathcal{T}^{\gamma}_{\alpha\beta} = -(G/c^4).\mathcal{F}^{\gamma}_{\beta}A_{\alpha}$ found in this later paper by Fernando and co-workers. Indeed, they have deduced that this the particular connection defined by $\Gamma^{\gamma}_{\alpha\beta} := \overline{\Gamma}^{\gamma}_{\alpha\beta} + \mathcal{T}^{\gamma}_{\alpha\beta}$ of the

Table 6.3 Theories of electromagnetism interacting with gravitation in curved spacetimes with torsion: Weitzenböck (non zero torsion, not curved) and Riemann–Cartan (non zero torsion and curved)

Weitzenböck		Teleparallel gravitation
Spacetime metric	$g_{\alpha\beta} := g_{ij} F_\alpha^i F_\beta^j$	$g := \sqrt{\lvert \mathrm{Det}\, g_{\alpha\beta}\rvert}$
Electromagnetic tensor	$\Gamma_{\alpha\beta}^\gamma := F_i^\gamma \partial_\alpha F_\beta^i = \overline{\Gamma}_{\alpha\beta}^\gamma + \mathscr{T}_{\alpha\beta}^\gamma$	$\mathscr{F}_{\alpha\beta} = \overline{\nabla}_\alpha A_\beta - \overline{\nabla}_\beta A_\alpha$
Constitutive laws	$\mathscr{L} := -\dfrac{1}{4} \mathscr{F}^{\alpha\beta} \mathscr{F}_{\alpha\beta}\, g$	$\mathscr{F}^{\alpha\beta} = \epsilon_0\, g^{\alpha\mu} g^{\beta\nu} \mathscr{F}_{\mu\nu}$
Conservation laws	$\overline{\nabla}_\beta \mathscr{F}^{\alpha\beta} = 0$	$g^{\nu\beta} \overline{\nabla}_\nu \overline{\nabla}_\beta A^\mu - g^{\mu\alpha} \mathscr{K}_{\alpha\gamma} A^\gamma = 0$
Riemann–Cartan		*Einstein–Cartan gravitation*
Spacetime metric	$g_{\alpha\beta} := g_{\alpha\beta}(x^\mu)$	$g := \sqrt{\lvert \mathrm{Det}\, g_{\alpha\beta}\rvert}$
Electromagnetic tensor	$\Gamma_{\alpha\beta}^\gamma$	$\mathscr{F}_{\alpha\beta} = \nabla_\alpha A_\beta - \nabla_\beta A_\alpha$
Constitutive laws	$\mathscr{L} := -\dfrac{1}{4} \mathscr{F}^{\alpha\beta} \mathscr{F}_{\alpha\beta}\, g$	$\mathscr{F}^{\alpha\beta} = \epsilon_0\, g^{\alpha\mu} g^{\beta\nu} \mathscr{F}_{\mu\nu}$
Conservation laws	$\nabla_\beta \mathscr{F}^{\alpha\beta} = 0$	$g^{\nu\beta} \nabla_\nu \nabla_\beta A^\mu + g^{\mu\alpha} \aleph_{\nu\alpha}^\gamma \nabla_\gamma A^\nu - g^{\mu\alpha} \Re_{\alpha\gamma} A^\gamma = 0$

Riemann–Cartan spacetime which allowed them to derive the two Maxwell's equations:

$$\nabla_\beta \mathscr{F}^{\alpha\beta} = \overline{\nabla}_\beta \mathscr{F}^{\alpha\beta} = 0, \qquad \nabla_{[\beta} \mathscr{F}_{\alpha\beta]} = \overline{\nabla}_{[\beta} \mathscr{F}_{\alpha\beta]} = \partial_{[\beta} \mathscr{F}_{\alpha\beta]} = 0$$

Remark 6.36 For this particular connection of the Riemann–Cartan spacetime, the model is Lorenz gauge invariant.

For instance, following another path Poplawski suggested to define the four-potential as a part of the trace of the torsion itself e.g. Poplawski (2010). All these aspects will certainly constitute future research topics. We observe that the both the torsion and the curvature influence the electromagnetic wave propagation in the Einstein–Cartan–Maxwell framework. Despite the crucial point on the Lorenz gauge invariance, this model seems to extend and thus include all previous models.

Remark 6.37 As a final remark of this chapter, the geometrization of gravity was first developed by Einstein. He recognized that gravity is due to the bending of the spacetime $\Re \neq 0$ and that gravity is indistinguishable from an accelerating inertial frame. The dependence of the Lagrangian on the curvature tensor is the starting point for deriving the gravity field equations (Einstein equations).

At a second step, from this finding and by introducing the tensor Faraday including electric and magnetic fields—within the Lagrangian, it is recognized that the light, which is a particular case of electromagnetic waves, bends if viewed from a uniformly accelerating frame and then accordingly that the gravity would therefore bend the light. Electromagnetism is governed by Maxwell's equations. The interaction of gravitation and electromagnetic waves are described Einstein–Maxwell's equations.

The geometrization of the electromagnetic fields constitute the third step when these fields are present in the spacetime. For that purpose, we have considered an extended spacetime where curvature and torsion are not present, Riemann–Cartan spacetime. By observing the third equation, we conclude that the gravitational and electromagnetic fields are respectively identified as geometric objects of t such a spacetime, namely the curvature $\Re^{\gamma}_{\alpha\beta\lambda}$ and the torsion $\aleph^{\gamma}_{\alpha\beta}$.

Chapter 7
General Conclusion

The axiomatic foundations of relativistic gravitation and electromagnetism were proposed by Hilbert by assuming two axioms. The first is the Mie axiom stating that the Lagrangian $\mathfrak{L}$ depends on both the space metric and their first and second derivatives, and on the electromagnetic potential and their first derivatives. The second is the covariance of the Lagrangian with respect to arbitrary transformations of coordinates of the spacetime. Most of field equations governing theoretical physics are obtained from a variational principle after defining a Lagrangian function and its arguments including both the generalized coordinates (x^μ), and the fields with their derivatives $(\Phi^i, \Phi^i_{\mu_1}, \cdots, \Phi^i_{\mu_1\cdots\mu_n})$. Invariance of Lagrangian and associated fields equations includes two aspects. First, passive diffeomorphism (covariance) is a mathematical requirement stating that it should be possible to use different coordinates to describe one physical situation (change of coordinates). Covariance dictates that physics laws (conservation laws, and constitutive laws) keep the same form, regardless of the coordinate system. Second, invariance with respect to an active diffeomorphism implies that any solution of the field equations can be transformed and still satisfies the same, untransformed field equations. The invariance group may be deduced from axioms of causality principle or the invariance of light velocity in relativistic theory. The invariance with respect the complete group of Poincaré including the Lorentz group (generated by three spatial rotations, and three boosts), and the spacetime translations together with dilations are considered throughout this book.

In the first part of the present book, we show that covariance imposes that Lagrangian depends only on metric, on torsion, and on curvature $\mathfrak{L}(\mathbf{g}, \aleph, \mathfrak{R})$. This result holds both for second strain gradient continuum, and for Einstein–Cartan gravity in relativistic mechanics. The arbitrariness of the connection choice is related to the frame-indifference principle. We do not investigate the concept of spinning particle by introducing explicitly metric and spin tensor as geometric arguments. Our method lies more on the Taylor expansion of the metric by means of independent connection on the continuum or on the spacetime manifolds.

© Springer International Publishing AG, part of Springer Nature 2018
L. R. Rakotomanana, *Covariance and Gauge Invariance in Continuum Physics*,
Progress in Mathematical Physics 73, https://doi.org/10.1007/978-3-319-91782-5_7

In a second part, we derive field equations by using the gauge invariance where Lie derivatives of metric, torsion, and curvature were the worth primal variables for that purpose. Extension to local translation for obtaining local Poincaré invariance is done. Dependence of Lagrangian on the three tensors, and not on any other non invariant arguments might be a guideline to design the Lagrangian density either for the spacetime for relativistic gravitation or the matter for strain gradient continuum. It seems that Einstein–Cartan manifold would be more convenient than Riemannian manifold. The knowledge of the metric, the torsion, and the curvature tensors allows us to define, at least locally, the entire geometry structure of a Einstein–Cartan manifold (continuum body and spacetime). We do not consider multipole models where additional arguments of the Lagrangian function are not torsion and curvature of the connection but rather some internal spin tensor, or angular momentum independently on the connection. When curved matter with torsion is in evolution within a curved spacetime with torsion too, the method for coupling both of them remains an open problem. We have suggested in the present work the concept of generalized deformation including the change of metric strain but also the change of internal topology, characterized by the contortion tensor $\mathcal{T}$ and its covariant derivative based on the Levi-Civita connection. This is different of the change of torsion with respect to the spacetime. In such a case the general form of the Lagrangian is modified to include the spacetime geometry and the generalized deformation $\mathfrak{G} := \int \mathfrak{L}(\hat{g}_{\alpha\beta}, \hat{\aleph}^{\gamma}_{\alpha\beta}, \hat{\mathcal{R}}^{\gamma}_{\alpha\beta\lambda}; \varepsilon_{\alpha\beta}, \mathcal{T}^{\gamma}_{\alpha\beta}, \mathcal{K}^{\gamma}_{\alpha\beta\lambda}) \, \omega_n$. Beyond the covariance formulation of equations which is mathematical requirement for any mathematical models. Invariance was also used to define both the constitutive laws for stress and the hypermomenta by means of the Lagrange variations $\Delta\varepsilon_{\alpha\beta}$, $\Delta\mathcal{T}^{\gamma}_{\alpha\beta}$, and $\Delta\mathcal{K}^{\gamma}_{\alpha\beta\lambda}$, and the associated conservation laws by means of the Lie derivative variations $\mathcal{L}_{\xi}\varepsilon_{\alpha\beta}$, $\mathcal{L}_{\xi}\mathcal{T}^{\gamma}_{\alpha\beta}$, and $\mathcal{L}_{\xi}\mathcal{K}^{\gamma}_{\alpha\beta\lambda}$. The arbitrariness of the vector field ξ^{α} induces the conservation laws. The last two chapters deal with some selected topics in the domain of continuum mechanics and electromagnetism interacting with gravitation. A system of equations involving both the gravitation, the electromagnetism and the presence of torsion is suggested at the end of the last chapter. The role of the torsion and the curvature in the derivation of conservation laws id highlighted particularly for elastic and electromagnetic wave propagation within a non homogeneous media modeled with Riemann–Cartan manifolds. It is illustrated by fundamental examples in continuum mechanics (5.23) and electromagnetism (6.88), namely on the influence of the geometric structures as torsion and curvature of the connection of the wave propagation. It may also give new insights to go further in the development of physics in the framework of Riemann–Cartan spacetime, namely the role of the torsion tensor related to the presence of cosmological constant in the gravitation theory.

Appendix

A.1 Lorentz Transformation

We remind the two postulates of the special relativity theory: (1) physics laws are the same (have the same shape) in two referential frames in relative constant motion (no rotation); (2) the speed of light c is finite and independent of the motion of its source in any referential frame e.g. Ryder (2009).

Let consider two inertial frames $\mathfrak{R}$ and $\tilde{\mathfrak{R}}$ with relative uniform and constant velocity $\mathbf{v}$, assumed to be along the x^1 and $\tilde{x}^1$ axis (without loss of the generality of our purpose). The length element within these two inertial frames takes the form of:

$$ds^2 := g_{\alpha\beta}dx^\alpha dx^\beta, \qquad d\tilde{s}^2 := \tilde{g}_{\mu\nu}d\tilde{x}^\mu d\tilde{x}^\nu \tag{A.1}$$

where the metric are $g_{\alpha\beta} := \{+1, -1, -1, -1\}$ and $\tilde{g}_{\mu\nu} := \{+1, -1, -1, -1\}$ respectively. We assume that the origin of the axis coincides at times $x^0 := ct = 0$ and $\tilde{x}^0 := c\tilde{t} = 0$. The linear transformation between the two coordinates of the frames $\mathfrak{R}$ and $\tilde{\mathfrak{R}}$ is given by:

$$\tilde{x}^\mu = \Lambda^\mu_\alpha x^\alpha$$

For the particular case we are interested in, this transformation takes the form of:

$$\begin{cases} \tilde{x}^0 = \Lambda^0_0 x^0 + \Lambda^0_1 x^1 \\ \tilde{x}^1 = \Lambda^1_0 x^0 + \Lambda^1_1 x^1 \\ \tilde{x}^2 = x^2 \\ \tilde{x}^3 = x^3 \end{cases} \tag{A.2}$$

© Springer International Publishing AG, part of Springer Nature 2018
L. R. Rakotomanana, *Covariance and Gauge Invariance in Continuum Physics*,
Progress in Mathematical Physics 73, https://doi.org/10.1007/978-3-319-91782-5

since there is no motion along the direction x^2 and x^3. The Lorentz transformation is defined as the transformation that verifies the invariance of the length:

$$g_{\alpha\beta} x^\alpha x^\beta = \tilde{g}_{\mu\nu} \tilde{x}^\mu \tilde{x}^\nu \quad \Longrightarrow \quad (x^0)^2 - (x^1)^2 \equiv (\tilde{x}^0)^2 - (\tilde{x}^1)^2 \tag{A.3}$$

Consider a particle at rest with respect to the inertial frame $\mathfrak{R}$ such that $x^1 = 0$, its velocity as seen by an observer at rest in the referential frame $\tilde{\mathfrak{R}}$ is $-v$ meaning that $\tilde{x}^1 = -v\tilde{x}^0$. We obtain:

$$\begin{cases} \tilde{x}^0 = \Lambda_0^0 \, x^0 \\ \tilde{x}^1 = \Lambda_0^1 \, x^0 \end{cases} \quad \Longrightarrow \quad \frac{\Lambda_0^1}{\Lambda_0^0} = -v$$

The same reasoning can be done with a particle at rest in the referential $\tilde{\mathfrak{R}}$ with $\tilde{x}^1 = 0$, its velocity as seen by an observer at rest in the referential frame $\mathfrak{R}$ is $+v$ meaning that $x^1 = vx^0$. We can write:

$$\begin{cases} \tilde{x}^0 = \Lambda_0^0 \, x^0 + \Lambda_1^0 \, x^1 \\ 0 = \Lambda_0^1 \, x^0 + \Lambda_1^1 \, x^1 \end{cases} \quad \Longrightarrow \quad \Lambda_0^1 + \Lambda_1^1 \, v = 0$$

We deduce $\Lambda_1^1 = \Lambda_0^0$. The transformation is written as follows:

$$\begin{cases} \tilde{x}^0 = \Lambda_0^0 \left(x^0 + \dfrac{\Lambda_1^0}{\Lambda_0^0} x^1 \right) \\ \tilde{x}^1 = \Lambda_0^0 \left(-v \, x^0 + x^1 \right) \\ \tilde{x}^2 = x^2 \\ \tilde{x}^3 = x^3 \end{cases} \tag{A.4}$$

Applying the second postulate of special relativity stating that a light pulse must propagate with the same speed in both referential frames, we have $\tilde{x}^1 = \tilde{x}^0$ and $x^1 = x^0$, then:

$$\frac{\Lambda_1^0}{\Lambda_0^0} = -v \tag{A.5}$$

Introducing (A.5) in (A.4) and in (A.3) gives the component:

$$\left(\Lambda_0^0 \right)^2 = \frac{1}{1 - v^2} \tag{A.6}$$

The Lorentz transformation thus takes the form of:

$$\begin{cases} \tilde{x}^0 = \dfrac{1}{\sqrt{1 - v^2}} \left(x^0 - v \, x^1 \right) \\ \tilde{x}^1 = \dfrac{1}{\sqrt{1 - v^2}} \left(-v \, x^0 + x^1 \right) \\ \tilde{x}^2 = x^2 \\ \tilde{x}^3 = x^3 \end{cases} \tag{A.7}$$

where the time coordinate is mixed with the spatial coordinate. They cannot be dissociated as in Newtonian mechanics (Galilean invariance). By writing the transformation as $\tilde{x}^\mu = \Lambda^\mu_\nu\, x^\nu$, the three matrices associated to the translation along the vector base $\mathbf{e}_1$, $\mathbf{e}_2$ and $\mathbf{e}_3$ are respectively:

$$\Lambda^{(1)} = \begin{bmatrix} \gamma & -\gamma v^1 & 0 & 0 \\ -\gamma v^1 & \gamma & 0 & 0 \\ 0 & 0 & 1 & 0 \\ 0 & 0 & 0 & 1 \end{bmatrix} \qquad \Lambda^{(2)} = \begin{bmatrix} \gamma & 0 & -\gamma v^2 & 0 \\ 0 & 1 & 0 & 0 \\ -\gamma v^1 & 0 & \gamma & 0 \\ 0 & 0 & 0 & 1 \end{bmatrix}$$

$$\Lambda^{(3)} = \begin{bmatrix} \gamma & 0 & 0 & -\gamma v^3 \\ 0 & 1 & 0 & 0 \\ 0 & 0 & 1 & 0 \\ -\gamma v^3 & 0 & 0 & \gamma \end{bmatrix}$$

For a uniform velocity along an arbitrary direction, the Lorentz transformation then holds:

$$\Lambda^i_j = \delta^i_j + \frac{(\gamma - 1)\, v^i v_j}{|\mathbf{v}|^2}, \qquad \Lambda^0_i = \Lambda^i_0 = -\gamma\, v^i, \qquad \Lambda^0_0 = \gamma \tag{A.8}$$

with $\gamma := \sqrt{1 - v^2}^{-1}$. We deduce the Poincaré-Lorentz transformation as:

$$\begin{cases} \tilde{x}^0 = \gamma \left(x^0 - \mathbf{v} \cdot \mathbf{x} \right) \\ \tilde{\mathbf{x}} = \mathbf{x} + (\gamma - 1)\, \dfrac{1}{\|\mathbf{v}\|^2}\, \mathbf{v} \otimes \mathbf{v}\, (\mathbf{x}) - \gamma \mathbf{v}\, x^0 \end{cases} \tag{A.9}$$

where $\mathbf{x}$ is the three-dimensional vector-position, and $\mathbf{v}$ is the relative velocity of the two referential frames (remind that $x^0 := ct$). The transformation rule where we factorize the matrix Λ is obtained accordingly[1]:

$$\begin{pmatrix} \tilde{x}^0 \\ \hline \tilde{\mathbf{x}} \end{pmatrix} = \left[\begin{array}{c|c} \gamma & -\gamma \mathbf{v} \\ \hline -\gamma \mathbf{v} & (\gamma - 1)\left(\mathbb{I} + \dfrac{\mathbf{v}}{\|\mathbf{v}\|} \otimes \dfrac{\mathbf{v}}{\|\mathbf{v}\|}\right) \end{array} \right] \begin{pmatrix} x^0 \\ \hline \mathbf{x} \end{pmatrix}$$

---

[1] For small relative velocity, Lorentz transformation reduces to Galilean transformation. Indeed, the relations become when $\|\mathbf{v}\| \ll c$ and when replacing $\tilde{x}^0 := c\tilde{t}$, and $x^0 := ct$ (owing that $\mathbf{v} \to \mathbf{v}/c$):

$$\begin{cases} c\tilde{t} = \gamma \left(ct - \dfrac{\mathbf{v}}{c} \cdot \mathbf{x} \right) \simeq t \\ \tilde{\mathbf{x}} = \mathbf{x} + (\gamma - 1)\, \dfrac{1}{\|\mathbf{v}\|^2}\, \mathbf{v} \otimes \mathbf{v}\, (\mathbf{x}) - \gamma \dfrac{\mathbf{v}}{c}\, ct \simeq \mathbf{x} - \mathbf{v}t \end{cases} \tag{A.10}$$

which gives the boost transformation along an arbitrary direction **v**. The rotation transformation does not present any difficulties since the rotation is based on the Euclidean spatial rotation as in three dimensions. Of course, spatial rotations alone are also Lorentz transformations since they leave the spacetime element invariant.

A.2   Some Relations for the Connection

Let (y^i) and (x^α) two coordinate systems associated to respectively the tangent bases $\{\mathbf{e}_i\}$ and $\{\mathbf{e}_\alpha\}$. Let ∇ be an affine connection and its coefficients $\nabla_{\mathbf{e}_i}\mathbf{e}_j = \Gamma_{ij}^k\mathbf{e}_k$, and $\nabla_{\mathbf{e}_\alpha}\mathbf{e}_\beta = \Gamma_{\alpha\beta}^\gamma\mathbf{e}_\gamma$. The coordinate transformation of ∇ holds

$$
\begin{aligned}
\Gamma_{\alpha\beta}^\gamma\mathbf{e}_\gamma = \nabla_{\mathbf{e}_\alpha}\mathbf{e}_\beta &= \nabla_{J_\alpha^i\mathbf{e}_i}\left(J_\beta^j\mathbf{e}_j\right) = J_\alpha^i\left[\nabla_{\mathbf{e}_i}\left(J_\beta^j\right)\mathbf{e}_j + J_\beta^j\nabla_{\mathbf{e}_i}\mathbf{e}_j\right] \\
&= J_\alpha^i\left[\nabla_{A_i^\alpha\mathbf{e}_\alpha}\left(J_\beta^j\right)\mathbf{e}_j + J_\beta^j\Gamma_{ij}^k\mathbf{e}_k\right] \\
&= J_\alpha^i\left(A_i^\alpha J_{\alpha\beta}^j\mathbf{e}_j + J_\beta^j\Gamma_{ij}^k\mathbf{e}_k\right),
\end{aligned}
$$

say

$$
\Gamma_{\alpha\beta}^\gamma = \left(J_\alpha^i J_\beta^j A_k^\gamma\right)\Gamma_{ij}^k + J_{\alpha\beta}^j A_j^\gamma. \tag{A.11}
$$

Let be the "double connection" $\nabla^2 = \nabla \circ \nabla$

$$
\begin{aligned}
\nabla_{\mathbf{e}_\lambda}\left(\nabla_{\mathbf{e}_\alpha}\mathbf{e}_\beta\right) = \nabla_{\mathbf{e}_\lambda}\left(\Gamma_{\alpha\beta}^\mu\mathbf{e}_\mu\right) &= \nabla_{\mathbf{e}_\lambda}\left(\Gamma_{\alpha\beta}^\mu\right)\mathbf{e}_\mu + \Gamma_{\alpha\beta}^\mu\left(\nabla_{\mathbf{e}_\lambda}\mathbf{e}_\mu\right) \\
&= \Gamma_{\alpha\beta,\lambda}^\mu\mathbf{e}_\mu + \Gamma_{\alpha\beta}^\mu\Gamma_{\lambda\mu}^\gamma\mathbf{e}_\gamma = \left(\Gamma_{\alpha\beta,\lambda}^\gamma + \Gamma_{\alpha\beta}^\mu\Gamma_{\lambda\mu}^\gamma\right)\mathbf{e}_\gamma.
\end{aligned}
$$

Without going into details, the coordinate transformation of ∇^2 holds

$$
\begin{aligned}
\Gamma_{\alpha\beta,\lambda}^\gamma + \Gamma_{\alpha\beta}^\mu\Gamma_{\lambda\mu}^\gamma &= \left[J_\alpha^i J_\beta^j J_\lambda^l A_k^\gamma\left(\Gamma_{ij,l}^k + \Gamma_{ij}^d\Gamma_{ld}^k\right)\right] \\
&\quad + \left(J_{\alpha\lambda}^i J_\beta^j A_k^\gamma + J_\lambda^i J_{\alpha\beta}^j A_k^\gamma + J_\alpha^i J_{\beta\lambda}^j A_k^\gamma\right)\Gamma_{ij}^k \\
&\quad + J_\mu^i J_\lambda^l J_{\alpha\beta}^j A_j^\gamma A_{il}^\mu + J_{\mu\lambda}^i J_{\alpha\beta}^j A_i^\mu A_j^\gamma \tag{A.12}
\end{aligned}
$$

Equations (A.11) and (A.12) show that the connection ∇ and the bi-connection ∇^2 are not tensors, according to (2.21). However, in the both equations, if the terms out of the square brackets vanish then the components behave as components of tensor. Torsion (2.22) and the curvature (2.23) are defined, with respect to the affine connection ∇, on the base $\{\mathbf{e}_a\}$ associated to coordinate system (y^a).

A.3 Algebraic Relations for Bi-connection

Again, we remind the guideline for the Quotient Theorem saying that a set of real numbers ($\mathbb{R}$) form the components of a tensor of a certain rank, if and only if its scalar product with another arbitrary tensor is again a tensor (practically, we attempt to obtain a scalar by a worth choice). For proving the Quotient Theorem, we extend in this section the Lemma 3.2 by adding terms $\Lambda^{ij,kl} h_{ij,kl}$. Adding fourth-order tensor is mandatory because we deal with the presence of curvature in the Lagrangian function.

A.3.1 Identification of Coefficients

Let us apply the Lemma 3.2 to the equation

$$\Lambda^{ij,kl} h_{ij,kl} + \Lambda^{ij,k} h_{ij,k} + \Lambda^{ij} h_{ij} = \Lambda^{ij,kl} h_{ij|k|l} + \Pi^{ij,k} h_{ij|k} + \Pi^{ij} h_{ij},$$

for the identification of the coefficients of $h_{ij,k}$ and h_{ij}. Let us remind that

$$\begin{aligned}
\Lambda^{ij,kl} h_{ij|k|l} = \Lambda^{ij,kl} [& h_{ij,kl} - \Gamma^a_{ik} h_{aj,l} - \Gamma^a_{jk} h_{ia,l} - \Gamma^a_{ik,l} h_{aj} - \Gamma_{ajk,l} h_{ia} \\
& - \Gamma^b_{il} h_{bj,k} + \Gamma^b_{il} (\Gamma^c_{bk} h_{cj} + \Gamma^{jk}_c h_{bc}) - \Gamma^b_{jl} h_{ib,k} \\
& + \Gamma^b_{jl} (\Gamma^c_{ik} h_{cb} + \Gamma^c_{bk} h_{ic}) - \Gamma^b_{kl} h_{ij,b} + \Gamma^b_{kl} (\Gamma^c_{ib} h_{cj} + \Gamma^c_{jb} h_{ic})]
\end{aligned}$$

and

$$\Pi^{ij,k} h_{ij|k} = \Pi^{ij,k} [h_{ij,k} - \Gamma^a_{ik} h_{aj} - \Gamma^a_{jk} h_{ia}].$$

The equation of the coefficients of "$h_{ij,k}$" is

$$\Lambda^{ij,k} h_{ij,k} = -\Lambda^{ij,kl} [\Gamma^a_{ik} h_{aj,l} + \Gamma^a_{jk} h_{ia,l} + \Gamma^b_{il} h_{bj,k} + \Gamma^b_{jl} h_{ib,k} + \Gamma^b_{kl} h_{ij,b}] + \Pi^{ij,k} h_{ij,k}.$$

The following permutations are necessary to explicitly write the common factor $h_{ij,k}$ in all the terms in right hand side (see above):

1. $a \longleftrightarrow i$ and $l \longleftrightarrow k$ for $\Lambda^{ij,kl} \Gamma^a_{ik} h_{aj,l}$
2. $a \longleftrightarrow j$ and $l \longleftrightarrow k$ for $\Lambda^{ij,kl} \Gamma^a_{jk} h_{ia,l}$
3. $b \longleftrightarrow i$ then $b \to a$ for $\Lambda^{ij,kl} \Gamma^b_{il} h_{bj,k}$
4. $b \longleftrightarrow j$ then $b \to a$ for $\Lambda^{ij,kl} \Gamma^b_{jl} h_{ib,k}$
5. $k \longleftrightarrow b$ for $\Lambda^{ij,kl} \Gamma^b_{kl} h_{ij,b}$.

Then, the symmetry of Λ and **h** reduces the identification of coefficients of $h_{ij,k}$

$$(1/2)(\Pi^{ij,k} + \Pi^{ji,k}) = \Lambda^{ij,k} + 2\Gamma^i_{al}\Lambda^{aj,kl} + 2\Gamma^j_{al}\Lambda^{ia,kl} + \Gamma^k_{bl}\Lambda^{ij,bl}.$$

In the same way, the equation of the coefficients of "h_{ij}" is

$$\Lambda^{ij}h_{ij} = \Lambda^{ij,kl}[-\Gamma^a_{ik,l}h_{aj} - \Gamma^a_{jk,l}h_{ia} + \Gamma^b_{il}(\Gamma^c_{bk}h_{cj} + \Gamma^c_{jk}h_{bc})$$

$$+ \Gamma^b_{jl}(\Gamma^c_{ik}h_{cb} + \Gamma^c_{bk}h_{ic})$$

$$+ \Gamma^b_{kl}(\Gamma^c_{ib}h_{cj} + \Gamma^c_{jb}h_{ic})] - \Pi^{ij,k}\Gamma^a_{ik}h_{aj} - \Pi^{ij,k}\Gamma^a_{jk}h_{ia} + \Pi^{ij}h_{ij}.$$

The following permutations are necessary to find again the common factor h_{ij} in all the terms in right hand side (see above):

1. $a \longleftrightarrow i$ for $\Lambda^{ij,kl}\Gamma^a_{ik,l}h_{aj}$ and $\Pi^{ij,k}\Gamma^a_{ik}h_{aj}$
2. $a \longleftrightarrow j$ for $\Lambda^{ij,kl}\Gamma^a_{jk,l}h_{ia}$ and $\Pi^{ij,k}\Gamma^a_{jk}h_{ia}$
3. $c \longleftrightarrow i$ for $\Lambda^{ij,kl}\Gamma^b_{il}\Gamma^c_{bk}h_{cj}$ and $\Lambda^{ij,kl}\Gamma^b_{kl}\Gamma^c_{ib}h_{cj}$
4. $c \longleftrightarrow j$ for $\Lambda^{ij,kl}\Gamma^b_{jl}\Gamma^c_{bk}h_{ic}$ and $\Lambda^{ij,kl}\Gamma^b_{kl}\Gamma^c_{jb}h_{ic}$
5. $c \longleftrightarrow i$ and $b \longleftrightarrow j$ for $\Lambda^{ij,kl}\Gamma^b_{il}\Gamma^c_{jk}h_{bc}$ and $\Lambda^{ij,kl}\Gamma^b_{jl}\Gamma^c_{ik}h_{cb}$.

Then, the symmetry of Λ and **h** reduces the identification of coefficients of h_{ij}

$$(1/2)(\Pi^{ij} + \Pi^{ji}) = \Lambda^{ij} + \Gamma^i_{ak,l}\Lambda^{aj,kl} + \Gamma^j_{ak,l}\Lambda^{ia,kl}$$

$$- \Gamma^b_{al}\Gamma^i_{bk}\Lambda^{aj,kl} - \Gamma^b_{cl}\Gamma^j_{bk}\Lambda^{ic,kl}$$

$$- \Gamma^i_{bl}\Gamma^j_{ck}\Lambda^{bc,kl} - \Gamma^j_{bl}\Gamma^i_{ck}\Lambda^{bc,kl}$$

$$- \Gamma^b_{kl}\Gamma^i_{cb}\Lambda^{cj,kl} - \Gamma^b_{kl}\Gamma^j_{cb}\Lambda^{ci,kl}$$

$$+ (1/2)\Gamma^i_{ak}(\Pi^{aj,k} + \Pi^{ja,k}) + (1/2)\Gamma^j_{ak}(\Pi^{ia,k} + \Pi^{ai,k}).$$

A.3.2   Coefficients of Bi-connection

We apply the permutation between i and l, such as

$$\mathbb{S}^k_{ij,l} + \mathbb{T}^k_{ij,l} + \mathbb{S}^m_{ij}\mathbb{S}^k_{lm} + \mathbb{S}^m_{ij}\mathbb{T}^k_{lm} + \mathbb{T}^m_{ij}\mathbb{S}^k_{lm} + \mathbb{T}^m_{ij}\mathbb{T}^k_{lm}$$

$$= (1/2)\left[\mathbb{S}^k_{ij,l} + \mathbb{S}^m_{ij}\mathbb{S}^k_{lm} + \mathbb{S}^m_{ij}\mathbb{T}^k_{lm} + \mathbb{S}^k_{lj,i} + \mathbb{S}^m_{lj}\mathbb{S}^k_{im} + \mathbb{S}^m_{lj}\mathbb{T}^k_{im}\right]$$

$$+ (1/2)\left[\mathbb{S}^k_{ij,l} + \mathbb{S}^m_{ij}\mathbb{S}^k_{lm} + \mathbb{S}^m_{ij}\mathbb{T}^k_{lm} - \mathbb{S}^k_{lj,i} - \mathbb{S}^m_{lj}\mathbb{S}^k_{im} - \mathbb{S}^m_{lj}\mathbb{T}^k_{im}\right]$$

$$+(1/2)\left[\mathbb{T}^k_{ij,l} + \mathbb{T}^m_{ij}\mathbb{T}^k_{lm} + \mathbb{T}^m_{ij}\mathbb{S}^k_{lm} + \mathbb{T}^k_{lj,i} + \mathbb{T}^m_{lj}\mathbb{T}^k_{im} + \mathbb{T}^m_{lj}\mathbb{S}^k_{im}\right]$$

$$+(1/2)\left[\mathbb{T}^k_{ij,l} + \mathbb{T}^m_{ij}\mathbb{T}^k_{lm} + \mathbb{T}^m_{ij}\mathbb{S}^k_{lm} - \mathbb{T}^k_{lj,i} - \mathbb{T}^m_{lj}\mathbb{T}^k_{im} - \mathbb{T}^m_{lj}\mathbb{S}^k_{im}\right].$$

Then we develop the expression in the right hand side. We have

$$\mathbb{S}^k_{ij,l} - \mathbb{S}^k_{lj,i} + \mathbb{T}^k_{ij,l} - \mathbb{T}^k_{lj,i} = \Gamma^k_{ij,l} - \Gamma^k_{lj,i},$$

and

$$\mathbb{S}^m_{ij}\mathbb{S}^k_{lm} - \mathbb{S}^m_{lj}\mathbb{S}^k_{im} = (1/4)\left(\Gamma^m_{ij}\Gamma^k_{lm} - \Gamma^m_{lj}\Gamma^k_{im}\right) + (1/4)\left(\Gamma^m_{ij}\Gamma^k_{ml} - \Gamma^m_{lj}\Gamma^k_{mi}\right)$$
$$+ (1/4)\left(\Gamma^m_{ji}\Gamma^k_{lm} - \Gamma^m_{jl}\Gamma^k_{im}\right) + (1/4)\left(\Gamma^m_{ji}\Gamma^k_{ml} - \Gamma^m_{jl}\Gamma^k_{mi}\right)$$

$$\mathbb{S}^m_{ij}\mathbb{T}^k_{lm} - \mathbb{S}^m_{lj}\mathbb{T}^k_{im} = (1/4)\left(\Gamma^m_{ij}\Gamma^k_{lm} - \Gamma^m_{lj}\Gamma^k_{im}\right) - (1/4)\left(\Gamma^m_{ij}\Gamma^k_{ml} - \Gamma^m_{lj}\Gamma^k_{mi}\right)$$
$$+ (1/4)\left(\Gamma^m_{ji}\Gamma^k_{lm} - \Gamma^m_{jl}\Gamma^k_{im}\right) - (1/4)\left(\Gamma^m_{ji}\Gamma^k_{ml} - \Gamma^m_{jl}\Gamma^k_{mi}\right)$$

$$\mathbb{T}^m_{ij}\mathbb{T}^k_{lm} - \mathbb{T}^m_{lj}\mathbb{T}^k_{im} = (1/4)\left(\Gamma^m_{ij}\Gamma^k_{lm} - \Gamma^m_{lj}\Gamma^k_{im}\right) - (1/4)\left(\Gamma^m_{ij}\Gamma^k_{ml} - \Gamma^m_{lj}\Gamma^k_{mi}\right)$$
$$- (1/4)\left(\Gamma^m_{ji}\Gamma^k_{lm} - \Gamma^m_{jl}\Gamma^k_{im}\right) + (1/4)\left(\Gamma^m_{ji}\Gamma^k_{ml} - \Gamma^m_{jl}\Gamma^k_{mi}\right)$$

$$\mathbb{T}^m_{ij}\mathbb{S}^k_{lm} - \mathbb{T}^m_{lj}\mathbb{S}^k_{im} = (1/4)\left(\Gamma^m_{ij}\Gamma^k_{lm} - \Gamma^m_{lj}\Gamma^k_{im}\right) + (1/4)\left(\Gamma^m_{ij}\Gamma^k_{ml} - \Gamma^m_{lj}\Gamma^k_{mi}\right)$$
$$- (1/4)\left(\Gamma^m_{ji}\Gamma^k_{lm} - \Gamma^m_{jl}\Gamma^k_{im}\right) - (1/4)\left(\Gamma^m_{ji}\Gamma^k_{ml} - \Gamma^m_{jl}\Gamma^k_{mi}\right).$$

The sum of the four previous equalities is equal to $\Gamma^m_{ij}\Gamma^k_{lm} - \Gamma^m_{lj}\Gamma^k_{im}$. Then we obtain

$$\mathbb{S}^k_{ij,l} + \mathbb{S}^k_{lj,i} + \mathbb{T}^k_{ij,l} + \mathbb{T}^k_{lj,i} = \Gamma^k_{ij,l} + \Gamma^k_{lj,i},$$

and

$$\mathbb{S}^m_{ij}\mathbb{S}^k_{lm} + \mathbb{S}^m_{lj}\mathbb{S}^k_{im} = (1/4)\left(\Gamma^m_{ij}\Gamma^k_{lm} + \Gamma^m_{lj}\Gamma^k_{im}\right) + (1/4)\left(\Gamma^m_{ij}\Gamma^k_{ml} + \Gamma^m_{lj}\Gamma^k_{mi}\right)$$
$$+ (1/4)\left(\Gamma^m_{ji}\Gamma^k_{lm} + \Gamma^m_{jl}\Gamma^k_{im}\right) + (1/4)\left(\Gamma^m_{ji}\Gamma^k_{ml} + \Gamma^m_{jl}\Gamma^k_{mi}\right)$$

$$\mathbb{S}_{ij}^m \mathbb{T}_{lm}^k + \mathbb{S}_{lj}^m \mathbb{T}_{im}^k = (1/4)\left(\Gamma_{ij}^m \Gamma_{lm}^k + \Gamma_{lj}^m \Gamma_{im}^k\right) - (1/4)\left(\Gamma_{ij}^m \Gamma_{ml}^k + \Gamma_{lj}^m \Gamma_{mi}^k\right)$$

$$+ (1/4)\left(\Gamma_{ji}^m \Gamma_{lm}^k + \Gamma_{jl}^m \Gamma_{im}^k\right) - (1/4)\left(\Gamma_{ji}^m \Gamma_{ml}^k + \Gamma_{jl}^m \Gamma_{mi}^k\right)$$

$$\mathbb{T}_{ij}^m \mathbb{T}_{lm}^k + \mathbb{T}_{lj}^m \mathbb{T}_{im}^k = (1/4)\left(\Gamma_{ij}^m \Gamma_{lm}^k + \Gamma_{lj}^m \Gamma_{im}^k\right) - (1/4)\left(\Gamma_{ij}^m \Gamma_{ml}^k + \Gamma_{lj}^m \Gamma_{mi}^k\right)$$

$$- (1/4)\left(\Gamma_{ji}^m \Gamma_{lm}^k + \Gamma_{jl}^m \Gamma_{im}^k\right) + (1/4)\left(\Gamma_{ji}^m \Gamma_{ml}^k + \Gamma_{jl}^m \Gamma_{mi}^k\right)$$

$$\mathbb{T}_{ij}^m \mathbb{S}_{lm}^k + \mathbb{T}_{lj}^m \mathbb{S}_{im}^k = (1/4)\left(\Gamma_{ij}^m \Gamma_{lm}^k + \Gamma_{lj}^m \Gamma_{im}^k\right) + (1/4)\left(\Gamma_{ij}^m \Gamma_{ml}^k + \Gamma_{lj}^m \Gamma_{mi}^k\right)$$

$$- (1/4)\left(\Gamma_{ji}^m \Gamma_{lm}^k + \Gamma_{jl}^m \Gamma_{im}^k\right) - (1/4)\left(\Gamma_{ji}^m \Gamma_{ml}^k + \Gamma_{jl}^m \Gamma_{mi}^k\right).$$

The sum of the four previous equalities is equal to $\Gamma_{ij}^m \Gamma_{lm}^k + \Gamma_{lj}^m \Gamma_{im}^k$. Then we have

$$\mathbb{S}_{ij,l}^k + \mathbb{T}_{ij,l}^k + \mathbb{S}_{ij}^m \mathbb{S}_{lm}^k + \mathbb{S}_{ij}^m \mathbb{T}_{lm}^k + \mathbb{T}_{ij}^m \mathbb{S}_{lm}^k + \mathbb{T}_{ij}^m \mathbb{T}_{lm}^k$$

$$= (1/2)\left(\Gamma_{ij,l}^k + \Gamma_{ij}^m \Gamma_{lm}^k - \Gamma_{lj,i}^k - \Gamma_{lj}^m \Gamma_{im}^k\right)$$

$$+ (1/2)\left(\Gamma_{ij,l}^k + \Gamma_{ij}^m \Gamma_{lm}^k + \Gamma_{lj,i}^k + \Gamma_{lj}^m \Gamma_{im}^k\right).$$

A.4   Lie Derivative and Exterior Derivative on Manifold $\mathfrak{M}$

A.4.1   Lie Derivative

Let ϕ, $\mathbf{u}$, and $\mathbf{S}$ a smooth real-valued scalar, vector and tensor fields on a finite-dimension manifold $\mathfrak{M}$. Let $\mathfrak{C}^n$ the set of sufficiently smooth sections of a tensor bundle of type (p, q) on $\mathfrak{M}$. The Lie derivative with respect to a vector field ξ is a map:

$$\mathcal{L}_\xi : \mathfrak{C}^n\left(T_q^p\right) \to \mathfrak{C}^n\left(T_q^p\right)$$

which is defined as follows. First, we define the Lie derivative of scalar field ϕ along the vector field ξ as the directional derivative:

$$\mathcal{L}_\xi \phi := \xi[\phi] = \xi^\mu(\mathbf{x})\, \partial_\mu \phi(\mathbf{x}) \tag{A.13}$$

Lie derivative of the scalar $\phi(\mathbf{x})$ coincides to covariant derivative on $\mathfrak{M}$. Second, we define the Lie derivative of a vector field $\mathbf{u}$ is equal to the Jacobi brackets as:

$$\mathcal{L}_\xi \mathbf{u} := [\xi, \mathbf{u}], \quad \mathcal{L}_{\xi(\mathbf{x})} u^\nu(\mathbf{x}) = \xi^\mu(\mathbf{x}) \partial_\mu u^\nu(\mathbf{x}) - u^\mu(\mathbf{x}) \partial_\mu \xi^\nu(\mathbf{x}) \quad \text{(A.14)}$$

In a Riemann–Cartan manifold endowed with an affine connection ∇ and torsion $\aleph$, the Lie derivative of a vector field takes the form of:

$$\mathcal{L}_\xi \mathbf{u} := \nabla_\xi \mathbf{u} - \nabla_\mathbf{u} \xi - \aleph(\xi, \mathbf{u}) \quad \text{(A.15)}$$

where the $\aleph$ denotes the torsion operator e.g. Nakahara (1996). Third, Lie derivative is defined to obey the following product rule of vector $\mathbf{u} \in T_P \mathfrak{M}$ and a covector $\omega \in T_P \mathfrak{M}^*$

$$\mathcal{L}_\xi [\omega(\mathbf{u})] := (\mathcal{L}_\xi \omega)(\mathbf{u}) + \omega(\mathcal{L}_\xi \mathbf{u}) \quad \text{(A.16)}$$

The Lie derivative of e 1-form ω is then a 1-form which respects:

$$\mathcal{L}_\xi \omega(\mathbf{u}) = \xi[\omega(\mathbf{u})] - \omega([\xi, \mathbf{u}]) \quad \text{(A.17)}$$

The Lie derivative is shown to obey the Leibniz rules for wedge (ω_1 and ω_2 are forms) and for tensor product ($\mathbf{T}_1$ and $\mathbf{T}_2$ are tensors):

$$\mathcal{L}_\xi (\omega_1 \wedge \omega_2) = \mathcal{L}_\xi(\omega_1) \wedge \omega_2 + \omega_1 \wedge \mathcal{L}_\xi(\omega_2) \quad \text{(A.18)}$$

$$\mathcal{L}_\xi (\mathbf{T}_1 \wedge \mathbf{T}_2) = \mathcal{L}_\xi(\omega_1) \wedge \omega_2 + \mathbf{T}_1 \wedge \mathcal{L}_\xi(\mathbf{T}_2) \quad \text{(A.19)}$$

For a tensor field of type (p, q) the component form of the Lie derivative is obtained accordingly. Lie derivative thus defines a derivative which measures the change of any tensor (for instance a 2-covariant tensor) on the manifold $\mathfrak{M}$ following a diffeomorphism $x^\mu \to x^\mu + \xi^\mu(x^\alpha)$ according to:

$$\mathcal{L}_{\xi(\mathbf{x})} S_{\mu\nu}(\mathbf{x}) = \xi^\alpha(\mathbf{x}) \partial_\alpha S_{\mu\nu}(\mathbf{x}) + S_{\alpha\nu}(\mathbf{x}) \partial_\mu \xi^\alpha(\mathbf{x}) + S_{\mu\alpha}(\mathbf{x}) \partial_\nu \xi^\alpha(\mathbf{x})$$

Knowledge of Lie derivative of vector fields allows us to calculate Lie derivatives on all tensor bundles over $\mathfrak{M}$. We then get a tool for computing Lie derivatives of all tensor fields. For other field than scalar, Lie derivative differs from covariant derivative. For a torsion-free connection, partial derivative ∂_α can be replaced with covariant derivative ∇_α, however the original definition is classically based only on partial derivative. There is no needs of metric and connection structures on the manifold $\mathfrak{M}$. For any p-form ω on $\mathfrak{M}$ the Lie derivative is defined as:

$$(\mathcal{L}_\xi \omega)(\mathbf{u}_1, \cdots, \mathbf{u}_p) := \mathcal{L}_\xi (\omega(\mathbf{u}_1, \cdots, \mathbf{u}_p))$$

$$- \sum_{i=1}^{i=p} \omega(\mathbf{u}_1, \cdots, [\xi, \mathbf{u}_i], \cdots, \mathbf{u}_p) \quad \text{(A.20)}$$

Definition A.1 For a volume-form ω_n on the n-dimensional manifold $\mathfrak{M}$, the Lie derivative $\mathcal{L}_\xi \omega_n$ is also a n-form then there is a scalar function on $\mathfrak{M}$ called divergence of the vector field $\xi(\mathbf{x})$ relative to ω_n denoted $\mathrm{div}_{\omega_n}\xi$ such that (Saa 1995):

$$\mathcal{L}_\xi \omega_n := \left(\mathrm{div}_{\omega_n}\xi\right)\omega_n \qquad (A.21)$$

No metric tensor is required for the divergence definition at this stage. For connected manifold endowed with affine connection ∇ (associated with non zero torsion and non zero curvature), the divergence reduces to $\nabla_\alpha \xi^\alpha$ e.g. Rakotomanana (2003), where the volume-form should be also compatible with the connection (Saa 1995). For practical calculus, we resume some properties of the Lie derivative. Let $\mathbf{A}$ and $\mathbf{B}$ be tensors of arbitrary types, φ a scalar function, ξ, $\mathbf{u}$ and $\mathbf{v}$ vector fields, ω a 1-form, and a and b constants. Then we have the following rules:

$$\begin{aligned}
\mathcal{L}_\xi\,(\mathbf{A}+\mathbf{B}) &= \mathcal{L}_\xi \mathbf{A} + \mathcal{L}_\xi \mathbf{B} \\
\mathcal{L}_\xi\,(\varphi\mathbf{A}) &= \mathbf{A}\mathcal{L}_\xi \varphi + \varphi\mathcal{L}_\xi \mathbf{A} \\
\mathcal{L}_\xi\,(\omega(\mathbf{u})) &= \left(\mathcal{L}_\xi\omega\right)(\mathbf{u}) + \omega\left(\mathcal{L}_\xi \mathbf{u}\right) \\
\mathcal{L}_\xi\,(\mathbf{A}\otimes\mathbf{B}) &= \left(\mathcal{L}_\xi \mathbf{A}\right)\otimes\mathbf{B} + \mathbf{A}\otimes\left(\mathcal{L}_\xi \mathbf{B}\right) \\
\mathcal{L}_{a\mathbf{u}+b\mathbf{v}}\mathbf{A} &= a\mathcal{L}_\mathbf{u}\mathbf{A} + b\mathcal{L}_\mathbf{v}\mathbf{A}
\end{aligned} \qquad (A.22)$$

A.4.2 Practical Formula for Lie Derivative

Component formulation of the Lie derivative of a type (p, q) tensor $\mathbf{T} = (T^{\alpha_1\cdots\alpha_p}_{\beta_1\cdots\beta_q})$ may be derived as (cf. Lovelock and Rund 1975):

$$\mathcal{L}_\xi T^{\alpha_1\cdots\alpha_p}_{\beta_1\cdots\beta_q} = \xi^\gamma \partial_\gamma T^{\alpha_1\cdots\alpha_p}_{\beta_1\cdots\beta_q} - \sum_{s=1}^{s=p} T^{\alpha_1\cdots\alpha_{s-1}\,\gamma\,\alpha_{s+1}\cdots\alpha_p}_{\beta_1\cdots\beta_q}\,\partial_\gamma\xi^{\alpha_s}$$

$$+ \sum_{s=1}^{s=q} T^{\alpha_1\cdots\alpha_p}_{\beta_1\cdots\beta_{s-1}\,\gamma\,\beta_{s+1}\cdots\beta_q}\,\partial_{\beta_s}\xi^\gamma \qquad (A.23)$$

A.4.3 Exterior Derivative

The exterior derivative of a form is the generalization of the total differential for function, of the differential operators as gradient of function, of rotational and divergence of vector field in the three dimensional space. We briefly remind the exterior differentiation. Let $\mathfrak{M}$ a differentiable manifold with or without metric and with or without connection. Let ω_0 a function of class $\mathcal{C}^1$ on $\mathfrak{M}$. The differential of function $d : \omega_0 \in \Omega^0(\mathfrak{M}) \to d\omega_0 \in \Omega^1(\mathfrak{M})$ is defined by the rule:

$$d\omega_0 := \partial_\mu \omega_0 \, dx^\mu \tag{A.24}$$

The exterior derivation is an extension of (A.24) to a map $d : \omega_p \in \Omega^p(\mathfrak{M}) \to d\omega_p \in \Omega^{p+1}(\mathfrak{M})$ a p-form ω_p to get a $(p+1)$-form $d\omega_p$.

Definition A.2 (Exterior Derivative) The exterior derivation is the unique set of maps $d : \omega \in \Omega^p(\mathfrak{M}) \to d\omega \in \Omega^{p+1}(\mathfrak{M})$ of p-forms which satisfy the following properties:

1. $d : \omega_0 \in \Omega^0(\mathfrak{M}) \to d\omega_0 \in \Omega^1(\mathfrak{M})$ is the differential of function (A.24),
2. $d(\omega_p \wedge \omega_q) = d\omega_p \wedge \omega_q + (-1)^p \omega_p \wedge \omega_q$, for all $\omega_p \in \Omega^p(\mathfrak{M})$, and $\omega_q \in \Omega^q(\mathfrak{M})$,
3. $d(\alpha \omega_p + \beta \omega'_p) = \alpha \omega_p + \beta \omega'_p$, for all $\alpha, \beta \in \mathbb{R}$, and $\omega_p, \omega'_p \in \Omega^p(\mathfrak{M})$,
4. $d(d\omega) = 0$, for all $\omega \in \Omega^p(\mathfrak{M})$.

For a p-form ω, the invariant expression of the exterior derivative takes the form of e.g. Nakahara (1996):

$$d\omega(\mathbf{u}_1, \cdots, \mathbf{u}_{p+1}) = \sum_{i=1}^{i=p} (-1)^{i+1} \mathbf{u}_i \left[\omega(\mathbf{u}_1, \cdots, \hat{\mathbf{u}}_i, \cdots, \mathbf{u}_{p+1}) \right]$$

$$+ \sum_{i<j}^{i=p} (-1)^{i+j} \omega \left([\mathbf{u}_i, \mathbf{u}_j], \mathbf{u}_1, \cdots, \hat{\mathbf{u}}_i, \cdots, \hat{\mathbf{u}}_j, \cdots, \mathbf{u}_{p+1} \right) \tag{A.25}$$

For a 1-form ω, we obtain for instance: $d\omega(\mathbf{u}_1, \mathbf{u}_2) = \mathbf{u}_1 [\omega(\mathbf{u}_2)] - \mathbf{u}_2 [\omega(\mathbf{u}_1)] - \omega([\mathbf{u}_1, \mathbf{u}_2])$ where $[,]$ is the Lie-Jacobi brackets. For a scalar field $f(x^\mu)$ on a manifold $\mathfrak{B}$ (with neither connection nor metric structure), the exterior derivatives can be derived as:

$$df = \partial_\alpha f \, dx^\alpha$$

$$d(df) = \left(\partial_\alpha \partial_\beta f - \partial_\beta \partial_\alpha f \right) dx^\alpha \wedge dx^\beta$$

where df is a 1-form, and $ddf \equiv 0$ if $f \in \mathcal{C}^2$. The component formulation of a p-form exterior derivative takes the form of:

$$d\omega = \sum_{j=1}^{j=n} \sum_{i_1 < \cdots < i_p} \frac{\partial \omega_{i_1 \cdots i_p}}{\partial x^j} \, dx^j \wedge dx^{i_1} \wedge dx^{i_2} \wedge \cdots \wedge dx^{i_p} \tag{A.26}$$

Definition A.3 Consider a field of p-form $\omega \in \Omega^p(\mathfrak{B})$ defined on a manifold $\mathfrak{B}$. If the exterior derivative of ω is zero, say $d\omega = 0$ then it is said to be closed.

Definition A.4 A field of $p + 1$-form $\omega \in \Omega^{p+1}(\mathfrak{B})$ defined on a manifold $\mathfrak{B}$ that can be written as the exterior derivative of a p-form η is said to be exact:

$$\omega := d\eta \tag{A.27}$$

For instance, for a scalar function $f \in \mathcal{C}^2$, the total differential df is a closed 1-form because we have $d(df) = 0$. For a non twice differentiable function, such is not the case.

A.4.4 Stokes' Theorem

One major reason to consider the exterior derivative is that the Stokes theorem holds. Form $\omega \in \Omega^{n-1}(\mathfrak{M})$ is basically a variable that can be integrated over an oriented manifold $\mathfrak{M}$ of dimension n, we assume with boundary $\partial\mathfrak{M}$. Then the Stokes theorem takes the form of:

$$\int_{\partial\mathfrak{M}} \omega = \int_{\mathfrak{M}} d\omega \tag{A.28}$$

which is an extension the most of formulae of Stokes theorem, divergence theorem on manifolds e.g. Rakotomanana (2003). Once again, the use of exterior derivative needs no structure as metric and connection, and then constitute a powerful tool for derivation of conservation laws on general manifolds. The particular case of Stokes' theorem in a three dimensional space are the following:

1. ω is a 0-form that is a scalar function f on a open interval of the real axis with $\mathfrak{B} := [a, b]$ and $\partial\mathfrak{B} := \{a, b\}$,

$$\int_{\partial\mathfrak{B}} \omega = \int_{\mathfrak{B}} d\omega \quad \Longrightarrow \quad f(b) - f(a) = \int_a^b df$$

which is the fundamental theorem of function integration,

2. ω is a 1-form defined on a compact oriented two-dimensional submanifold $\mathfrak{B}$ in a plane, with $\omega := \omega_1 dx^1 + \omega_2 dx^2 + \omega_3 dx^3$ isomorph to the vector field $\mathbf{w} = (\omega_1, \omega_2, \omega_3)$:

$$\int_{\partial\mathfrak{B}} \omega = \int_{\mathfrak{B}} d\omega \quad \Longrightarrow \quad \int_{\partial\mathfrak{B}} \mathbf{w} \cdot d\mathbf{s} = \int_{\mathfrak{B}} \text{rot}(\mathbf{w}) \cdot d\mathbf{S}$$

which states that the circulation of the vector field $\mathbf{w}$ along the curve-boundary $\partial\mathfrak{B}$ is equal to the flux of the curl $\text{rot}(\mathbf{w})$ (axial vector across the surface $\mathfrak{B}$),

3. ω is a 2-form on a compact oriented $3D$ submanifold $\mathfrak{B} \in \mathbb{R}^3$, with $\omega :=$ $\omega_{12}dx^1 \wedge dx^2 + \omega_{23}dx^2 \wedge dx^3 + \omega_{31}dx^3 \wedge dx^1$, isomorph to the vector field $\mathbf{w} = (\omega_{23}, \omega_{31}, \omega_{12})$:

$$\int_{\partial\mathfrak{B}} \omega = \int_{\mathfrak{B}} d\omega \qquad \Longrightarrow \qquad \int_{\partial\mathfrak{B}} \mathbf{w} \cdot d\mathbf{S} = \int_{\mathfrak{B}} \mathrm{div}(\mathbf{w}) \cdot dv$$

expressing that the flux of the vector field $\mathbf{w}$ through the surface (boundary) $\partial\mathfrak{B}$ is equal to the divergence of the vector field $\mathbf{w}$ within the volume $\mathfrak{B}$. Here $\mathbf{w}$ is a an axial vector.

References

Abbott et al (2016) Observation of gravitational waves from binary black hole merger. Phys Rev Lett 116:061102/1-16

Acedo L (2015) Autoparallel vs. Geodesic trajectories in a model of torsion gravity. Universe 1:422–445

Agiasofitou EK, Lazar M (2009) Conservation and balance laws in linear elasticity. J Elast 94:69–85

Aldrovandi R, Pereira JG (2007) Gravitation: on search of the missing torsion. Ann Fond Louis de Broglie 32(2–3):229–251

Aldrovandi R, Pereira JG (2013) Teleparallel gravity: an introduction. Fundamental theories of physics, vol 173. Springer, Dordrecht

Ali SA, Cafaro C, Capozziello S, Corda C (2009) On the Poincaré gauge theory of gravitation. Int J Theor Phys 48:3426–3448

Amendola L, Enqvist K, Koivisto T (2011) Unifying Einstein and Palatini gravities. Phys Rev D 83:044016(1)–044016(14)

Anderson JL (1971) Covariance, invariance, and equivalence. Gen Relativ Gravit 2(2):161–172

Anderson IM (1978) On the structure of divergence-free tensors. J Math Phys 19(12):2570–2575

Anderson IM (1981) The principle of minimal gravitational coupling. Arch Ration Mech Anal 75:349–372

Andringa R, Bergshoeff E, Panda S, de Roo M (2011) Newtonian gravity and the Bargmann algebra. Classical Quantum Gravity 28:105011 (12pp)

Antonio TN, Rakotomanana L (2011) On the form-invariance of Lagrangian function for higher gradient continuum. In: Altenbach H, Maugin G, Erofeev V (eds) Mechanics of generalized continua. Springer, New York, pp 291–322

Antonio TN, Buisson M, Rakotomanana L (2011) Wave propagation within some non-homogeneous continua. Académie des Sciences de Paris: Comptes Rendus Mécanique 339:779–788

Appleby PG (1977) Inertial frames in classical mechanics. Arch Ration Mech Anal 67(4):337–350

Askes H, Aifantis EC (2011) Gradient elasticity in statics and dynamics: an overview of formulations, length scale identification procedures, finite element implementations and new results. Int J Solids Struct 48:1962–1990

Bain J (2004) Theories of Newtonian gravity and empirical indistinguishability. Stud Hist Philos Mod Phys 35:345–376

Baldacci R, Augusti V, Capurro M (1979) A micro relativistic dislocation theory. Lincei Memoria Sc Fisiche, ecc S VIII, vol XV, Sez II 2:23–68

© Springer International Publishing AG, part of Springer Nature 2018   317
L. R. Rakotomanana, *Covariance and Gauge Invariance in Continuum Physics*,
Progress in Mathematical Physics 73, https://doi.org/10.1007/978-3-319-91782-5

Bamba K, Capoziello S, De Laurentis M, Nojiri S, Sáez-Gómez (2013) No further gravitational modes in $F(T)$ gravity. Phys Lett B 727:194–198

Banerjee R, Roy D (2011) Poincaré gauge, Hamiltonian symmetries, and trivial gauge transformations. Phys Rev D 84:124034-1/8

Barra F, Caru A, Cerda MT, Espinoza R, Jara A, Lund F, Mujica N (2009) Measuring dislocations density in aluminium with resonant ultrasound spectroscopy. Int J Bifurcation Chaos 19(10):3561–3565

Bernal AN, Sánchez M (2003) Leibnizian, Galilean, and Newtonian structures of spacetime. J Math Phys 44(3):1129–1149

Betram A, Svendsen B (2001) On material objectivity and reduced constitutive equations. Arch Mech 53(6):653–675

Bideau N, Le Marrec L, Rakotomanana L (2011) Influence of finite strain on vibration of a bounded Timoshenko beam. Int J Solids Struct 48:2265–2274

Bilby BA, Bullough R, Smith E (1955) Continuous distributions of dislocations: a new application of the methods of non-Riemannian geometry. Proc R Soc Lond A 231:263–273

Birkhoff GD, Langer RE (1923) Relativity and modern physics. Harvard University Press, Boston

Boehmer CG, Downes RJ (2014) From continuum mechanics to general relativity. Int J Mod Phys D 23(12):1442015/1-6

Brading KA, Ryckman TA (2008) Hilbert's "Foundations of Physics": gravitation and electromagnetism within the axiomatic method. Stud Hist Philos Mod Phys 39:102–153

Bruzzo U (1987) The global Utiyama theorem in Einstein–Cartan theory. J Math Phys 28(9):2074–2077

Bull P et al (2016) Beyond ΛCDM: problems, solutions, and the road ahead. Phys Dark Universe 12:56–99

Capoziello S, De Laurentis D (2009) Gravity from local Poincaré gauge invariance. Int J Geom Meth Mod Phys 6(1):1–24

Capoziello S, De Laurentis D (2011) Extended theories of gravity. Phys Rep 509:167–321

Capoziello S, Cardone VF, Piedipalumbo E, Sereno M, Troisi A (2003) Matching torsion Lambda-term with observations. Int J Mod Phys D 12:381–394

Capozziello S, De Laurentis M, Francaviglia M, Mercadante S (2009) From dark energy and dark matter to dark metric. Found Phys 39:1161–1176

Cartan E (1922) Sur les équations de la gravitation d'Einstein. J Math Pures Appl 1:141–203

Cartan E (1986) On manifolds with affine connection and the theory of general relativity (translated by A. Magon and A. Ashtekar). Monographs and textbooks in physical science, vol 1. Bibliopolis, Naples

Carter B (1973) Elastic perturbation theory in general relativity and a variation principle for a rotating solid star. Commun Math Phys 30:261–286

Carter B, Quintana H (1977) Gravitational and acoustic waves in an elastic medium. Phys Rev D 16(10):2928–2938

Challamel N, Rakotomanana L, Le Marrec L (2009) A dispersive wave equation using nonlocal elasticity. Académie des Sciences de Paris: Comptes Rendus Mécanique 337(8):591–595

Charap JM, Duff MJ (1977) Gravitational effects on Yang-Mills topology. Phys Lett 69B(4):445–447

Cho YM (1976a) Einstein Lagrangian as the translational Yang-Mills Lagrangian. Phys Rev D 14(10):2521–2525

Cho YM (1976b) Gauge theory of Poincaré symmetry. Phys Rev D 14(12):3335–3340

Cho YM (1976c) Gauge theory, gravitation, and symmetry. Phys Rev D 14(12):3341–3344

Chrusciel PT (1984) On the unified affine electromagnetism and gravitation theories. Acta Phys Polon B15:35–51

Clayton JD, Bammann DJ, McDowell DL (2004) Anholonomic configuration spaces and metric tensors in finite elastoplasticity. Int J Non-Linear Mech 39:1039–1049

Clayton JD, Bammann DJ, McDowell DL (2005) A geometric framework for the kinematics of crystals with defects. Philos Mag 85(33–35):3983–4010

Clifton T, Ferreira PG, Padilla A, Skordis C (2012) Modified gravity and cosmology. Phys Rep 513:1–189

Cordero NM, Forest S, Busso EP (2016) Second strain gradient elasticity of nano-objects. J Mech Phys Solids 97:92–124

Curnier A, Rakotomanana LR (1991) Generalized strain and stress measures: critical survey and new results. Eng Trans (Pol Acad Sci) 39(3–4):461–538

Dadhich N, Pons JM (2012) On the equivalence if the Einstein-Hilbert and the Einstein-Palatini formulations of general relativity for an arbitrary connection. Gen Relativ Gravit 44:2337–2352

Darabi F, Mousavi M, Atazadeh K (2015) Geodesic deviation equation in $f(T)$ gravity. Phys Rev D 91:084023/1–11

de Andrade VC, Pereira JG (1999) Torsion and the electromagnetic field. Int J Mod Phys D 8(2):141–151

Defrise P (1953) Analyse géométrique de la cinématique des milieux continus. Institut Royal Météorologique de Belgique – Publications Série B 6:5–63

Dias L, Moraes F (2005) Effects of torsion on electromagnetic fields. Braz J Phys 35(3A):636–640

Dirac PAM (1974) An action principle for the motion of particles. Gen Relativ Gravit 5(6):741–748

Dixon WG (1975) On the uniqueness of the Newtonian theory as a geometric theory of gravitation. Commun Math Phys 45:167–182

Duval C, Kunzle HP (1978) Dynamics of continua and particles from general covariance of Newtonian gravitation theory. Rep Math Phys 13(3):351–368

Duval C, Burdet G, Kunzle HP, Perrin M (1985) Bargmann structures and Newton–Cartan theory. Phys Rev D 31(8):1841–1853

Earman J (1974) Covariance, invariance, and the equivalence of frames. Found Phys 4(2):267–289

Ehlers J (1973) The nature and concept of spacetime. In: Mehra J (ed) The Physicist's concept of nature. Reidel Publishing Company, Dordrecht, pp 71–91

Ehlers J, Geroch R (2004) Equation of motion of small bodies in relativity. Ann Phys 309:232–236

Exirifard Q, Sheikh-Jabbari MM (2008) Lovelock gravity at the crossroads of Palatini and metric formulations. Phys Lett B 661:158–161

Fernado J, Giglio T, Rodrigues WA Jr (2012) Gravitation and electromagnetism as Geometrical objects of Riemann–Cartan Spacetime structure. Adv Appl Clifford Algebr 22:640–664

Fernandez-Nunez I, Bulashenko O (2016) Anisotropic metamaterial as an analogue of a black hole. Phys Lett A 380:1–8

Ferraro R, Fiorini F (2011) Spherical symmetric static spacetimes in vacuum $f(T)$ gravity. Phys Rev D 84:083518-1/8

Fiziev P, Kleinert H (1995) New action principle for classical particle trajectories in spaces with torsion. Europhys Lett 35(4):241–246

Flügge W (1972) Tensor analysis and continuum mechanics. Springer, Berlin

Forger M, Römer H (2004) Currents and the energy-momentum tensor in classical field theory: a fresh look at an old problem. Ann Phys 309:306–389

Frankel T (1997) The geometry of physics: an introduction. Cambridge University Press, Cambridge

Frewer M (2009) More clarity on the concept of material frame-indifference in classical continuum mechanic. Acta Mech 202:213–246

Fuchs H (1990) Deviation of circular geodesics in static spherically symmetric space-times. Astron Nachr 311(5):271–276

Fumeron S, Pereira E, Moraes F (2015) Generation of optical vorticity from topological defects. Physica B 476:19–23

Futhazar G, Le Marrec L, Rakotomanana-Ravelonarivo L (2014) Covariant gradient continua applied to wave propagation within defective material. Arch Appl Mech 84(9–11):1339–1356

Garcia De Andrade LC (2004) Non-Riemannian geometry of vortex acoustics. Phys Rev D 70:064004-1/064004-5

Garcia De Andrade LC (2005) On the necessity of non-Riemannian acoustic spacetime in fluids with vorticity. Phys Lett A 346:327–329

Gelman H (1966) Generalized conversion of electromagnetic units, measures, and equations. Am J Phys 34(191):291–295

Goenner HFM (1984) A variational principle for Newton–Cartan theory. Gen Relativ Gravit 16(6):513–526

Gonseth F (1926) Les fondements des mathématiqes: De la géométrie d'Euclide à la relativité générale et à l'intuitionisme Ed. Albert Blanchard, Paris

Greenberg PJ (1974) The equation of geodesic deviation in Newtonian theory and the oblatness of the earth. Il Nuovo Cimento 24(2):272–286

Griffiths DJ (2011) Resource letter EM-1: electromagnetic momentum. Am J Phys 80(1):7–18

Hackmann E, Lämmerzahl C (2008) Geodesic equation in Scharzschild-(anti-)de Sitter space-times. Phys Rev D 78:024035-1, 16

Hammond RT (1987) Gravitation, torsion, and electromagnetism. Gen Relativ Gravit 20(8):813–827

Hammond RT (1989) Einstein-Maxwell theory from torsion. Classical Quantum Gravitation 6:195–198

Hammond RT (1990) Second order equations and quadratic Lagrangians. J Math Phys 31:2221–2224

Hammond RT (2002) Torsion gravity. Rep Prog Phys 65:599–649

Hartle JB, Sharp DH (1967) Variational principle for the equilibrium of a relativistic, rotating star. Astrophys J 147:317–333

Havas P (1964) Four-dimensional formulations of Newtonian mechanics and their relation to the special and the general theory of relativity. Rev Mod Phys 36:938–965

Hayashi K, Shirafuji T (1979) New general relativity. Phys Rev D 19:3524–3553

Hehl FW (1971) Dow does one measure torsion of space-time? Phys Lett 36A(3):225–226

Hehl FW (1985) On the kinematics of the torsion of spacetime. Found Phys 15(4):451–471

Hehl FW (2008) Maxwell's equations in Minkowski's world: their premetric generalization and the electromagnetic energy-momentum tensor. Ann Phys 17(0–10):691–704

Hehl FW, Kerlick GD (1976/1978) Metric-affine variational principles in general relativity. I. Riemannian spacetime. Gen Relativ Gravit 9(8):691–710

Hehl FW, Obukhov YN (2003) Foundations of classical electrodynamics: charge, flux, and metric. Birkhäuser, Boston

Hehl FW, von der Heyde P (1973) Spin and the structure of spacetime. Ann Inst Henri Poincaré Sect A 19(2):179–196

Hehl FW, von der Heyde P, Kerlick GD (1974) General relativity with spin and torsion and its deviation from Einstein's theory. Phys Rev D 16(4):1066–1069

Hehl FW, von der Heyde P, Kerlick GD, Nester JM (1976) General relativity with spin and torsion: foundations and prospects. Rev Mod Phys 48(3):393–416

Hehl FW, McCrea JD, Mielke EW, Ne'eman Y (1995) Metric-affine gauge theory of gravity: field equations, Noether identities, world spinors, and breaking of dilation invariance. Phys Rep 258:1–173

Hehl FW, Obukhov YN, Puetzfield D (2013) On Poincaré gauge theory of gravity, its equation of motions, and gravity Probe B. Phys Lett A 377:1775–1781

Hojman S (1976/1978) Lagrangian theory of the motion of spinning particles in torsion gravita-tional theories. Phys Rev D 18(8):2741–2744

Itin Y (2012) Covariant jump conditions in electromagnetism. Ann Phys 327359–375

Javili A, dell'Isola F, Steinmann P (2013) Geometrically nonlinear higher-gradient elasticity with energetic boundaries. J Mech Phys Solids 61:2381–2401

Kadianakis ND (1996) The kinematics of continua and the concept of connection on classical spacetime. Int J Eng Sci 34(3):289–298

Katanaev MO, Volovich IV (1992) Theory of defects in solids and three-dimensional gravity. Ann Phys 216:1–28

Kempers LJTM (1989) The principle of material indifference and the covariance principle. Il Nuovo Cimento 103B(3):227–236

Kibble TWB (1961) Lorentz invariance and gravitational field. J Math Phys 3(2):212–221

Kijowski J, Magli G (1992) Relativistic elastomechanics as a Lagrangian field theory. J Geom Phys 9:207–223

Kleinert H (1987) Gravity as a theory of defects in a crystal with only second gradient elasticity. Ann Phys 499(2):117–119

Kleinert H (1999) Universality principle for orbital angular momentum and spin in gravity with torsion. Gen Relativ Gravit 32(7):1271–1280

Kleinert H (2000) Nonholonomic mapping principle for classical and quantum mechanics in spaces with curvature and torsion. Gen Relativ Gravit 32(5):769–839

Kleinert H (2008) Multivalued fields: in condensed matter, electromagnetism, and gravitation. World Scientific, Singapore

Kleman M, Friedel J (2008) Disclinations, dislocations, and continuous defects: a reappraisal. Rev Mod Phys 80:61–115

Knox E (2013) Effective spacetime geometry. Stud Hist Phil Sci B 44(3):346–356

Kobelev V On the Lagrangian and instability of medium with defects. Meccanica. https://doi.org/10.1007/s11012-011-9480-7

Koivisto T (2011) New variational principles as alternatives to the Palatini method. Phys Rev D 83:101501/4

Kovetz A (2000) Electromagnetic theory. Oxford Science Publications, New York

Krause J (1976) Christoffel symbols and inertia in flat spacetime theory. Int J Theor Phys 15(11):801–807

Kroener E (1981) Continuum theory of defects. In: Balian et al (ed) Physique des défauts. Les Houches 28 July–29 August. North-Holland, Amsterdam, pp 219–315

Landau L, Lifchitz EM (1971) The classical theory of fields: course of theoretical physics volume 2. Third Revised English edn. Pergamon Press, Oxford

Lazar M (2002) An elastoplastic theory of dislocations as a physical field with torsion. J Phys A: Math Gen 35:1983–2004

Lazar M, Anastassiadis C (2008) The gauge theory of dislocations: conservation and balance laws. Philos Mag 88(11):1673–1699

Le KC, Stumpf H (1996) On the determination of the crystal reference in nonlinear continuum theory of dislocations. Proc R Soc Lond A 452:359–37

Leclerc M (2005) Mathisson-Papapetrou equations in metric and gauge theories of gravity in a Lagrangian formulation. Classical Quantum Gravitation 22:3203–3221

Lehmkuhl D (2011) Mass-energy-momentum in general relativity. Only there because of spacetime? Br J Philos Sci 62(3):453–488

Leonhardt U, Philbin TG (2006) General relativity in electrical engineering. New J Phys 8/247:1–18

Leonhardt U, Piwnicki P (2000) Relativistic effects of light in moving media with extremely low group velocity. Phys Rev Lett 84/5:822–825

Levi-Civita T (1927) Sur l'écart géodésique. Math Ann 97:291–320

Li B, Sotiriou TP, Barrow JD (2011) $f(T)$ gravity and local Lorentz invariance. Phys Rev D 83:064035/1-pp 064035/5

Lichnerowicz A (1955) Théories relativistes de la gravitation et de léléctromagnétisme. Masson, Paris

Logunov AA, Mestvirishvili MA (2012) Hilbert's causality principle and equations of general relativity exclude the possibility of black hole formation. Theor Math Phys 170(3):413–419

Lompay RR (2014) On the energy-momentum and spin tensors in the Riemann–Cartan space. Gen Relativ Gravit 46(1692):1–123

Lovelock D (1969) The uniqueness of the Einstein field equations in a four-dimensional space. Arch Ration Anal Mech 33:54–70

Lovelock D (1971) The Einstein tensor and its generalizations. J Math Phys 12:498–501

Lovelock D, Rund H (1975) Tensors, differential forms, and variational principles, chap 8. Wiley, New York

Maier R (2014) Static vacuum solutions in non-Riemannian gravity. Gen Relativ Gravit 46(1830):1–15

Maldacena J (1998) The Large N limit of superconformal field theories and supergravity. Adv Theor Math Phys 2:231–252

Maluf JW, da Rocha-Neto JF, Toríbio TML, Castello-Branco KH (2002) Energy and angular momentum of gravitational field in the tele parallel geometry. Phys Rev D 65:124001/1 - 12

Malyshev C (2000) The $T(3)$-gauge model, the Einstein-like gauge equation, and Volterra dislocations with modified asymptotics. Ann Phys 286:249–277

Manoff S (1999) Lagrangian theory of tensor fields over spaces with contra variant and covariant affine connections and metrics and its application to Einstein's theory of gravitation in $\overline{V}_4$ spaces. Acta Appl Math 55:51–125

Manoff S (2001a) Frames of reference in spaces with affine connections and metrics. Classical Quantum Gravity 18:1111–1125

Manoff S (2001b) Deviation operator and deviation equations over spaces with affine connections and metrics. J Geom Phys 39:337–350

Mao Y, Tegmark M, Guth AH, Cabi S (2007) Constraining torsion with Gravity Probe B. Phys Rev D 76:104029/1-26

Marsden JE, Hughes TJR (1983) Mathematical foundations of elasticity. Prentice-Hall, Englewood Cliffs

Mathisson M (2010) New mechanics of material system. In: General relativity and gravitation, vol 42, pp 1011–1048/Translated by A Ehlers from the Original Paper: Neue mechanik materieller Systeme, Acta Physica Polonica 6, 1937, pp 163–200

Maugin GA (1978) Exact relativistic theory of wave propagation in prestressed nonlinear elastic solids. Ann Inst Henri Poincaré Sect A 28(2):155–185

Maugin GA (1993) Material inhomogeneities in elasticity. Chapman and Hall, London

McKellar RJ (1981) The uniqueness of gravity as a Poincaré or Lorentz gauge theory. J Math Phys 22 (12):2934–2942

Metrikine AV (2006) On causality of the gradient elasticity models. J Sound Vib 297:727–742

Milonni PW, Boyd RW (2010) Momentum of light on a dielectric medium. Adv Opt Photon 2:519–553

Minazzoli O, Karko T (2012) New derivation of the Lagrangian of a perfect fluid with a barotropic equation of state. Phys Rev D 86:087502/1-4

Mindlin RD (1964) Micro-structure in linear elasticity. Arch Ration Mech Anal 16:51–78

Mindlin RD (1965) Second gradient of strain and surface-tension in linear elasticity. Int J Solids Struct 1:417–438

Nakahara (1996) Geometry, topology, and physics. In: Brower D (ed) Graduate student series in physics. Institute of Physics Publishing, Bristol

Ni WT, Zimmermann M (1978) Inertial and gravitational effects in the proper reference frame of an accelerated, rotating observer. Phys Rev D 17(6):1473–1476

Nieto JA, Saucedo J, Villanueva VM (2003) Relativistic top deviation equation and gravitational waves. Phys Lett A 312:175–186

Nieto JA, Saucedo J, Villanueva VM (2007) Geodesic deviation equation for relativistic tops and the detection of gravitational waves. Rev Mex Fís S 53(2):141–145

Noll W (1967) Materially uniform simple bodies with inhomogeneities. Arch Ration Mech Anal 27:1–32

Norton JD (1993) General covariance and the foundations of general relativity: eight decades of disputes. Rep Prog Phys 56:791–858

Obukhov YN (2008) Electromagnetic energy and momentum in moving media. Ann Phys Berlin 17(9–10):830–851

Obukhov YN, Hehl FW (2003) Electromagnetic energy-momentum and forces in matter. Phys Lett A 311:277–284

Obukhov YN, Puetzfeld D (2014) Conservation laws in gravity: a unified framework. Phys Rev D 90(02004):1–10

Obukhov YN, Ponomariev VN, Zhytnikov VV (1989) Quadratic Poincaré gauge theory of gravity: a comparison with the general relativity theory. Gen Relativ Gravit 21(11):1107–1142

Olmo GJ, Rubiera-Garcia D (2013) Importance of torsion and invariant volumes in Palatini theories of gravity. Phys Rev D 88(084030):1–11

Oprisan CD, Zet G (2006) Gauge theory on a space-time with torsion. Rom J Physiol 51(5–6):531–540

Padmanabhan T (2003) Cosmological constant-the weight of vacuum. Phys Rep 380:235–320

Papapetrou A (1951) Spinning test-particles in general relativity I. Proc R Soc Lond A 209:248–258

Pellegrini YP (2012) Screw and edge dislocations with time-dependent core width: from dynamical core equations to an equation of motion. J Mech Phys Solids 60:227–249

Petrov AN, Lompay RR (2013) Covariantized Noether identities and conservation laws for perturbations in metric theories of gravity. Gen Relativ Gravit 45:545–579

Pettey D (1971) One-one-mappings onto locally connected generalized continua. Pac J Math 50(2):573–582

Philipp D, Perlick V, Lammerzahl C, Deshpande K (2015) On geodesic deviation in Schwarzschild spacetime. In: Metrology for aerospace IEEE, pp 198–203

Plebanski J (1960) Electromagnetic waves in gravitational fields. Phys Rev 118(5):1396–1408

Polizzotto C (2012) A gradient elasticity theory for second-grade materials and higher order inertia. Int J Solids Struct 49:2121–2137

Polizzotto C (2013a) A second strain gradient elasticity theory with second velocity gradient inertia- Part I: constitutive equations and quasi-static behavior. Int J Solids Struct 50:3749–3765

Polizzotto C (2013b) A second strain gradient elasticity theory with second velocity gradient inertia- Part I: dynamic quasi-static behavior. Int J Solids Struct 50:37–3777

Polyzos D, Fotiadis DI (2012) Derivation of Mindlin's first and second strain gradient elastic theory via simple lattice and continuum models. Int J Solids Struct 49:470–480

Pons JM (2011) Noether symmetries, energy-momentum tensors, and conformal invariance in classical field theory. J Math Phys 52:012904-1/21

Poplawski NJ (2009) A variational formulation of relativistic hydrodynamics. Phys Lett A 373:2620–2621

Poplawski NJ (2010) Torsion as electromagnetism and spin. Int J Theor Phys 49(7):1481–1488

Prasanna AR (1975a) Maxwell's equations in Riemann–Cartan space U_4. Phys Lett A 54(1):17–18

Prasanna AR (1975b) Static fluid spheres in Einstein–Cartan theory. Phys Rev D 11(8):2076–2082

Puetzfeld D, Obukhov YN (2008) Probing non-Riemannian spacetime geometry. Phys Lett A 372:6711–6716

Puetzfeld D, Obukhov YN (2013a) Covariant equations of motion for test bodies in gravitational theories with general nonminimal coupling. Phys Lett D 87:044045/1–7

Puetzfeld D, Obukhov YN (2013b) Equations of motion in gravity theories with nonminimal coupling: a loophole to detect torsion macroscopically. Phys Lett D 88:064025/1–9

Puntigam RA, Lämmerzahl C, Hehl FW (1997) Maxwell's theory on a post-Riemannian spacetime and the equivalence principle. Classical Quantum Gravitation 14:1347–1356

Rakotomanana RL (1997) Contribution à la modélisation géométrique et thermodynamique d'une classe de milieux faiblement continus. Arch Ration Mech Anal 141:199–236

Rakotomanana RL (2003) A geometric approach to thermomechanics of dissipating continua. Progress in Mathematical Physics Series. Birkhaüser, Boston

Rakotomanana RL (2005) Some class of SG continuum models to connect various length scales in plastic deformation. In: Steinmann P, Maugin GA (ed) Mechanics of material forces, chap 32. Springer, Berlin

Rakotomanana RL (2009) Élements de dynamiques des structures et solides déformables. Presses Polytechniques et Universitaires Romandes, Lausanne

Ramaniraka NA, Rakotomanana LR (2000) Models of continuum with microcrack distribution. Math Mech Solids 5:301–336

Riles K (2013) Gravitational waves sources, detectors and searches. Prog Part Nucl Phys 68:1–54

Romero JM, Bellini M, Aguilar JEM (2016) Gravitational waves and magnetic monopoles during inflation with Weitzenböck torsion. Phys Dark Univ 13:121–125

Rosen G (1972) Galilean invariance and the general covariance of nonrelativistic laws. Am J Phys 40:683–687

Ross DK (1989) Planck's constant, torsion, and space-time defects. Int J Theor Phys 28(11):1333–1340

Rousseaux G (2008) On the electrodynamics of Minkowski at low velocities. Europhys Lett 84:20002/p1-p3

Ruedde C, Straumann N (1997) On Newton–Cartan cosmology. Helv Phys Acta 71(1–2):318–335

Ruggiero ML, Tartaglia A (2003) Einstein–Cartan as theory of defects in spacetime. Am J Phys 71(12):1303–1313

Ryder L (2009) Introduction to general relativity. Cambridge University Press, New York

Ryskin G (1985) Misconception which led to the "material frame-indifference" controversy. Phys Rev A 32(2):1239–1240

Saa A (1995) Volume-forms and minimal action principles in affine manifolds. J Geom Phys 15:102–108

Schmidt H-J (2007) Fourth-order gravity: equations, history, and applications to cosmology. Int J Geom Meth Mod Phys 4(2):209–248

Schücker T, Tilquin A (2012) Torsion, an alternative to the cosmological constant? Int J Mod Phys D 21(13):1250089

Schutz BF (1970) Perfect fluids in general relativity: velocity potentials and a variational principle. Phys Rev D 2(12):2762–2773

Schutzhold R, Plunien G, Soff G (2002) Dielectric black hole analogs. Phys Rev Lett 88(6):061101/1-061101/4

Sciama DW (1964) The physical structure of general relativity. In: Reviews of modern physics, Contributed papers for relativity, nuclear physics, and post deadline papers, January, pp 463–469

Shapiro IL (2002) Physical aspects of spacetime torsion. Phys Rep 357:113–213

Sharma P, Ganti S (2005) Gauge-field-theory solution of the elastic state of a screw dislocation in a dispersive (non-local) crystalline solid. Proc R Soc Lond 461:1–15

Shen W, Moritz H (1996) On the separation of gravitation and inertia and the determination of the relativistic gravity field in the case of free motion. J Geod 70:633–644

Smalley LL (1986) On the extension of geometric optics from Riemaniann to Riemann–Cartan spacetime. Phys Lett A 117(6):267–269

Smalley LL, Krisch JP (1992) Minimal coupling of electromagnetic fields in Riemann–Cartan space-times for perfect fluids with spin density. J Math Phys 33(3):1073–1081

Söderholm L (1970) A principle of objectivity for relativistic continuum mechanics. Arch Ration Mech Anal 39(2):89–107

Sotiriou TP (2008) The viability of theories with matter coupled to the Ricci scalar. Phys Lett B 664:225–228

Sotiriou TP (2009) $f(R)$ gravity, torsion and non-metricity. Classical Quantum Gravitation 26:152001

Sotiriou TP, Faraoni V (2010) $f(R)$ theories of gravity. Rev Mod Phys 82:451–497

Sotiriou TP, Liberati S (2007) Metric-affine $f(R)$ theories of gravity. Ann Phys 322:935–966

Sotiriou TP, Li B, Barrow JD (2011) Generalizations of tele parallel gravity and local Lorentz symmetry. Phys Rev D 83:104030/1-104030/6

Svendsen B, Betram A (1999) On frame-indifference and form-invariance in constitutive theory. Acta Mech 132:195–207

Synge JL (1934) On the deviation of geodesics and null-geodesics, particularly in relation to the properties of spaces of constant curvatures and indefinite line-element. Ann Math 35(4):705–713

Tamanini N (2012) Variational approach to gravitational theories with two independent connections. Phys Rev D 86:024004/1-9

Taub AH (1954) General relativistic variational principle for perfect fluids. Phys Rev 94(6):1468–1470

Terrier A, Miyagaki J, Fujie H, Hayashi K, Rakotomanana L (2005) Delay of intracortical bone remodelling following a stress change: a theoretical and experimental study. Clin Biomech 20(9):998–1006

Tiwari RN, Ray S (1997) Static spherical charged dust electromagnetic mass models in Einstein–Cartan theory. Gen Relativ Gravit 29(6):683–690

Toupin RA (1962) Elastic materials with couple stresses. Arch Ration Mech Anal 11:385–414

Truesdell C, Noll W (1991) The non-linear field theories of mechanics, 2nd edn. Springer, Berlin

Utiyama R (1956) Invariant theoretical interpretation of interaction. Phys Rev 101:1597–1607

Vandyck MA (1996) Maxwell's equations in spaces with non-metricity and torsion. J Phys A: Math Gen 29:2245–2255

Verçyn A (1990) Metric-torsion gauge theory of continuum line defects. Int J Theor Phys 29(1):7–21

Vitagliano V, Sotiriou TP, Liberati S (2011) The dynamics of metric-affine gravity. Ann Phys 326:1259–1273

Wang CC (1967) Geometric structure of simple bodies, or mathematical foundation for the theory of continuous distributions of dislocations. Arch Ration Mech Anal 27:33–94

Westman H, Sonego S (2009) Coordinates, observables and symmetry in relativity. Ann Phys 324:1585–1611

Weyl H (1918) Gravitation and electricity. Sitzungsber Preuss Akad Berlin 465:24–37

Weyl H (1929) Gravitation and the electron. Proc Natl Acad Sci 15:323–334

Wigner E (1939) On unitary representations of the inhomogeneous Lorentz group. Ann Math 40(1):149–204

Williams G (1973) A discussion of causality and the Lorentz group. Int J Theor Phys 7(6):415–421

Williams DN (1989) The elastic energy-momentum tensor in special relativity. Ann Phys 196:345–360

Yang CN, Mills RL (1954) Conservation of isotopic spin and isotopic gauge invariance. Phys Rev 96:191–201

Yang G, Duan Y, Huang Y (1998) Topological invariant in Riemann–Cartan manifold and spacetime defects. Int J Theor Phys 37(12):2953–2964

Yasskin PB, Stoeger WR (1980) Propagation equations for test bodies with spin and rotation in theories of gravity with torsion. Phys Rev D 21(8):2081–2093

Yavari A, Goriely A (2012) Weyl geometry and the nonlinear mechanics of distributed point defects. Proc R Soc A 468:3902–3922

Zeeman EC (1964) Causality implies the Lorentz group. J Math Phys 5(4):490–493